Modern Approaches in Solid Earth Sciences

Volume 25

Background and motivation

Earth Sciences are going through an interesting phase as the traditional disciplinary boundaries are collapsing. Disciplines or sub-disciplines that have been traditionally separated in the past have started interacting more closely, and some new fields have emerged at their interfaces. Disciplinary boundaries between geology, geophysics and geochemistry have become more transparent during the last ten years. Geodesy has developed close interactions with geophysics and geology (tectonics). Specialized research fields, which have been important in development of fundamental expertise, are being interfaced in solving common problems.

In Earth Sciences the term System Earth and, correspondingly, Earth System Science have become overall common denominators. Of this full System Earth, Solid Earth Sciences – predominantly addressing the Inner Earth - constitute a major component, whereas others focus on the Oceans, the Atmosphere, and their interaction. This integrated nature in Solid Earth Sciences can be recognized clearly in the field of Geodynamics. The broad research field of Geodynamics builds on contributions from a wide variety of Earth Science disciplines, encompassing geophysics, geology, geochemistry, and geodesy. Continuing theoretical and numerical advances in seismological methods, new developments in computational science, inverse modelling, and space geodetic methods directed to solid Earth problems, new analytical and experimental methods in geochemistry, geology and materials science have contributed to the investigation of challenging problems in geodynamics. Among these problems are the high-resolution 3D structure and composition of the Earth's interior, the thermal evolution of the Earth on a planetary scale, mantle convection, deformation and dynamics of the lithosphere (including orogeny and basin formation), and landscape evolution through tectonic and surface processes. A characteristic aspect of geodynamic processes is the wide range of spatial and temporal scales involved. An integrated approach to the investigation of geodynamic problems is required to link these scales by incorporating their interactions. Scope and aims of the new series

The book series "Modern Approaches in Solid Earth Sciences" provides an integrated publication outlet for innovative and interdisciplinary approaches to problems and processes in Solid Earth Sciences, including Geodynamics.

It acknowledges the fact that traditionally separate disciplines or sub-disciplines have started interacting more closely, and some new fields have emerged at their interfaces. Disciplinary boundaries between geology, geophysics and geochemistry have become more transparent during the last ten years. Geodesy has developed close interactions with geophysics and geology (tectonics). Specialized research fields (seismic tomography, double difference techniques etc), which have been important in development of fundamental expertise, are being interfaced in solving common problems.

Accepted for inclusion in Scopus.

Prospective authors and/or editors should consult one of the Series Editors or the Springer Contact for more details. Any comments or suggestions for future volumes are welcomed.

Xuanxue Mo • Jinfu Deng • Zengqian Hou •
Zhidan Zhao • Guochen Dong •
Dicheng Zhu • Yaoling Niu

Magmatism and Crustal Evolution of the Tibetan Plateau

Springer

Xuanxue Mo
China University of Geosciences
Beijing, China

Zengqian Hou
Institute of Geology
Chinese Academy of Geological Sciences
Beijing, China

Guochen Dong
Earth Sciences and Resources
China University of Geosciences
Beijing, China

Yaoling Niu
Laoshan Laboratory
Qingdao, China

Jinfu Deng
China University of Geosciences
Beijing, China

Zhidan Zhao
China University of Geosciences
Beijing, China

Dicheng Zhu
China University of Geosciences
Beijing, China

ISSN 1876-1682 ISSN 1876-1690 (electronic)
Modern Approaches in Solid Earth Sciences
ISBN 978-981-97-2107-8 ISBN 978-981-97-2109-2 (eBook)
https://doi.org/10.1007/978-981-97-2109-2

This Springer imprint is published by the registered company Springer Nature Singapore Pte Ltd.
The registered company address is: 152 Beach Road, #21-01/04 Gateway East, Singapore 189721, Singapore

Preface

The Tibetan Plateau is the unique and most prominent surface construct on Earth with an average elevation of ~ 4500 m and a vast area of ~ 2.5 million km^2. Despite the longstanding debate, it has become clear that the plateau formation is recent in response to the India-Asia continental collision at ~55 ± 5 Ma and continued convergence to this day. Nevertheless, the Greater Tibetan Plateau is not a coherent massive terrane, but a geological amalgamation developed as a result of several northward drift and collision of Gondwana-derived land masses from the Early Paleozoic to the Cenozoic, marked by progressively younger suture zones from Northeast to Southwest. These suture zones (SZs), from northeast to southwest, are Qilian SZ (Early Paleozoic), Kunlun SZ (Paleozoic), Jinsha SZ (Triassic), Bangong SZ (Jurassic), and Indus-Yarlung Zangbo SZ (Early Cenozoic) (e.g., Tapponnier et al. 2001; Metcalfe 2013). The exact timings of these collision/suture zones remain debated, yet the more confusing uses/discussions are the *Tethys Oceans* through which the aforementioned continental drift and collision events took place, including their opening, size, longevity, and closing.

"Tethys Ocean" has been widely and loosely used in all studies of the magmatism, metamorphism, and geological evolution of the Greater Tibetan Plateau, but the "Tethys Ocean" was originally used to describe the "new ocean" resulting from the breakup of the Pangea to form continental masses of Gondwana to the south and Laurasia to the north. Derived from "Tethys," researchers coined terms of "Proto-Tethys" (600–460 Ma), "Paleo-Tethys" (440–200 Ma), and "Neo-Tethys" (200–50 Ma) used to describe the land masses with volcanic arcs and lost oceans with crustal remnants in the west (e.g., the Black, Caspian Aral seas, and much of the present-day Mediterranean).

The eastern part of the Tethys Ocean is sometimes referred to as Eastern Tethys. In Asia, the Paleo-Tethys (Devonian–Triassic), Meso-Tethys (late Early Permian–Late Cretaceous), and Ceno-Tethys (Late-Triassic–Cenozoic) are used to describe land masses with volcanic arcs and the suture zones preserved in the Greater Tibetan Plateau and the adjacent regions (e.g., Metcalfe 2013). These are also meticulously summarized in several recent studies (e.g., Pan et al. 2022; Wu et al. 2020; Zhu et al. 2022) although a rigorous use of the terms is warranted for effective communication.

Tapponnier et al. (2001) and Song et al. (2014) argue that the Longmuco-Shuang Hu SZ may actually represent the final suture of Laurasia-Gondwana convergence preserved in this part of the world with composite terranes north of the suture belonging to Laurasia land mass whereas the composite terranes south of the suture belonging to Gondwana land mass based on interpreted subduction polarity. This is possible, but to be consistent and for convenience we prefer the following: The Kunlun SZ representing the Paleo-Tethys (e.g., Fan et al. 2021a), the Bangong-Nujiang SZ representing the Meso-Tethys (Fan et al. 2021b), and the Indus-Yarlung Zangbo SZ representing the Ceno-Tethys (Mecalfe 2013). This is in general consistent with the data analysis-based division that will be used throughout this book.

The Tibetan Plateau has abundant igneous rocks of varying ages, types and compositions, ophiolite suites and related mineral resources, which contain rich information on crustal and mantle processes and mineralization in the course of the plateau formation and its continued development. This book aims to summarize the state-of-the-art research on and further exploration for the petrogenesis of all these igneous rocks and mineralization as a means to discussing the crustal and mantle processes associated with the formation of the Tibetan Plateau in the context of the evolution of the Eastern Tethys realm.

This book represents decades of dedicated research on the Tibetan Plateau by our research group since the 1970s. The authors of the book and all chapters are, in alphabetic order, Jinfu Deng, Guochen Dong, Zengian Hou, Xuanxue Mo, Yaoling Niu, Zhidan Zhao, and Dicheng Zhu. The chapters are written by the following lead authors with the assistance of others. Chapter 1, by Guochen Dong, provides an overview on the distribution of igneous rocks on the Tibetan Plateau in space and time. Chapter 2, by Dicheng Zhu, discusses pre-Cenozoic igneous rocks of the Tibetan Plateau. Chapter 3, by Guochen Dong, discusses early Cenozoic collisional igneous rocks on the Tibetan Plateau. Chapter 4, by Zhidan Zhao, discusses Cenozoic post-collisional igneous rocks on the Tibetan Plateau. Chapter 5, by Jinfu Deng, investigates the tectonic evolution from Pangea to the Neo-Tethys, and to the Tibetan Plateau. Chapter 6, by Xuanxue Mo, discusses the timing, process, and mechanism of India-Asia collision and the uplift of the Tibetan Plateau. Chapter 7, by Dicheng Zhu, discusses accretional and collisional processes of the Lhasa Terrane to the southern Tibetan Plateau. Chapter 8, by Yaoling Niu, discusses juvenile crustal growth in continental collision zones and crustal "destruction" in continental interiors by granitoid magmatism. Chapter 9, by Zhidan Zhao, discusses the diversity and evolution of the Tibetan Plateau lithosphere. Chapter 10, by Zengqian Hou, discusses the formation, preservation, metallogenic fertility of juvenile crust in the Eastern Tethyan collisional orogens, and lithospheric architecture control on the metallogenic systems.

We thank many who have helped and supported us since the 1970s and Professor Yildirim Dilek for his interest, enthusiasm, and support of this book project. We thank Zeguang Chang, Yongjun Di, Pengsheng Dong, Hangsheng Lei, Shi-Min Li, An-Lin Liu, Cui Liu, Dong Liu, Li Liu, Yanbin Liu, Yongjun Lu, Shujun Lü, Yunhe Lü, Chenguang Sun, Changming Wang, Qing Wang, Tao Wang, Ying Xia, Qinghui

Xiao, Wei Xu, Qiong-Yao Zhan, Hongrui Zhang, and Su Zhou (in alphabetical order) for their support and contributions in various forms to the book. Our gratitude also goes to the Springer and the project coordinators for their diligent work towards the publication of the book *Magmatism and the Crustal Evolution of the Tibetan Plateau*.

Beijing, China Xuanxue Mo
Beijing, China Zengqian Hou
Durham, UK Yaoling Niu
1 December 2022

References

Fan JJ, Niu YL, Luo AB, Xie CM, Hao YJ, Liu HY (2021a) Timing of the Meso-Tethys Ocean opening: evidence from Permian sedimentary province changes in the South Qiangtang Terrane, Tibetan Plateau. Paleogeogr Paleoclimatol Paleoecol 567:110265.

Fan JJ, Niu YL, Liu YM, Hao YJ (2021b) Timing of closure of the Meso-Tethys Ocean: constraints from remnants of a 141-135 ocean island within the Bangong-Nujiang suture zone, Tibetan Plateau. Geol Soc Am Bull 133:1875–1889.

Metcalfe I (2013) Gondwana dispersion and Asian accretion: tectonic and palaeogeographic evolution of eastern Tethys. J Asian Earth Sci 66:1–33.

Pan GT, Wang LP, Yim FG, Geng QR, Li GM, Zhu DC (2022) Researches on geological-tectonic evolution of Tibetan Plateau: a review, recent advances and directions in the future. Sediment Geol Tethyan Geol 42:151–175. (Chinese with English Abstract).

Song SG, Niu YL, Su L, Zhang C, Zhang LF (2014) Continental orogenesis from seafloor subduction, continent collision/subduction, to orogen collapse, and orogen recycling: The example of the North Qaidam UHPM belt, NW China. Earth-Sci Rev 129:59–84.

Tapponnier P, Xu Z, Roger F, Meyer B, Arnaud N, Wittlinger G, Yang JS (2001) Oblique stepwise rise and growth of the Tibet Plateau. Science 294:1671–1677.

Wu FY, Wan B, Zhao L, Xiao WJ, Zhu RX (2020) Tethys geodynamics. Acta Petrologica Sinica 36:1627–1674.

Zhu RX, Zhao P, Zhao L (2022) The evolution and dynamic process of Neo-Tethys. Sci China Ser D Earth Sci 52:1–25.

Contents

Abbreviations

AC	Arc crust
AFC	Assimilation and fractional crystallization
AFM	Alkali-FeO-MgO (diagram)
ACM	Active continental margin
ACM-PCM collision	Collision between active continental margin and passive continental margin
ARC	Island arc
ARC-PCM collision	Island arc-passive continental margin collision
BCC	Bulk continental crust
BMA	Bangonghu magmatic arc
BNS	Bangong-Nujiang suture
BNSZ	Bangong-Nujiang Suture Zone
CA	Calc-alkaline
CAOB	Central Asian Orogenic Belt
CC	Continental crust
CFB	Continental flood basalt
CHUR	Chondrite Uniform Reservoir
CLS	Central Lhasa subterrane)
DM	Depleted mantle
DMM	Depleted MORB mantle (radiogenic isotopes)
DUPAL	An Indian Ocean basalts Pb-Sr mixing scenario proposed by Dupre and Allegre (1983)
EMII	Enriched mantle type 2 (radiogenic isotopes)
EMI	Enriched mantle type 1 (radiogenic isotopes)
E-MORB	Enriched mid-ocean ridge basalt
EUR	Europe
E-W	East-west
GMG	Granodiorite-monzonitic granite
GGL	Great gradient line in eastern continental China
GMA	Gangdese magmatic arc
GMB	Gangdese magmatic belt

HFSE	High field strength element
HKCA	High-K calc-alkaline
HMB	High-Mg Basalt
HOT	Harzburgite ophiolite
HREE	Heavy rare earth element
HSE	Highly siderophile element
IAT	Island arc tholeiite
INC	Indochina
IUGS	International Union of Geological Sciences
IYZO	Indus-Yarlung Zangbo Ocean
IYZS	Indus-Yarlung Zangbo suture
IYZSZ	Indus-Yarlung Zangbo suture zone
JDMA	Jiangda magmatic arc
KAR	Karakorum
KAZ	Kazakhstan
KARF	Karakorum fault
LDF	Lower Dianzhong Formation of the LVS
LILE	Large ion lithophile element
LIP	Large igneous province
LLF	Luonan-Luanchuan fault
LLWF	Lingbao-Lushan-Wuyang fault
LMF	Luobadui-Milashan fault
LREE	Light rare earth element
LST	Lashsa Terrane
LSSZ	Longmu Tso-Shuanghu Suture Zone
LVS	Linzizong volcanic secession
MASH	Melting-assimilation-storage-homogenization
MBT	Main boundary thrust
MCT	Main central thrust
MLS	Mianlue suture
MME	Mafic magmatic enclave
MME	Mafic microgranular enclave
MNF	Middle Nianbo Formation of the LVS
MON	Mongolia
MON-NC	Mongolia-North China plate
MORB	Mid-ocean ridge basalt
MSL	Mechanically strong lithospheric boundary layer
MTM	Malaysia-Thailand Massif
MVT	Mississippi Valley-type
NCB	North China Block
NCC	North China Craton
NCI	Neodymium (isotope) crust index
NHRL	Northern hemisphere reference line
NLS	Northern Lhasa subterrane
N-MORB	Normal mid-ocean ridge basalt

NQS	Northern Qiangtang subterrane
NQD	North Qinling terrane
NQT	North Qinling terrane
NRC	National Research Council
N-S	North-south
NW	Northwest
OIB	Ocean island basalt
OPS	Ocean plate stratigraphy
PCD	Porphyry Cu deposit
PCM	Passive continental margin
PMD	Porphyry Mo deposit
PREMA	Prevalent mantle
PVR	Potassic volcanic rock
QD	Qaidam
QTT	Qiangtang terrane
REE	Rare earth element
SCB	South China Block
SCLM	Sub-continental lithospheric mantle
SDS	Shangdan suture
SPGB	Songpan-Ganzi Block
SGB	South Gangdese batholith
SGMB	South Gangdese magmatic belt
STB	Shan Tai block
SHO	Shoshonite
SHRIMP	Sensitive high-resolution ion microprobe
SIB	Siberia
SIMS	Secondary ionization mass spectrometry
SLS	Southern Lhasa subterrane
S-NCB	Southern North China Block
SNMZ	Shiquanhe-Nam Tso melange zone
SNMZ	Shiquan River-Nam Tso Melange zone
SQST	Southern Qiangtang subterrane
SQT	South Qinling terrane
SSZ	Supra-subduction zone
SSMZ	Sanandaj-Sirjan magmatic arc zone
STD	Southern Tibetan detachment
STDS	South Tibetan detachment system
STSA	Southern Tibet syncollisional andesite
TAR	Tarim
TAS	Total alkali-silica (diagram)
TTG	Tonalite–trondhjemite–granodiorite
UDMA	Urumieh-Dokhtar magmatic arc
UPF	Upper Pana Formation of the LVS
UPVR	Ultrapotassic volcanic rocks

Authors and Contributors

About the Authors

Di-Cheng Zhu is a professor of petrology and geochemistry at China University of Geosciences (Beijing) (CUGB). He received his B.Sc. (1997) and M.Sc. (2000) from Chengdu University of Technology, China, and his Ph.D. (2003) from the Chinese Academy of Geological Sciences, China. He joined the Chengdu Institute of Geology and Mineral Resources in 2003 and moved to CUGB in 2007, before becoming a professor in 2009. He devoted his enthusiasm for studying the formation of the Tibetan Plateau from a magmatic perspective since 2000. His research interests include (1) magma generation and crustal evolution in collision zones, (2) magmatism and the formation of the Tibetan Plateau, and (3) paleogeographic reconstruction of microcontinents. He has published more than 180 research papers, with a total of more than 15,000 citations (Google Scholar h-index 59). He is currently serving as a Co-Editor-in-Chief of Lithos (2022 to present).

Guochen Dong is a professor at the China University of Geosciences (Beijing). He has been both an educator and a researcher in the field of Earth Science. He conducted geological mapping and mineral exploration at the Hebei Geological Bureau (1983–1998) and taught optical mineralogy, mineralogy, petrology, and field geology at both undergraduate and postgraduate levels at the China University of Geosciences (2003–2023). His research includes: (1) petrology and geochemistry of granites, granitoid rocks, and volcanic rocks associated with the collision between the Indian and Asia continentals; (2) magmatism associated with continental collision and continental crust accretion; (3) felsic and granitic rocks and related ore deposits; and (4) genetic and prospecting mineralogy, geochronology, and geological applications.

Jinfu Deng (1935–2021) was a professor at the China University of Geosciences, Beijing. He had served as the dean of the Department of Geology, the vice-chairman of the Academic Committee of the University, the executive director of the Chinese Society of Mineralogy, Petrology, and Geochemistry, among other roles. He was a visiting professor in the Department of Geology and Planetary Sciences at Caltech,

USA (1986–1988). His research interests included igneous petrology, petrotectonics, petrophysical chemistry, and mineralization. (1) He was the first to introduce physicochemistry into the study of genesis of igneous rocks, crust-mantle deep processes, regional petrogeotectonics, and mineralization. (2) He proposed a theoretical model of "continental roots–plume tectonics" to explain the geodynamic evolution of the lithosphere of China and Eurasia, based on tectonic division at the lithosphere scale and crust-mantle deep processes. (3) He proposed a composition profile and thermal structure model for the upper mantle in eastern China, revealing the huge thinning of the North China Craton lithosphere. (4) He proposed that the cataclysm of the lithosphere–asthenosphere system was the fundamental power source behind the metallogenic "great explosion," which was efficiently applied to explaining metallogenic mechanism and evaluating the potential of mineral resources.

Xuanxue Mo is an academician of the Chinese Academy of Sciences and a professor at the China University of Geosciences. He conducted research at UC Berkeley and LBL Berkeley, USA (1981–1983). His research interests are igneous petrology, petrophysical chemistry, petrotectonics, and mineralization, including (1) the magmatism and tectonic evolution of the Tibetan Plateau and its surrounding regions; (2) the geodynamic background of mineralization and distribution of large to super-large deposits in China, especially in the Tibetan Plateau and the southwest China; and (3) experiments and modeling of phase equilibrium and thermodynamics of magma system. He has been teaching igneous and metamorphic petrology, thermodynamics of magmatic processes, and advanced igneous petrology at both undergraduate and postgraduate levels.

Yaoling Niu is currently an honorary professor at the Laoshan Laboratory, China. He has been both an educator and a researcher in the field of Earth Science. He taught crystallography, optical mineralogy, mineralogy, igneous and metamorphic petrology, elemental and isotope geochemistry, ore deposits, thermodynamics for geologists, global tectonics, and field geology at both undergraduate and postgraduate levels in China (1982–1985), Australia (1993–2001), the USA (1986–1987 and 2003–2004), and the UK (2005–2022). His research includes: (1) petrology and geochemistry of mantle peridotites, basalts, basaltic rocks, granites, granitoid rocks, high- and ultra-high-pressure metamorphic rocks, and ore deposits; (2) ocean ridge magmatism, intraplate magmatism, subduction-zone processes, magmatism associated with continental collision, and continental crust accretion; (3) causes and effects of seafloor subduction, global tectonics, chemical geodynamics, and mantle circulation; and (4) elemental and isotope geochemistry, geochronology, and geological applications.

Zengqian Hou is an academician of the Chinese Academy of Science and a senior research geologist at the Chinese Academy of Geological Sciences (CAGS). He has held various leadership positions, including Director-General of the Institute of Mineral Resources (2000–2005) and the Institute of Geology (2005–2018), CAGS. He has also served as vice-president of NSFC (2018–2023), chairman of the

IGCP Committee of China (2016–2023), and regional vice president of the Society of Geology Applied to Mineral Deposits (2011–2013). He has led several major scientific projects, including an IGCP project and two Chinese State Key 973 Fundamental Research Projects, focusing on the lithospheric architecture and relevant mineral systems in collisional orogenies. These researches have led to the establishment of a new theoretical framework and genetic models for collision-related metallogenesis (REE, Cu, Pb-Zn, and Au) and major discoveries of world-class copper and other metallic deposits in China. He was thus awarded the Grand Prize and First Prize of the National Scientific and Technical Advancement Award as well as the Second Prize of the National Natural Science Award. He also served as Regional Vice President of the Society of Geology Applied to Mineral Deposits (2011–2013).

Zhidan Zhao is a professor of petrology and geochemistry at the China University of Geosciences in Beijing (CUGB). He received his BS degree (1990) and PhD (1995) degree in geochemistry from the China University of Geosciences in Wuhan (CUGW). He completed his postdoctoral research at the Institute of Geochemistry, Chinese Academy of Sciences in Guiyang in 1998, before joining the faculty at CUGB, where he remains to this day. In the intervening years, he served as a visiting professor at UC Berkeley (2002–2003 and 2012–2014), where he collaborated with Professor Don DePaolo on the isotopic geochemistry of postcollisional rocks from the Tibetan Plateau. He taught optical mineralogy, petrology, and petrogeochemistry at both undergraduate and postgraduate levels. His research includes: (1) regional geochemistry of various rocks along the Qinling orogenic belts, (2) high pressure and high temperature experimental study of elastic wave velocity in rocks, and (3) petrology and geochemistry of the postcollisional magmatism on the Tibetan plateau and in the adjacent regions.

Contributors

Jinfu Deng China University of Geosciences, Beijing, China

Guochen Dong China University of Geosciences, Beijing, China

Zengqian Hou Institute of Geology, Chinese Academy of Geological Sciences, Beijing, China

Xuanxue Mo China University of Geosciences, Beijing, China

Yaoling Niu Laoshan Laboratory, Qingdao, China

China University of Geosciences, Beijing, China

Institute of Oceanology, Qingdao, China

Zhidan Zhao China University of Geosciences, Beijing, China

Dicheng Zhu China University of Geosciences, Beijing, China

Chapter 1
Spatial Distribution of Igneous Rocks in the Tibetan Plateau

Abstract The Tibetan Plateau contains abundant igneous rocks of varying kind, including ultramafic, mafic, intermediate and felsic volcanic and intrusive rocks of Paleozoic to Cenozoic age, which record the origin and evolution of the Tibetan Plateau since the Paleozoic. They are distributing in five giant magmatic belts from the north to south as follows: Qinling-Qilian-Kunlun giant magmatic belt (QQKB), Qiangtang giant magmatic belt (QTB), Gangdese giant magmatic belt (GDB), Himalaya giant magmatic belt (HMLB) and Sanjiang giant magmatic belt (SJB). Magmatic rocks in west Qinling formed during late Permian to early middle Triassic, and late Triassic to early Jurassic epochs, associated with seafloor subduction and continental collision. The Qilian magmatic belt contains ocean ridge basalts and oceanic island basalts of early Cambrian to early Ordovician age, arc volcanic rocks and granitoids association and back arc basin magmatic association of the Ordovician age with locally para-aluminum granite and A-type granite association of late Silurian to Devonian age. The Kunlun magmatic belt widely developed vast intrusive (mainly granitoids) and some volcanic rocks from Precambrian, through early Paleozoic (Caledonian), late Paleozoic to early Mesozoic (Hercynian to Indosinian), late Mesozoic (Yanshanian) to Cenozoic (Himalaya) epoch. In Qiangtang area volcanic rocks formed mainly in Cenozoic epoch, forming the largest Cenozoic volcanic belt, while intrusive rocks developed mainly in the southern Qiangtang, including late Paleozoic mafic rocks and late Mesozoic granite-granodiorite-diorite and some sporadic Cenozoic porphyry, ultramafic rocks and diabase dikes. In Himalaya area, the main intrusive rock granitoids with pale color distribute in the High Himalayan Belt and Northern Himalayan Belt during the Proterozoic, Ordovician and Miocene epoch and also a large number of volcanic interbeds dominated by basalts of varying ages are found in the Himalayan Belt. The Sanjiang region developed vast intrusive and volcanic rocks with complicated lithologies forming several magmatic belts, which contain important information of the history and evolution of pre-Paleo-Tethys, the Paleo-Tethys, the Neo-Tethys and Indo-Eurasian collision.

Prof. Guochen Dong has fully contributed to this chapter.

X. Mo et al., *Magmatism and Crustal Evolution of the Tibetan Plateau*, Modern Approaches in Solid Earth Sciences 25,
https://doi.org/10.1007/978-981-97-2109-2_1

Keywords The Tibetan Plateau · Magmatic rocks · Intrusive rocks · Volcanic rocks · Spatial distribution · Giant magmatic belts

The Tibetan Plateau is the youngest, largest and highest plateau on earth and contains abundant and widespread igneous rocks of varying kind, including ultramafic, mafic, intermediate and felsic volcanic and intrusive rocks of Paleozoic to Cenozoic age. They record the origin and evolution of the Tibetan Plateau since the Paleozoic. According to tectonic regimes, histories and locations as well as lithological associations, five giant magmatic belts are recognized from the north to south as follows (Fig. 1.1): Qinling-Qilian-Kunlun giant magmatic belt (QQKB), Qiangtang giant magmatic belt (QTB), Gangdese giant magmatic belt (GDB), Himalaya giant magmatic belt (HMLB) and Sanjiang giant magmatic belt (SJB).

1.1 Igneous Rocks in Qinling-Qilian-Kunlun Orogenic Belt

1.1.1 West Qinling Magmatic Belt

The west Qinling magmatic belt, the main part of the Qinling magmatic belt, is located on the northeastern margin of the Tibetan Plateau. Magmatic rocks in the belt formed during late Permian to early middle Triassic, and late Triassic to early Jurassic epochs. It is considered that the west Qinling magmatic belt was associated with seafloor subduction during from the late Permian to early middle Triassic, and followed by continental collision in the timeframe of the late middle Triassic to early Jurassic.

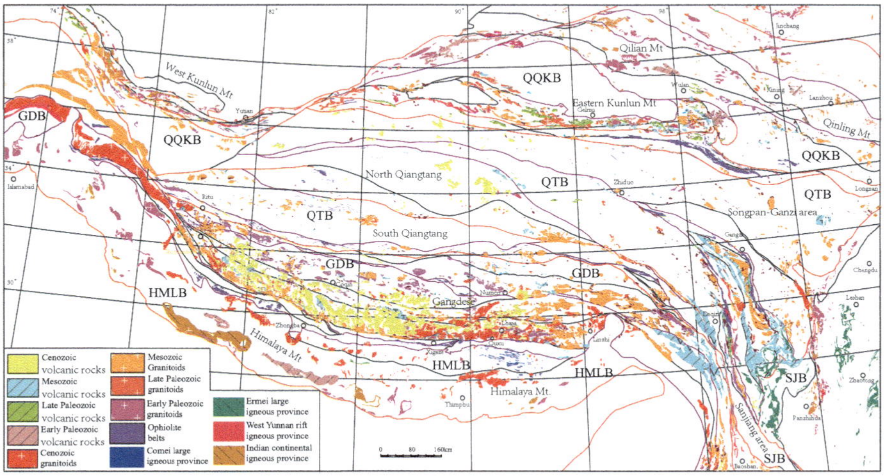

Fig. 1.1 Igneous rock distribution in the Tibetan Plateau. *QQKB* Qinling-Qilian-Kunlun giant magmatic belt, *QTB* Qiangtang giant magmatic belt, *GDB* Gangdese giant magmatic belt, *HMLB* Himalaya giant magmatic belt, *SJB* Sanjiang giant magmatic belt

There are widespread granitoid intrusions of intermediate and felsic composition in the west Qinling magmatic belt. These granitoids are of varying age mostly associated with the Indosinian and Yanshanian orogenesis and lesser with the Caledonian, Hercynian and Himalayan orogenesis.

The Caledonian intrusions are characterized by high Ba–Sr, and can be divided into K-rich high Ba–Sr intrusion represented by the Yangjiazhuang quartz diorites and Na-rich high Ba–Sr intrusion represented by the Tangzang quartz diorites. The zircon U–Pb dating gives 439–438 Ma for the K-rich dioritic intrusions, and 429–425 Ma for the Na-rich intrusion. Also, these intrusions are metaluminous to weakly peraluminous with relatively uniform Sr ($^{87}Sr/^{86}Sr = 0.7032–0.7059$) and Nd ($\varepsilon_{Nd}(t) = -0.65$ to -2.78) isotopic compositions.

The Indosinian intrusions are mainly found in eastern and western segments of the west Qinling belt with lithologies of monzogranite, granodiorite, diorite, and alkaline granite. They generally occur as batholith and stocks of complex lithologies in the eastern segment and as small stocks with porphyritic texture in the western segment. There were two periods of Indosinian granitic magmatism in the West Qinling belt, i.e., early and late magmatic activities with a gap of ~6 Myrs (233–227 Ma) in between. The early magmatic activity (249–233 Ma) formed a series of magmatic rocks in the western and northern segments of the West Qinling belt, with a peak at ~242 Ma. The late magmatic activity (227–203 Ma) with the peak at ~218 Ma mainly developed in the middle and southern segments of the West Qinling Belt with little in the western segments. There are also some Indosinian mafic-felsic magmatic rocks such as pyroxenite-syenite and adakite in the west Qinling belt. The pyroxenite-syenite is mainly composed of biotite pyroxenite, gabbro, quartz monzonite and syenite. High Sr/Y granitoids of varying genesis in the early and late periods are also found in the eastern segment of the west Qinling belt. Among the late Triassic high Sr/Y granitoids, the Miba granodiorite as an example was considered to have resulted from partial melting of mafic lower crust of normal thickness, while the Dashan granodiorite as another example was generated by partial melting of the thickened mafic lower crust.

The volcanic rocks were mainly formed since the late Triassic concentrated with characteristics in the following areas: (1) those of the late Triassic are of high K calc-alkaline intermediate to felsic composition distributed in the Ela Mt. area as a NNW belt, and are of intercontinental mountain building origin with a thickness >6000 m, (2) those of the late Triassic to Jurassic are mainly distributed in Gansu and Qinghai as a NW-SE belt, and are dominated by basaltic andesite, andesite and dacite of high K calc-alkaline feature, (3) those of the Jurassic mainly occur in the boundary area between Gansu and Sichuan with an outcrop area of tens of square kilometers, and are mostly of high K_2O shoshonite association consisting of mafic, intermediate and felsic rocks, (4) those of the Cretaceous consist of alkaline olivine basalt and basaltic andesite and occurs locally in Gansu and Qinghai with small outcrop and thickness, (5) those of the Cenozoic are mainly found in a Cenozoic faulted basin in Lixian, Gansu and are ultrapotassic magmatic rocks of kamafugite and carbonate association. The kamafugite gives varying Ar–Ar age of 23–7.1 Ma, and commonly contains abundant inclusions of clinopyroxene and phlogopite crystals and mantle xenoliths, which is a good site to study the lithosphere and asthenosphere.

1.1.2 Qilian Magmatic Belt

The Qilian tectonic magmatic belt is located on the northern margin of the Tibetan Plateau, connected with the Qaidam block in the south, with the North China craton in the north, with the west Qinling in the east, and cut as a boundary in the west. The magmatic belt contains ocean ridge basalts and oceanic island basalts of early Cambrian to early Ordovician age, arc volcanic rocks and granitoid association and back arc basin magmatic association of the Ordovician age. There is a para-aluminum granite and A-type granite association of late Silurian to Devonian age (Song et al. 2013). Generally, the belt develops different magmatism in the north, middle and southern sub-belts of the tectonic units.

The north part of the Qilian is consisting of Cambrian to early Ordovician mafic to ultramafic volcanic rocks in Tuolai Mt. to Daban Mt. area, lenticular mafic ultramafic blocks and some tholeiitic to calc alkaline series of volcanic rocks in the south mountain of Zoulang. The tholeiitic to calc-alkaline volcanic rocks was considered as a product of volcanic arc with eruption age of 445–486 Ma (Xia et al. 1998). There are some late Cenozoic (Pleistocene) basaltic volcanic rocks in Hongliuxia and Hanxia area of Yumen area, which is located in the contact area between Qilian Mt. and Aerhchin Mt. (Pan et al. 1990). The lithology of the volcanic rocks is composed of alkaline basalt and trachytic basalt with olivine phenocryst and in the matrix (Lai and Zhang 1999).

The middle part of the Qilian was developed based on the Archean to early Proterozoic tonalite-trondhjemite-granodiorite (shortly TTG) series (Pan et al. 2004). The early Proterozoic metamorphic rocks include migmatite, garnet quarts shist, amphibolite and marble. Neoproterozoic metasandstone, phyllite and slate also occur in the area. There are only a few mafic volcanic rocks as interlayers in the Ordovician strata.

The south part of the Qilian developed middle Cambrian to late Ordovician abyssal volcanic rocks and clastic sedimentary rock association on the base of Proterozoic metamorphic basement. The lithology of the association consists of andesite, basaltic andesite, basalt and the related pyroclastic rocks associated with Caledonian intermediate to felsic intrusions of biotite granite, monzogranite, granodiorite and diorite. There are locally basaltic layers in middle to upper Ordovician carbonate strata, andesite in Silurian dolomite strata, and a few Hercynian granite intrusions as stocks.

1.1.3 Kunlun Magmatic Belt

The Kunlun magmatic belt as part of the central orogen of the key tectonic-magmatic active belt can be divided into east Kunlun and west Kunlun (Jiang et al. 2000). Vast intrusive and some volcanic rocks are widely distributed in the Kunlun belt from Precambrian, through early Paleozoic (Caledonian), late Paleozoic to early Mesozoic

(Hercynian to Indosinian), late Mesozoic (Yanshanian) to Cenozoic (Himalaya) epoch. Especially the granitoids are most abundant in the late Paleozoic to early Mesozoic (Hercynian to Indosinian), with the total granitoid outcrop area in excess of 47,000 km^2.

1.1.3.1 East Kunlun

There are various magmatic rocks in different geological time periods in the east Kunlun belt.

Precambrian Magmatic Rocks in the Metamorphic Basement

Metamorphic gabbro was considered as the oldest rocks with zircon U–Pb age of 2468 ± 46 Ma (Luo et al. 2002). There are also tonalite-trondhjemite association (shortly TT) (Deng et al. 1995) or tonalite-trondhjemite-granodiorite (shortly TTG) and granitoids with Rb–Sr isotopic age of 1846 Ma and zircon U–Pb age of 1955 ± 6 Ma (Luo et al. 2002) associated with the metamorphic gabbro.

Early Paleozoic Magmatic Rocks

The Cambrian gabbro occurs locally with zircon U–Pb age of 518 Ma and hornblende Ar–Ar age of 578.5 ± 6.1 Ma (Jiang et al. 2000) in the Qingshuiquan ophiolite. There are some magmatic arc-type granitoids in the east Kunlun, such as Alake intrusion with bulk-rock Rb–Sr isochron age of 508 Ma, Wanbaogou intrusion with hornblende Ar–Ar age of 450 Ma and Shihuigou intrusion with zircon U–Pb age of 471 and 485 Ma. In the belt are also two mica granitoid intrusion in wangaogou with zircon U–Pb age of 412.6 Ma and Elgunsi intrusion with Rb–Sr isochron age of 379.6 Ma. All these data indicate that the ocean basin was closed in the late Silurian to middle Devonian before the continental collision.

Late Paleozoic to Middle Mesozoic Magmatic Rocks

Some basalts with Rb–Sr age of 347 ± 9 Ma (Bian et al. 2001) and zircon U–Pb age of 308.2 ± 4.9 Ma (Yang et al. 2004) occur along the ophiolite suites, e.g., the Buqingshan ophiolite and Derni ophiolite.

Great volume of middle-late Permian to early Triassic granitoids are widespread in the W-E belt and occur mainly as elongated batholith in the northern part of the east Kunlun and tend to become less southwards. For example, the Xialamusong intrusion near Dulan is a granitoid batholith with an outcrop area of 240 km^2. It is the granitoids that constitute the Kunlun magmatic belt. MMEs are also common in these granitoids. Ultramafic or mafic intrusive complexes also occur in the granitic batholith, especially at the margins. Such as in Jialuhe of east Kunlun, the mafic to ultramafic rocks consist of serpentine peridotite, hornblende gabbro, pyroxenite and the hornblende in gabbro gave an Ar–Ar age of 226.4 Ma (Mo et al. 2004; Liu et al. 2004), showing magma mixing with the granodiorite host rocks (Liu et al. 2004).

The late Paleozoic volcanic rocks include mostly Carboniferous basaltic andesite and andesitic pyroclastic rocks and few upper Devonian and Permian pyroclastic rocks. There are also few mafic volcanic rocks in some ophiolite suite and few arc type volcanic rocks locally. The Mesozoic volcanic rocks abundant in the Triassic, forming two volcanic strata formations (Babao Mt. Formation and Ela Mt. Formation) with thickness of >3500 M. The volcanic rocks in the Babao Mt. Formation are distributed as an E-W belt and consist of andesite, rhyolite, the related pyroclastic rocks with a few basalts. The Ela Mt. Formation occurs along Ela Mountain and is composed of intermediate to mafic lavas and sedimentary pyroclastic rocks in the lower part and intermediate to felsic lavas and sedimentary pyroclastic rocks in the upper part.

1.1.3.2 West Kunlun

The magmatic activity is very intensive from the Proterozoic to the Cenozoic with diverse lithologies in the west Kunlun. Especially, the granitoids are abundant and exposed over a large area in excess of ~50% of the total area.

The Granitoids in the West Kunlun

The Precambrian granitoids distribute along the main ridge of west Kunlun and consist of batholiths of gneissic granodiorite and gneissic syenitic granite. The Caledonian granitoids are the main components of the west Kunlun and comprise of diorite, tonalite, granodiorite and biotite monzogranite with LA-ICP-MS zircon U–Pb isotopic age of 540–400 Ma (Zhang et al. 2000). The Hercynian to Indosinian granitoids mainly occur along the Kangxiwa fault in the E-W belt and consist of batholiths and stocks of diorite, tonalite, granodiorite, monzogranite, syenitic granite and alkaline granite with zircon U–Pb age of 260–200 Ma. The Cenozoic granitoids are minor with a few occurrences of late Miocene (10.5 Ma) tourmaline-bearing two mica granite overlain in unconformity by the upper Miocene (4.6 Ma) quartz sanidine tuff.

Volcanic Rocks in the West Kunlun

The early Proterozoic volcanic rocks occur mainly along the Muji-Bulunkou area with only a few outcrops. The middle Proterozoic strata (Changcheng system Sailajiazi group and Jixian system Daxidaban group) contains some oceanic faces spilite-keratophyre suites. Paleozoic volcanic rocks are widespread, including magmatic arc, back arc basin and rifted basin assemblages with different characteristics in eastern and western parts of the west Kunlun. The volcanic rocks in the eastern part formed a Silurian to middle Devonian volcanic-sedimentary construction of 6000 m thick, which is mafic in the low section and more felsic in the upper section, forming a spilite-keratophyre suite (Jia et al. 1999). And the volcanic rocks in the western part occur as an E-W belt of >30 km long and up to 2000 m wide, composed of hornblende basalt, basaltic andesite, andesite and rhyolite and ignimbrite along with coeval hornblende gabbro and dolerite.

There are some potassic–ultra-potassic volcanic rocks, considered as post collisional products. There are also >10 m thick olivine basalt, trachyte and pyroclastic rocks in Kanxiwa, 1–2 m thick volcanic braccias and pyroxene trachyte in Dahongliutan (Deng et al. 1998), >100 m thick olivine basalt and andesite in Quanshuigou (Arnaud et al. 1992), and 10–15 and 7 m thick two layers of grey olivine trachyte and shaoshonite in the Pulu etc. with LA-ICP-MS zircon U–Pb isotopic age of 0.12–6.4 Ma (Mo et al. 2006).

1.2 Igneous Rocks in Qiangtang: The Central Belt

1.2.1 Hoh Xil-Bayan Har Tectonic Magmatic Rock Belt

The belt refers to the vast region south of the Kangxiwar-muzitag-Maqin suture zone and north of the Xijir Ulan Lake-Jinshajiang suture zone. In general, the magmatic rocks in this belt are very distinctive but less abundant. The intrusive rocks are mainly intermediate-felsic composition of late Indosinian-Yanshanian age. The volcanic rocks are mainly potassic basalt series of younger age from the tertiary to present. The most representative intrusive rocks of late Indosinian-early Yanshanian age (~172–201 Ma) are the Aksai Chin intermediate-felsic composition in the west of the belt with an exposed area of ~3000 m^2. The main rock types include granodiorite, monzogranite, syenitic granite, diorite and granite porphyry. The Bayan Har intermediate-felsic intrusion is emplaced during the late Indosinian period (~180–204 Ma) with a total outcrop area of ~4000 m^2, including diorite, quartz monzonite, granodiorite, monzogranite, syenogranite and syenite etc. The Yanshanian (100–130 Ma) intermediate-felic intrusive rocks have an exposed area of ~2800 km^2, mostly distributed south of the East and West Kunlun Mountains, dominated by granodiorite with some quartz porphyry.

The Cenozoic potassic volcanic rocks make the Hoh Xil magmatic belt special. From west to East, a large volume of Cenozoic volcanic rocks are distributed along the southern margin of the Kunlun Mountain and Hoh Xil, concentrated to form volcanic fields such as Yanghu, Qiangbaqian, Yongbocuo, Muztag, Jingyuhu-Xiongyingtai, Xiangyang Lake, Heituofeng, Wuxuefeng, Kekaohu and Damaoshan etc.

1.2.2 North Qiangtang Magmatic Belt

This belt includes the large region between the Xijir Ulan Lake-Jinshajiang suture zone and the Longmucuo-Shuanghu suture zone. Intrusions only crop out in a few places, but Cenozoic volcanic rocks are widespread, forming the largest Cenozoic volcanic belt in the northern regions of the Greater Tibetan Plateau.

1.2.2.1 Intrusive Rocks in the North Qiangtang

The intrusive rocks in the North Qiangtang are widespread in the east and southeast area of the belt, comprising middle Late Mesozoic granites, including mainly of monzogranite, syenogranite, granodiorite, quartz diorite, diorite, granodiorite porphyry, and monzogranite porphyry and so on. All the intrusive rocks are concentrated to form the Nyainqentanglha granite belt, the Chasang-Shuanghu granite belt, the Margog Caka intermediate-felic intrusive rock area and the Wuqi-Changmaoling small porphyry (dyke and branch) area. There are also Cenozoic Geladandong monzogranite and Laoriteco granite porphyry.

There are also small intrusions in the belt, such as Cenozoic syenite porphyry, lamprophyre and alkaline rocks. Syenite porphyry associated with quartz syenite porphyry and monzonite porphyry mostly occur in the northeast of Wangxing Mt.-Mazhangcuoqin, Quemocuo-Wuguo, the middle reaches of Bancheng River and north of Mt. Jurhen Ul in the eastern section of Mt. Ulan Wula. In addition to the above three porphyry belts, syenite porphyry is also found in Laxiongcuodong, Badaohu and other places. Alkaline rocks are only found in the northeast of Cha Wula in the east of the belt. There are 4–5 intrusive bodies crop out in a small area of <0.5 km^2. The lamprophyre intruding along the fault as dikes and veins is distributed on the side of new North-South faults such as the upper reach of the Tuotuo River, peninsula lake and MayiGangri-Bamaoqiongzong. The lithology is mainly biotite lamprophyre and olivine-biotite lamprophyre.

A few mafic and ultramafic intrusions are also exposed, reflecting activities of mantle magmatism. There are two small peridotite-gabbro intrusions in the west Quemoco mountain, altered pyroxene peridotite, diabase stocks and two N-S striking dykes in the Margog Caka mountain, and a number of diabase dykes in the northwest of Chaduo Gangri.

1.2.2.2 Volcanic Rocks in the North Qiangtang

In addition to a small amount of Late Triassic, Jurassic and early Cretaceous volcanic rocks, the volcanic rocks in the North Qiangtang region are mainly of Cenozoic age.

The late Triassic volcanic rocks in the North Qiangtang region are mainly distributed in two areas (Yang and Jin 2001). The first with name of Rola Kangri volcanic rocks is mainly distributed in the MargaiCaka, Jinshi mountain, Rola Kangri, Yongbo lake and Yinchunkou in the north Qiangtang. They occur as interlayers in the flysch volcanic sedimentary formation of the Rola Kangri group. The rocks include alkaline olivine basalt and basaltic amphibole-pyroxene andesite, resembling ocean island alkaline basalts, emplaced in flysch formation of an island arc front slope basin. The second, named Xiaochaka-Panqiong volcanic rocks, occurs between Xiaochaka and Xiangqiong, and is distributed in an E-W belt along the southern margin of the central uplift of Qiangtang.

In addition, the Jurassic and lower Cretaceous strata contain an assemblage of andesite/dacite and an assemblage of nepheline alkali basalt and andesite south of the Guozhaco-Chazuri-Nadigangri-Maweishan area.

The Cenozoic volcanic rocks in the North Qiangtang region include those in the Bamaoqiongzong area, the Dogai Coring area, the Taiping Lake-Botao Lake area, the Luanqing Mt. area, the Yongbo Lake area and the Zhentouya area.

The Bamaoqiongzong lavas that cover a large area are olivine leucite porphyry, phonolite, pseudo-leucite porphyry and trachyte. The rocks are ultra-alkaline series, giving an average K–Ar age of 26.54 Ma (Wangming 1998).

The Dogai Coring lavas consisting of two eruption episodes cover a large area. Each episode is successively transformed upward from the thick layer of tuffaceous brecciaed basalt in the lower section to molten slag basalt, trachyte basalt and pyroxene andesite. These first episode volcanic rocks give a K–Ar age 39.7 Ma (Chi et al. 1999).

Taiping Lake-Botuo Lake volcanic rocks were developed in the eastern extension of the Dogai Coring volcanic belt. The north part of the volcanic belt is dominated by trachyte andesite and andesite, and some volcanic breccias. Nepheline basanite, leucite basanite, basalt, pyroxene andesite and biotite trachyte andesite are exposed in the area of the Potao Lake and the Yanzi Lake. Leucite tephrite is exposed 10 km away from the Potuo lake. Dacite samples in the south of the Taiping Lake give a K–Ar age of 34.55 Ma.

The Luanqing Mt. volcanic rock area includes those erupted around the Luanqing Mt., the Bandao lake, the Heihu range, Donghuyuejinla and other places. The volcanic rocks are unconformably over the upper Jurassic in the form of lava beds. The lavas are porphyritic, massive and rhyolitic with vesicular and amygdale structures. The phenocrysts consist mainly of feldspar, pyroxene, hornblende and biotite and the matrix are interwoven and cryptocrystalline. The main rock types are trachyte, dacite, shoshonite and andesite. The thickness of volcanic rocks varies in different from one place to another, up to 244 m. There are more pyroclastic materials in the Yuejinla area than other places with thickness of >780 m. To the east of Mt. Jurhen Ul, the lavas and pyroclastic rocks develop alternately with a thickness of more than 1000 m. There are calderas on the North Bank of the Haobo Lake and Luanqingshan. The K–Ar dating gives an age of 43.7 Ma on trachyte of Luanqing Mt., 33.8 Ma on trachy andesite, 39.8 Ma on andesite (Tan et al. 2000), 40.7 Ma on the trachyte of Bandao lake, 35.8 Ma on dacite porphyry of Heihuling, and 33.9–32.6 Ma on the trachy andesite of Yuejinla (Li et al. 2000).

Volcanic rocks in the Yongbo lake region are lava flows dominated by amygdaloidal pyroxene andesite and amygdaloidal andesite with a K–Ar age of 9.4 Ma.

The Zhentouya volcanic rock is mainly Paleogene lava flows. The lower section is massive trachy andesite, and the upper section is vesicular lavas with obvious flow structure. The K–Ar dating gives the lower trachy andesite an age of 44.66 Ma, and the top vesicular trachy andesite an age of 4.27 Ma (Zhang et al. 1994, 1996).

1.2.2.3 Volcanic Rocks in West Qiangtang

West Qiangtang refers to the region of North Qiangtang west of the Altun fault. In this region, there are several Cenozoic volcanic field, such as Tongtianqiao, Hongshan lake and bangdacuo. Deng et al. (2001) defined these rocks as a sodium rich series, which differ from the potassium volcanic rocks widespread in other areas of the Qiangtang.

The Tongtianqiao volcanic rocks consist mainly of mafic and intermediate-mafic lavas between Cretaceous purple clastic limestone of Tielongtan group on both sides. The rocks are porphyritic in texture, in which phenocrysts are subhedral augite or altered plagioclase and the matrix are phanerocrystalline or cryptocrystalline. The rocks were also altered with epidotization, sericitization and carbonation.

The Hongshanhu volcanic rocks are composed mainly of diabase with a small amount phenocryst of altered clinopyroxene and anorthite, intruding the Cretaceous clastic limestone of the Tielongtan group with baked margins.

Bangdacuo volcanic rocks consist of olivine basalt and basalt with olivine or clinopyroxene phenocrysts and glassy matrix (Deng et al. 2001). They occur along an E-W trending fault, in which the north side is the gravel-bearing slate of the Late Carboniferous Guliya group, and the south side is the clastic limestone of the Cretaceous Tielongtan group.

1.2.3 South Qiangtang Magmatic Belt

1.2.3.1 Intrusive Rocks

The intrusive rocks in this zone mainly include late Paleozoic mafic rocks and late Mesozoic granite-granodiorite-diorite. There are also sporadic Cenozoic porphyry and some ultramafic rocks and diabase dikes, with a total exposed area of ~4000 km^2.

The late Paleozoic mafic intrusive rocks are concentrated in the Alucuo-Dayouri-Jiaomuri area in the middle of the Qiangtang central uplift belt. They intruded the metamorphic rocks of the Middle Carboniferous Mushirebuka formation, including diabase and gabbro (Li et al. 1996).

The Mesozoic granitoids are mostly distributed in three parts, i.e., the Zecuo-Qingche lake, Chabu-Luguxi-Yulinshan and northwest of Yibuchaka in the middle and western part of the belt, as well as the southern edge of the Chawula-Lalongong granitoid belt, with a total area of ~1700 km^2. The rock types are mostly granite, biotite granite, monzogranite, granodiorite and quartz diorite.

There are only minor Cenozoic intrusive rocks, mostly hypabyssal intrusive vein bodies. They consist of granite porphyry, syenite porphyry and Muscovite pegmatite vein, all of which intruded the Jurassic system strata.

In addition, there are a small amount of mafic and ultramafic dykes. The mafic dykes include diabase and diabase porphyrite, which occur on the south bank of Zaga Zangbu. Ultramafic cumulate rocks of varying composition occur as small stocks in the western side of a NW-SE strike slip fault southwest of Bulaco. The rocks are unconformably overlain by Neogene Kangtuo formation northwest of the fault.

1.2.3.2 Volcanic Rocks

The south Qiangtang is another important area of Cenozoic volcanic rock distribution in northern Tibet. The Cenozoic volcanic rocks are mainly distributed in several areas including Yulinshan, Nadingcuo, lagala, zogouyouchagou, Kangtuo and other areas.

Yulinshan volcanic rock as a medium-sized lava flow covers unconformably the Jurassic, Cretaceous and Paleogene strata. It is mainly composed of leucite porphyry, phonolite, aegirine trachyte, neon nepheline trachyte, tetrahedrite trachyte etc. The Nadingcuo volcanic rock consists mainly of andesite, basalt, andesite, rhyolite, dacite and pyroclastic rock. The lithology of Lagala volcanic rock is mainly composed of basalt with olivine phenocryst. The Zogouyouchagou volcanic rock is mainly High potassium calc-alkaline andesite-trachytic andesite (Chi et al. 1999).

1.2.4 Karakoram Area

Karakoram refers to the region to the west of the Arkin fault, south of Karakax-Bolike-Muztag fault and north of Hongshanhu-Qiao'ertianshan-Jinshajiang fault zone.

The magmatic rocks in the Karakoram region are mainly intrusive rocks, including Jurassic and Cretaceous granitoids and Neogene alkaline intrusive rocks. The Jurassic granitoids are mainly distributed in three areas, i.e., Taxkorgan, Mazha-Tatulugou and Guding-Kalatage mountain. The rock types include biotite granite, monzogranite, porphyritic monzogranite and granodiorite and diorite.

The Cretaceous granitic intrusions, occurring in NWW-SEE direction in the Khunjerab-karachigu and its west area, include Waugiri, Baldarum, Chaengtomu, Taike, Mingtiegai, Attigial and Khunjerab intrusions. They include those of east Khunjerab, Tatulugou, the upper reaches of keleqing River, east Ageledaban and sharekle, forming the North Karakoram belt and distributing in southeast direction from Karachigu to the

Hongshan Lake through jogori peak and its north slope. The rock types are tonalite, biotite monzogranite, biotite syenogranite and biotite alkali feldspar granite.

Cenozoic alkaline intrusions are located south of the west Kunlun belt and east of the Pamir tectonic syntaxis. They are distributed along the Karakoram fault zone on both sides of the Taxkorgan valley and consists of several Cenozoic alkaline complexes and granitoid intrusions. The Taxkorgan intrusion west of the Tashkurgan county is the largest Cenozoic alkaline complex in the area. It is distributed N-S along the Taxkorgan fault and is composed of Kuzigan and Karibasheng intrusions. A series of alkaline extrusive to hypabyssal complexes with an age of 11 Ma are developed from the Taxkorgan complex northwest towards the Dunkeldik Valley in Tajikistan and are composed of alkaline basalt to trachyte, syenite and carbonatite with xenoliths of deep crust and mantle such as eclogite, garnet pyroxenite and granulite (Hacker et al. 2005).

1.3 Igneous Rocks in the Gangdese

The Gangdese magmatic belt (also known as the Lhasa block or Lhasa terrane) is a giant tectono-magmatic belt located between the Bangong Lake-Nujiang ophiolite belt, which is ~2500 km long along E-W strike and 100–300 km wide across the strike. It is the largest magmatic belt in the Tibetan Plateau.

1.3.1 The Granitoids in the Gangdese

The Gangdese magmatic belt can be divided into three magmatic sub-belts from north to south.

1.3.1.1 Gaize-Bange-Boshula Ridge Granite Belt

The granite belt is distributed in the north magmatic belt of the Gangdese between the Bangong Lake-Nujiang suture zone to the north and the Yongzhu-Jiali-Bomi suture zone to the south. The intrusions are discontinuously distributed in Xingdukushi, Bangebengcuo, Nierong, Jiali, Bomi, and Boshula Ridge and other places with slightly difference between the east and the west cross the Selincuo as a boundary. There are 18 complex batholith and stock intrusions in the western section, which consists of granodiorite, quartz diorite, quartz monzonite, monzonite granite, porphyry granite, and alkali-feldspar granite porphyry and a few tonalite. The rocks are generally calc-alkaline and tend to become more alkaline. These rocks coexist with coeval intermediate-mafic volcanic rocks, and altogether form the magmatic complex.

There are 36 intrusions in the eastern section, intruding the Mesozoic strata in the form of batholith and stocks. The rock types are mainly biotite granite, biotite monzonite granite, granodiorite, quartz diorite and a few two-mica granite. There are alkali feldspar granites (porphyry) formed later.

The SHRIMP zircon U–Pb isotopic dating on rocks of the granite belt are concentrated on ages of late Jurassic-Cretaceous with minor of Triassic. We can roughly divide them into three age groups with different tectonic properties: (1) Triassic Island arc granites of 227–195 Ma (Mo et al. 2004) throughout the extension of the E-W belt, indicating the presence and significance of an island arc orogeny in the Gangdese in the late Triassic. (2) Late Jurassic-Early Cretaceous island arc granites with an age range of 170–99 Ma (Guo et al. 1991; Qiu et al. 2002), dominated by quartz diorite, tonalite and granodiorite. It is I-type granite, and presumed to be related to the southward subduction of the Nujiang Tethys Ocean. (3) Early Cretaceous to the Late Cretaceous orogenic granites with an age range of of 99–74 Ma, dominated by monzogranite, and evolved into peraluminum granite in the late period. These younger granitoids are considered as the product after closure of the Nujiang Tethys ocean orogen.

1.3.1.2 Shiquanhe-Cuoqin-Namtso Igneous Belt

This belt is the central belt of the Gangdese, also named as Geji-Cuoqin back-arc belt and the Longgeer-Gongbujiangda fault-uplift belt (Pan et al. 2004), located north of the Gaer-Asuo-Yongzhu-Jiali fault and south of the Longgerer-Cuomai fault. There are dominantly 150 small isolated intrusions, mainly concentrated in the areas of Cuoqin, Bangduo and Shenzha. They intrude the late Paleozoic and Mesozoic strata, including Laga Formation (C_2–P_1), Jienu Formation (J_{2-3}), Zenong Group (K_1) and can be divided into three stages, including Triassic to late Early Jurassic (195–227 Ma), Late Jurassic to Cretaceous (178–65 Ma) and Paleogene to Miocene (62–15 Ma).

The late Triassic to Early Jurassic granites are distributed in the Bomi area, south of the Polongzangbo ophiolitic melange, and are I-type with island arc properties. They intrude the Mesoproterozoic Nianqingtanggula group, Devonian Songzong Formation and Carboniferous Nuocuo Formation and Laigu Formation. Li and Wan (2003) recognized a porphyry granodiorite called the Luoza intrusion with zircon U–Pb age of 217 ± 3.4 Ma (age) in Nanmulin, the middle part of the belt, which is considered as recording the ocean closure event of the Paleotethys orogeny.

The Cretaceous granitoids include 39 intrusions, most of which are composite intrusion complexes. Their lithologies are granodiorite, quartz diorite and biotite monzogranite. Among them, the larger-scale complexes, such as the Demula batholiths, can be recognized as representing a time sequence of early phase granodiorite, second phase and third phase coarse-grained porphyritic biotite monzogranite and medium-coarse grained biotite syenite granite, and last stage of medium grained two-mica syenitic granite. The plutons intruded the Late Paleozoic and Cretaceous strata. The LA-ICP-MS zircon U–Pb isotopic ages are of 128–60.7 Ma, ranging from the Cretaceous to the Paleogene.

Some small-scale Cenozoic granite (<40 Ma) associated with post-collisional potassic-ultrapotassic volcanic rocks (25 Ma) is a feature of the middle belt. The rock types are mainly biotite syenogranite and monzogranite.

1.3.1.3 Southern Gangdese-Nimu-Shangchayu Granite Belt (South Belt)

This belt is the Gangdese magma belt in a narrow sense, adjacent to the Yalungzangbu ophiolitic mélange belt in the south, and the Shiquanhe-Longggeer-Cuomai fault in the north. It was mainly made up of batholith and large granitic complexes, and accompanied by widespread coeval medium-felsic volcanic rocks, both together accounting for more than 60% of the total area of the Gangdese magmatic belt. The granitoids are formed at Triassic, Cretaceous, Paleogene and Neogene.

The Triassic intrusions in this belt were recognized for the first time by the 1:200,000 scale Woka sheet geological survey. Then 1:250,000 scale Shenzha sheet mapping (2003) recognized two porphyric granodiorite intrusions from the batholith in the area of Kangbowuri and Xiabo in Nanmulin. The zircon U–Pb age is 217 ± 3.4 Ma (Li and Wan 2003). There are also coeval volcanic rocks in the same period, indicating an island arc orogeny taking place in the belt in the late Triassic (Pan et al. 2004).

The Cretaceous-Paleogene intrusions are widespread and make up the main body of this belt. The intrusions as batholith or large plutonic stocks tend to become slightly younger from west to east. The largest Qushui batholith in this belt is distributed in the Nanmulin-Nimu-Qushui area on the north side of the IYSZ, intruding the Mesozoic strata. The zircon U–Pb dating gives ages in the range of 47–52.5 Ma (Mo et al. 2005; Dong et al. 2005), which are coeval with the Linzizong volcanic succession (LVS) rocks in the belt, which is interpreted as magmatic response to the India-Asia collision. The batholiths are lithologically quartz diorite, granodiorite, monzonite and potassic feldspar granite with abundant mafic microgranular enclaves (MMEs). Some mafic and ultramafic small stocks can be found south of the batholith. These observations indicate the importance of mantle contribution to the petrogenesis of the granitoids. A large number of Cretaceous-Paleogene granitoid intrusions are also outcropped in the Bomi and Chayu areas consisting of diorite, quartz diorite, tonalite, granodiorite, monzogranite, and syenogranite. Zircon U–Pb dating show two groups: 127–78.3 and 22–17 Ma (Liu and Wang 1994).

1.3.2 Volcanic Rocks in the Gangdese

The volcanic rocks in the Gangdese are known as the Paleozoic-Mesozoic volcanic rocks, the Paleogene LVS rocks, the Miocene potassic and ultra-potassic volcanic rocks, constituting the following volcanic belts:

1.3.2.1 Late Paleozoic-Early Mesozoic Composite Arc Volcanic Belt

The volcanic belt is located south of the Cuoqin-Dingrenla fault and north of the Shamole-Maila-Luobadui-Milashan fault (Pan et al. 2004). In addition to the Late Mesozoic and Cenozoic volcanic rocks, a large amount of Late Paleozoic and Early Mesozoic arc volcanic rocks have been recognized in this belt, including of the Early Carboniferous Nuocuo Formation, the Early Permian Angjie Formation, and the Late Permian Dibucuo Formation. The Paleozoic volcanic rocks are abundant and volumetrically only subordinate to the Late Mesozoic to Neogene volcanism, and is composed of basalt, andesite basalt, andesite, dacite, rhyolite and pyroclastic rocks.

1.3.2.2 Early-Middle Jurassic Yeba Arc Volcanic Belt

The volcanic belt is located south of the Luobadui-Milashan fault and north of the Xietongmen-Angang-Zhigang ductile shear zone (Pan et al. 2004). From Dazi in the west to the Sangri in the east, the belt with a width of ~30 km extends for ~80 km trending NWW-SEE. The zircon U–Pb dating on dacites near Dazi indicates the volcanism before the Middle Jurassic.

1.3.2.3 Middle-Late Jurassic-Early Cretaceous Zenong Arc Volcanic Belt

The belt is located south of the Shiquanhe-Asuo-Yongzhu-Juzila-Jiali-Bomi suture zone and north of the Dawacuo-Taerma fault (Pan et al. 2004). It is a long and narrow lenticular shape pinching out at both ends. The volcanic rocks include the Middle and Late Jurassic Jienu Group and the Late Jurassic-Early Cretaceous Zenong Group. In addition, there are also volcanic rocks in the Early Cretaceous Woronggou and Jiega Formations.

The Jienu Group volcanic rocks are less exposed and consist of gray andesitic tuff, altered andesitic tuff, gray-green andesitic sedimentary tuff, and tuffaceous greywacke. All the rocks are distributed in the Langdingmaji-Jiarang area southwest of Shenzha with the assemblage of basalt-dacite-rhyolite.

The Zenong group is widely distributed south of Bangduo, Nima and Shenzha area in trending east-west and is composed of rhyolite, dacite, dacitic volcanic breccia, dacitic tuff, andesite and basalt, showing volcanic arc volcanic arc characteristics. The LA-ICP-MS zircon dating in the 1:250,000 Nima area gave 83.35 ± 1.33 Ma, 95.4–104.6 Ma, and whole rock Rb–Sr age of 110–115 Ma in the Cuoqin area. The $^{40}Ar/^{39}Ar$ age for dark grey massive hornblende dacite at the top of the Zhaliena Formation is 126.37 ± 1.61 Ma in the Bangduo area. Therefore, the main age of the Zenong Group volcanic rocks should be Early Cretaceous.

The volcanic rocks of the Woronggou Formation are generally distributed in Bange trending east-west. The rock types are basalt, basaltic andesite, andesite, pyroxene-bearing trachyandesite and rhyolite, as well as tuffs. The early felic volcanic rocks of the Woronggou Formation belong to the potassic basalt series. The pyroxene andesite from top of the Ronggou formation in Deqing gives zircon U–Pb age of 114.2 ± 1.1 Ma.

1.3.2.4 Late Jurassic-Early Cretaceous Sangri Arc Volcanic Belt

This belt is located south of the Xietongmen-Angang-Zhigang ductile shear zone and north of the Yaluzangbu suture zone (Pan et al. 2004). The volcanic rocks are grouped as the Mamuxia Formation in the lower part and the Bima Formation in the upper part. The former is scattered along Mamen and Nanluo south of the Yaluzangbu River, and the latter is mainly distributed in Xietongmen, Nimu, Qushui, Jiangxiang and Sangri areas on the north bank of the Yaluzangbu River. The area reaches out >400 km in east-west extension with a width of ~20–40 km.

1.3.2.5 Middle-Late Jurassic-Cretaceous Bange-Bengcuo Arc Volcanic Belt

This belt is located south of the Banggong Lake-Nujiang suture zone, and north of the Shiquanhe-Asuo-Yongzhu-Jiuzila-Jiali-Bomi suture zone (Pan et al. 2004). In addition to the volcanic rock interlayers of the Middle and Late Jurassic Lagongtang formation, there are also volcanic interlayers of the Early Cretaceous Duoni Formation, Langshan Formation, and Late Cretaceous Jingzhushan Formation.

The Lagontang Formation, consisting of volcanic rocks, is exposed in the Banglang to Mayue of Bange to the south of the Baila-Jueweng ophiolite belt. The lithology is rhyolite of ~214 km thick.

The Early Cretaceous Duoni Formation, in which the middle and upper parts contain volcanic rocks that are, trending east-west, distributed in the Bange, Banduo, Nima and Shenzha areas. The volcanic lithology includes welded rhyolitic tuffs and glassy ignimbrite in middle part and complex in the upper part, including altered andesite, pyroxene andesite, hornblende andesite, dacite, altered dacite, breccia-bearing dacite, rhyolite, and related pyroclastic rocks. All the rocks belong to the middle-potassium calc-alkaline series and are produced under tectonic environment of a continental margin volcanic arc. In addition, there are two thin layers of volcanic rocks exposed in the Early Cretaceous Langshan Formation.

The Late Cretaceous Jingzhushan Formation contains a small amount of volcanic rock interlayers, and mainly exposed in Jiangqiong, Shenzha, Biru of Nima County. Two layers of volcanic rocks of ~110 km thick are mainly andesitic tuffs and breccia-bearing quartz trachyte with K–Ar age of 90.24 ± 6.88 Ma.

In this belt are also granitoids with subduction-related arc characteristics are distributed in nearly east-west direction. The rocks include quartz diorite, tonalite and granodiorite of 124–103 Ma age (Jiang et al. 2003).

1.4 Igneous Rocks in the Himalaya

The Himalayan Magmatic Belt in the Himalayan region, south of the IYSZ, can be divided into the Northern Himalayan Belt and the High Himalayan Belt.

1.4.1 Intrusive Rocks of the Himalayan Belt

The intrusive rocks in the Himalayan Belt are mainly granitoids with pale color. They were formed mainly in the Proterozoic, Ordovician and Miocene and mostly concentrated in the High Himalayan Belt and Northern Himalayan Belt (Lhagoi Kangri Tectonic Belt), with Jurassic granitoids and Cretaceous gabbro-diabase found in places (Pan et al. 2004).

The Proterozoic granitoids are mainly distributed in the metamorphic rock series of the Middle-Neoproterozoic Nyalam Group in the High Himalayan Belt, covering an area of ~555 km^2. Seven typical intrusions studied in detail are gneissic porphyritic biotite monzogranite with peraluminous and S-type characteristics, and contain abundant xenoliths of marble, quartz schist, garnet amphibolite, amphibolite and other host rocks.

The Ordovician granites are primarily concentrated in the Northern Himalayan Belt and the High Himalayan Belt. The gneissose monzogranite represented by the Kangmar intrusion dominates the North Himalayan Belt, forming a series of domes in response to east-west extension. The isotopic ages of the granitic rocks are 562–451 and 266 Ma for zircon U–Pb dating, and 31.8–12.5 Ma for biotite K–Ar ages in the Kangmar Gneissic Dome (Wang et al. 1981; Scharer et al. 1986).

There are Jurassic porphyritic biotite monzogranite only in the area between Saga and Ngamring in the Yarlung Zangbo Belt. They are peraluminous and S-type characteristics with an outcrop area of ~72 km^2, and intrude the Upper Triassic Xiukang group and crosscut by the Neogene Xiaru granites. More than 10 small gabbro-diabase intrusions have also been seen in Lhaza with possible forming age of early Cretaceous, the largest which has a exposure of 4 km^2.

The occurrence of Cenozoic leucogranites is characteristic of the intrusive activity in the Himalayan Belt. It is mainly distributed in the Lhagoi Kangri Tectonic Belt and the High Himalayan Belt. The age of intrusion is mostly less than 40 Ma and predominately 23 Ma and younger.

The leucogranites in the Lhagoi Kangri Belt are distributed north of the Gyirong-Tingri-Kamba fault, and exposed at nearly equal intervals along the east-west direction. Some of them are associated with the Ordovician granitic intrusions, and some individual intrusions themselves. They are mainly the strongly peraluminous two-mica granites, gneissic two-mica monzogranite, biotite monzogranite, granites with tourmaline, intruding in the light metamorphic rocks of Triassic, Carboniferous-Permian, Jurassic and Cambrian ages. The granitic intrusions are mostly elongated in NE and NEE trending and are often accompanied by strike-slip faults of same trending. They are in general small in size and simple in lithology with gneissose foliation with radiometric age of 43.3–20.4 Ma, and often contain accessory minerals such as tourmaline, garnet, fluorite, topaz and beryl, making them distinct from older rocks.

The Cenozoic leucogranite intrusions are distributed nearly 2000 km long as a leucogranite belt along the High Himalaya (Le Fort 1981). The lithologies of the intrusions are tourmaline leucogranite, two-mica granites, and gneissose two-mica granites. They occur mostly as small intrusive sheets, layers or stocks, intruding old rocks such as Early Paleozoic or Archean metasedimentary strata. The leucogranites of Miocene give radiometric ages of 20.6–10.5 Ma.

1.4.2 Volcanic Rocks of the Himalayan Belt

A large number of volcanic interbeds dominated by basalts of varying age are found in the Himalayan Belt (Pan et al. 2002; Zhu et al. 2003, 2004a, b).

The Permian volcanic rocks are exposed in the Jilong valley east of the Selong Mt. with a thickness of several hundreds of meters (Zhu et al. 2004a). The lithology is mainly massive basalt, vesicular and amygdaloidal basalts as well as volcanic breccias (Zhu et al. 2002). These basalts show characteristics of OIB-type tholeiite in continental tensional environment, emplaced in response to continental extension and rifting.

The Triassic volcanic rocks occur as interlayers of tens of meters thick in the Gyangzê Basin, Pumoyum Co and other places. They are compositionally E-MORB type tholeiite formed in a continental margin rifting background (Zhu et al. 2004a).

The Jurassic volcanic rock layers are outcropped in Sa'gya and Nagarzê, and the typical volcanic rocks of the Zhela Formation show similar major and trace element compositions to that of the present-day OIB, characterized by high Ti (TiO_2 averaging 2.57 wt.%), high Fe (average TFeO = 10.50 wt.%), low Mg (MgO averaging 4.06 wt.%), enrichment of LREEs and HFSEs with high Ti/V, Ti/Y and Zr/Y ratios, and high $^{87/86}Sr$ and $\varepsilon_{Nd}(t)$ values of −0.29 to 0.50, indicating alkali basalts produced in a continental margin rifting background with OIB-type geochemical characteristics (Zhu et al. 2004b).

The Cretaceous volcanic interlayers of ~133 Ma are found in the Yamdrok Tso area. The representative volcanic rocks of early Cretaceous Sangxiu Formation are basalts in the lower part and felsic varieties in the upper part, distributed in the southeast region of YamdrokTso, The thickness and occurrence of the basalt intercalation varies from one outcrop to another. There are massive, pillow and amygdaloidal basalts with thicknesses varying from a few tens to hundreds meters, and are alkali basalts with OIB-type geochemistry formed under continental margin rift setting (Zhu et al. 2005b). Felsic volcanic rocks have weak negative Eu anomalies, which may be related to the remelting of siliciclastic crust caused by mantle plume or hot spots caused by basaltic magmatism (Zhu et al. 2005a).

1.5 Igneous Rocks in Sanjiang Area

Magmatic rocks in the Sanjiang region can be divided temporally into three suites: pre-Paleo-Tethys magmatic rocks, magmatic rocks associated with the open, subduction and closure of the Paleo-tethys, magmatic rocks associated with the open, subduction and closure of the Neo-Tethys and Indo-Eurasian collision. These magmatic rocks can spatially overlap in places. The Sanjiang region can be divided into several magmatic belts in terms of spatial distribution. The following discussion concentrates on the main magmatic belts.

1.5.1 Yidun Magmatic Belt

The Yidun magmatic belt is located west of the Ganzi-Litang ophiolite belt, and east of the Jinshajiang suture zone and the Zhongza micro-continent. The Garzê-Litang Ocean, a branch of the Paleo-Tethys, opened during the Late Permian (e.g., Deng et al. 2014; Hou et al. 2003, 2007). It is generally believed that the Garzê-Litang seafloor subduction beneath the Yidun Terrane in the Triassic (e.g., Hou et al. 2003, 2007; Reid et al. 2005, 2007). Extensive Triassic magmatism temporally and spatially associated with the Garzê-Litang suture is well-represented in the Yidun Terrane (Dong et al. 2021). The Yidun belt is mainly occupied by late Triassic volcanic-sedimentary rock series and Indosinian-Yanshanian (Triassic-Cretaceous) granitic batholith. The Yidun belt contains two volcanic sedimentary basins in the southern and northern parts, Baiyu Basin and Xiangcheng Basin, respectively.

Volcanic activity in the Yidun belt can be divided into pre-arc, main arc, and post-arc periods. The main arc period can be divided into three stages: early arc-formation stage, intra-arc rift stage, and late arc formation stage. Before the formation of the volcanic arc during the early Carnic, a series of grabens and horsts of extensional structures developed in this area, resulting in a rifting type high TiO_2 alkaline-transition series of basalts or bimodal volcanic assemblages that have similar geochemical characteristics to the Emeishan basalts in the Yangtze plate. In the Baiyu Basin in the northern section, the volcanic rocks are symmetrically distributed with the old on both sides and the young in the middle part. In the Xiangcheng Basin in the southern section, they are mainly distributed in the western part of the basin near the Zhongza microcontinent. During the transition from the end of the pre-arc stage to the early main-arc stage, high MgO tholeiite, massive low-K tholeiite, and high MgO and SiO_2 and extremely low TiO_2 Boninite were formed. These rocks have obvious arc features, and also retain some rift features. This set of volcanic rocks is only exposed in the Chizhong area. The main arc volcanic rocks are dominated by andesite in the two arc-forming stages, including a small amount of calc-alkaline basalt and dacite-rhyolite, typical arc volcanic rocks. Bimodal volcanic assemblage developed in the intra-arc rifting stage is the main feature that distinguishes the Yidun belt from other magmatic arcs, and it is also the mafic structure-volcanic construct for the polymetallic massive sulfide mineralization such as the Gacun deposits.

The intermediate-felsic intrusive rocks can be divided into the east and west two sub-belts. The west belt mainly comprises intermediate-felsic porphyries associated with the assemblage of calc-alkaline volcanic rocks in the main arc, which are concentrated in the Xuejiping area of Zhongdian in the southern Yidun belt. The rock type is dioritic-monzonitic porphyry, which are characterized by low SiO_2, high CaO, MgO, rich in sodium and poor in potassium. They belong to I-type granite and are products of arc magmatism, with porphyry-type copper and polymetallic mineralization.

The east belt mainly comprises monzogranite-granodiorite batholith, distributed in the northern Yidun belt. The main magmatic period is the Indosinian (237–195 Ma). Some Hercynian granites also exist, belonging to syn-collisional granite. In addition, there are some Yanshanian–Himalayan post-collisional granites in this area, overprinting the pre-arc and syn-collisional granites.

The Changtai-Xiangcheng magmatic arc is closely related in space and time to the Ganzi-Litang belt. From east to west, the following spatial configuration can be seen in the northern section of the volcanic arc (Zengke-Changtai area): the Ganzi-Litang suture (Garze-Litang ridge-type volcanic rock-ophiolite belt), the Queershan granite belt and Triassic sedimentary rocks, outer arc (eastern andesite belt), intra-arc rift (Bimodal volcanic rock belt), inner arc (west Andesite belt), a back-arc area (Miange rhyolite-high-K basalt belt). The back-arc area and the main arc area are roughly divided by the Keludong-Xiangcheng fault, along which there are many serpentinites as structural diapirs. In general, volcanic rocks are younger from east to west, and the volcanic activity center migrates from east to west. In the southern segment (Xiangcheng area) of the Yidun belt, the spatial distribution of volcanic rocks are roughly comparable with that in the northern segment. The difference is that no back-arc volcanic rocks are found, and the back-arc position is occupied by volcanic rocks from the fore-arc period. The center has a tendency to migrate from west to east, which may be due to the differences in the dynamic boundary conditions or subduction mechanisms of the southern and northern sections.

It can be seen from the foregoing that the Ganzi-Litang belt and the Yidun volcanic arc are coeval. They are a pair of connected "double rock belts", clearly marking the relative positions of the ancient Ganzi-Litang Ocean and the active continental margin of the Zhongza micro-continent, indicating the westward seafloor subduction.

In summary, the Yidun belt was developed on the basis of the continental rift graben-barrier system during the short time interval of the late Jurassic due to westward subduction of the Garze-Litang seafloor plate. Since its history of tension-compression alternation and the existence of inter-arc rifts, it is also an extensional volcanic arc.

1.5.2 Jiangda-Weixi Arc/Collisional Magmatic Belt

This belt is located west of the Jinshajiang ophiolites belt. They are closely related in time and space, and share a common petrogenesis, forming an ophiolite-arc volcanic belt. The Jiangda-Weixi rock belt can be further divided into four sub-belts: the early-later Permian Jubalon-Gongka arc volcanic sub-belt; collisional-post-collisional-arc volcanic sub-belt in the Jijiading-Shizzhongshan area during the early to later Triassic; Post-collisional arc volcanic sub-belt later Triassic in The the Jiangda Area; Mojiang luchun Arc volcanic-collisional volcanic sub-belt during middle to later Permian.

Jubalon-Gongka arc volcanic sub-belt is located in the Mangkang-Zhubalong, Xiquheqiao-Deqingongka and Dongzhulindasi. The main arc volcanic rocks are calc-alkaline andesite, basalt, basaltic andesite and sodium dacite, which are characterized by high Al_2O_3 and low TiO_2. In the Gongka and Dongzhulindashi area, there are tectonic mélanges include lithologies of serpentinites, gabbros, diabase dikes and radiolarite siliceous rocks. The back-arc volcanic rocks are only exposed between Jubalong and Xiquheqiao, located on the west side of the main arc belt. These rocks include low TiO_2 tholeiite, dense diabase wall, radiolarite siliceous rock and flysch.

The Jijiading-Shizzhongshan volcanic sub-belt is located in the Shusong-Erjiading-Pantiange-Shizhongshan ares, which is composed of three sets of volcanic rocks, partially overlapping in space. The first set is the early Triassic (Marathduo Formation in mangkang Jiaolongqiao area) and Middle Triassic (Pantiange Formation, mainly in Jijiading-Pantiange-Shizongshan area) rhyolite dacite with high SiO_2 and Al_2O_3 contents, which is produced in the southern Deqin together with collisional granitoid intrusions. The second set is the late Triassic (Xiaodingxi Formation, represented by Shusong) post-collisional arc volcanic andesite-dacite-rhyolite porphyry assemblage. The third set, located in the northern section of Susong-Yijiading, is a combination of basalt, basaltic andesite and radiolarian siliceous slate, which is also associated with serpentinites and gabbros in Baimaxueshan. In the southern section, the Cuibi-Jigai rocks are composed of late Triassic biotite, quartz-keratophyre and radiolarian siliceous rocks.

The Jiangda volcanic sub-belt is located in Jiangda-Tongpu-Dongdu-Jiadoling-Chesuo-Dengke. The arc volcanic rocks are calc-alkaline basalt-andesite-rhyolite-dacite assemblage with age of T1–T31. The T31 (Jiangda Formation) is the most abundant. The T1 and T2 volcanic rocks were emplaced after the molasse accumulation at the bottom of The Lower Triassic Pushuiqiao Formation, while the T31 volcanic rocks were formed after the Molasse accumulation at the bottom of the Upper Triassic Jiangda Formation. They are typical post-collisional volcanic rocks. The back-arc volcanic rocks in the Chesuo area are low-TiO_2 tholeiite with pillow structure in Marine facies.

The Mojiang-Luchun volcanic sub-belt is located in the area of Mojiang-Luchun in western Yunnan. It is composed of two sets of volcanic rocks. The first set is the late Permian calc-alkaline basaltic andesite-dacite and the corresponding pyroclastic rocks, which are distributed in the Queliyuan-Caopiba-Yaxuanqiao area. The second is the late Triassic high SiO_2 and high Al_2O_3 collisional dacite-rhyolite assemblage (Gaoshanzhai Formation), distributed in the Luchun-Yuanyang area. The felsic intrusive rocks are also distributed in the southern Luchun area, with radiometric ages ranging from 230 to 211 Ma. They are mainly I-type biotite monzogranite and moyite, with minor two-mica granites, which are syncollisional products. There is a 30 km long mafic volcanic belt in the Bulong-Dalongkai-Wusu-Wannianqing close to the eastern side of the late Permian arc volcanic belt. The lithologies are mainly pillow basalts, containing a small amount of felsic lavas in the marine sand and argillaceous flysch stratum, and may be of Carboniferous age.

The late Indosinian collisional granites include Renda, Anmexi, Jiarene, Baimanxueshan, Ludian, Datuan, Xinanzhai and Bader intrusions from north to south. They are aluminum-saturation series rocks, with initial $^{87}Sr/^{86}Sr$ value up to 0.7175.

1.5.3 Magmatic Rocks in the Changdu-Simao Microcontinent

The magmatic rocks in the Changdu-Simao microcontinent are mainly scattered volcanic rocks. The opposite subduction between the Lancangjiang oceanic slab and the Jinshajiang oceanic slab resulted in the generation of arc volcanic rocks, including the middle Carboniferous intermediate-felsic volcanic rocks in the Qingnidong, P2 andesite and dacite in the Jiaseding of Mangkang and Lingzhiheqiao, P1–P2 andesite-dacite-rhyolite assemblage in the Simao basin. However, the intraplate basalts are also found in the baoshan block and have no outcrop in Zhongza block during the same period. After the collision between the Changdu block and the Zuogong-Baoshan block in the Middle Triassic (T2), the intraplate volcanic rocks of the T3 Ze'e Formation and the later Paleogene trachyte were generated. The Jinshajiang oceanic plate subducted under the Changdu block and retained in the mantle may provide the materials for the Himalayan Yulong porphyry copper deposits.

1.5.4 Yunxian-Jinghong Arc/Collision Post-collisional Magmatic Belt

The Yunxian-Jinghong magmatic belt is located between the Changning-Menglian ophiolite belt and the Simao magmatic belt. The arc volcanic group are mainly distributed in Yunxian-Jinghong area in the east of the Lincang granite, along the Lancang river. This belt is mainly composed of two sets of spatially overlapping volcanic series. One set is the Permian andesitic-dacitic-rhyolitic volcanic rock assemblage, with some late Carboniferous intermediate-felsic volcanic rocks, mainly distributed in the southern part of the volcanic arc and the western margin of the Simao basin. The other set is Triassic collisional volcanic rocks, post-collisional arc volcanic rocks and post-collisional extensional volcanic rocks, including middle Triassic (Huaimang Formation) collisional dacite, rhyolite and late Triassic (Xiaodingxi Group) collisional potassic trachybasalt, latite (Yunxian), basalt, andesite, dacite, rhyolite and late Triassic (Manshuihe Group) syn-collisional potassic basalt, and bimodal volcanic rocks.

The intermediate-felsic intrusive rocks are also well developed in this belt, including Lincang granitic batholith, the largest batholith in the Sanjiang region, extending more than 300 km from north to south and an E-W width of 15–45 km, which is a type syncollisional granites in the Sanjiang region. In northern Tibet, there are Dongdashan and Youxi plutons with the same origin, age and rock type, which together constitute the huge Indosinian syncollisional granite belt. There are also Yanshanian granite superimposed in this belt.

1.5.5 Magmatic Rocks in the Baoshan Microcontinent

The volcanic rocks are mainly distributed in the Baoshan-Zhenkang area, composed of Carboniferous continental tholeiite, which is characterized by containing standard quartz-normative composition resembling the Deccan basalt in India, reflecting similar mantle source.

1.5.6 Bomi-Tengchong Arc Magmatic Belt

This belt is located west of the Dingqing-Basu ophiolite (mélange) belt, which is generally considered to be the southeast extension of the Gangdese-NyenqingTanggula. But there are obvious differences in the Tengchong-Lianghe area.

The volcanic rocks are mainly J-K2 calc-alkaline basalt-andesite-rhyolite and the corresponding pyroclastic assemblage, which coincides with the subduction of the Nujiang seafloor slab. This belt extends westward to the Gangdese-NyenchenTangula volcanic arc. The temporal and spatial relationship between this belt and the Dingqing-Basu ophiolite (mélange) belt reveal that the opening time of the Nujiang ocean is T3–J3, and the closure time is K1–K2. The Bomi-Tengchong arc may be a continental margin arc formed by westward subduction of the Nujiang oceanic slab under the Lhasa block. In addition, there are also Cenozoic syn-collisional potassic volcanic rocks exposed in this belt, represented by the Tengchong volcanic group.

Magmatism is widely developed in this belt as represented by abundant Mesozoic-Early Cenozoic intrusions and minor Quaternary volcanic rocks (Xie et al. 2016; Zhao and Liu 2010). The northern segment of the granite belt is composed of four large batholith belts, which have varying age of early cretaceous to Paleogene, i.e., 140–40 Ma. The ages show a younging trend from northeast to southwest. The I- and S-type granite batholith in Chayu is quite different from the typical continental margin arc granites. The S-type alaskite-moyite is the host of Sn and W mineralization in the region (Dong et al. 2022; Sun et al. 2017, 2020; Wang et al. 2020).

References

Arnaud N, Vidal P, Tapponnier P, Matte P, Deng W (1992) The high K2O volcanism of northwestern Tibet: geochemistry and tectonic implications. Earth Planet Sci Lett 111:351–367

Bian Q, Gao S, Li D, Ye Z, Chang C, Luo X (2001) A study of the Kunlun-Qilian-Qinling suture system. Acta Geologica Sinica-English Edition 4:364–374

Chi X, Li C, Jin W, Liu S, Yang R (1999) Spatial-temporal evolution of Cenozoic volcanism and uplifting in North Tibet. Geol Rev 45:978–986

Deng J, Zhao H, Lai S, Li Y, Liu H, Mo X, Di Y (1995) Intracontinental orogenic igneous rocks and orogenic processes in Qinghai - Xizang -Himalaya. J China Univ Geosci 2:5–12
Deng W, Huang X, Zhong D (1998) Petrological characteristics and genesis of Cenozoic alkali rich porphyry in West Yunnan, China. Sci Geol Sin 4:31–44
Deng J, Mo X, Luo Z, Wang Y, Zhao H, Zhao Z, Su S, Yu X (2001) Inhomogeneity of the lithosphere of the Tibetan Plateau and implications for geodynamics. Sci China Earth Sci 201:56–63
Deng J, Wang QF, Li GJ, Li CS, Wang CM (2014) Tethys tectonic evolution and its bearing on the distribution of important mineral deposits in the Sanjiang region, SW China. Gondwana Res 26:419–437
Dong GC, Mo XX, Zhao ZD, Wang LL, Chen T (2005) Geochronologic constraints by SHRIMP II zircon U-Pb dating on magma underplating in the Gangdese Belt following India-Eurasia collision. Acta Geol Sin 79:787–794. (in Chinese with English abstract)
Dong P, Dong G, Santosh M, Mo X, Wang P, Li H, Tang J, Su L, Sun Z, Ketchaya YB (2021) Multiple sources and magmatic evolution of the late Triassic Daocheng batholith in the Yidun Terrane: implications for the Paleo-Tethys Ocean evolution in the eastern Tibetan Plateau. Geol Soc Am Bull 134:1660–1680. https://doi.org/10.1130/B35970.1
Dong P, Dong G, Santosh M, Mo X, Sun Z, Ketchaya YB, Pan Y, Lemdjou YB (2022) Eocene magmatism in the western Tengchong block: implications for crust-mantle interaction associated with the slab rollback of the Neo-Tethys Ocean. Gondwana Res 106:259–280
Guo TY, Liang DY, Zhang YZ, Zhao CH (1991) Geology of Arli, Tibet. China University of Geosciences Press, Wuhan, pp 1–464
Hacker B, Luffi P, Lutkov V, Minaev V, Ratschbacher L, Plank T, Ducea M, Patino-Douce A, McWilliams M, Metcalf J (2005) Near-ultrahigh pressure processing of continental crust: miocene crustal xenoliths from the pamir. J Petrol 46(8):1661–1687
Hou ZQ, Wang LQ, Zaw K, Mo XX, Wang MJ, Li DM, Pan GT (2003) Postcollisional crustal extension setting and VHMS mineralization in the Jinshajiang orogenic belt, southwestern China. Ore Geol Rev 22:177–199
Hou ZQ, Zaw K, Pan GT, Mo XX, Xu QM, Hu Y, Li X (2007) Sanjiang Tethyan metallogenesis in S.W. China: tectonic setting, metallogenic epochs and deposit types. Ore Geol Rev 31:48–87
Jia D, Lu HF, Jia CZ, Wei GQ (1999) Structural features of northern tarim basin: implications for regional tectonics and petroleum traps: reply. AAPG Bull 83(8):1284–1286
Jiang CS, Wang SJ, Zhou RQ, Zhou ZH, Long XF, Wang V (2000) Dynamic study of the Tengchong volcanic active structure. Seismol Res 23:179–187
Jiang YS, Zhou YY, Wang MG, Xie YX, Li JB, Peng B (2003) Characteristics and geological significance of Quaternary volcanic rocks in the central segment of the Gangdise area. Geol Bull China 22(1):16–20
Lai SC, Zhang GW (1999) Ophiolites and their tectonic significance in the Mianlue Suture Zone, Qinling belt. Geol Rev 45(Sup):1062–1071. (in Chinese with English abstract)
Le Fort P (1981) Manaslu Leucogrante: a collision signature of the Himalaya. A Model for its genesis and emplacement. J Geophys Res 86:10545–10568
Li GB, Wan XQ (2003) Eocene microfossils and the extinction of Tethys in Gamba-Tingri region, Southern Tibet. J Stratigr 27(2):99–108
Li ZX, Metcalfe I, Powell CM (1996) Breakup of Rodinia and Gondwana land and assembly of Asia. Aust J Earth Sci 43(6):591–592
Li G, Feng X, Huang Z, Gao D (2000) The multiple island arc-basin systems and their evolution in the Gangdise tectonic belt, Xizang. Sediment Geol Tethyan Geol 20(4):38–46. (in Chinese with English abstract)
Liu ZS, Wang JM (1994) Geology and Geochemistry of Granites in the South Tibetan Plateau. Southwest Jiaotong University Press, Chengdu
Liu CD, Mo X, Luo ZH, Yu XH, Chen HW, Li SW, Zhao X (2004) Mixing events between the crust- and mantle-derived magmas in eastern kunlun: evidence from zircon SHRIMP II chronology. Chin Sci Bull 49(8):828–834

Luo ZH, Ke S, Cao YQ, Deng JF, Zhan HW (2002) Late Indosinian mantle-derived magmatism in the east Kunlun. Reg Geol China 21(6):292–297. (in Chinese with English abstract)
Mo X, Dong G, Zhao Z, Guo TY, Wang LL, Chen T (2004) Timing of magma mixing in the Gangdisê Magmatic Belt during the India-Asia Collision: zircon SHRIMP U-Pb dating. Acta Geologica Sinica (English Edition) 79(1):66–76
Mo X, Dong GC, Zhao ZD, Zhou S, Wang LL, Qiu RZ, Zhang FQ (2005) Temporal- spatial distribution and characteristics of granites in the Gangdese belt, Tibet and implications for crustal growth and evolution. Geol J China Univ 11(3):281–290
Mo X, Zhao Z, Deng J, Flower M, Yu X, Luo Z, Li Y, Zhou S, Dong G, Zhu D, Wang L (2006) Petrology and geochemistry of post-collisional volcanic rocks from the Tibetan Plateau: implications for lithosphere heterogeneity and collision-induced asthenospheric mantle flow. In: Yildirim D, Spyros P (eds) Postcollisional tectonics and magmatism in the mediterranean region and Asia. Geological Society of America Special Paper 409, pp 507–530
Pan GT, Wang PS, Xu YR, Jiao SP, Xiang TX (1990) Cenozoic tectonic evolution of the Tibetan plateau. Geological Publishing House, Beijing, p 165
Pan GT, Li XZ, Wang LQ, Ding J, Chen ZL (2002) Preliminary division of tectonic units of the Qinghai-Tibet Plateau and its adjacent regions. Geol Bull China 21:701–707. (in Chinese with English abstract)
Pan GT, Ding J, Yao DS, Wang LQ, Luo JN, Yan YJ, Yong YY, Zheng JK, Liang XZ, Qin DH, Jiang XS, Wang QH, Li RS, Geng QR, Liao ZL, Zhu DC (2004) Geological map of the Tibetan plateau and adjacent areas (1:1500 000) and the manual. Chengdu Map Publishing House, Chengdu, p 133
Qiu R, Xiao R, Zhou S, Li S, Cai Z, Zhang W (2002) Geochro- nology and Nd and Sr isotope characteristcs of gabbro dikes from the Shemalagou area in the middle sector of the Bangongco-Nujiang ophiolite belt, north Tibet. Geol Rev 48:Supp. 64-68
Reid AJ, Wilson CJL, Liu S (2005) Structural evidence for the Permo-Triassic tectonic evolution of the Yidun arc, eastern Tibetan Plateau. J Struct Geol 27:119–137
Reid AJ, Wilson CJL, Shun L, Pearson N, Belousova E (2007) Mesozoic plutons of the Yidun arc, SW China: U/Pb geochronology and Hf isotopic signature. Ore Geol Rev 31:88–106
Scharer U, Rong-Hua X, Allègre Claude J (1986) U(Th)Pb systematics and ages of Himalayan leucogranites, South Tibet. Earth Planet Sci Lett 77(1):35–48
Song SG, Niu YL, Su L, Xia XH (2013) Tectonics of the North Qilian orogen, NW China. Gangwana Res 23:1378–1401
Sun ZR, Dong GC, Zhao ZX, Wang WQ, Liu SQ (2017) Petrological, geochemical and geochronological features of Lailishan granitoids in western Yunnan and their genesis of partial melting of crustal source. Geol China 44:1140–1158 (in Chinese with English abstract)
Sun Z, Dong G, Santosh M, Mo X, Dong P, Wang W, Fu B (2020) A late Cretaceous felsic magmatic suite from the Tengchong block, western Yunnan: integrated geochemical and isotopic investigation and implications for Sn mineralization. Geol Mag 157(8):1316–1332
Wang JW, Cheng ZL, Gui XT, Xu RH, Zhang YQ (1981) Rb-Sr Isotopic studies of some intermediate-acidic plutons in southern Xizang. Geochimica 3:242–246. (in Chinese with English abstract)
Wang W, Dong G, Sun Z, Dong P, Pan Y, Ketchaya YB, Lemdjou YB, Geng J (2020) The genesis of Eocene granite-related Lailishan tin deposit in western Yunnan, China: constraints from geochronology, geochemistry, and S-Pb-H-O isotopes. Geol J 56:508–524
Wangming D (1998) Cenozoic Intraplate Volcanic Rocks in the Northern Qinghai-Tibetan plateau [M]. Geological Publishing House, Beijing, pp 1–180. (in Chinese with English abstract)
Xia LX, Xia ZC, Xu XY (1998) Early Palaeozoic mid-ocean island and back-arc basin volcanism in the north Qinlian Mountains. Acta Geol Sin 72(4):301–312. (in Chinese with English abstract)
Xie J-C, Zhu D-C, Dong G, Zhao Z-D, Wang Q, Mo X (2016) Linking the Tengchong Terrane in SW Yunnan with the Lhasa Terrane in southern Tibet through magmatic correlation. Gondwana Res 39:217–229

Yang XS, Jin ZM (2001) Studies on Rb-Sr and Sm-Nd isotope of Yadong leucogranite in Tibet: constrait on its age and source material. 47(3):294–300

Yang DM, He ZH, Zheng CQ, Wang TW (2004) Zircon SHRIMP U-Pb dating of Dezong granitoid gneisses and its geological implications in Menba area, Tibet. J Jilin Univ (Earth Sci Ed) 34:12–15

Zhang HF, Luo TC, Zhang BR, Ling WL (1994) Geochemical study of compositional polarity and causes of late Proterozoic Island-Arc granitoids from northern margin of Yangtze Craton. Earth Sci J China Univ Geosci 19(2):219–226. (in Chinese with English abstract)

Zhang Q, Zhou DJ, Zhao DS, Peng XJ, Luo WL, Liu XP (1996) Wilson Cycle of the Paleo-Tethyan orogenic belt in western Yunnan: record of magmatism and discussion on Mantle processes. Acta Petrol Sin 12(1):17–28. (in Chinese with English abstract)

Zhang ZW, Zhu BQ, Chang XY (2000) Nd, Sr, Pb isotopic geochemisty of the alkali-rich intrusive rocks in East Qinling, central China and its tectonic significance. Geochimica 29(5):460–461. (in Chinese with English abstract)

Zhao D, Liu L (2010) Deep structure and origin of active volcanoes in China. Geosci Front 1(1):31–44

Zhu DC, Duan LP, Pan GT, Liao ZL, Xiong YZ (2002) Some tectonic motions and magmatic response events of important influence on Qinghai-Tibet. J Chengdu Univ Technol 29(4):405–409. (in Chinese with English abstract)

Zhu DC, Pan GT, Mo XX, Duan LP, Liao ZL, Wang LQ (2003) Sr-Nd-Pb isotopic variations of the Cenozoic volcanic rocks from Qinghai-Xizang plateau and its adjacent area. Sediment Geol Ththyan Geol 23(3):1–11. (in Chinese with English abstract)

Zhu DC, Pan GT, Mo X, Duan LP, Liao ZL (2004a) Timing of collision between India and Eurasia. Adv Earth Science 19(4):564–571

Zhu DC, Pan GT, Mo XX, Liao ZL, Wang XS, Wang LQ (2004b) Permian to Cretaceous volcanic activities in the central segment of the Tethyan Himalayas (I): distribution characteristics and significance. Geol Bull China 23(7):645–654

Zhu B, Kidd WSF, Rowley DB et al (2005a) Age of initiation of the India-Asia collision in the east-central Himalaya. Geology 113:265–285

Zhu DC, Pan GT, Mo XX, Wang LQ, Liao ZL, Jiang XS, Geng QR (2005b) SHRIMP U-Pb zircon dating for the dacite of the Sangxiu Formation in the central segment of Tethyan Himalaya and its implications. Chin Sci Bull 50(6):563–568

Chapter 2
Pre-Cenozoic Igneous Rocks of the Tibetan Plateau

Abstract Magmatism of multiple types records the histories of origin, drift, and accretion of various terranes comprising of the present-day Tibetan Plateau. This chapter provides a synoptic overview of the pre-Cenozoic igneous rocks in the Tibetan Plateau in terms of spatial and temporal distributions, rock associations, geochemical characteristics, petrogenesis, and geodynamic interpretations. The Early Paleozoic (Cambrian-Ordovician) igneous rocks are poorly exposed in the Southern Qiangtang subterrane, Central Lhasa subterrane, and Tethyan Himalaya. These rocks are mostly high-K calc-alkalic to shoshonitic, as well as minor peraluminous, resulting from the southward subduction, roll-back, and break-off of the proto-Tethys slab beneath the northern Gondwana margin. The Late Paleozoic (Permian) igneous rocks commonly occur in the Northern Qiangtang subterrane, Southern Qiangtang subterrane, and Tethyan Himalaya. These rocks are composed mainly of basalts and ultramafic-mafic dykes, showing geochemical affinity to ocean island basalts (OIB), alkaline, and tholeiitic basalts. These features indicate an extensional tectonic setting most likely related to a mantle plume represented by the coeval Panjal Traps in NW India. The Mesozoic igneous rocks are widespread in the entire Tibetan Plateau. The Triassic peraluminous granites widespread in the Northern Qiantang subterrane are attributed to the collision between southern and northern Qiangtang. The Jurassic to Early Cretaceous igneous rocks in the Southern Qiangtang subterrane are medium-K to high-K calc-alkalic and show arc signatures, which form an E-W belt that was attributed to the northward subduction of the Bangong-Nujiang Tethys. The Late Triassic to Early Cretaceous are widespread in the Northern and Central Lhasa subterranes, representing a magmatic arc related to either the northward subduction of the Neo-Tethys or the southward subduction of the Bangong-Nujiang Tethys. The Late Cretaceous igneous rocks in the Southern Qiangtang, Northern, and Central Lhasa subterranes show geochemical features of high-Mg andesites or adakites and are generally interpreted to originate from thickened or delaminated lower crust following the Lhasa-Qiangtang collision. The igne-

Prof. Di-Cheng Zhu has fully contributed to this chapter.

X. Mo et al., *Magmatism and Crustal Evolution of the Tibetan Plateau*, Modern Approaches in Solid Earth Sciences 25,
https://doi.org/10.1007/978-981-97-2109-2_2

ous rocks in the Southern Lhasa subterrane were emplaced from the Late Triassic to Late Cretaceous and were dominated by medium-K calc-alkalic gabbros, diorites, and granodiorite. These rocks show arc signatures related to the northward subduction of the Neo-Tethys. The Mesozoic igneous rocks in the Tethyan Himalaya consist mainly of basalts and show OIB compositions, in which the Triassic and Jurassic rocks likely indicate a rift setting at passive continental margin, whereas the Early Cretaceous rocks that have been named the Comei large igneous province, were the products of the Kerguelen mantle plume.

Keywords Pre-Cenozoic igneous rocks · Geochronology · Spatial-temporal distribution · Rock association · Geochemistry · Geodynamic interpretation · Tibetan Plateau

The Tibetan Plateau, from north to south comprises: the Qiangtang Terrane, Lhasa Terrane, and Himalayas, separated by the Bangong–Nujiang suture zone (BNSZ) and the Indus–Yarlung Zangbo suture zone (IYZSZ), respectively (Fig. 2.1; Zhu et al. 2013; Kapp and DeCelles 2019). The Longmu Tso–Shuanghu suture zone (LSSZ) subdivides the Qiangtang Terrane into the Northern Qiangtang (NQ) and Southern Qiangtang (SQ) subterranes (Li et al. 1995). The Shiquan River-Nam Tso mélange zone (SNMZ) and Luobadui-Milashan Fault (LMF) subdivides the Lhasa Terrane into Northern (NL), Central (CL), and Southern (SL) Lhasa subterranes (Zhu et al. 2013).

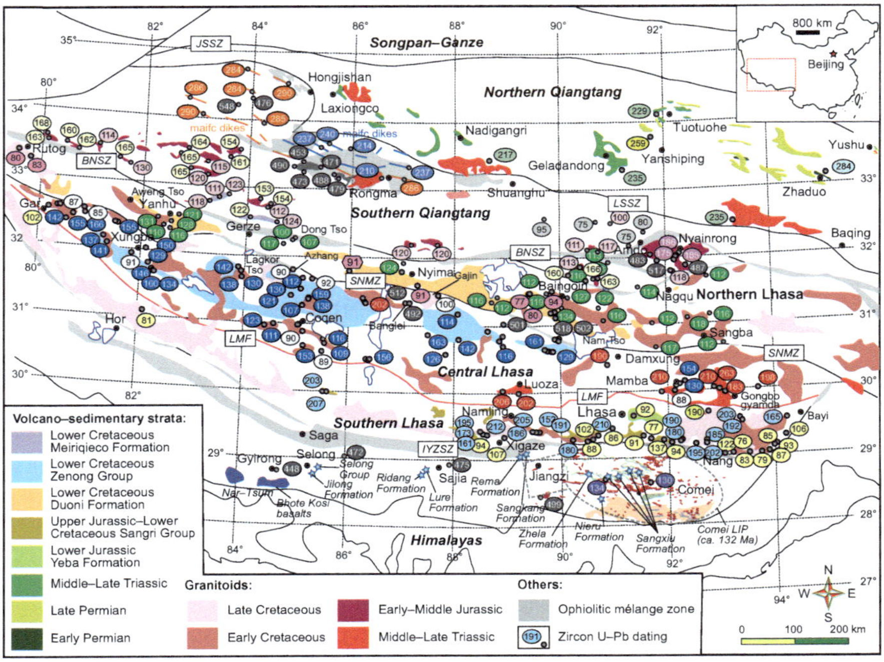

Fig. 2.1 Distributions of the pre-Cenozoic igneous rocks of the Tibetan Plateau (modified from Zhu et al. 2013). See text for details

Pre-Cenozoic igneous rocks in the Tibetan Plateau are of Early Paleozoic, Late Paleozoic and Mesozoic ages. The Early Paleozoic (Cambrian-Ordovician) igneous rocks are poorly exposed in the SQ, CL, and Himalayas; whereas the Late Paleozoic (Permian) igneous rocks commonly occur in the NQ, SQ, and Himalayas. The Mesozoic igneous rocks are widespread in the entire Tibetan Plateau. This chapter presents a synoptic overview of the pre-Cenozoic magmatic rocks in terms of spatial and temporal distributions, rock associations, geochemical characteristics, petrogenesis and geodynamic interpretations.

2.1 Qiangtang Terrane

2.1.1 Northern Qiangtang Subterrane

The Northern Qiangtang (NQ) subterrane lies between the Jinshajiang suture zone (JSSZ) to the north and the Longmu Co-Shuanghu suture zone (LSSZ) to the south (Fig. 2.1) (Zhu et al. 2013). It consists mainly of Devonian to Lower Cretaceous rock units. Magmatism in this subterrane can be divided into three episodes, including early Permian, late Permian, and Middle to Late Triassic.

2.1.1.1 Early Permian Igneous Rocks

Early Permian igneous rocks are volumetrically minor and are distributed in the Zhaduo to Yanshiping area (Duan et al. 2006, 2010; Wang et al. 2007b; Li et al. 2009; Liu 2014). They are dominated by basalts with minor ultramafic and mafic dykes. The ultramafic/mafic dykes consist of pyroxene-peridotite, olivine-pyroxenite, gabbro, and diabase and intrude the Carboniferous strata. The basalts and ultramafic-mafic dykes have zircon U–Pb ages of ca. 284 Ma (Li et al. 2009; Liu 2014).

The majority of the basalts and mafic dykes show geochemical affinity to ocean island basalts (OIB) or intraplate alkaline basalts (Fig. 2.2a, b) and are characterized by high TiO_2 contents, enrichment of high field strength elements (HFSEs), and enrichment of light rare earth elements (LREEs) relative to heavy rare earth elements (HREEs). These rocks also have depleted bulk-rock Nd isotopic compositions with $\varepsilon_{Nd}(t)$ of +2.2 to +5.5. Two models have been proposed for the tectonic setting of these mafic rocks including intraplate rifting (Duan et al. 2006, 2010; Li et al. 2009; Liu 2014) and back arc extension (Wang et al. 2007b).

2.1.1.2 Late Permian Igneous Rocks

Late Permian igneous rocks are exposed in the Zhiduo-Yushu and Tuotuohe areas (Liu et al. 2016a; Wang et al. 2018a). These igneous rocks consist of volcanic rocks (Tuotuohe area) and mafic dykes (Zhiduo-Yushu area). The mafic dykes include gabbro and diabase and have zircon U–Pb age of ca. 258 Ma (Yong et al. 2011; Liu et al. 2016a). The volcanic rocks are bimodal including basalts and rhyolites with zircon U–Pb age of ca. 256–259 Ma (Wang et al. 2018a).

The mafic dykes are divided into three groups based on geochemistry (Fig. 2.2a; Liu et al. 2016a). The first group consists of gabbros in the Yushu area and dolerites in the Zhiduo area showing high TiO_2 contents, enrichment of large ion lithophile elements (LILEs) and HFSEs, and enrichment of LREEs relative HREEs, which are comparable to the present-day OIB or the Emeishan high Ti-basalts in southwest China. They have weakly depleted bulk-rock Nd isotopic compositions with $\varepsilon_{Nd}(t)$ of 0.0 to +0.7. The second group consists of hornblende gabbros between the Zhiduo and Yushu area and reveals relatively low TiO_2 contents, weak LREE enrichment, moderate enrichment of LILEs, and obvious depletion of HFSEs. They have weakly enriched bulk-rock Nd isotopic compositions with $\varepsilon_{Nd}(t)$ of −0.6 to 0.0. The third group consists of gabbros in the Zhiduo area show moderate TiO_2, enrichment of LREEs and LILEs, and weak depletion of HFSEs. They have slightly enriched bulk-rock Nd isotopic compositions with $\varepsilon_{Nd}(t)$ of −1.0 to −1.2.

Some mafic rocks were originally interpreted as ophiolites formed in either a subduction zone or an oceanic island (Li et al. 2010; Yong et al. 2011). However, a model involving plume-subduction interaction was proposed to explain the geodynamics and generation of the three groups of mafic rock (Liu et al. 2016a). This model suggests that the activity of the late Permian Emeishan mantle plume caused extension and rifting of the Yushu-Zhiduo area in association with rollback of the northward subduction of the Longmuco-Shuanghu oceanic lithosphere. Group 1 of the mafic rock is interpreted to be derived from a plume-related mantle source. Group 2 and 3 are thought to be related to the subduction of the Longmuco-Shuanghu oceanic lithosphere under the Northern Qiangtang terrane. Group 2 and 3 are derived from a subduction-metasomatized asthenospheric mantle and a subduction-metasomatized lithospheric mantle, respectively.

The basalts of the bimodal volcanic rocks are characterized by enrichment of LILEs and LREEs relative to HREEs, and moderate depletion of HFSEs (Fig. 2.2a, b). They have depleted bulk-rock Nd and Hf isotopic compositions with $\varepsilon_{Nd}(t)$ of

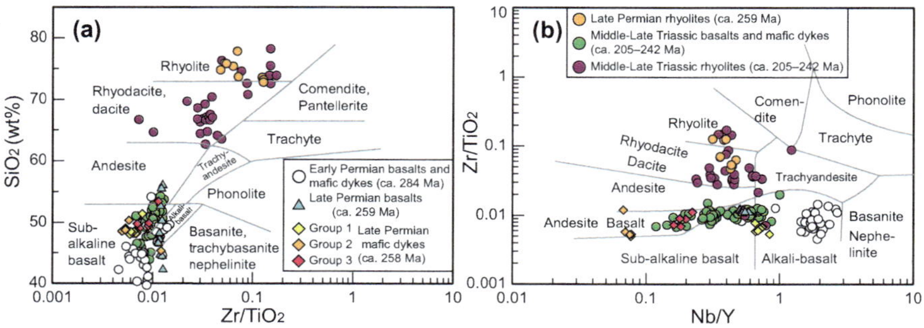

Fig. 2.2 Rock classification diagrams of (**a**) SiO2 vs. Zr/TiO2, and (**b**) Zr/TiO2 vs. Nb/Y (Winchester and Floyd 1976) for early Permian to Late Triassic volcanic rocks and mafic dykes in the Northern Qiangtang subterrane

+3.1 to +3.5 and $\varepsilon_{Hf}(t)$ of +10.9 to +11.7. The rhyolites in the bimodal volcanic rocks are characterized by high SiO_2 (Fig. 2.2a), strongly peraluminous, enrichment of LILEs, and depletion of HFSEs. They have depleted bulk-rock Nd and Hf isotopic compositions with $\varepsilon_{Nd}(t)$ of +2.7 to +2.8 and $\varepsilon_{Hf}(t)$ of +9.5 to +9.7. A subduction-plume interaction model was also proposed to elucidate the petrogenesis of the bimodal volcanic rocks (Wang et al. 2018a). This model suggests that rollback of northward subduction slab beneath the Northern Qiangtang terrane induced flowing of the Emeishan mantle plume to the back-arc extension region. The basalts are interpreted to be derived from partial melting of the plume material with minor contributions from subducted sediment-derived fluids. The rhyolites were produced by partial melting of juvenile mafic crustal rocks as a result of mafic magma underplating.

2.1.1.3 Middle to Late Triassic Igneous Rocks

Triassic igneous rocks are common in the Northern Qiantang terrane including both plutonic and volcanic rocks. The plutonic rocks are mainly located at the southern margin, broadly parallel to the Longmuco-Shuanghu suture zone. From west to southeast, the plutonic bodies include those in the Laxiongco and Shuangshu area and the Tanggula batholith in the Baqing area. The plutons in the Shuanghu area consist of amphibole-bearing granites which have dioritic enclaves and the granodiorites with zircon U–Pb age of ca. 210–220 Ma (Li et al. 2015a; Jiang et al. 2021). The Tanggula batholith consists of biotite-bearing tonalite, granodiorite, monzogranite, and syenogranite and has zircon U–Pb age of ca. 223–235 Ma (Lu et al. 2017; Song et al. 2021a). These Middle to Late Triassic granitoids dominantly belong to high-K calc-alkaline series (Fig. 2.3a) and are characterized by enrichment of LILEs but depletion of HFSEs, enrichment of LREEs relative to HREEs, and weakly to strongly peraluminous (Fig. 2.3b). They have enriched bulk-rock Nd isotopic compositions with $\varepsilon_{Nd}(t)$ of −11.1 to −8.5 and enriched to depleted zircon Hf isotopic compositions with large variation of $\varepsilon_{Hf}(t)$ from −29.9 to +5.6. These granitoids are interpreted to be derived from partial melting of middle or lower crust of the Qiangtang terrane in a collisional setting (Li et al. 2015a; Jiang et al. 2021; Song et al. 2021b). Alternatively, the granitoids of the Tanggula batholith are suggested to have formed in a subduction setting and generated by partial melting of the flysch from the Songpan-Ganzi complex underthrusting beneath the Qiangtang terrane along the Jinshajiang suture zone (Lu et al. 2017).

The Triassic mafic plutonic rocks are only reported in the Yushu area close to the Jinshajiang suture zone (Liu et al. 2020a). The mafic rocks are composed of gabbroic and doleritic dykes intruding into Triassic sedimentary rocks and have zircon U–Pb age of ca. 230 Ma. These mafic rocks are characterized by the enrichment of LILEs but depletion of HFSEs, LREEs enrichment relative to HREEs, and slightly depleted bulk-rock Nd isotopic compositions with $\varepsilon_{Nd}(t)$ of +0.5 to +2.0. They are

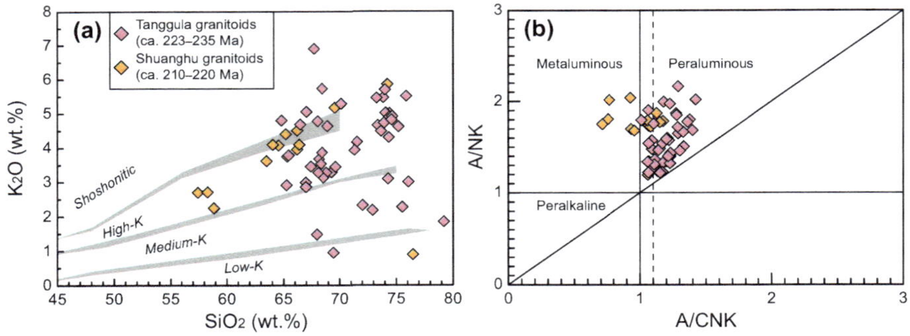

Fig. 2.3 (**a**) Rock series (Rickwood 1989) and (**b**) alumina saturation indices of Late Triassic granitoids in the Northern Qiantang subterrane

interpreted to be generated in a back-arc basin setting associated with a northward subduction system beneath the Northern Qiangtang terrane.

The volcanic rocks extend from the Laxiongco area in the west, through Nadigangri area, Geladandong area, Yanshiping area, and Tuotuohe area, to the Yushu area in the east. These volcanic rocks consist of basalts, andesites, dacites, and rhyolites (Fig. 2.2a, b) and show a progressive younger age towards the west. The volcanic rocks in the Yanshiping area to Tuotuohe area are the oldest with zircon U–Pb ages of ca. 230–242 Ma (Wang et al. 2008b; Chen et al. 2016). The volcanic rocks in the Geladandong area have slightly younger zircon U–Pb ages of ca. 220–237 Ma (Fu et al. 2010; Ou et al. 2021). The volcanic rocks in the Laxiongco area to Nadigangri area are the youngest with zircon U–Pb ages of ca. 205–220 Ma (Zhai and Li 2007; Zhai et al. 2013b; Li et al. 2007; Wang et al. 2007a, 2008a).

Among the volcanic rocks, some basalts are classified as Nb-enriched basalts and are limited to the Yanshiping area to Tuotuohe area. They show variable Nb enrichment with Nb/La ratios mostly >0.6 and have depleted bulk-rock Nd isotopic compositions with variable $\varepsilon_{Nd}(t)$ of +1.71 to +5.16 (Wang et al. 2008a; Chen et al. 2016). These Nb-enriched basalts are suggested to be derived from a mantle source metasomatized by a slab melt in a subduction setting related to either south-dipping subduction of the Jinshajiang suture zone (e.g., Ou et al. 2021) or north-dipping subduction of the Bangonghu-Nujiang suture zone (Chen et al. 2016).

In contrast to the Nb-enriched basalts, andesites, dacites, rhyolites and other basalts are all characterized by enrichment of LILEs, depletion of HFSEs, and enrichment to slightly depleted bulk-rock Nd isotopic compositions with large variation of $\varepsilon_{Nd}(t)$ (e.g., −9.60 to +1.21). Different tectonic setting and geodynamic background are proposed to explain the genesis of these basalts to rhyolites, including: (1) arc magmatism related to both a northward subduction system along the Longmuco-Shuanghu suture zone or Bangonghu-Nujiang suture zone and a southward subduction system along the Jinshajiang suture zone (e.g., Ou et al. 2021) and (2) within-plate magmatism formed in a continental rift setting (Fu et al. 2010).

2.1.2 *Southern Qiangtang Subterrane*

The Southern Qiangtang subterrane (SQ) lies to the south of the Longmu Co-Shuanghu suture zone (LSSZ) and north of the Bangong-Nujiang suture zone (BNSZ) (Fig. 2.1). It is primarily composed of Upper Carboniferous to Middle Permian strata, and Upper Triassic to Lower Cretaceous strata (Zhu et al. 2016; Kapp and DeCelles 2019). It likely has an Archean basement as inferred by zircon Hf isotopes from Mesozoic igneous rocks (e.g., Li et al. 2014b). The Amdo microterrane (also termed Amdo microcontinent; Zhu et al. 2013) within the BNSZ show a geological affinity (e.g., detrital zircon ages and magmatism) to the southern Qiangtang subterrane (Zhu et al. 2013). Thus, it is also discussed here.

Pre-Cenozoic magmatism in the SQ and Amdo microterrane was active from Early Paleozoic to Late Cretaceous, and can be divided into seven periods: (1) Early Paleozoic (ca. 502–465 Ma), (2) Early Permian (ca. 291–283 Ma), (3) Middle-Late Triassic (ca. 247–210 Ma), (4) Early Jurassic (ca. 194–173 Ma), (5) Middle-Late Jurassic (ca. 170–148 Ma), (6) Early Cretaceous (ca. 125–100 Ma), and (7) Late Cretaceous (ca. 95–76 Ma).

2.1.2.1 Early Paleozoic Igneous Rocks

The Early Paleozoic igneous rocks are volumetrically minor and have experienced lower-greenschist-facies metamorphism. In the SQ, Early Paleozoic magmatism was active at ca. 502–439 Ma and are exposed around the Bensong Co, Gemuri, Duguer, Dabure, and Dawashan areas (Fig. 2.4a; Pullen et al. 2011; Liu

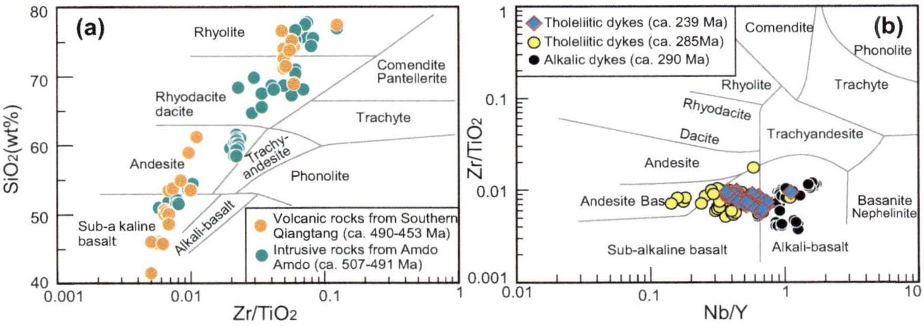

Fig. 2.4 Rock classification diagrams of (**a**) SiO2 vs. Zr/TiO2, and (**b**) Zr/TiO2 vs. Nb/Y (Winchester and Floyd 1976) for Early Paleozoic, Early Permian to Late Triassic igneous rocks in the Southern Qiangtang subterrane

et al. 2019; Wang et al. 2020). Around the Duguer area, granitic gneisses are dated at ca. 502–492 Ma (Liu et al. 2016b) and ca. 465 Ma (Dan et al. 2020). These rocks belong to high-K calc-alkaline series and geochemically resemble S-type granites. Similar granitic gneisses are also reported around the Bensong Co (ca. 480 Ma; Dan et al. 2020). These S-type granites yield high zircon O isotopes and negative zircon Hf isotopes, and are likely generated by partial melting of sedimentary rocks. Some mafic rocks, including both basalt and gabbro, are also reported around the Duguer area at ca. 490–483 Ma. The mafic rocks are geochemically similar to enriched mid-ocean-ridge basalt with no depletions in Nb and Ta and are interpreted to originate from high degree partial melting of an enriched mantle (Liu et al. 2019). Slightly younger (ca. 475 Ma) andesites around the Duguer area show negative anomalies in Nb and Ta (Liu et al. 2019). These rocks were replaced by a bimodal volcanic suite of meta-basalt and meta-rhyolite (470–455 Ma; Xie et al. 2017) and A-type granite gneiss (ca. 453 Ma; Liu et al. 2019). The meta-basalt is characterized by depletion of Rb, Ba, and Sr and enrichments in TiO_2, Th, U, Nb and Ta. The meta-rhyolites show high-K calc-alkaline signatures. The A-type granites belong to high-K calc-alkaline and shoshonitic series. A magma source of ancient basement is suggested for the A-type granites (Liu et al. 2019). The last episodes of magmatism were emplaced at ca. 438 Ma represented by the granitic gneisses (Wang et al. 2020). The Early Paleozoic igneous rocks in the Amdo microterrane indicate active magmatism from ca. 532 to 483 Ma (Guynn et al. 2012; Zhang et al. 2012c; Hu et al. 2021a, b). These rocks are granitoids, quartz monzonites and mafic enclaves, interpreted as resulting from mixing between mantle-derived mafic and crust-derived felsic magmas (Hu et al. 2021a). Two end-member models have been proposed for geodynamic settings of the early Paleozoic igneous rocks, including a post-collision extensional setting of the Pan-African orogenic cycle (e.g., Pan et al. 2012) and an Andean-type arc-related setting during southward subduction of the proto-Tethys beneath the northern Gondwana margin (e.g., Cawood et al. 2007; Zhu et al. 2012; Li et al. 2018b).

2.1.2.2 Early Permian Igneous Rocks

The Early Permian igneous rocks are exposed as E-W trending mafic dyke swarms in the western segment of the SQ. Previous studies suggested that these dykes emplaced during ca. 320–280 Ma (Zhai et al. 2013a; Wang et al. 2019) based on LA-ICPMS zircon U–Pb dating. Recently, Dan et al. (2021a) documented that these mafic dykes likely formed at ca. 291–283 Ma with two peaks of ca. 290 and 285 Ma using SIMS and SHRIMP zircon U–Pb dating methods. These Early Permian rocks are basaltic and include both alkalic and tholeiitic series (Fig. 2.4b). The ca. 290 Ma phase igneous rocks are mainly tholeiitic basaltic in composition (or classified as low-Ti suite) with minor alkalic basalts (or classified as high-Ti suite); whereas rocks formed at ca. 285 Ma phase are widespread and are mainly tholeiitic basalts (low-Ti suite) (Wang et al. 2019; Dan et al. 2021a). The tholeiitic rocks are enriched in light REEs and LILEs, and depleted in heavy REEs and HFSEs (e.g., Nb and Ta). The alkalic rocks show enrichment in light REEs, LILEs, and HFSEs (e.g., Nb and Ta), similar to the present-day OIB. All these rocks yield positive bulk-rock $\varepsilon_{Nd}(t)$ values and are suggested to be derived from different parts of a single mantle plume.

2.1.2.3 Middle-Late Triassic Igneous Rocks

Triassic magmatic rocks consist of mafic dyke swarms and granitic plutons on the northern margin of the southern Qiangtang subterrane. The Triassic mafic dykes are recently identified (Dan et al. 2021b) along the margin of the Longmu Co-Shuanghu suture zone. These mafic dykes were mapped as part of the Permian dykes described above (Zhai et al. 2013a; Dan et al. 2021a). These dykes are E-W trending with no more than a few meters in width. The zircon grains from seven samples yield ages ranging from ca. 247 to 232 Ma, with a mean age of 239 ± 1 Ma (Dan et al. 2021b). These mafic dykes have tholeiitic compositions (Fig. 2.4b), showing enrichment in light REEs, negative Nb–Ta–Ti anomalies, mantle-like zircon $\delta^{18}O$ values, and large range of $\varepsilon_{Nd}(t)$ values (−4 to +3). These features are different from subduction-related basaltic rocks, the Permian dikes (ca. 291–283 Ma), and typical OIB. These rocks are interpreted to originate from the metasomatized lithospheric mantle because of regional extension on a passive margin caused by enhanced slab-pull forces (Dan et al. 2021b). A younger period magmatism was active between ca. 220 and 210 Ma, including the Rongma and Shuanghu batholiths (Li et al. 2015a). These rocks belong to high-K calc-alkaline granites and are enriched in light REEs and LILEs and depleted in HFSEs. They show negative $\varepsilon_{Nd}(t)$ values, resembling typical I-type granite. These rocks also occurred on a passive margin and are considered to be products of slab break-off during collision between southern and northern Qiangtang subterranes (Zhai et al. 2013b; Li et al. 2015a).

2.1.2.4 Early Jurassic Igneous Rocks

The Early-Middle Jurassic igneous rocks are exposed as intermediate-felsic intrusions in the Amdo microterrane (Kapp and DeCelles 2019). The phase of magmatism ranges from ca. 194 to 173 Ma (Zhu et al. 2011; Liu 2012; Yan et al. 2016; Liu et al. 2017) and can be subdivided into two stages based on rock types and geochemical data (Fig. 2.5a, b).

The early stage rocks (ca. 194–180 Ma) consist of diorite, granodiorite, monzogranite, and syenogranite (Liu 2012). These rocks belong to the calc-alkalic series and are characterized by varying K_2O and thus range from medium-K to shoshonitic series (Fig. 2.5a, b). They are enriched in light REEe and LILEe compared with HFSEe and yield evolved $\varepsilon_{Nd}(t)$ (−11.6 to −9.2) and zircon $\varepsilon_{Hf}(t)$ (−17.8 to −0.9) values (Liu 2012). These rocks are I-type granitoids because of the presence of amphibole. They are interpreted to originate from partial melting of the ancient

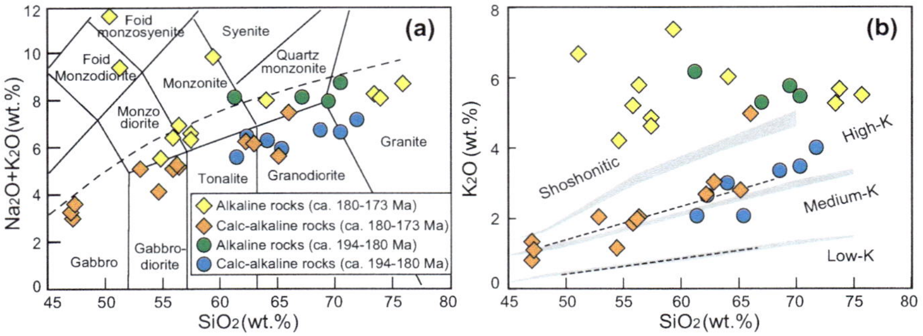

Fig. 2.5 Plots of (**a**) $Na_2O + K_2O$ vs. SiO_2 (Middlemost 1994) and (**b**) K_2O vs. SiO_2 (Rickwood 1989) of the Early Jurassic igneous rocks in the Southern Qiangtang subterrane

lower crust (Liu 2012) and are generally ascribed to northward subduction of the Bangong-Nujiang Tethyan lithosphere (e.g., Guynn et al. 2006; Zhu et al. 2013; Tian et al. 2021).

The late stage rocks (ca. 180–173 Ma) are composed of various rock types, including calc-alkalic gabbro, diorite, and granodiorite (Fig. 2.5a, b; Yan et al. 2016) and alkalic series of monzonite diorite, monzonite, and alkali feldspar syenite (Fig. 2.5a, b; Liu 2012; Yan et al. 2016). The gabbros have flat REE patterns and are geochemically similar to MORB. They yield positive $\varepsilon_{Nd}(t)$ values of 4.4–5.5, identical to those of the SSZ-type ophiolites within the BNSZ (Wang et al. 2016a). The calc-alkalic diorite and granodiorite are very similar to the early-stage rocks (ca. 194–180 Ma) rocks. The alkalic diorite are characterized by high Mg#, Cr, and Ni and have similar isotopic features to the early stage rocks; whereas the alkali feldspar syenite have low Mg# (~45) and more negative $\varepsilon_{Nd}(t)$ (-14.4 to -12.8) values (Liu 2012; Yan et al. 2016). The 180–173 Ma igneous rocks were likely generated by partial melting of various sources, including asthenospheric mantle, lithospheric mantle, and ancient lower crust (Yan et al. 2016). Rocks in this stage are either attributed to slab roll-back of the north-dipping subducting Bangong-Nujiang Tethyan lithosphere (Yan et al. 2016; Xie et al. 2020) or to slab break-off during the Amdo-Qiangtang collision (Liu 2012).

2.1.2.5 Middle-Late Jurassic Igneous Rocks

Four groups of magmatic-rock associations (Fig. 2.6) have been identified in the Southern Qiangtang subterrane and BNSZ during Middle-Late Jurassic (see Li et al. 2020, and references therein).

1. Low-Mg series: Normal calc-alkaline intermediate-felsic intrusive rocks are broadly exposed around the Gerze, Qingcaoshan, Liqunshan, and Rutog areas (ca. 170–148 Ma; Li et al. 2014a, b; Zhang et al. 2017; Liu et al. 2017), extending over 400 km along strike in the western segment of the Southern Qiangtang

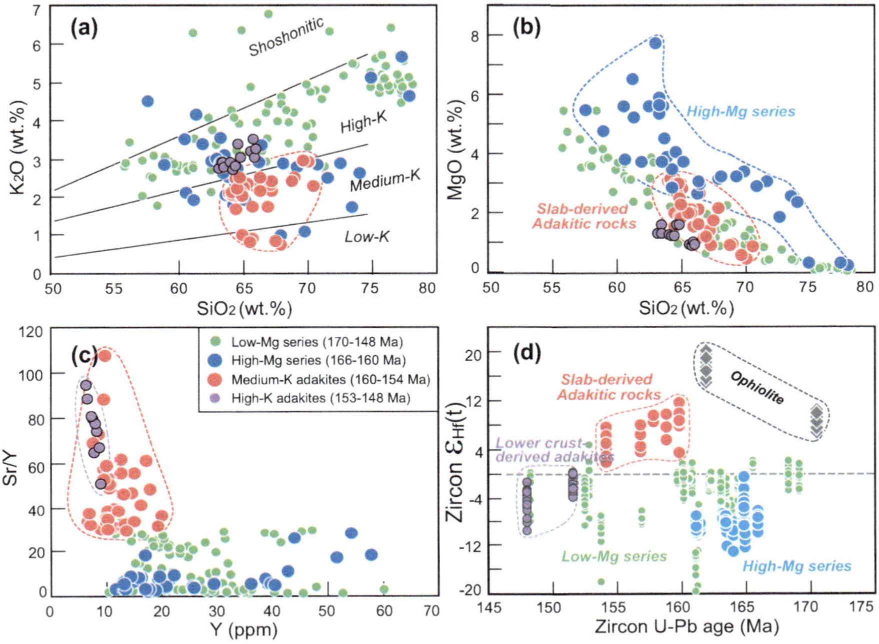

Fig. 2.6 Plots of (**a**) K_2O vs. SiO_2 (Le Maitre 2002), (**b**) MgO vs. SiO_2, (**c**) Sr/Y vs. Y (Defant and Drummond 1990), and (**d**) zircon $\varepsilon_{Hf}(t)$ vs. zircon U–Pb ages for Middle-Late Jurassic igneous rocks in the Southern Qiangtang subterrane

subterrane (Li et al. 2020). There are also minor exposures near Daru Tso (ca. 165–163 Ma; Li et al. 2020) in the eastern BNSZ. These igneous rocks are linked to the Rutog-Gerze magmatic rocks, indicating the development of an over 1200 km magmatic belt. These rocks are high-K calc-alkaline rocks (Fig. 2.6a) and belongs to I-type granitoids and highly fractionated I-type granitoids (Li et al. 2014b). They belong to low-Mg series and are characterized by low Sr/Y, high Y, negative $\varepsilon_{Nd}(t)$ and zircon $\varepsilon_{Hf}(t)$ values (Fig. 2.6b–d), and high zircon $\delta^{18}O$ values of 6.7–8.0‰. Besides, mafic enclaves and dikes have been reported in these intrusions, which show geochemical affinity of high-Al basalt and low-Mg andesite (Li et al. 2014b). All these rocks are common in modern magmatic arcs and can be interpreted by a MASH process, i.e., a combined process of *m*elting, *a*ssimilation, *s*torage, and *h*omogenization in lower crust above subduction zones (Li et al. 2014b).

2. High-Mg series: High-Mg andesites, dacites, and rhyolites are reported near Daru Tso (ca. 166–160 Ma) (Li et al. 2015b, 2020; Zeng et al. 2016) in the eastern BNSZ, stretching approximately 40 km from east to west. These rocks are characterized by have high MgO, Mg# (>42), Cr, and Ni, differing significantly from the low-Mg series. They also yield low Sr/Y, low Y, negative $\varepsilon_{Nd}(t)$ and zircon $\varepsilon_{Hf}(t)$ values (Fig. 2.6c, d), and high zircon $\delta^{18}O$ values of 7.2–8.4‰, simi-

lar to the low-Mg series. The trace element features of the high-Mg series are identical to those of the low-Mg series, exhibiting a magmatic arc affinity. They are thought to either originate from the mantle wedge that has been modified by sediment-derived melts during initial subduction of a separate branch of the Bangong Tethys (Zeng et al. 2016) or be derived from partial melting of mélange diapirs in the mantle wedge during northward subduction of the Bangong Tethys (Li et al. 2020).

3. Medium-K adakite and OIB-type rocks: Slightly younger medium-K adakitic dacites are reported near Gerze-Rena Tso (ca. 160–154 Ma; Li et al. 2016a; Fan et al. 2016) and Kangqiong (ca. 148 Ma; Li et al. 2016b). These rocks are characterized by high Sr/Y, low Y, high MgO and Mg#, small negative to positive $\varepsilon_{Nd}(t)$, and positive zircon $\varepsilon_{Hf}(t)$, distinct from the high-K intermediate-felsic rocks (i.e., both low-Mg and high-Mg series described above) (Fig. 2.6). These features indicate an ocean crust origin (Li et al. 2016a, b). Coeval ocean island basalt (OIB) type mafic rocks have also been identified around Gerze (ca. 166 and 158 Ma; Li et al. 2016a) and Rutog (ca. 163 Ma, Shiquanhe; Li et al. 2018a) (Fig. 2.6). These rocks are enriched in light REEs and Nb–Ta, and have positive $\varepsilon_{Nd}(t)$, suggesting a sub-slab asthenosphere source (Li et al. 2016a). This rock association of coeval adakite and OIB rock is likely generated in a slab window setting, which is suggested to be related to ridge subduction (Li et al. 2020).
4. High-K adakite: Diorites and granodiorites (ca. 153–148 Ma) from the Rena plutons are slightly younger than the medium-K adakitic dacites (ca. 156 Ma) around the Gerze-Rena Tso. These intrusive rocks belong to high-K series (Hao et al. 2016; Song et al. 2021b) and have high Sr/Y and La/Yb, akin to adakites (Fig. 2.6). However, these adakites have negative $\varepsilon_{Nd}(t)$ and zircon $\varepsilon_{Hf}(t)$ values, similar to the low-Mg series and differing from the adakitic dacites described above. Therefore, these adakitic granodiorites could indicate crustal thickening (Hao et al. 2016).

2.1.2.6 Early Cretaceous Igneous Rocks

The Early Cretaceous magmatism was active from ca. 125 to 100 Ma and is exposed along the southern margin of the Southern Qiangtang subterrane, in the Amdo microterrane of western segment (Zhu et al. 2011; Liu et al. 2017), Xiabie batholith of middle segment (Yang et al. 2019), and along Gerze-Rutog belt of western segment (Li et al. 2013a, 2014a; Hao et al. 2016; Liu et al. 2017). Exposures are dominantly intermediate-felsic intrusive rocks (Li et al. 2014a, 2017) with minor volcanic rocks of basalt to rhyolite (the Meiriqiecuo Formation) (Zhang et al. 2017). These rocks belong to medium-K to high-K series (Fig. 2.7) and are enriched in light REEs and LILEs relative to HFSEs and yield positive $\varepsilon_{Nd}(t)$ and zircon $\varepsilon_{Hf}(t)$ values, indicating a juvenile lower crust origin. These rocks are thought to record the collision processes between Lhasa and Qiangtang terranes, although the collision timing remains controversial, ranging from ca. 166 to 100 Ma (Zhang et al. 2012a; Zhu et al. 2016).

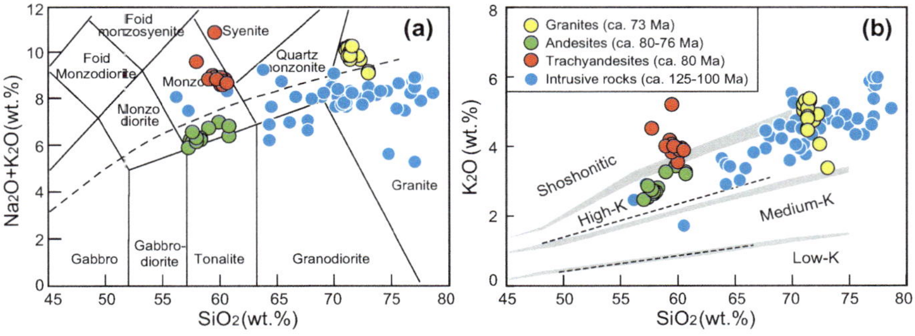

Fig. 2.7 Plots of (**a**) $Na_2O + K_2O$ vs. SiO_2 (Middlemost 1994) and (**b**) K_2O vs. SiO_2 (Rickwood 1989) of the Early and Late Cretaceous igneous rocks in the Southern Qiangtang subterrane

2.1.2.7 Late Cretaceous Igneous Rocks

Late Cretaceous igneous rocks are volumetrically minor and including two stages. The first stage rocks (ca. 95 Ma) are exposed as andesites around Biluoco. They have high MgO, Mg#, Cr and Ni, termed the magnesian andesite (He et al. 2018). These rocks have high Y, Sr, and high Sr/Y, differing from both adakites (high Sr, low Y) and normal arc rocks (low Sr, high Y) and similar to the bajaites. The $\varepsilon_{Nd}(t)$ values of these rocks are positive. These rocks are interpreted to be derived from the mantle wedge metasomatized by slab- and sediment-derived melts and represent the last phase of the Lhasa-Qiangtang collision (He et al. 2018). The second stage rocks (ca. 80–76 Ma) consist of alkali granite around the Cuona lake (He et al. 2019), trachy-andesite around the Amdo (ca. 80 Ma; Bai et al. 2009; Chen et al. 2017), and K-rich andesite and dacite around the Abushan (ca. 80–76 Ma; Li et al. 2013b) (Fig. 2.7). These rocks yield varying Mg# and Sr/Y and negative $\varepsilon_{Nd}(t)$. They likely record post-collisional geodynamic processes and are commonly interpreted to originate from delaminated lower crust or lithospheric mantle (Li et al. 2013b; Chen et al. 2017).

2.2 Lhasa Terrane

2.2.1 Northern Lhasa Subterrane

The Northern Lhasa subterrane (NL) lies to the south the Bangong-Nujiang suture zone (BNSZ) and north of the Shiquan River-Nam Tso mélange zone (SNMZ) (Fig. 2.1). This subterrane likely has a juvenile lower crust based on zircon Hf isotope mapping (Zhu et al. 2011). Exposures are dominated by extensive Lower Cretaceous volcanic rocks (Duoni and Qushenla formations) and coeval plutons with minor Triassic-Jurassic strata in the eastern segment. The NL consists mainly of Early Cretaceous igneous rocks, forming an E-W trending belt. Minor Late Cretaceous igneous are also reported in this subterrane.

2.2.1.1 Early Cretaceous Igneous Rocks

Early Cretaceous igneous rocks are widespread in the NL and are exposed as batholiths and plutons in the eastern segment and the Qushenla Formation in the western segment. Based on rock associations, the Early Cretaceous magmatism can be divided into two stages (Fig. 2.8).

Stage 1 ranges from ca. 139 to 116 Ma. In the eastern segment, the earliest rocks are weakly peraluminous I-type granites (ca. 139–127 Ma) with depleted isotopes (zircon $\varepsilon_{Hf}(t) = -1.9$ to $+11.0$) in the Baingoin batholith (Zhu et al. 2016). They are replaced by peraluminous S-type granites (ca. 125–120 Ma) with negative zircon $\varepsilon_{Hf}(t)$ (-7.0 to -2.1) in the Baingoin batholith (Zhu et al. 2016), Sangba batholith (He et al. 2006) and coeval S-type rhyolites (Sun et al. 2015b). Therefore, the zircon $\varepsilon_{Hf}(t)$ values decrease from ca. 132 to 120 Ma. In the western segment, the volcanic rocks from the Qushenla Formation are medium-K calc-alkaline basalt and andesite (ca. 131–116 Ma; Fig. 2.8a; Kang et al. 2010; Li et al. 2018c). The basalts have low K_2O, Th, TiO_2, moderate MgO, and high Al_2O_3, resembling typical high-alumina basalt, which is widespread in magmatic arcs. The andesites also show low K_2O and Th, belonging to medium-K series. The andesites are considered as Mg-rich andesite because of higher MgO, Mg#, Cr and Ni contents when compared to typical arc andesites (Li et al. 2018c). Overall, rocks in this stage are common above subduction zones and can be generated by partial melting of the mantle wedge and lower crust. Therefore, rocks in this stage likely represent an E-W trending magmatic arc along the northern Lhasa subterrane (Zhu et al. 2016; Li et al. 2018c).

Stage 2 magmatism was active from ca. 116 to 100 Ma with a peak at ca. 110 Ma. In the eastern segment, peraluminous granites with negative zircon $\varepsilon_{Hf}(t)$ values of -12.6 to -0.5 and high zircon $\delta^{18}O$ values of 8–11‰ are identified around Nagqu (Sun et al. 2015b), and metaluminous to strongly peraluminous granites with varying zircon $\varepsilon_{Hf}(t)$ values of -13.5 to $+6.4$ are reported in the Baingoin batholith (Zhu et al. 2016). These rocks resemble S-type granites. In the middle segment, a rock association of A_2-type granites and rhyolites to the west of Baingoin (ca.

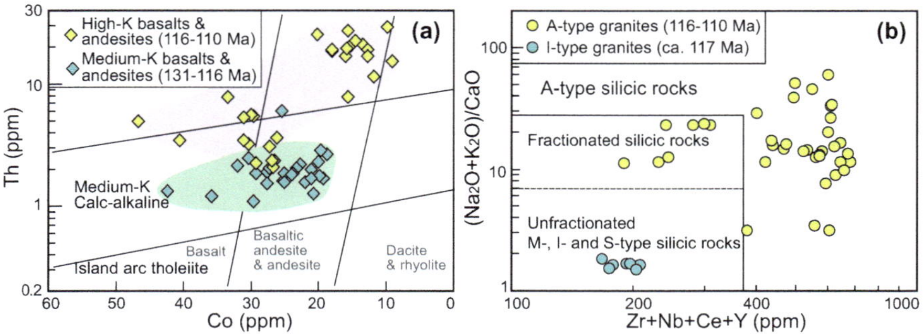

Fig. 2.8 Plots of (**a**) Th vs. Co for basalts and andesites (Hastie et al. 2007) and (**b**) $(Na_2O + K_2O)/CaO$ vs. Zr + Nb + Ce + Y for granitoids and rhyolites (Whalen et al. 1987) of the Early Cretaceous igneous rocks in the northern Lhasa subterrane

116–110 Ma; Fig. 2.8b; Chen et al. 2014) and high-K calc-alkaline basalts and andesites form Mayue (ca. 116 Ma; Fig. 2.8a) has been identified. In the western segment, diverse rock types at ca. 110 ± 2 Ma have been reported, including high-K basalt (Fig. 2.8a; Li et al. 2018c), a bimodal suite of basalt and rhyolite and adakitic diorite around Yanhu (Sui et al. 2013). The coexistence of A-type granite, S-type granite, adakite, high-K basalt and the bimodal suite suggests partial melting of differing sources, including deep asthenosphere mantle, mantle wedge, ocean crust, and lower continental crust (Li et al. 2018c). Rocks in this stage are interpreted as products of slab break-off during Lhasa-Qiangtang collision (Zhu et al. 2009b, 2011, 2016; Sui et al. 2013; Li et al. 2018c). Slightly younger rocks are minor, including low-K adakitic granodiorites (ca. 105 Ma) (Wu et al. 2015a), high-Mg medium-K monzonites (100 ± 1 Ma) (Wu et al. 2015b), and high-K rhyolites (ca. 107–103 Ma) (Wu et al. 2013) around Gerze in the western segment. These rocks likely represent the last magmatic responses to slab break-off.

2.2.1.2 Late Cretaceous Igneous Rocks

The late Cretaceous magmatism is volumetrically minor and is exposed as dacitic and rhyolitic dykes to the west of Baingoin (ca. 94 Ma; Yi et al. 2018), andesite and dacite around the Zhuogapu (ca. 91 Ma; Wang et al. 2014b), and granitoids around the Rutog (ca. 81–79 Ma; Zhao et al. 2008) (Fig. 2.9a, b). The Baingoin dacite and rhyolite show high Al_2O_3 contents, enrichment in light REEs, depletion in heavy REEs, high Sr, and low Y, positive bulk-rock $\varepsilon_{Nd}(t)$ and zircon $\varepsilon_{Hf}(t)$ values, resemble adakite (Yi et al. 2018). The dacite is divided into low-Mg group (Mg# = 37–45; Fig. 2.9a) and high-Mg group (Mg# = 48–52; Fig. 2.9a), which were likely generated by partial melting of thickened lower crust and foundering lower crust followed by melt–mantle interaction, respectively (Yi et al. 2018). Differing from the Baingoin adakites, the Zhuogapu andesite and dacite have low light REEs and Sr (Fig. 2.9b). All the Zhuogapu rocks are characterized by high MgO contents and

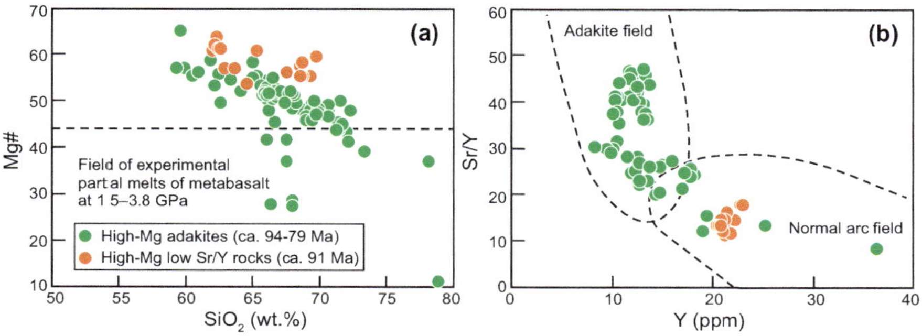

Fig. 2.9 Plots of (**a**) Mg# vs. SiO_2 and (**b**) Sr/Y vs. Y (Defant and Drummond 1990) of Late Cretaceous igneous rocks in the Northern Lhasa subterrane

Mg# values, negative bulk $\varepsilon_{Nd}(t)$ and positive zircon $\varepsilon_{Hf}(t)$ values (Wang et al. 2014b). They are thought to be derived from a mixed source of juvenile and ancient lower crust followed by melt–mantle interaction and high extent of magma fractionation (Wang et al. 2014b). The Rutog granitoids are geochemically similar to the Zhuogapu dacites with high Sr/Y, moderate Mg#, and positive bulk $\varepsilon_{Nd}(t)$ values and also can be explained by partial melting of a thickened lower crust (Zhao et al. 2008).

2.2.2 Central Lhasa Subterrane

The Central Lhasa subterrane (CL) lies between the Shiquan River-Nam Tso mélange zone (SNMZ) to the north and the Luobadui-Milashan fault (LMF) to the south (Fig. 2.1). It likely has a Proterozoic and Archean crystalline basement as indicated by zircon Hf isotopic mapping of magmatic rocks (Zhu et al. 2011). These magmatic rocks are covered by widespread Permian to Cretaceous meta-sedimentary rocks (Zhu et al. 2013).

The Cambrian magmatism (ca. 518–492 Ma) is volumetrically minor and only occurs on the northern margin of the CL (Fig. 2.1). The Mesozoic magmatism (ca. 236–84 Ma) is common in the CL (Fig. 2.1), as documented by the presence of Late Triassic to Late Cretaceous mafic, intermediate and felsic rocks. This section further introduces the research progress of these magmatic rocks in terms of spatial and temporal distributions, rock assemblages, geochemical characteristics, petrogenesis, and geodynamic interpretation based on four timeframes of major phase of magmatic activities: Cambrian (ca. 518–492 Ma), Late Triassic to Early Jurassic (ca. 236–183 Ma), Middle Jurassic to Early Cretaceous (ca. 172–107 Ma), and Late Cretaceous (ca. 101–84 Ma).

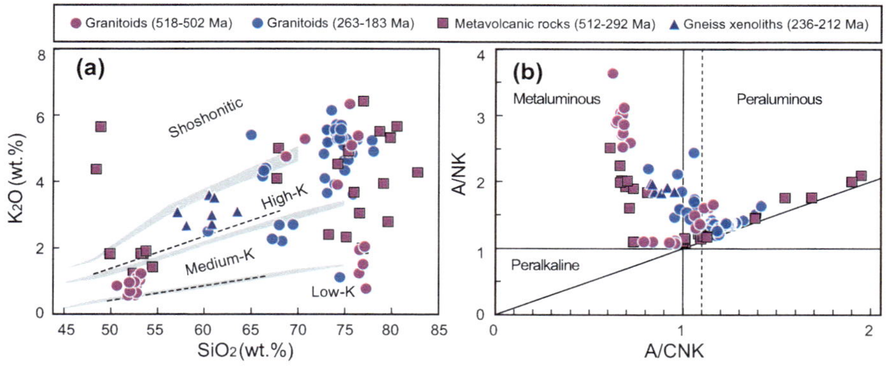

Fig. 2.10 (**a**) K_2O vs. SiO_2 diagram (Rickwood 1989), and (**b**) molar $Al_2O_3/(Na_2O + K_2O)$ (A/NK) vs. molar $Al_2O_3/(CaO + Na_2O + K_2O)$ (A/CNK) diagram (Maniar and Piccoli 1989) for the Cambrian to Early Jurassic magmatic rocks from the Central Lhasa subterrane

2.2.2.1 Cambrian Igneous Rocks

The Cambrian igneous rocks are exposed as silicic metavolcanic rocks and metabasalts around Banglei County (ca. 512–492 Ma, Liu et al. 2020b; Zhu et al. 2012), and are exposed as gabbros and granitoids in the Ren Co area (ca. 518–502 Ma, Hu et al. 2021a) along the north CL margin. Geochemically, these metaigneous rocks are high-K calc-alkaline and shoshonite (Fig. 2.10a). The silicic metavolcanic rocks have negative $\varepsilon_{Hf}(t)$ values (−13.9 to −0.6), suggesting that these rocks were generated by partial melting of ancient Lhasa basement (Liu et al. 2020b; Zhu et al. 2012). The metabasalts yield varying zircon $\varepsilon_{Hf}(t)$ values (−0.7 to +7.5), indicating an enriched mantle origin (Liu et al. 2020b; Zhu et al. 2012). The Nd-Hf isotopic data suggest that the gabbros and granitoids exposed in the Ren Co area were probably generated by mixing of mantle-derived mafic and crust-derived felsic melts (Hu et al. 2021a). Two opposite views exist on the development of the Cambrian magmatism. Some researchers have argued that the generation of magmatism at this stage (ca. 512–492 Ma) may be the result of slab roll-back and subsequent break-off of the proto-Tethyan oceanic lithosphere, indicating an early Paleozoic Andean-type northern margin of the Tibet Plateau (Hu et al. 2021a; Zhu et al. 2012). However, Liu et al. (2020b) preferred to suggest that these magmatic rocks correlate with intraplate tectonics, indicating a passive northern margin of the Tibet Plateau in the Early Paleozoic.

2.2.2.2 Late Triassic to Early Jurassic Igneous Rocks

The intrusive rocks of the Late Triassic to Early Jurassic have been documented in the CL, including Luozha (ca. 231–202 Ma, Zhang et al. 2007a, 2019; Zhu et al. 2011; Wang et al. 2014a; Bu et al. 2015; Zhou et al. 2017b), Mamba (ca. 215–195 Ma, Ji 2010; Zhu et al. 2011), Gongbogyamda (215–183 Ma, Ji 2010; Zhu et al. 2011) and Ningzhong (ca. 193–190 Ma, Liu et al. 2006) areas. These rocks contain a small amount of dioritic genesis xenoliths (Zhou et al. 2017b), granodiorite, biotite monzogranite, muscovite monzogranite, two-mica granite and syenogranite.

These granitoids fall in the high-K calc-alkaline and shoshonite fields (Fig. 2.10a) and are metaluminous to peraluminous with varying A/CNK of 0.8–1.4 (Fig. 2.10b) (molar $Al_2O_3/CaO + Na_2O + K_2O$). These rocks are enriched in LREEs and LILEs (e.g., U, Th and Pb), and are depleted in HFSEs (e.g., Nb, Ta, Ti), showing typical affinities of arc magmatism. Most of these rocks are S-type granitoids with enriched zircon Hf isotopic compositions (Zhang et al. 2007a, 2019; Zhu et al. 2011; Zhou et al. 2017b). Therefore, the Late Triassic to Early Jurassic magmatism is generally thought to be related to re-melting of ancient Lhasa basement (Zhang et al. 2019; Zhu et al. 2011).

Two popular geodynamic models have been proposed to explain the development of the Late Triassic to Early Jurassic magmatism. Some researchers have argued that these magmatic rocks were the products related to the northward subduction of the Neo-Tethys ocean lithosphere (Allègre et al. 1984; Yin and Harrison 2000; Zhou et al. 2017b), which may have started since the Late Triassic (Zhang et al. 2007b; Wu et al. 2010; Guo et al. 2013; Kang et al. 2014) (model 1). However, other researchers suggested that these magmatic activities were generated during the southward subduction of the Bangong-Nujiang Tethyan lithosphere (Pan et al. 2012), which opened before the Early Cretaceous (Zhu et al. 2009b, 2011, 2013) (model 2).

Recently, Zhang et al. (2019) reported an integrated dataset of bulk-rock geochemistry, zircon U–Pb age and Hf isotope of Luoza batholith in the CL, suggesting that these granodiorites with low SiO_2 content (64–67 wt%) were crystallized under high-temperature and low-pressure conditions. In combination with sedimentary records, these unique temperature-pressure conditions are interpreted as a consequence of back-arc extensional setting triggered by the southward subduction of the Bangong-Nujiang Tethyan lithosphere.

2.2.2.3 Middle Jurassic to Early Cretaceous Igneous Rocks

The Middle Jurassic intrusive rocks are exposed as granodiorite around Geji County (ca. 172–170 Ma, Jiang et al. 2010; Wang et al. 2017b). The Late Jurassic to Early Cretaceous intrusive rocks are widely exposed as intermediate-felsic intrusions in the Shiquanhe, Geji, Xiongba, Zhabuye, Coqein, Xuru Tso, Xainza and Mamba regions of the CL from west to east (166–107 Ma, Cao et al. 2016; Du et al. 2011; Jiang et al. 2010; Qu et al. 2009; Qi et al. 2019; Volkmer et al. 2007; Wang et al. 2017b, 2018b; Yan et al. 2017; Yan 2019; Zhou et al. 2008; Zhang et al. 2010a, 2011, 2012b; Zhu et al. 2009b, 2011; Zheng et al. 2018). The dioritic enclaves are developed in these intrusivons. The igneous rocks of Late Jurassic to Early Cretaceous are mainly exposed as Zenong Group in southern Xiongba, Coqein, Xainza, and Mamba regions (ca. 161–107 Ma, Chen et al. 2014; Kang et al. 2008; Tang et al. 2020; Zhu et al. 2009b, 2011), which consist of basalt, andesite, dacite, and rhyolite (Fig. 2.11a).

The intrusive rocks are metaluminous to peraluminous with varying A/CNK (molar $Al_2O_3/CaO + Na_2O + K_2O$; 0.7–2.0), plotting in the middle-K to high-K calc-alkaline fields (Fig. 2.11b). These rocks are mainly I-type granitoids given the presence of amphibole. In the Zr + Nb + Ce + Y vs. FeO*/MgO diagram, some granitic samples plot in the fractionated silicic rocks field (Fig. 2.11e). The volcanic rocks belong to calc-alkalic to shoshonitic series (Chen et al. 2014; Wang et al. 2016b). They are enriched in LREEs and LILEs (e.g., U, Th and Pb) relative to HFSEs and yield enriched bulk-rock Nd isotope and zircon Hf isotope compositions. It should be noted that the zircon Hf isotopes become more depleted from Middle Jurassic to Early Cretaceous, which indicates that the composition of the magma source regions changed significantly at ca. 110 Ma (Zhu et al. 2009b, 2011; Cao et al. 2016). The igneous rocks (ca. 172–107 Ma) were likely generated

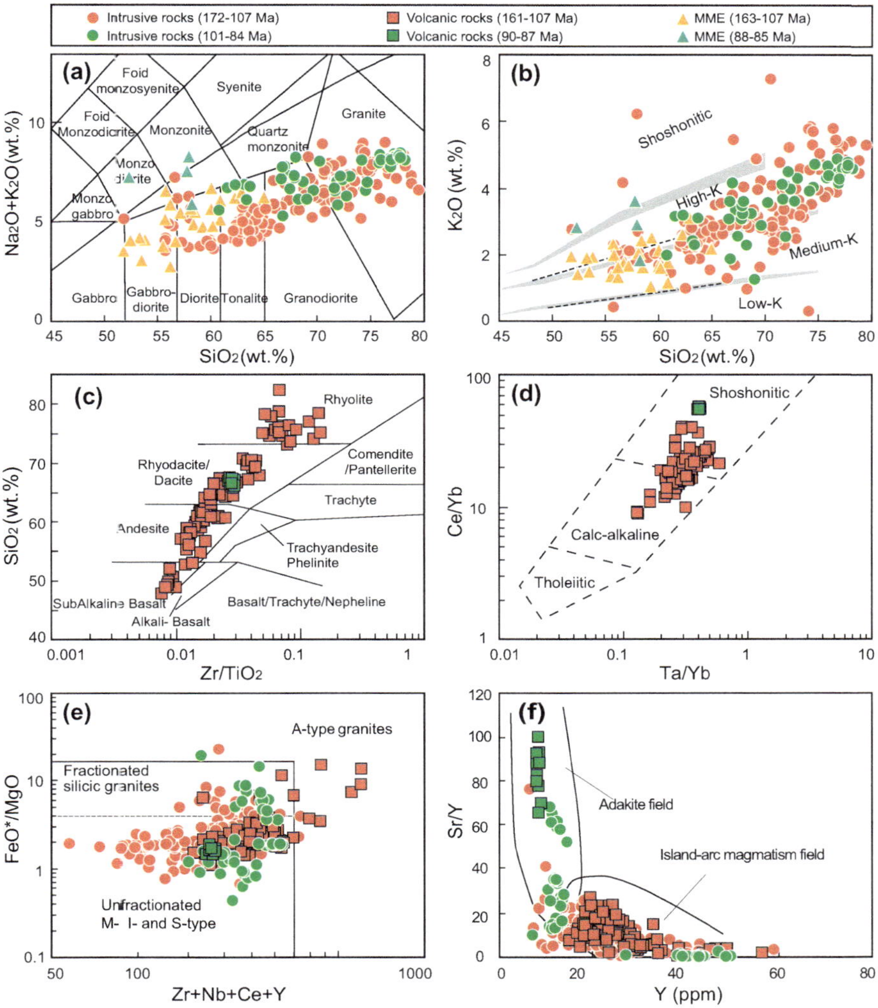

Fig. 2.11 Classification diagrams for the Middle Jurassic to Late Cretaceous magmatic rocks from the Central Lhasa subterrane. (**a**) $Na_2O + K_2O$ vs. SiO_2 (Middlemost 1994); (**b**) K_2O vs. SiO_2 (Rickwood 1989); (**c**) SiO_2 vs. Zr/TiO_2 (Winchester and Floyd 1976); (**d**) Ce/Yb vs. Ta/Yb (Pearce 1982); (**e**) FeO*/MgO vs. Zr + Nb + Ce + Y (Whalen et al. 1987); and (**f**) Sr/Y vs. Y (Defant and Drummond 1990)

by partial melting of various sources, including lithospheric mantle (e.g., Wang et al. 2018b), ancient lower crust (e.g., Wang et al. 2018b), and mixed sources (e.g., Chen et al. 2014). Rocks formed at this stage are either attributed to slab roll-back and subsequent slab break-off of the south-dipping subducting Bangong-Nujiang Tethyan lithosphere (e.g., Zhu et al. 2011) or to low-angle or flat north-dipping subduction of the Neo-Tethys Ocean lithosphere (e.g., Kapp et al. 2005, 2007; Zheng et al. 2018; Zhang et al. 2012a).

2.2.2.4 Late Cretaceous Igneous Rocks

The Late Cretaceous magmatism was active from ca. 101 Ma to ca. 84 Ma and is exposed in Gaerqiong (ca. 89–82 Ma, Lei et al. 2015; Yao et al. 2013), Xiongba (ca. 90 Ma, Liu et al. 2014), Azhang (ca. 90–87 Ma, Sun et al. 2015a), Zhongcang (94–88 Ma, Chen et al. 2015; Yu et al. 2011), and Gajin areas (ca. 100 Ma, Shuai et al. 2021) along the north CL margin, and in Coqein (ca. 90–87 Ma, Qu et al. 2006; Yu et al. 2015) and Memba (ca. 88–84 Ma, Meng et al. 2010, 2014) areas along the southern margin of the CL from west to east. Exposures are dominantly intermediate-felsic intrusive rocks (Fig. 2.11a) with minor volcanic rocks of dacite (Fig. 2.11c, Sun et al. 2015a).

The Late Cretaceous intrusive rocks belong to medium-K to high-K calc-alkaline series (Fig. 2.11b), whereas volcanic rocks belong to the shoshonitic series (Fig. 2.11d). Apart from peraluminous granites exposed in Coqein, Zhongcang and Gajin areas (Shuai et al. 2021), other intrusive and volcanic rocks are characterized by enrichments of LREEs, weak negative Eu anomalies (Eu/Eu* = 0.7–1.0), high Sr contents (Sr = 836–305 ppm) and high Sr/Y ratios (Sr/Y = 100–23) (Fig. 2.11f) (Chen et al. 2015; Lei et al. 2015; Meng et al. 2010, 2014; Sun et al. 2015a; Yu et al. 2011), resembling typical adakites (Defant and Drummond 1990).

These rocks with geochemical characteristics of adakites are likely originated from partial melting of varying sources which include a delaminated thickened lower crust (Chen et al. 2015; Lei et al. 2015), a thickened juvenile lower crust (Sun et al. 2015a), and mixed sources of partial melts from ancient thickened lower crust and enriched fluid-metasomatized mantle sources (Meng et al. 2014). Rocks in this stage are either attributed to the slab rollback and break-off of the southward subducting Bangong-Nujiang Tethyan lithosphere after the Lhasa-Qiangtang initial collision or to the back-arc extension of the Neo-Tethyan Ocean (Meng et al. 2014). In summary, these adakitic rocks provide robust evidence for the Late Cretaceous crustal thickening prior to the India-Asia collision (Chen et al. 2015; Lei et al. 2015).

2.2.3 Southern Lhasa Subterrane

The Southern Lhasa subterrane (SL) lies between the Luobadui-Milashan fault (LMF) to the north and the Indus–Yarlung Zangbo suture zone (IYZSZ) to the south (Fig. 2.1). It is composed primarily of the Gangdese batholith and the Linzizong volcanic succession. Magmatism in the southern Lhasa subterrane has an over 200 Myr history (237–10 Ma). Prior to the India-Asia continental collision in the Cenozoic (Zhu et al. 2015; Hu et al. 2016), the magmatism was formed in an Andean-type active continental arc. It can be divided into three stages: (1) Late

Triassic–Middle Jurassic magmatism, (2) Late Jurassic–Early Cretaceous magmatism, and (3) Late Cretaceous magmatism. Among them, the Late Cretaceous magmatism is most intense.

2.2.3.1 Late Triassic to Jurassic Igneous Rocks

Late Triassic to Middle Jurassic intrusive rocks (220–170 Ma) in the SL is distributed along a semi-continuous, east-west-trending, magmatic belt, which is over 800 km in length (Fig. 2.1) (e.g., Chu et al. 2006; Ji et al. 2009; Zhu et al. 2011; Guo et al. 2013; Wang et al. 2017a; Xu et al. 2019). In the Gayca area, it intrudes the Late Devonian–Early Carboniferous gneissic granites with clear intrusive boundaries to the south, and is intruded in turn by late Cretaceous tonalitic dikes (~100 Ma) and Eocene granitoids (~49 Ma) to the north (Xu et al. 2019). It mainly consists of diorite, tonalite, granodiorite and granite with minor hornblende gabbro. Recently, ultramafic-mafic cumulates (clinopyroxene hornblendite, hornblendite and hornblende gabbro) with abundant amphibole and felsic cumulates (tonalite) have been found in the ca. 200 Ma Cuijiu Igneous Complex (Xu et al. 2019). In addition, mafic microgranular enclaves commonly occur as dark blobs in the diorites to tonalites (Wang et al. 2017a; Shui et al. 2017; Xu et al. 2020). Middle Jurassic (ca. 173–165 Ma) granitic magmatism in the SL western segment was accompanied by mineralization, forming the famous Xiongcun Cu–Au deposit (Tafti et al. 2009; Tang et al. 2010). Corresponding to the intrusive rocks, coeval volcanic rocks (ca. 237–170 Ma) are also locally documented in the SL eastern and western segments (Zhu et al. 2008b; Kang et al. 2014; Wang et al. 2016a; Wei et al. 2017; Liu et al. 2018). These volcanic rocks are dominated by basalt and rhyolite with minor andesite (Xu et al. 2019).

The Late Triassic to Middle Jurassic intrusive and volcanic rocks span a similar, wide and continuous range of bulk-rock compositions (SiO_2 = 45–85 wt%) and

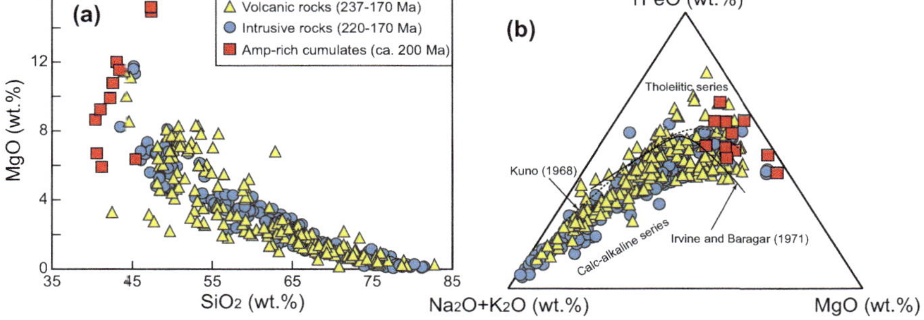

Fig. 2.12 (**a**) MgO (wt %) vs. SiO_2 (wt%) and (**b**) AFM diagram (Wager and Deer 1939) for the Late Triassic to Middle Jurassic magmatic rocks in the Southern Lhasa subterrane. Discriminatory lines in (**b**) classifying the calc-alkaline and tholeiitic magmatic series are from Irvine and Baragar (1971) and Kuno (1968)

define a typical calc-alkaline trend in the AFM diagram (Fig. 2.12). Both show a bimodal compositional distribution with minor intermediate compositions (Xu et al. 2019). Most of these magmatic rocks have positive zircon Hf and depleted bulk-rock Sr–Nd–Pb isotopic compositions (Xu et al. 2019). These features, along with minor outcrops of mafic plutonic rocks and the absence of ultramafic–mafic cumulates, have convinced most researchers that the felsic rocks were formed through partial melting of pre-existing juvenile lower crust with intermediate rocks resulting from magma mixing between mantle-derived basaltic and crust-derived felsic magmas (e.g., Zhu et al. 2008b; Guo et al. 2013; Wang et al. 2017a). Mixing of basaltic and felsic magmas invariably generates linear arrays in elemental variation diagrams (Keller et al. 2015). However, the intrusive and volcanic rocks define a curve line in the MgO vs. SiO_2 diagram (Fig. 2.12). In addition, with increasing SiO_2, these rocks show a bend in P_2O_5 contents and continually decreasing $(Dy/Yb)_N$ ratios (Xu et al. 2019). These observations have two direct implications: (1) they indicate that amphibole-dominated fractional crystallization without garnet involvement explains the main features of the rocks and their geochemistry (Davidson et al. 2007; Lee and Morton 2015); (2) fractional crystallization rather than the 'partial melting followed by magma mixing' model was the main magmatic process controlling the magmatism. This is consistent with the newly found Cuijiu ultramafic-mafic cumulates that contain abundant amphibole and are characterized by low SiO_2 and high Fe_2O_{3tot} contents (Fig. 2.12). It should be noted that the magmatic rocks located near the Central Lhasa subterrane (CL), which has an ancient basement, show more enriched zircon Hf and bulk-rock Nd isotopic compositions than magmatic rocks located away from the CL (Xu et al. 2019). These trends are considered to reflect minor involvement of older crustal materials in their formation.

Although two different geodynamic models have been developed to explain the formation of the SL Late Triassic to Middle Jurassic magmatism: (1) northward subduction of the Neo-Tethyan oceanic slab beneath the southern Lhasa subterrane (Chu et al. 2006; Ji et al. 2009), and (2) southward subduction of Bangong–Nujiang Tethyan oceanic slab beneath the Southern Lhasa subterrane (Zhu et al. 2011, 2013, 2019), it is widely accepted that the magmatism was formed in a continental arc. Based on the Cuijiu ultramafic-mafic and felsic cumulates (Xu et al. 2019), an evolved AFC model has been proposed by Xu et al. (2019) to explain the formation of the continental arc magmatism. The descending oceanic slab and associated sediment dehydration and releasing hydrous fluids, enriched in incompatible elements, into the mantle wedge, triggered melting of mantle peridotite (Elliott et al. 1997). Partial melting of this metasomatized mantle wedge generated primitive, hydrous basaltic magmas, which intruded the lower crust where they underwent amphibole-dominated fractional crystallization. This was accompanied by minor wall-rock assimilation and produced, after 50–70% crystallization, evolved andesitic to dacitic liquids and complementary ultramafic–mafic cumulates. The andesitic to dacitic liquids extracted from crustal magma reservoirs either erupted at the surface to form abundant andesite lava flows or were trapped in the middle to upper crust to form crystal mushes. These mushes, with intermediate compositions, either would eventually freeze to form diorite and tonalite plutons or would further fractionate and assimilate minor volumes of wall-rock, to form high-silica rhyolitic liquids and complementary cumulates after 50–70% crystallization. Like the andesitic liquids, the high-silica rhyolitic liquids were extracted from their mush source and also erupted at the surface to form voluminous rhyolites.

2.2.3.2 Early Cretaceous Igneous Rocks

The Early Cretaceous igneous rocks, mainly as intrusions, are rarely exposed in the Gangdese belt from southern Lhasa subterrane (Fig. 2.1). They include, in the WE direction, diabase and diorite in Xigaze to Quxu (Wen et al. 2008b; Ji et al. 2009; Zeng et al. 2017), diorite, granodiorite and mafic enclave, and granitoid in Nang (Quidelleur et al. 1997; Wen et al. 2008b; Wang et al. 2013; Zheng et al. 2014), and metamorphosed gabbro, diorite, and granite in Milin. These mafic to felsic intrusive rocks yielded crystallization ages ranging from ca. 122 to 101 Ma (e.g., Quidelleur et al. 1997; Wen et al. 2008b). Besides, andesites that emplaced at ca. 137 Ma were reported in the Mamen area, Zedong (Zhu et al. 2009c).

Limited geochemical data for the Early Cretaceous magmatism shows that these rocks mainly have subalkaline and calc-alkaline compositions (Fig. 2.13). Gabbroic

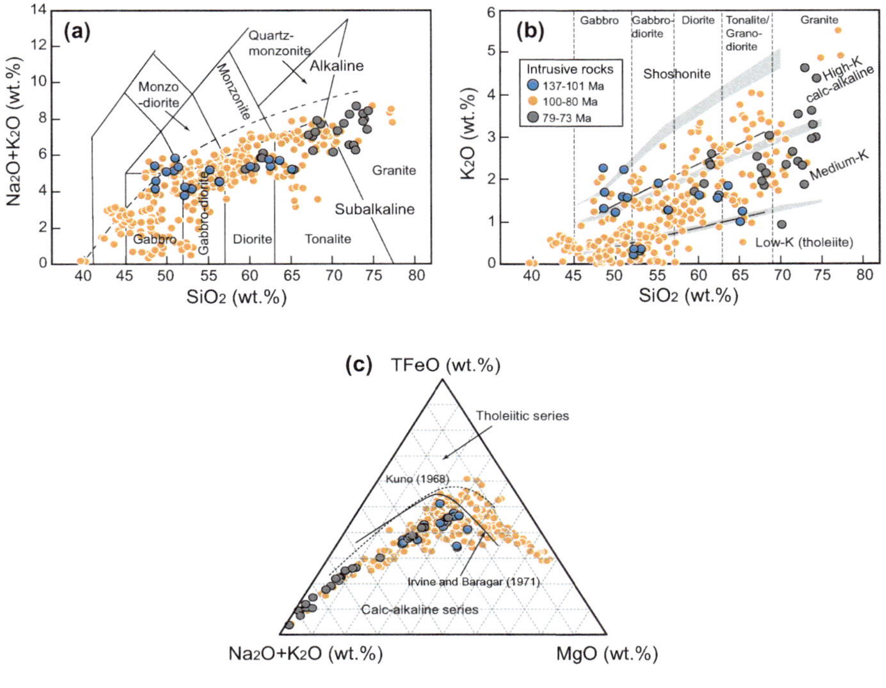

Fig. 2.13 Plots of (**a**) $Na_2O + K_2O$ vs. SiO_2 (Middlemost 1994), (**b**) K_2O vs. SiO_2 (Rickwood 1989), and (**c**) AFM diagram (Wager and Deer 1939) for the Early Cretaceous igneous rocks in the Southern Lhasa subterrane

to dioritic rocks are characterized by high Al_2O_3 and Mg# and their sodic nature. They are enriched in LREEe and LILEe relative to HFSEs and have positive zircon $\varepsilon_{Hf}(t)$ and bulk-rock $\varepsilon_{Nd}(t)$ values (e.g., Zhu et al. 2009c; Wang et al. 2013). Consequently, more mafic rocks are interpreted to originate from partial melting of metasomatized depleted mantle (Wang et al. 2013) and dioritic rocks with high-Mg adakitic signatures are thought to derive from partial melting of subducting Neo-Tethyan slab and subsequent hybridization by peridotite in the mantle wedge (Zhu et al. 2009c). Fractional crystallization of more mafic magmas has been proposed for the generation of adakitic granitoids with similar isotopic compositions (Zheng et al. 2014). Overall, the formation of the Early Cretaceous magmatism at ca. 137–101 Ma is generally attributed to north-dipping subduction of the Neo-Tethyan oceanic lithosphere, most likely at high-angle (Zhu et al. 2009c).

2.2.3.3 Late Cretaceous Igneous Rocks

The exposure of the Late Cretaceous magmatism is restricted to 89–94° E in the eastern segment of the Southern Lhasa subterrane (Fig. 2.1). Age data indicate an intense magmatic pulse at ca. 100–80 Ma and the other phase at ca. 79–73 Ma (Zhu et al. 2019). The early phase magmatism consists of gabbro, gabbroic diorite, diorite, and granitoids from Namling to Quxu (Xu et al. 2015; Ye et al. 2015; Ma et al. 2017), diabase from Dagze (Ma et al. 2015), diorite, quartz monzonite, and granitoids from Zedong (Jiang et al. 2012, 2014; Zheng et al. 2014; Zhang et al. 2021), Gabbro from Zhengga (Ma et al. 2013b), and ultramafic rocks of mainly hornblendite, mafic-intermediate rocks of hornblende gabbro, norite, gabbroic diorite, charnockite, diorite, and quartz diorite, and granitoids of tonalite, granodiorite and granite from Lilong to Milin (e.g., Wen et al. 2008a; Zhang et al. 2010b, 2014, 2020; Ma et al. 2013a, c; Ji et al. 2014; Guo et al. 2020). Magmatic rocks of the late stage are volumetrically minor and are always associated with the magmatism of the early phase. They are exposed as intrusions ranging from gabbro, diorite to monzogranite, and two-mica granite (Zhang et al. 2014, 2020; Ji et al. 2014; Jiang et al. 2014; Zheng et al. 2014).

The early phase magmatic rocks (ca. 100–80 Ma) have subalkaline and calc-alkaline compositions according to the geochemistry (Fig. 2.13a–c). Based on their variable contents of SiO_2 and K_2O, they display as low-K series for some mafic-intermediate rocks, medium-K series for most rocks, and even high-K to shoshonitic series for some mafic to felsic rocks (Fig. 2.13b). Gabbroic rocks generally show high Al_2O_3, low MgO and Mg#, and are enriched in LILEs and HFSEs, suggesting an origin from partial melting of melt- or fluid-metasomatized mantle wedge with minor subsequent fractionation of mafic minerals (e.g., Guan et al. 2011; Ma et al. 2013a, c). The ultramafic lithologies associated with these gabbroic rocks are interpreted as cumulates from fractional crystallization (e.g., Ma et al. 2013a). Intermediate-felsic rocks normally have adakitic affinities and their origins are explained by several possible processes: (a) partial melting of thickened juvenile lower crust (Wen et al. 2008a), (b) partial melting of subducting Neo-Tethyan slab (Zhang et al. 2010b), and (c) hornblende-dominated fractional crystallization of mantle wedge derived basaltic magmas (Xu et al. 2015). A recent study proposed both a damp (low H_2O) and a wet igneous fractional crystallization (of arc magmas) models to explain the composition and formation of mafic-intermediate sequence, ultramafic rocks, and the granitoids (Guo et al. 2020). Two popular geodynamic models have been proposed to explain the generation of the magmatic pulse at ca. 100–80 Ma: (1) Neo-Tethyan oceanic ridge subduction (e.g., Zhang et al. 2010b) and (2) rollback of subducting Neo-Tethyan slab (Ma et al. 2013a).

The late phase magmatism (ca. 79–73 Ma) mainly consists of intermediate-felsic rocks and mostly displays medium-K calc-alkaline compositions (Fig. 2.13). These rocks have adakitic geochemical characteristics and show I-type geochemical signatures with positive zircon Hf isotopic compositions. Therefore, they are interpreted as partial melting of thickened mafic lower crust during the subduction of Neo-Tethyan oceanic slab (Ji et al. 2014; Jiang et al. 2014; Zheng et al. 2014). However, Guo et al. (2020) proposed that granitoids of this phase could derive from fractional crystallization of wet arc magmas and experienced crustal assimilation during emplacement.

2.3 Tethyan Himalaya

The Himalaya is located between the Indus–Yarlung Zangbo suture zone (IYZSZ) to the north and the India shield to the south (Fig. 2.1). The Himalaya consists of Lesser, Greater, and Tethyan Himalaya from south to north. They are separated by three north-dipping Late Cenozoic fault systems including the Main Boundary Thrust (MBT), the Main Central Thrust (MCT), and the South Tibetan Detachment System (STDS) (Fig. 2.1). The Tethyan Himalaya is characterized by Proterozoic-Cenozoic siliciclastic and carbonate rocks interbedded with Paleozoic and Mesozoic volcanic rocks. This rock association can be divided into four sequences: (1) Proterozoic to Devonian pre-rift sequence, (2) Carboniferous–Lower Jurassic rift and post-rift sequence, (3) Jurassic–Cretaceous passive continental margin sequence, and (4) Cenozoic syn-collisional (the India-Asia collision) sequence (Liu and Einsele 1996; Yin 2006). The Paleozoic and Mesozoic igneous rocks in the Tethyan Himalaya are discussed in the following sections.

2.3.1 Early Paleozoic Igneous Rocks

Early Paleozoic granite and granitic gneiss (ca. 510–430 Ma) have been discovered as an important component of a series of well-defined gneiss domes including the Xiaru, Kangri, Mabja, Kangmar, and Yalashangbo (Fig. 2.1). The Paleozoic granites are of high potassium calc-alkaline to shoshonite and peraluminous type. They are enriched in light rare-earth elements (REE) and relatively depleted in HFSEs (i.e., Nb, Ta, and Ti) in primitive mantle-normalized multi-element diagrams. They have enriched initial Nd and Hf isotopic ratios with $\varepsilon_{Nd}(t)$ values of −12.6 to −7.7 and $\varepsilon_{Hf}(t)$ values of −15.3 to +1.0, respectively (Gao et al. 2019). Such geochemical signatures suggest that the Early Paleozoic granites were derived from partial melting of metasedimentary rocks with input of mantle-derived materials under a passive crustal extension environment (Gao et al. 2019).

2.3.2 Permian Igneous Rocks

Permian volcanic rocks are locally exposed in Gyirong to Selong area, within the Early Permian Jilong Formation and Late Permian Selong Group (Zhu et al. 2010). Permian volcanic rocks are mainly basalts with thickness of ~130 m in the Jilong Formation and ~55 m in Selong Group. The timing of basalts within the Jilong Fomration is inferred to be Sakamarian-Artinskian age (Zhu et al. 2010), coeval with the Bhote Kosi basalts south of the Gyirong County (Garzanti et al. 1999) and the Panjal Traps in NW India (Gaetani and Garzanti 1991; Garzanti et al. 1999). Abundant brachiopods within slate overlying the basalts in the Selong Group indicate a Capitanian-Lopingian age (Zhu et al. 2010).

The basalts in Jilong and Selong Formations are characterized by moderate TiO_2, high MgO and Mg# showing geochemical affinity to high-MgO basalts (HMB) (SiO_2 ≤54 wt%, MgO ≥7 wt%, and Al_2O_3 <16.5 wt%; Crawford et al. 1987; Kersting and Arculus 1994). The chondrite-normalized, REE patterns of the Jilong Formation and Selong Group basalts are characterized by light REE enrichment with or without a weak Eu anomaly (Fig. 2.14c). The basalts in the Jilong Formation and Selong Group show similar patterns on primitive mantle-normalized multi-element diagrams, characterized by moderately negative Nb–Ta anomalies and the absence of Ti anomalies (Fig. 2.14d). These basalts are distinct from arc basalts (e.g., Andean arc basalts) in terms of Nb and Ti anomalies as typical arc basalts showing a marked depletion in Nb and Ti (Hawkesworth et al. 1991). Initial isotopic ratios were corrected to ca. 280 Ma for the Jilong Formation basalts and ca. 260 Ma for the Selong Group basalts. The initial Sr isotopic ratios of these basalts are very high, ranging from 0.7160 to 0.7185 at relatively constant $\varepsilon_{Nd}(t)$ values (+0.7 to +1.2) (Fig. 2.14b).

The basalts in the Jilong Formation and Selong Group show relatively high Ti/Y values and related data were plotted mostly in the within-plate tholeiitic basalt field of the Ti/Y vs. Nb/Y diagram (Fig. 2.15a). Their low Th/Ta values (5.4–5.9) are consistent with the formation in a within-plate setting (Gorton and Schandl 2000). These results favor a Permian extensional tectonic regime for the generation of the basalts in the Jilong Formation and Selong Group and Bhote Kosi basalts in the Tethyan Himalaya (Garzanti et al. 1999). Such extensional tectonic setting may relate to the Panjal Traps generated by mantle plume.

2.3.3 Triassic Igneous Rocks

Triassic volcanic rocks, locally exposed in the Jiangzi Basin, occurred in the Early Triassic Rema, Middle Triassic Sangkang, and Late Triassic Nieru formations (Fig. 2.1). These volcanic rocks are mainly basalts and andesite basalts. Basalts of the Rema and Sangkang Formations have high TiO_2 content. They are enriched in LREEs with or without weak Nb anomaly which is similar to high-Ti basalts in Emeishan large igneous province (LIP), Hawaii oceanic island basalts (OIB), and alkaline basalt in Rajmahal Traps of India (Fig. 2.14c; Zhu et al. 2004). They plot in the within-plate basalt field in the Ti/Y vs. Nb/Y diagram (Fig. 2.15a). The basalts

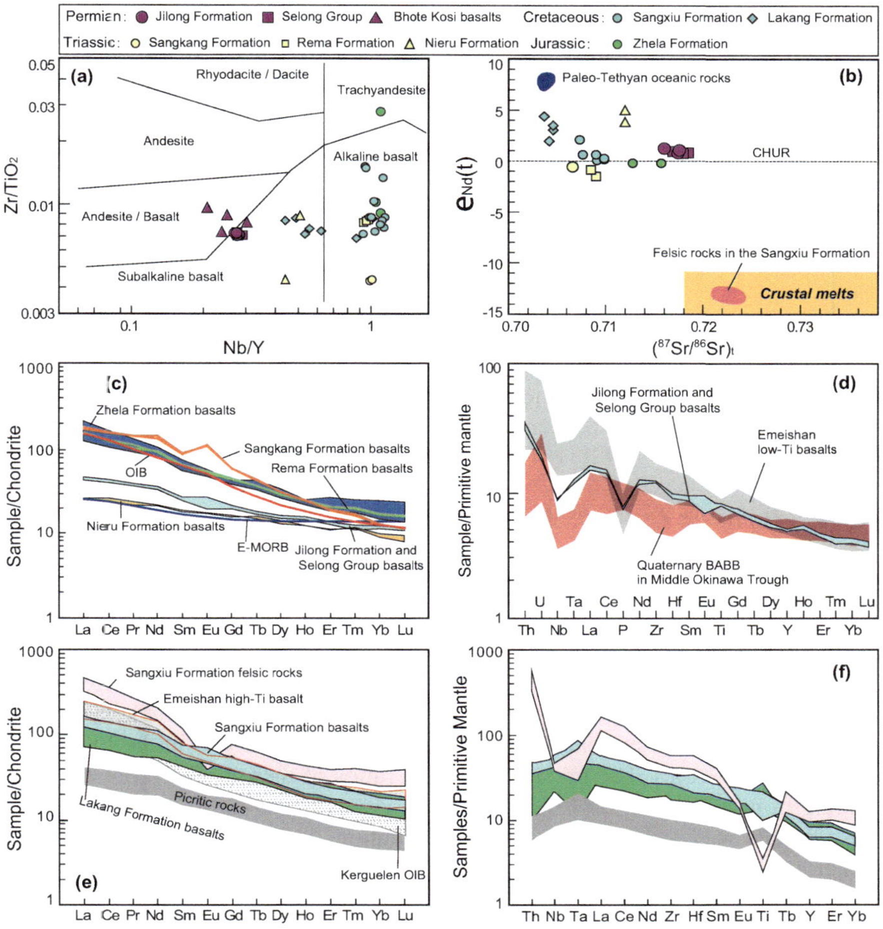

Fig. 2.14 Diagram for volcanic rocks in the Tethyan Himalaya. (**a**) Zr/TiO_2 vs. Nb/Y (Winchester and Floyd 1976); (**b**) Nd–Sr isotopic diagram; (**c**, **e**) Chondrite-normalized rare-earth element (REE) patterns; and (**d**, **f**) primitive mantle-normalized multi-element diagrams. Normalization values are from Sun and McDonough (1989). (Data sources: Zhu et al. 2004, 2007, 2008a, 2010; Xia et al. 2014)

in the Nieru Formation have lower TiO_2 content and $(La/Yb)_N$ than those in the Rema and Sangkang Formations and show geochemical affinity to enriched middle ocean ridge basalts (E-MORB). The basalts in the Nieru Formation also exhibit high Th/Yb and Nb/Y, similar to within-plate basalt (Fig. 2.15b). Initial isotopic ratios were corrected to ca. 245 Ma for the Rema Formation basalts, ca. 230 Ma for the Sangkang Group basalts, and ca. 215 Ma for the Nieru Formation basalts. Basalts of the Rema and Sangkang Formations have similar initial Sr isotopes of 0.7067 to 0.7091 and $\varepsilon_{Nd}(t)$ values of −0.6 to −1.5. For comparison, basalts of the Nieru Formation have more enriched initial Sr isotopes (0.7120) but depleted $\varepsilon_{Nd}(t)$ (+3.8 to +5.0) (Fig. 2.14b) (Zhu et al. 2004). These results suggest that the Triassic volcanic rocks are generated at a continental rift margin.

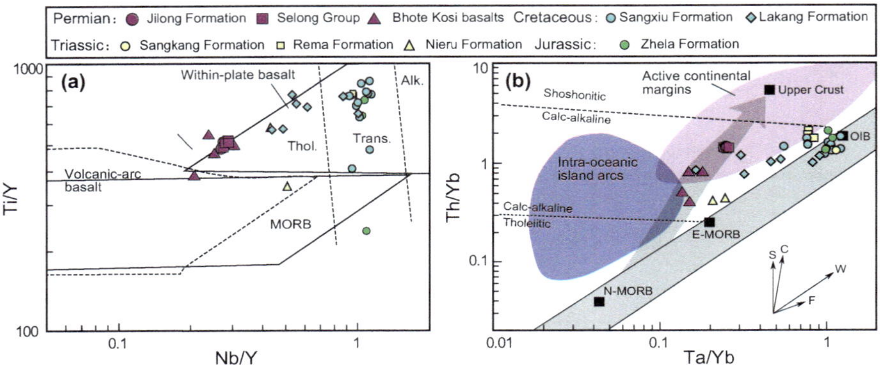

Fig. 2.15 (**a**) Ti/Y vs. Nb/Y (Pearce 1982); (**b**) Th/Yb vs. Ta/Yb (Pearce 1983). Abbreviations: *S* subduction components, *C* crustal contamination, *W* within-plate component, *F* fractional crystallization. Values of OIB, E-MORB, and N-MORB are from Sun and McDonough (1989)

2.3.4 *Jurassic Igneous Rocks*

Jurassic volcanic rocks occur in the Early Jurassic Ridang Formation, Middle Jurassic Lure and Zhela formations. However, the data of Jurassic volcanic rocks are rarely reported except for the basalts in the Zhela Formation (Zhu et al. 2004). These basalts have moderate-to-high TiO_2 content. They are characterized by high Zr/TiO_2 and Nb/Y, suggesting that they are alkaline basalt to trachyandesite (Fig. 2.14a). They are enriched in LREEs and HFSEs, which is similar to present-day OIB (Fig. 2.14c). The initial Sr isotopic ratios of these basalts are very high, ranging from 0.71357 to 0.71577 (values are corrected to 142 Ma), suggesting that these basalts were subjected to seawater alteration. Their $\varepsilon_{Nd}(t)$ values (−0.3) are similar to basalts in the Triassic Rema and Sangkang formations (Fig. 2.14b) (Zhu et al. 2004). Basalts in the Zhela Formation show relatively high Ti/Y values and plot mostly in the within-plate tholeiitic basalt field on the Ti/Y vs. Nb/Y diagram (Fig. 2.15a). These results indicate that they may have derived from the same mantle source of basalts in the Triassic Rema and Sangkang formations in the continental rift margin.

2.3.5 *Early Cretaceous Igneous Rocks*

The Early Cretaceous igneous rocks are mainly exposed in the eastern Tethyan Himalaya, namely the Comei LIP (Zhu et al. 2009a) (Fig. 2.1). The Comei LIP is mainly composed of mafic rocks with subordinate silicic and ultramafic rocks with areal extent of ~50,000 km^2 (Zhu et al. 2009a; Qiu et al. 2010). The volcanic remnants are interbedded with the Early Cretaceous siliciclastic deposit in the Sangxiu and Lakang formations. The volcanic rocks in the Sangxiu Formation consist of an average thickness of ~100 m basalts in the middle-lower parts and ~130 m dacites/rhyolites in the upper part (Zhu et al. 2007). Zircon U–Pb ages of the silicic lavas suggest that they were erupted during ca. 137–131 Ma (Zhu et al. 2005, 2009a; Zhou et al. 2017a). In the Lakang Formation, basaltic lavas occur in the middle-lower parts, ranging from several meters to ~600 m thick (Zhu et al. 2008a). Recent zircon U–Pb age data indicate that they were emplaced between ca. 147 and 144 Ma (Zhu et al. 2008a; Shi et al. 2018; Bian et al. 2019). Abundant E-W trending dolerite dykes/sills and gabbro intrusions intruded the Mesozoic strata in the eastern segment of the Tethyan Himalaya with similar zircon U–Pb ages from ca. 145 to 130 Ma (Zhu et al. 2009a).

The mafic rocks in the Comei LIP are characterized by high TiO_2 content. They are enriched in LREEs and HFSEs, similar to the Kerguelen OIB (Fig. 2.14e, f). The picritic rocks show sub-parallel trace element patterns to the mafic rocks. The $\varepsilon_{Nd}(t)$ values of the Sangxiu basalts (−1.7 to 2.0) are identical to the high-Ti basalts in the Emeishan LIP (−1.17 to 0.43; Xiao et al. 2004). The felsic rocks in the Sangxiu Formation are characterized by high HFSE content with moderate Eu anomaly showing geochemical affinity to A-type granite (Fig. 2.14e, f; Zhu et al. 2007). They have enriched Sr isotopes of 0.72165 to 0.72357 and $\varepsilon_{Nd}(t)$ of −13.4 to −13.1, which is different from the basalts in the Sangxiu Formation. The picritic rocks and mafic rocks of the Comei LIP were likely derived from the Kerguelen plume during the breakup of India and Australia, and the contemporaneous felsic rocks are generated by remelting of crust during underplating of mafic rocks (Zhu et al. 2007, 2008b, 2009a; Xia et al. 2014).

Acknowledgements Discussions with Sun-Lin Chung, Fu-Yuan Wu, Xiumian Hu, and Qiang Wang over the years are gratefully appreciated. This work was financially supported by the Second Tibetan Plateau Scientific Expedition and Research Program (2019QZKK0702), the Chinese National Natural Science Foundation (grants 42121002 and 91755207), and the 111 project (B18048).

References

Allègre CJ, Courtillot V, Tapponnier P, Hirn A, Mattauer M, Coulon C, Jaeger JJ, Achache J, Scharer U, Marcoux J, Burg JP, Girardeau J, Armijo R, Gariepy C, Gopel C, Li TD, Xiao XC, Chang CF, Li GQ, Lin BY, Teng JW, Wang NW, Cheng GM, Han TL, Wang XB, Den WM, Sheng HB, Cao YG, Zhou J, Qiu HR, Bao PS, Wang SC, Wang BX, Zhou YX, Xu RH (1984) Structure and evolution of the Himalaya-Tibet orogenic belt. Nature 307:17–22

Bai Z-D, Xu DB, Chen MJ, Sun LX (2009) Characteristics and zircon SHRIMP U–Pb dating of the Amdo trachyte, Tibet, China. Geol Bull China 28(9):1229–1235

Bian W, Yang T, Ma Y, Jin J, Gao F, Wang S, Peng W, Zhang S, Wu H, Li H, Cao L, Shi Y (2019) Paleomagnetic and geochronological results from the Zhela and Weimei formations lava flows of the eastern Tethyan Himalaya: new insights into the breakup of eastern Gondwana. J Geophys Res Solid Earth 124(1):44–64

Bu T, Wei QR, Ding PF, Wang C, Liu XN, Zhang XQ, Sun J, Wang JY (2015) LA-ICPMS zircon U-Pb ages and petrogenesis of Rindu intrusion near Namling, middle Gangdese magmatic belt. Geol Bull China 34(12):2254–2265 (in Chinese with English abstract)

Cao MJ, Qin KZ, Li GM, Li JX, Zhao JX, Evans NJ, Hollings P (2016) Tectono-magmatic evolution of late Jurassic to early Cretaceous granitoids in the west central Lhasa subterrane, Tibet. Gondwana Res 39:386–400

Cawood PA, Johnson MRW, Nemchin AA (2007) Early Palaeozoic orogenesis along the Indian margin of Gondwana: tectonic response to Gondwana assembly. Earth Planet Sci Lett 255(1):70–84

Chen Y, Zhu DC, Zhao ZD, Meng FY, Wang Q, Santosh M, Wang LQ, Dong GC, Mo XX (2014) Slab breakoff triggered ca. 113 Ma magmatism around Xainza area of the Lhasa terrane, Tibet. Gondwana Res 26(2):449–463

Chen JL, Xu JF, Yu HX, Wang BD, Wu JB, Feng YX (2015) Late Cretaceous high-Mg# granitoids in southern Tibet: implications for the early crustal thickening and tectonic evolution of the Tibetan Plateau? Lithos 232:12–22

Chen S-S, Shi R-D, Yi G-D, Zou H-B (2016) Middle Triassic volcanic rocks in the northern Qiangtang (central Tibet): geochronology, petrogenesis, and tectonic implications. Tectonophysics 666:90–102

Chen S-S, Fan W-M, Shi R-D, Gong X-H, Wu K (2017) Removal of deep lithosphere in ancient continental collisional orogens: a case study from central Tibet, China. Geochem Geophys Geosyst 18(3):1225–1243

Chu MF, Chung SL, Song B, Liu DY, O'Reilly SY, Pearson NJ, Ji JQ, Wen DJ (2006) Zircon U-Pb and Hf isotope constraints on the Mesozoic tectonics and crustal evolution of southern Tibet. Geology 34(9):745–748

Crawford AJ, Falloon TJ, Eggins S (1987) The origin of Island arc high-alumina basalts. Contrib Mineral Petrol 97(3):417–430

Dan W, Wang Q, Zhang X-Z, Tang G-J (2020) Early Paleozoic S-type granites as the basement of southern Qiantang Terrane, Tibet. Lithos 356–357:105395

Dan W, Wang Q, Murphy JB, Zhang X-Z, Xu Y-G, White WM, Jiang Z-Q, Ou Q, Hao L-L, Qi Y (2021a) Short duration of early Permian Qiangtang-Panjal large igneous province: implications for origin of the neo-Tethys Ocean. Earth Planet Sci Lett 568:117054

Dan W, Wang Q, White WM, Li X-H, Zhang X-Z, Tang G-J, Ou Q, Hao L-L, Qi Y (2021b) Passive-margin magmatism caused by enhanced slab-pull forces in central Tibet. Geology 49(2):130–134

Davidson J, Turner S, Handley H, Macpherson C, Dosseto A (2007) Amphibole "sponge" in arc crust? Geology 35(9):787–790

Defant MJ, Drummond MS (1990) Derivation of some modern arc magmas by melting of young subducted lithosphere. Nature 347:662–665

Du DD, Qu XM, Wang GH, Xin HB, Liu ZB (2011) Bidirectional subduction of the middle Tethys oceanic basin in the west segment of Bangonghu-Nujiang suture, Tibet: evidence from zircon U-Pb LAICPMS dating and petrogeochemistry of arc granites. Acta Petrol Sin 27:1993–2002 (in Chinese with English abstract)

Duan QF, Yang ZQ, Wang JX, Bai YS, Niu ZJ, Yao HZ (2006) Geochemical characteristics of Permian high-Ti basalt in the eastern part of the northern Qiangtang basin, Qinghai-Tibet Plateau. Geol Bull China 25(1–2):156–15n

Duan QF, Wang JX, Bai YS, Niu ZJ, He LQ, Yao HZ, Kou XH, Zhang KX (2010) Geochemistry and mantle source characteristics of the Permian basalts in Moyun area, eastern Tanggula range. Acta Petrol Mineral 29(2):125–138

Elliott T, Plank T, Zindler A, White W, Bourdon B (1997) Element transport from slab to volcanic front at the Mariana arc. J Geophys Res Solid Earth 102(B7):14991–15019

Fan J-J, Li C, Wu H, Zhang T-Y, Wang M, Chen J-W, Xu J-X (2016) Late Jurassic adakitic granodiorite in the Dong Co area, northern Tibet: Implications for subduction of the Bangong–Nujiang oceanic lithosphere and related accretion of the Southern Qiangtang terrane. Tectonophysics 691:345–361

Fu X-G, Wang J, Tan F-W, Chen M, Chen W-B (2010) The late Triassic rift-related volcanic rocks from eastern Qiangtang, northern Tibet (China): age and tectonic implications. Gondwana Res 17(1):135–144

Gaetani M, Garzanti E (1991) Multicyclic history of the northern India continental margin (northwestern Himalaya). Am Assoc Pet Geol Bull 75(9):1427–1446

Gao L-E, Zeng L, Hu G, Wang Y, Wang Q, Guo C, Hou K (2019) Early Paleozoic magmatism along the northern margin of East Gondwana. Lithos 334-335:25–41

Garzanti E, Le Fort P, Sciunnach D (1999) First report of lower Permian basalts in South Tibet: tholeiitic magmatism during break-up and incipient opening of Neotethys. J Asian Earth Sci 17(4):533–546

Gorton MP, Schandl ES (2000) From continents to Island arcs: a geochemical index of tectonic setting for arc-related and within-plate felsic to intermediate volcanic rocks. Can Mineral 38(5):1065–1073

Guan Q, Zhu DC, Zhao ZD, Dong GC, Mo XX, Liu YS, Hu ZC, Yuan HL (2011) Zircon U-Pb chronology, geochemistry of the late Cretaceous mafic magmatism in the southern Lhasa Terrane and its implications. Acta Petrol Sin 27(7):2083–2094

Guo L, Liu Y, Liu S, Cawood PA, Wang Z, Liu H (2013) Petrogenesis of early to middle Jurassic granitoid rocks from the Gangdese belt, southern Tibet: implications for early history of the neo-Tethys. Lithos 179:320–333

Guo L, Jagoutz O, Shinevar WJ, Zhang HF (2020) Formation and composition of the late Cretaceous Gangdese arc lower crust in southern Tibet. Contrib Mineral Petrol 175(6):58. https://doi.org/10.1007/s00410-020-01696-y

Guynn JH, Kapp P, Pullen A, Heizler M, Gehrels G, Ding L (2006) Tibetan basement rocks near Amdo reveal "missing" Mesozoic tectonism along the Bangong suture, central Tibet. Geology 34(6):505–508

Guynn J, Kapp P, Gehrels GE, Ding L (2012) U–Pb geochronology of basement rocks in central Tibet and paleogeographic implications. J Asian Earth Sci 43(1):23–50

Hao L-L, Wang Q, Wyman DA, Ou Q, Dan W, Jiang Z-Q, Yang J-H, Li J, Long X-P (2016) Andesitic crustal growth via mélange partial melting: evidence from early Cretaceous arc dioritic/andesitic rocks in southern Qiangtang, central Tibet. Geochem Geophys Geosyst 17(5):1641–1659

Hastie AR, Kerr AC, Pearce JA, Mitchell S (2007) Classification of altered volcanic Island arc rocks using immobile trace elements: development of the Th–Co discrimination diagram. J Petrol 48(12):2341–2357

Hawkesworth CJ, Hergt JM, Ellam RM, McDermott F, Tarney J, Pickering KT, Knipe RJ, Dewey JF (1991) Element fluxes associated with subduction related magmatism. Philos Trans R Soc Lond A Phys Eng Sci 335(1638):393–405

He Z, Yang D, Wang T (2006) The determination of early Cretaceous post/collision granites in Sangba area of Ganddese tectonic belt and its tectonic significance. Acta Petrol Mineral 25(3):185

He H, Li Y, Wang C, Zhou A, Qian X, Zhang J, Du L, Bi W (2018) Late Cretaceous (ca. 95 Ma) magnesian andesites in the Biluoco area, southern Qiangtang subterrane, central Tibet: Petrogenetic and tectonic implications. Lithos 302-303:389–404

He H, Li Y, Wang C, Han Z, Ma P, Xiao S (2019) Petrogenesis and tectonic implications of late Cretaceous highly fractionated I-type granites from the Qiangtang block, central Tibet. J Asian Earth Sci 176:337–352

Hu XM, Garzanti E, Wang JG, Huang WT, An W, Webb A (2016) The timing of India-Asia collision onset—facts, theories, controversies. Earth-Sci Rev 160:264–299

Hu PY, Zhai QG, Cawood PA, Zhao GC, Wang J, Tang Y, Zhu ZC, Wang W, Wu H (2021a) Cambrian magmatic flare-up, central Tibet: magma mixing in proto-Tethyan arc along north Gondwanan margin. GSA Bull 133(9–10):2171–2188

Hu P-Y, Zhai Q-G, Zhao G-C, Wang J, Tang Y, Zhu Z-C, Wang W, Wu H (2021b) Cambrian and Cryogenian tectonothermal events in the Amdo microcontinent, central Tibet: implications for paleogeographic reconstruction and tectonic evolution of northern Gondwana. Palaeogeogr Palaeoclimatol Palaeoecol 569:110332
Irvine TN, Baragar WRA (1971) A guide to the chemical classification of the common volcanic rocks. Can J Earth Sci 8:523–548
Ji WQ (2010) Geochronology and petrogenesis of granitic rocks from east segment of the Gangdese batholith, southern Tibet. Institute of Geology and Geophysics, Chinese Academy of Sciences
Ji WQ, Wu FY, Chung SL, Li JX, Liu CZ (2009) Zircon U–Pb geochronology and Hf isotopic constraints on petrogenesis of the Gangdese batholith, southern Tibet. Chem Geol 262(3):229–245
Ji WQ, Wu FY, Chung SL, Liu CZ (2014) The Gangdese magmatic constraints on a latest Cretaceous lithospheric delamination of the Lhasa terrane, southern Tibet. Lithos 210:168–180
Jiang X, Zhao ZD, Zhu DC, Zhang FQ, Dong GC, Mo XX, Guo TY (2010) Zircon U-Pb geochronology and Hf isotopic geochemistry of Jiangba, and Xiongba granitoids in western Gangdese, Tibet. Acta Petrol Sin 26:2155–2164 (in Chinese with English abstract)
Jiang ZQ, Wang Q, Li ZX, Wyman DA, Tang GJ, Jia XH, Yang YH (2012) Late Cretaceous (ca. 90 Ma) adakitic intrusive rocks in the Kelu area, Gangdese Belt (southern Tibet): slab melting and implications for Cu–Au mineralization. J Asian Earth Sci 53:67–81
Jiang ZQ, Wang Q, Wyman DA, Li ZX, Yang JH, Shi X-B, Ma L, Tang GJ, Gou GN, Jia XH, Guo HF (2014) Transition from oceanic to continental lithosphere subduction in southern Tibet: evidence from the late Cretaceous–early Oligocene (~91–30 Ma) intrusive rocks in the Chanang–Zedong area, southern Gangdese. Lithos 196–197:213–231
Jiang Q, Dan W, Wang Q, Zhang X, Tang G (2021) Neoproterozoic basement information revealed in the Triassic granites in the northern Qiangtang, Tibetan Plateau: constraints from zircon SIMS U-Pb age and Hf-O isotopes. Geotecton Metallog 45(2):389–400
Kang ZQ, Xu JF, Dong YH, Wang BQ (2008) Cretaceous volcanic rocks of Zenong Group in north-middle Lhasa Block: products of southward subducting of the Slainajap ocean? Acta Petrol Sin 24:303–314 (in Chinese with English abstract)
Kang Z, Xu J, Wang B, Chen J (2010) Qushenla formation volcanic rocks in north Lhasa block: products of Bangong Co-Nujiang Tethy's southward subduction. Acta Petrol Sin 26(10):3106–3116
Kang ZQ, Xu JF, Wilde SA, Feng ZH, Chen JL, Wang BD, Fu WC, Pan HB (2014) Geochronology and geochemistry of the Sangri Group Volcanic Rocks, Southern Lhasa Terrane: implications for the early subduction history of the neo-Tethys and Gangdese Magmatic Arc. Lithos 200–201:157–168
Kapp P, DeCelles PG (2019) Mesozoic–Cenozoic geological evolution of the Himalayan-Tibetan orogen and working tectonic hypotheses. Am J Sci 319(3):159–254
Kapp P, Yin A, Harrison TM, Ding L (2005) Cretaceous-tertiary shortening, basin development, and volcanism in central Tibet. Geol Soc Am Bull 117(7):865
Kapp P, DeCelles PG, Gehrels GE, Heizler M, Ding L (2007) Geological records of the Lhasa-Qiangtang and Indo-Asian collisions in the Nima area of central Tibet. Geol Soc Am Bull 119(7–8):917–933
Keller CB, Schoene B, Barboni M, Samperton KM, Husson JM (2015) Volcanic-plutonic parity and the differentiation of the continental crust. Nature 523:301–307
Kersting AB, Arculus RJ (1994) Klyuchevskoy Volcano, Kamchatka, Russia: the role of high-flux recharged, tapped, and fractionated magma chamber(s) in the genesis of high-Al_2O_3 from high-MgO basalt. J Petrol 35(1):1–41
Kuno H (1968) Differentiation of basalt magmas. In: Hess HH, Poldervaart AA (eds) Basalts: the Poldervaart treatise on rocks of basaltic composition, vol 2. Interscience, New York, pp 623–688
Lee CTA, Morton DM (2015) High silica granites: terminal porosity and crystal settling in shallow magma chambers. Earth Planet Sci Lett 409:23–31

Lei M, Chen JL, Xu JF, Zeng YC (2015) Geochemistry of early late Cretaceous Gaerqiong high-Mg# diorite porphyry in midnorthern Lhasa terrane: partial melting of delaminated lower continental crust? Geol Bull China 34(2/3):337–346 (in Chinese with English abstract)
Le Maitre RW (2002) Igneous rocks: a classification and glossary of terms: recommendations of international union of geological sciences subcommission on the systematics of igneous rocks. Cambridge University Press, Cambridge, p 236. https://doi.org/10.1017/CBO9780511535581
Li C, Cheng LR, Hu K, Yang ZR, Hong YR (1995) Study on the paleo-Tethys suture zone of Longmu Co-Shuanghu, Tibet. Geological Publishing House, Beijing, pp 1–131
Li C, Zhai QG, Chen W, Yan QR, Dong YS, Yu JJ (2007) Geochronology evidence of the closure of Longmu Co-Shuanghu suture, Qinghai-Tibet plateau; Ar-Ar and zircon SHRIMP geochronology from ophiolite and rhyolite in Guoganjianian. Acta Petrol Sin 23(5):911–918
Li L, Bai YS, Ma LY, Niu ZJ, Duan QF (2009) Geochemical characteristics and tectonic significance of Qixianian volcanic rocks in Zhigengaka area, Zhiduo County, eastern Qiangtang. Geol China 36(6):1289–1301
Li SP, Pan T, Li YX, Wang L, Wen DY, Wang SL, Chang YY, Huang QH, Wang QY (2010) Geochemistry of the Duocai ophiolite in north Qiangtang Basin, Qinghai-Tibet plateau: environments for tectonics. Geol China 37(6):1592–1606
Li JX, Qin KZ, Li GM, Xiao B, Zhao JX, Cao M, Chen L (2013a) Petrogenesis of ore-bearing porphyries from the Duolong porphyry Cu-Au deposit, central Tibet: evidence from U-Pb geochronology, petrochemistry and Sr-Nd-Hf-O isotope characteristics. Lithos 160-161:216–227
Li Y, He J, Wang C, Santosh M, Dai J, Zhang Y, Wei Y, Wang J (2013b) Late Cretaceous K-rich magmatism in central Tibet: evidence for early elevation of the Tibetan Plateau? Lithos 160-161:1–13
Li J-X, Qin K-Z, Li G-M, Richards JP, Zhao J-X, Cao M-J (2014a) Geochronology, geochemistry, and zircon Hf isotopic compositions of Mesozoic intermediate–felsic intrusions in central Tibet: petrogenetic and tectonic implications. Lithos 198–199:77–91
Li SM, Zhu DC, Wang Q, Zhao ZD, Sui QL, Liu SA, Liu D, Mo XX (2014b) Northward subduction of Bangong–Nujiang Tethys: Insight from late Jurassic intrusive rocks from Bangong Tso in western Tibet. Lithos 205:284–297
Li G-M, Li J-X, Zhao J-X, Qin K-Z, Cao M-J, Evans NJ (2015a) Petrogenesis and tectonic setting of Triassic granitoids in the Qiangtang terrane, central Tibet: evidence from U–Pb ages, petrochemistry and Sr–Nd–Hf isotopes. J Asian Earth Sci 105:443–455
Li X, Wang B, Liu H, Wang L, Chen L (2015b) The late Jurassic high-Mg andesites in the Daru Tso area, Tibet: evidence for the subduction of the Bangong Co–Nujiang River oceanic lithosphere. Geol Bull China 34(2/3):251–261 (in Chinese with English abstract)
Li SM, Zhu DC, Wang Q, Zhao Z, Zhang LL, Liu SA, Chang QS, Lu YH, Dai JG, Zheng YC (2016a) Slab-derived adakites and subslab asthenosphere-derived OIB-type rocks at 156 ± 2 Ma from the north of Gerze, central Tibet: records of the Bangong–Nujiang oceanic ridge subduction during the late Jurassic. Lithos 262:456–469
Li Y, He J, Han Z, Wang C, Ma P, Zhou A, Liu S-A, Xu M (2016b) Late Jurassic sodium-rich adakitic intrusive rocks in the southern Qiangtang terrane, central Tibet, and their implications for the Bangong–Nujiang Ocean subduction. Lithos 245:34–46
Li C, Li Z, Cao H, Zeng M (2018a) Newly discovered ca. 163 Ma OIB-type diabase dike from the Shiquanhe Ophiolites, Western Tibet. Acta Geol Sin 92(6):2443–2444
Li S, Zhao S, Liu X, Cao H, Yu S, Li X, Somerville I, Yu S, Suo Y (2018b) Closure of the proto-Tethys ocean and early Paleozoic amalgamation of microcontinental blocks in East Asia. Earth Sci Rev 186:37–75
Li S-M, Wang Q, Zhu D-C, Stern RJ, Cawood PA, Sui Q-L, Zhao Z (2018c) One or two early Cretaceous arc systems in the Lhasa Terrane, southern Tibet. J Geophys Res Solid Earth 123(5):3391–3413
Li S-M, Wang Q, Zhu D-C, Cawood PA, Stern RJ, Weinberg R, Zhao Z, Mo X-X (2020) Reconciling orogenic drivers for the evolution of the Bangong-Nujiang Tethys during middle-late Jurassic. Tectonics 39(2):e2019TC005951

Liu M (2012) Petrogenesis and tectonic significance of early Jurassic alkalic Pluton in Nyainrong microcontinent, central Tibet. Doctoral thesis, China University of Geosciences (in Chinese with English abstract)
Liu B (2014) Petrogenesis and geodynamic setting of Permian to Triassic mafic rocks in the Yushu area, central Qinghai-Tibetan Plateau. China University of Geosciences
Liu G, Einsele G (1996) Sedimentary history of the Tethyan basin in the Tibetan Himalayas. Geol Rundsch 83(1):32–61
Liu QS, Jiang W, Jian PS, Wu ZH, Hu DG (2006) Zircon SHRIMP U-Pb age and petrochemical and geochemical features of Mesozoic muscovite monzonitic granite at Ningzhong, Tibet. Acta Petrol Sin 22(3):643–652
Liu D, Zhao Z, Zhu DC, Niu Y, Harrison TM (2014) Zircon xenocrysts in Tibetan ultrapotassic magmas: imaging the deep crust through time. Geology 42(1):43–46
Liu B, Ma C-Q, Guo P, Sun Y, Gao K, Guo Y-H (2016a) Evaluation of late Permian mafic magmatism in the central Tibetan Plateau as a response to plume-subduction interaction. Lithos 264:1–16
Liu Y, Li C, Xie C, Fan J, Wu H, Jiang Q, Li X (2016b) Cambrian granitic gneiss within the central Qiangtang terrane, Tibetan Plateau: implications for the early Palaeozoic tectonic evolution of the Gondwanan margin. Int Geol Rev 58(9):1043–1063
Liu D, Shi R, Ding L, Huang Q, Zhang X, Yue Y, Zhang L (2017) Zircon U–Pb age and Hf isotopic compositions of Mesozoic granitoids in southern Qiangtang, Tibet: implications for the subduction of the Bangong–Nujiang Tethyan Ocean. Gondwana Res 41(Suppl C):157–172
Liu ZC, Ding L, Zhang LY, Wang C, Qiu ZL, Wang JG, Shen XL, Deng XQ (2018) Sequence and petrogenesis of the Jurassic volcanic rocks (Yeba formation) in the Gangdese arc, southern Tibet: implications for the neo-Tethyan subduction. Lithos 312–313:72–88
Liu Y, Xie C, Li C, Li S, Santosh M, Wang M, Fan J (2019) Breakup of the northern margin of Gondwana through lithospheric delamination: evidence from the Tibetan Plateau. Bulletin 131(3–4):675–697
Liu B, Xu Y, Li Q, Sun Y, Zhao S-Q, Huang J, Dong H, Yan R (2020a) Origin of Triassic mafic magmatism in the north Qiangtang terrane, central Tibetan Plateau: implications for the development of a continental back-arc basin. J Geol Soc 177(4):826–842
Liu Y, Li S, Santosh M, Cao H, Yu S, Wang Y, Guo R, Xu L, Zhou J, Zhou Z (2020b) The passive margin of northern Gondwana during early Paleozoic: evidence from the central Tibet Plateau. Gondwana Res 78:126–140
Lu L, Zhang KJ, Yan LL, Jin X, Zhang YX (2017) Was late Triassic Tanggula granitoid (central Tibet, western China) a product of melting of underthrust Songpan-Ganzi flysch sediments? Tectonics 36(5):902–928
Ma L, Wang Q, Li ZX, Wyman DA, Jiang ZQ, Yang JH, Gou GN, Guo HF (2013a) Early late Cretaceous (ca. 93Ma) norites and hornblendites in the Milin area, eastern Gangdese: lithosphere–asthenosphere interaction during slab roll-back and an insight into early late Cretaceous (ca. 100–80Ma) magmatic "flare-up" in southern Lhasa (Tibet). Lithos 172–173:17–30
Ma L, Wang Q, Wyman DA, Jiang ZQ, Yang JH, Li QL, Gou GN, Guo HF (2013b) Late Cretaceous crustal growth in the Gangdese area, southern Tibet: petrological and Sr–Nd–Hf–O isotopic evidence from Zhengga diorite–gabbro. Chem Geol 349–350:54–70
Ma L, Wang Q, Wyman DA, Li ZX, Jiang ZQ, Yang JH, Gou GN, Guo HF (2013c) Late Cretaceous (100–89 Ma) magnesian charnockites with adakitic affinities in the Milin area, eastern Gangdese: partial melting of subducted oceanic crust and implications for crustal growth in southern Tibet. Lithos 175:315–332
Ma L, Wang Q, Wyman DA, Jiang ZQ, Wu FY, Li XH, Yang JH, Gou GN, Guo HF (2015) Late Cretaceous back-arc extension and arc system evolution in the Gangdese area, southern Tibet: geochronological, petrological, and Sr-Nd-Hf-O isotopic evidence from Dagze diabases. J Geophys Res Solid Earth 120(9):6159–6181
Ma X, Xu Z, Meert JG (2017) Syn-convergence extension in the southern Lhasa terrane: evidence from late Cretaceous adakitic granodiorite and coeval gabbroic-dioritic dykes. J Geodyn 110:12–30

Maniar PD, Piccoli PM (1989) Tectonic discrimination of granitoids. Geol Soc Am Bull 101:635–643
Meng FY, Zhao ZD, Zhu DC, Zhang LL, Guan Q, Liu M, Yu F, Mo XX (2010) Petrogenesis of late Cretaceous adakite-like rocks in Mamba from the eastern Gangdese, Tibet. Acta Petrol Sin 26:2180–2192 (in Chinese with English abstract)
Meng FY, Zhao Z, Zhu DC, Mo X, Guan Q, Huang Y, Dong G, Zhou S, DePaolo DJ, Harrison TM, Zhang Z, Liu J, Liu Y, Hu Z, Yuan H (2014) Late Cretaceous magmatism in Mamba area, central Lhasa subterrane: products of back-arc extension of Neo-Tethyan Ocean? Gondwana Res 26(2):505–520
Middlemost EA (1994) Naming materials in the magma/igneous rock system. Earth Sci Rev 37:215–224
Ou Q, Wang Q, Zeng J-P, Yang J-H, Zhang H-X, Xia X-P, Chen Y-W (2021) Petrogenesis and tectonic implications of middle Triassic basalts and rhyolites in the northern Qiangtang Block, central Tibet. J Asian Earth Sci 206:104573
Pan GT, Wang LQ, Li RS, Yuan SH, Ji WH, Yin FG, Zhang WP, Wang BD (2012) Tectonic evolution of the Qinghai-Tibet Plateau. J Asian Earth Sci 53:3–14
Pearce JA (1982) Trace element characteristics of lavas from destructive plate boundaries. In: Thorpe RS (ed) Andesites, orogenic andesites and related rocks. John Wiley, New York, pp 525–548
Pearce J (1983) Role of the sub-continental lithosphere in magma genesis at active continental margin. In: Hawkesworth CJ, Norry MJ (eds) Continental basalts and mantle xenoliths. Shiva, Nantwich, pp 230–249
Pullen A, Kapp P, Gehrels GE, Ding L, Zhang Q (2011) Metamorphic rocks in central Tibet: lateral variations and implications for crustal structure. Geol Soc Am Bull 123(3–4):585–600
Qi NY, Zhao ZD, Tang Y, Yang JJ, Shi QS, Liu D, Zhu DC, Sheikh L (2019) Geochronology, geochemistry and petrogenesis of the late Jurassic-early Cretaceous granitoids in Zuozuo, western central Lhasa Terrane, Tibet. Acta Petrol Sin 35:405–422 (in Chinese with English abstract)
Qiu BB, Zhu DC, Zhao ZD, Wang LQ (2010) The westward extension of Comei fragmented large igneous province in southern Tibet and its implications. Acta Petrol Sin 26(7):2207–2216
Qu XM, Xin HB, Xu WY, Yang ZS, Li ZQ (2006) Discovery and significance of copperbearing bimodal rocks series in Coqin area of Tibet. Acta Petrol Sin 22:707–716 (in Chinese with an English abstract)
Qu XM, Wang RJ, Xin HB, Zhao YY, Fan XT (2009) Geochronology and geochemistry of igneous rocks related to the subduction of the Tethys oceanic plate along the Bangong Lake arc zone, the western Tibetan Plateau. Geochimica 38:523–535 (in Chinese with English abstract)
Quidelleur X, Grove M, Lovera OM, Harrison TM, Yin A, Ryerson FJ (1997) Thermal evolution and slip history of the Renbu Zedong Thrust, southeastern Tibet. J Geophys Res Solid Earth 102(B2):2659–2679
Rickwood PC (1989) Boundary lines within petrologic diagrams which use oxides of major and minor elements. Lithos 22:247–263
Shi Y, Hou C, Anderson JL, Yang T, Ma Y, Bian W, Jin J (2018) Zircon SHRIMP U–Pb age of late Jurassic OIB-type volcanic rocks from the Tethyan Himalaya: constraints on the initial activity time of the Kerguelen mantle plume. Acta Geochim 37(3):441–455
Shuai X, Li SM, Zhu DC, Wang Q, Zhang LL, Zhao ZD (2021) Tetrad effect of rare earth elements caused by fractional crystallization in high-silica granites: an example from central Tibet. Lithos 384–385:105968
Shui XF, He ZY, Klemd R, Zhang ZM, Lu TY, Yan LL (2017) Early Jurassic adakitic rocks in the southern Lhasa sub-terrane, southern Tibet: petrogenesis and geodynamic implications. Geol Mag 155(1):132–148
Song S-W, Zhu DC, Wang Q, Cawood PA, Zhan Q-Y, Li S-M, Zhang L-L, Zhao Z-D (2021a) Generation of syn-collisional S-type granites in collision zones: an example from the late Triassic Tanggula batholith in northern Tibet. Gondwana Res 104:185–198

Song Y, Chen W, Wei S, Ma X, Sun M, Liu Q (2021b) Evolution of continental arc at root caused by igneous garnet and amphibole fractionation: evidence from Jurassic intermediate–felsic intrusive rocks in southern Qiangtang, Tibet. Lithos 382–383:105935

Sui QL, Wang Q, Zhu DC, Zhao ZD, Chen Y, Santosh M, Hu ZC, Yuan HL, Mo XX (2013) Compositional diversity of ca. 110 Ma magmatism in the northern Lhasa Terrane, Tibet: implications for the magmatic origin and crustal growth in a continent-continent collision zone. Lithos 168:144–159

Sun SS, Mcdonough WF (1989) Chemical and isotopic systematics of oceanic basalts: implications for mantle composition and processes. Geol Soc Lond Spec Publ 42(1):313–345

Sun G, Hu X, Sinclair HD, BouDagher-Fadel MK, Wang J (2015a) Late Cretaceous evolution of the Coqen Basin (Lhasa terrane) and implications for early topographic growth on the Tibetan Plateau. Geol Soc Am Bull 127(7/8):1001–1020

Sun S-J, Sun W-D, Zhang L-P, Zhang R-Q, Li C-Y, Zhang H, Hu Y-B, Zhang Z-R (2015b) Zircon U–Pb ages and geochemical characteristics of granitoids in Nagqu area, Tibet. Lithos 231:92–102

Tafti R, Mortensen JK, Lang JR, Rebagliati M, Oliver JL (2009) Jurassic U–Pb and Re-Os ages for the newly discovered Xietongmen Cu-Au porphyry district, Tibet, PRC: implications for metallogenic epochs in the southern Gangdese belt. Econ Geol 104(1):127–136

Tang JX, Li FJ, Li ZJ, Zhang L, Tang XQ, Deng Q, Lang XH, Huang Y, Yao XF, Wang Y (2010) Period of time for the formation of main geologic bodies in Xiongcun copper–gold deposit, Xietongmen County, Tibet: evidence from Zircon U–Pb ages and Re–Os ages of molybdenite. Mineral Deposits 29(3):461–475 (in Chinese with English abstract)

Tang Y, Zhai QG, Hu PY, Chung SL, Xiao XC, Wang HT, Zhu ZC, Wang W, Wu H, Lee HY (2020) Southward subduction of the Bangong-Nujiang Tethys Ocean: insights from ca. 161–129 Ma arc volcanic rocks in the north of Lhasa terrane, Tibet. Int J Earth Sci 109(2):6314–6647

Tian Z, Zhang Z, Piccoli PM, Dong X (2021) Early Jurassic accretion of retrograde eclogites and granulites in the Amdo complex, Bangong–Nujiang suture zone, central Tibet. Gondwana Res 104:70

Volkmer JE, Kapp P, Guynn JH, Lai Q (2007) Cretaceous-tertiary structural evolution of the north central Lhasa terrane, Tibet. Tectonics 26:TC6007

Wager LR, Deer WA (1939) Geological investigation in East Greenland, part III, the petrology of the Skaergaard intrusion, Kangerdluqssuaq, East Greenland. Meddelser Gronland 105:1

Wang J, Wang ZJ, Chen WX, Fu XG, Chen M (2007a) New evidences for the age assignment of the Nadi Kangri formation in the north Qiangtang basin, northern Tibet, China. Geol Bull China 26(4):404–409

Wang YZ, Qi SS, An SW, Xu CQ (2007b) Characteristics and Ar-Ar dating of ultramafic-mafic rocks in the Zadoi area, southern Qinghai, China. Geol Bull China 26(6):668–674

Wang J, Fu X, Chen W, Tan F, Wang Z, Chen M, Zhuo J (2008a) Geochronology and geochemistry of the volcanic rocks in the Woruoshan area, northern Qiangtang. Sci China (Ser D) 38:33–43

Wang Q, Wyman DA, Xu J, Wan Y, Li C, Zi F, Jiang Z, Qiu H, Chu Z, Zhao Z (2008b) Triassic Nb-enriched basalts, magnesian andesites, and adakites of the Qiangtang terrane (central Tibet): evidence for metasomatism by slab-derived melts in the mantle wedge. Contrib Mineral Petrol 155(4):473–490

Wang L, Zeng LS, Gao LE, Chen ZY (2013) Early Cretaceous high Mg# and high Sr/Y clinopyroxene-bearing diorite in the southeast Gangdese batholith, southern Tibet. Acta Petrol Sin 29(6):1977–1994

Wang C, Wei QR, Liu XN, Ding PF, Bu T, Sun J, Zhang XQ, Wang JY (2014a) Post-collision related late Indosinian granites of Gangdise Terrane: evidences from zircon U-Pb geochronology and petrogeochemistry. Earth Sci J China Univ Geosci 39(09):1277–1300 (in Chinese with English abstract)

Wang Q, Zhu DC, Zhao ZD, Liu SA, Chung SL, Li SM, Liu D, Dai JG, Wang LQ, Mo XX (2014b) Origin of the ca. 90Ma magnesia-rich volcanic rocks in SE Nyima, central Tibet: products of lithospheric delamination beneath the Lhasa-Qiangtang collision zone. Lithos 198:24–37

Wang B-D, Wang L-Q, Chung S-L, Chen J-L, Yin F-G, Liu H, Li X-B, Chen L-K (2016a) Evolution of the Bangong–Nujiang Tethyan ocean: insights from the geochronology and geochemistry of mafic rocks within ophiolites. Lithos 245:18–33
Wang C, Ding L, Liu ZC, Zhang LY, Yue YH (2016b) Early Cretaceous bimodal volcanic rocks in the southern Lhasa terrane, south Tibet: age, petrogenesis and tectonic implications. Lithos 268–271:260–273
Wang RQ, Qiu JS, Yu SB, Zhao JL (2017a) Crust–mantle interaction during early Jurassic subduction of neo-Tethyan oceanic slab: evidence from the Dongga gabbro–granite complex in the southern Lhasa subterrane, Tibet. Lithos 292–293:262–277
Wang Y, Tang J, Wang L, Duan J, Danzhen W, Li S, Li Z (2017b) Petrogenesis of Jurassic granitoids in the west central Lhasa subterrane, Tibet, China: the Geji example. Int Geol Rev 60(9):1155–1171
Wang J, Wang Q, Zhang C, Dan W, Qi Y, Zhang XZ, Xia XP (2018a) Late Permian bimodal volcanic rocks in the northern Qiangtang Terrane, central Tibet: evidence for interaction between the Emeishan plume and the Paleo-Tethyan subduction system. J Geophys Res Solid Earth 123(8):6540–6561
Wang Q, Zhu DC, Liu AL, Cawood PA, Liu SA, Xia Y, Chen Y, Wang H, Zhang LL, Zhao ZD (2018b) Survival of the Lhasa Terrane during its collision with Asia due to crust-mantle coupling revealed by ca. 114 Ma intrusive rocks in western Tibet. Lithos 304–307:200–210
Wang M, Li C, Zeng X-W, Li H, Fan J-J, Xie C-M, Hao Y-J (2019) Petrogenesis of the southern Qiangtang mafic dykes, Tibet: link to a late Paleozoic mantle plume on the northern margin of Gondwana? GSA Bull 131(11–12):1907–1919
Wang H, Zhai Q, Hu P, Zeng L, Tang Y, Zhu Z (2020) Late Cambrian to early Silurian granitic rocks of the Gemuri area, central Qiangtang, North Tibet: new constraints on the tectonic evolution of the northern margin of Gondwana. Acta Geol Sin Eng Ed 94(4):1007–1019
Wei YQ, Zhao ZD, Niu YL, Zhu DC, Liu D, Wang Q, Mo XX, Wei JC (2017) Geochronology and geochemistry of the early Jurassic yeba formation volcanic rocks in southern Tibet: initiation of back-arc rifting and crustal accretion in the southern Lhasa terrane. Lithos 278–281:477–490
Wen DR, Chung SL, Song B, Iizuka Y, Yang HJ, Ji J, Liu D, Gallet S (2008a) Late Cretaceous Gangdese intrusions of adakitic geochemical characteristics, SE Tibet: petrogenesis and tectonic implications. Lithos 105(1):1–11
Wen DR, Liu D, Chung SL, Chu MF, Ji J, Zhang Q, Song B, Lee TY, Yeh MW, Lo CH (2008b) Zircon SHRIMP U–Pb ages of the Gangdese batholith and implications for Neotethyan subduction in southern Tibet. Chem Geol 252(3):191–201
Whalen JB, Currie KL, Chappell BW (1987) A-type granites: geochemical characteristics, discrimination and petrogenesis. Contrib Mineral Petrol 95:407–419
Winchester JA, Floyd PA (1976) Geochemical magma type discrimination: application to altered and metamorphosed basic igneous rocks. Earth Planet Sci Lett 28(3):459–469
Wu FY, Ji WQ, Liu CZ, Chung SL (2010) Detrital zircon U-Pb and Hf isotopic data from the Xigaze fore-arc basin: constraints on trans Himalayan magmatic evolution in southern Tibet. Chem Geol 271:13–25
Wu H, Li C, Hu P, Fan J, Zhang H, Li J (2013) The discovery of Qushenla volcanic rocks in Tasepule area of Nyima County, Tibet, and its geological significance. Geol Bull China 32(7):1014–1026
Wu H, Li C, Hu P, Li X (2015a) Early Cretaceous (100–105 Ma) Adakitic magmatism in the Dachagou area, northern Lhasa terrane, Tibet: implications for the Bangong–Nujiang Ocean subduction and slab break-off. Int Geol Rev 57(9–10):1172–1188
Wu H, Li C, Xu M, Li X (2015b) Early Cretaceous adakitic magmatism in the Dachagou area, northern Lhasa terrane, Tibet: implications for slab roll-back and subsequent slab break-off of the lithosphere of the Bangong–Nujiang Ocean. J Asian Earth Sci 97(Part A):51–66
Xia Y, Zhu DC, Wang Q, Zhao ZD, Liu D, Wang LQ, Mo XX (2014) Picritic porphyrites and associated basalts from the remnant Comei Large Igneous Province in SE Tibet: records of mantle-plume activity. Terra Nova 26(6):487–494

Xiao L, Xu YG, Mei HJ, Zheng YF, He B, Pirajno F (2004) Distinct mantle sources of low-Ti and high-Ti basalts from the western Emeishan large igneous province, SW China: implications for plume–lithosphere interaction. Earth Planet Sci Lett 228(3):525–546
Xie C-M, Li C, Fan J-J, Su L (2017) Ordovician sedimentation and bimodal volcanism in the southern Qiangtang terrane of northern Tibet: implications for the evolution of the northern Gondwana margin. Int Geol Rev 59(16):2078–2105
Xie C, Song Y, Duan M, Su L, Hao Y (2020) Geochronology, geochemistry, origin, and tectonic implications of high-pressure mafic granulites of the Amdo region, central Tibet. Geol J 55(11):7458–7473
Xu WC, Zhang HF, Luo BJ, Guo L, Yang H (2015) Adakite-like geochemical signature produced by amphibole-dominated fractionation of arc magmas: an example from the late Cretaceous magmatism in Gangdese belt, south Tibet. Lithos 232(Suppl C):197–210
Xu W, Zhu DC, Wang Q, Weinberg RF, Wang R, Li SM, Zhang LL, Zhao ZD (2019) Constructing the Early Mesozoic Gangdese crust in southern Tibet by hornblende-dominated magmatic differentiation. J Petrol 60(3):515–552
Xu W, Zhu DC, Wang Q, Weinberg RF, Wang R, Li SM, Zhang LL, Zhao ZD (2020) Mafic microgranular enclaves formed by gas-driven filter pressing during rapid cooling: an example from the Gangdese batholith in southern Tibet. J Petrol 61(11–12):1–21
Yan JJ (2019) Geochronology, geochemistry and petrogenesis of Mesozoic magmatic rocks in Shiquanhe area, Tibetan Plateau. China University of Geosciences, Beijing
Yan H, Long X, Wang X-C, Li J, Wang Q, Yuan C, Sun M (2016) Middle Jurassic MORB-type gabbro, high-mg diorite, calc-alkaline diorite and granodiorite in the Ando area, central Tibet: evidence for a slab roll-back of the Bangong-Nujiang Ocean. Lithos 264:315–328
Yan JJ, Zhao ZD, Liu D, Wang ZZ, Tang Y (2017) Geochemistry and petrogenesis of the late Jurassic Xuru Tso batholith in central Lhasa Terrane, Tibet. Acta Petrol Sin 33:2437–2453 (in Chinese with English abstract)
Yang Z-Y, Wang Q, Yang J-H, Dan W, Zhang X-Z, Ma L, Qi Y, Wang J, Sun P (2019) Petrogenesis of early Cretaceous granites and associated microgranular enclaves in the Xiabie Co area, central Tibet: crust-derived magma mixing and melt extraction. Lithos 350–351:105199
Yao XF, Tang JX, Li ZJ, Deng SL, Ding S, Hu ZH, Zhang Z (2013) The redefinition of the ore-forming porphyry's age in Gaerqiong skarn-type gold-copper deposit, western Bangong Lake-Nujiang River metallogenic belt, Xizang (Tibet). Geol Rev 59:193–200 (in Chinese with English abstract)
Ye LJ, Zhao ZD, Liu D, Zhu DC, Dong GC, Mo XX, Hu ZC, Liu YS (2015) Late Cretaceous disbase and granite dike in Namling, Tibet: petrogenesis and implications for extension. Acta Petrol Sin 31(5):1298–1312
Yi J-K, Wang Q, Zhu D-C, Li S-M, Liu S-A, Wang R, Zhang L-L, Zhao Z-D (2018) Westward-younging high-Mg adakitic magmatism in central Tibet: record of a westward-migrating lithospheric foundering beneath the Lhasa–Qiangtang collision zone during the late Cretaceous. Lithos 316–317:92–103
Yin A (2006) Cenozoic tectonic evolution of the Himalayan orogen as constrained by along-strike variation of structural geometry, exhumation history, and foreland sedimentation. Earth Sci Rev 76(1–2):1–131
Yin A, Harrison TM (2000) Geologic evolution of the Himalayan Tibetan Orogen. Annu Rev Earth Planet Sci 28:211–280
Yong Y, Chen W, Zhang Y, Liu XY (2011) Zircon SHRIMP U-Pb dating and geochemistry of the Rangnianggongba gabbro in Yushu area, Qinghai Province. Acta Petrol Mineral 30(3):419–426
Yu HX, Chen JL, Xu JF, Wang BD, Wu JB, Liang HY (2011) Geochemistry and origin of late Cretaceous (90 Ma) ore-bearing porphyry of Balazha in mid-northern Lhasa terrane, Tibet. Acta Petrol Sin 27:2011–2022 (in Chinese with an English abstract)
Yu YS, Yang ZS, Dai PY, Tian SH, Gao Y, Liu YC, Xiu D (2015) Geochronology and genesis of the magmatism in the Ria copper polymetallic deposit of the Nixiong orefield, Coqen, Tibet. Geol China 42:118–133 (in Chinese with an English abstract)

Zeng Y-C, Chen J-L, Xu J-F, Wang B-D, Huang F (2016) Sediment melting during subduction initiation: geochronological and geochemical evidence from the Darutso high-mg andesites within ophiolite melange, central Tibet. Geochem Geophys Geosyst 17(12):4859–4877

Zeng LS, Gao LE, Guo CL, Hou KJ, Wang Q (2017) Early Cretaceous forearc extension of the Gangdese continental arc, southern Tibet. Acta Petrol Sin 33(8):2377–2394

Zhai QG, Li C (2007) Zircon SHRIMP dating of volcanic rock from the Nadigangri Formation in Juhuashan, Qiangtang, northern Tibet and its geological significance. Acta Geol Sin 81(6):795–800

Zhai Q-G, Jahn B-M, Su L, Ernst RE, Wang K-L, Zhang R-Y, Wang J, Tang S (2013a) SHRIMP zircon U–Pb geochronology, geochemistry and Sr–Nd–Hf isotopic compositions of a mafic dyke swarm in the Qiangtang terrane, northern Tibet and geodynamic implications. Lithos 174:28–43

Zhai Q-G, Jahn B-M, Su L, Wang J, Mo X-X, Lee H-Y, Wang K-L, Tang S (2013b) Triassic arc magmatism in the Qiangtang area, northern Tibet: zircon U–Pb ages, geochemical and Sr–Nd–Hf isotopic characteristics, and tectonic implications. J Asian Earth Sci 63:162–178

Zhang HF, XuWC GJQ, Zong KQ, Cai HM, Yuan HL (2007a) Indosinian orogenesis of the Gangdese terrane: evidences from zircon U-Pb dating and petrogenesis of granitoids. Earth Sci J China Univ Geosci 32:155–166 (in Chinese with English abstract)

Zhang HF, Xu WC, Guo JQ, Zong KQ, Cai HM, Yuan HL (2007b) Zircon U-Pb and Hf isotopic composition of deformed granite in the southern margin of the Gangdese belt, Tibet: evidence for early Jurassic subduction of neo-Tethyan oceanic slab. Acta Petrol Sin 23:1347–1353 (in Chinese with English abstract)

Zhang XQ, Zhu DC, Zhao ZD, Wang LQ, Hang JC, Mo XX (2010a) Petrogenesis of the Nixiong pluton in Coqen, Tibet and its potential significance for the Nixiong Fe-rich mineralization. Acta Petrol Sin 26(6):1793–1804 (in Chinese with English abstract)

Zhang ZM, Zhao GC, Santosh M, Wang JL, Dong X, Shen K (2010b) Late Cretaceous charnockite with adakitic affinities from the Gangdese batholith, southeastern Tibet: evidence for Neo-Tethyan mid-ocean ridge subduction? Gondwana Res 17(4):615–631

Zhang LL, Zhu DC, Zhao ZD, Liao ZL, Wang LQ, Mo XX (2011) Early granitoids in Xainza, Tibet: evidence of slab break-off. Acta Petrol Sin 27:1938–1948 (in Chinese with English abstract)

Zhang KJ, Zhang YX, Tang XC, Xia B (2012a) Late Mesozoic tectonic evolution and growth of the Tibetan Plateau prior to the indo-Asian collision. Earth Sci Rev 114:236–249

Zhang XQ, Zhu DC, Zhao ZD, Sui QL, Wang Q, Yuan SH, Hu ZC, Mo XX (2012b) Geochemistry, zircon U-Pb geochronology and in-situ Hf isotope of the Maiga batholith in Coqen, Tibet: constraints on the petrogenesis of the early Cretaceous granitoids in the central Lhasa terrane. Acta Petrol Sin 28:1615–1634 (in Chinese with English abstract)

Zhang ZM, Dong X, Liu F, Lin YH, Yan R, Santosh M (2012c) Tectonic evolution of the Amdo terrane, central Tibet: petrochemistry and zircon U-Pb geochronology. J Geol 120(4):431–451

Zhang ZM, Dong X, Xiang H, He Z, Liou JG (2014) Metagabbros of the Gangdese arc root, south Tibet: implications for the growth of continental crust. Geochim Cosmochim Acta 143(Suppl C):268–284

Zhang Y-X, Li Z-W, Yang W-G, Zhu L-D, Jin X, Zhou X-Y, Tao G, Zhang K-J (2017) Late Jurassic–early Cretaceous episodic development of the Bangong Meso-Tethyan subduction: evidence from elemental and Sr–Nd isotopic geochemistry of arc magmatic rocks, Gaize region, central Tibet, China. J Asian Earth Sci 135:212–242

Zhang LX, Wang Q, Zhu DC, Li SM, Zhao ZD, Zhang LL, Chen Y, Liu SA, Zheng YC, Wang R, Liao ZL (2019) Generation of leucogranites via fractional crystallization: a case from the late Triassic Luoza batholith in the Lhasa terrane, southern Tibet. Gondwana Res 66:63–76

Zhang ZM, Ding H, Palin RM, Dong X, Tian Z, Chen Y (2020) The lower crust of the Gangdese magmatic arc, southern Tibet, implication for the growth of continental crust. Gondwana Res 77:136–146

Zhang L, Hu Y, Deng J, Wang J, Wang K, Sui Q, Chen Y, Wu K, Sun W (2021) Genesis and mineralization potential of the late Cretaceous Chemen granodioritic intrusion in the southern Gangdese magmatic belt, Tibet. J Asian Earth Sci 217:104829

Zhao T-P, Zhou M-F, Zhao J-H, Zhang K-J, Chen W (2008) Geochronology and geochemistry of the c. 80 Ma Rutog granitic pluton, northwestern Tibet: implications for the tectonic evolution of the Lhasa Terrane. Geol Mag 145(6):845–857

Zheng YC, Hou ZQ, Gong YL, Liang W, Sun QZ, Zhang S, Fu Q, Huang KX, Li QY, Li W (2014) Petrogenesis of Cretaceous adakite-like intrusions of the Gangdese Plutonic Belt, southern Tibet: implications for mid-ocean ridge subduction and crustal growth. Lithos 190:240–263

Zheng H, Huang QT, Cai ZR, Zhang KJ, Liu HC, Cheng C, Lu LJ, Yang P, Yu SR, Yang G (2018) Early Cretaceous arc granitoids from the central Lhasa subterrane: production of the northward subduction of Yarlung Zangbo Neo-Tethyan Ocean? Geol J 54(6):4001–4013

Zhou CY, Zhu DC, Zhao ZD, Xu JF, Wang LQ, Chen HH, Xie LW, Dong GC, Zhou S (2008) Petrogenesis of Daxiong pluton in western Gangdese, Tibet: zircon U-Pb dating and Hf isotopic constraints. Acta Petrol Sin 24:348–358 (in Chinese with English abstract)

Zhou Q, Liu Z, Lai Y, Wang G-C, Liao Z, Li Y-X, Wu J-Y, Wang S, Qing C-S (2017a) Petrogenesis of mafic and felsic rocks from the Comei large igneous province, South Tibet: implications for the initial activity of the Kerguelen plume. GSA Bull 130(5–6):811–824

Zhou X, Zheng JP, Xiong Q, Yang JS, Wu YB, Zhao JH, Griffin WL, Dai HK (2017b) Early Mesozoic deep-crust reworking beneath the central Lhasa terrane (South Tibet): evidence from intermediate gneiss xenoliths in granites. Lithos 274–275:225–239

Zhu DC, Pan GT, Mo XX, Liao ZL, Wang XS, Wang LQ (2004) Permian to Cretaceous volcanic activities in the central segment of the Tethyan Himalayas (I): distribution characteristics and significance. Geol Bull China 23(7):645–654

Zhu DC, Pan GT, Mo XX, Wang LQ, Liao ZL, Jiang XS, Geng QR (2005) SHRIMP U-Pb zircon dating for the dacite of the Sangxiu formation in the central segment of Tethyan Himalaya and its implications. Chin Sci Bull 50(6):563–568

Zhu DC, Pan GT, Mo XX, Liao ZL, Jiang XS, Wang LQ, Zhao ZD (2007) Petrogenesis of volcanic rocks in the Sangxiu formation, central segment of Tethyan Himalaya: a probable example of plume–lithosphere interaction. J Asian Earth Sci 29(2):320–335

Zhu DC, Mo XX, Pan GT, Zhao ZD, Dong GC, Shi YR, Liao ZL, Wang LQ, Zhou CY (2008a) Petrogenesis of the earliest early Cretaceous mafic rocks from the Cona area of the eastern Tethyan Himalaya in South Tibet: interaction between the incubating Kerguelen plume and the eastern greater India lithosphere? Lithos 100(1–4):147–173

Zhu DC, Pan GT, Chung SL, Liao ZL, Wang LQ, Li GM (2008b) SHRIMP zircon age and geochemical constraints on the origin of lower Jurassic volcanic rocks from the Yeba formation, southern Gangdese, South Tibet. Int Geol Rev 50(5):442–471

Zhu DC, Chung SL, Mo XX, Zhao ZD, Niu YL, Song B, Yang YH (2009a) The 132 Ma Comei-Bunbury large igneous province: remnants identified in present-day southeastern Tibet and southwestern Australia. Geology 37(7):583–586

Zhu DC, Mo XX, Niu YL, Zhao ZD, Wang LQ, Liu YS, Wu FY (2009b) Geochemical investigation of early Cretaceous igneous rocks along an eastwest traverse throughout the central Lhasa terrane, Tibet. Chem Geol 268:298–312

Zhu DC, Zhao ZD, Pan GT, Lee HY, Kang ZQ, Liao ZL, Wang LQ, Li GM, Dong GC, Liu B (2009c) Early Cretaceous subduction-related adakite-like rocks of the Gangdese Belt, southern Tibet: products of slab melting and subsequent melt–peridotite interaction? J Asian Earth Sci 34(3):298–309

Zhu DC, Mo XX, Zhao ZD, Niu YL, Wang LQ, Chu QH, Pan GT, Xu JF, Zhou CY (2010) Presence of Permian extension- and arc-type magmatism in southern Tibet: paleogeographic implications. GSA Bull 122(7–8):979–993

Zhu DC, Zhao ZD, Niu YL, Mo XX, Chung SL, Hou ZQ, Wang LQ, Wu FY (2011) The Lhasa terrane: record of a microcontinent and its histories of drift and growth. Earth Planet Sci Lett 301:241–255

Zhu D-C, Zhao Z-D, Niu Y, Dilek Y, Wang Q, Ji W-H, Dong G-C, Sui Q-L, Liu Y-S, Yuan H-L, Mo X-X (2012) Cambrian bimodal volcanism in the Lhasa terrane, southern Tibet: record of an early Paleozoic Andean-type magmatic arc in the Australian proto-Tethyan margin. Chem Geol 328:290–308

Zhu DC, Zhao ZD, Niu YL, Dilek Y, Hou ZQ, Mo XX (2013) The origin and pre-Cenozoic evolution of the Tibetan Plateau. Gondwana Res 23:1429–1454

Zhu DC, Wang Q, Zhao ZD, Chung SL, Cawood PA, Niu YL, Liu SA, Wu FY, Mo XX (2015) Magmatic record of India-Asia collision. Sci Rep 5:14289

Zhu D-C, Li S-M, Cawood PA, Wang Q, Zhao Z-D, Liu S-A, Wang L-Q (2016) Assembly of the Lhasa and Qiangtang terranes in central Tibet by divergent double subduction. Lithos 245(Suppl C):7–17

Zhu DC, Wang Q, Chung SL, Cawood PA, Zhao ZD (2019) Gangdese magmatism in southern Tibet and India–Asia convergence since 120 Ma. Geol Soc Lond Spec Publ 483(1):583–604

Chapter 3
Early Cenozoic Igneous Rocks in the Tibetan Plateau

Abstract Vast amounts of Paleocene-Eocene volcanic rocks and Eocene intrusions had developed as a large approximately west-east plutonic-volcanic belt in southern Tibet, forming the Gangdese Magmatic Belt (GMB). The Paleocene-Eocene volcanic rocks consist mainly of Linzizong volcanic succession (LVS), which is lithologically intermediate-rhyolitic lavas and pyroclastic rocks with interlayers of sedimentary rocks. The LVS sequence has been well characterized in total thickness of 5200 m thickness in the Linzhou basin about 30 km north of the Lhasa city and can be divided into three formations upwards, i.e., the Dianzhong, the Nianbo and the Pana formation with continuous between-formation relation contacts and a clear unconformity over the highly deformed Cretaceous meta-sedimentary rocks. Geochemically, the LVS volcanic rocks are mostly calc-alkaline although some are more alkalic, and can be further classified into several rock types in TAS diagram with variable SiO_2 content (SiO_2 = 47–79 wt.%). Detail chronology on the andesite and rhyolite places the LVS eruptions in the Paleocene to Early Eocene with age of 63.5–43.9 Ma, suggesting a continuous volcanic activity for ~20 Myrs. The Gangdese magmatic belt (GMB) contains numerous granitoid batholiths and can be roughly divided into three sub-belts, which are getting younger southwards. The South Gangdese Batholith (SGB) is dominated by 60–40 Ma composite plutons, extends E-W for up to 2000 km, and accounts for ~45% of all the batholiths exposed in the Gangdese Belt. The lithologies of SGB are diverse, ranging from mafic to felsic rocks, including granodiorite, quartz diorite, quartz monzonite and monzogranite with abundant mafic micro-granular enclaves (MME) in these granites, indicating magma mixing process and interaction between mafic and felsic magmas. They are alkalic to calc alkalic and mostly plot in the field of shoshonitic rock series. These granitic intrusions are coeval with the LVS, and were previously interpreted as resulting from Tethyan seafloor subduction beneath the Andean-type Eurasian continental margin. It is accepted that the Cenozoic magmatic rocks have formed as a result of northward subduction of the neo-Tethyan seafloor beneath the Asian continent and the subsequent India-Asia continental collision and provide constraints on the timing and processes of the collision.

Prof. Guochen Dong has fully contributed to this chapter.

X. Mo et al., *Magmatism and Crustal Evolution of the Tibetan Plateau*, Modern Approaches in Solid Earth Sciences 25,
https://doi.org/10.1007/978-981-97-2109-2_3

Keywords Early Cenozoic igneous rocks · Gangdese magmatic belt · Linzizong volcanic sequence · India-Asia continental collision

The large magmatic belt of abundant plutonic and volcanic rocks in southern Tibet has long been interpreted as result of northward subduction of the Neo-Tethyan seafloor beneath the Asian continent and the subsequent India-Asia continental collision (Chung et al. 2003; Ji et al. 2009; Lee et al. 2012; Mo et al. 2007, 2008; Zhu et al. 2011, 2013, 2015, 2017). This magmatic belt is also called Gangdese Magmatic Belt (GMB), and the Gangdese is also known as Lhasa Terrane in the more recent literature, which is the southernmost Eurasian block immediately north of the Yarlung–Zangbo suture (YZSZ) of the India–Asia collision (Yin and Harrison 2000; Flower et al. 2001; Mo et al. 2003, 2006, 2007). The Cenozoic magmatic rocks widespread in the Lhasa Terrane, as a large approximately west-east plutonic-volcanic belt mainly comprised large amounts of Paleocene-Eocene volcanic rocks and Eocene granitoid intrusions. It is generally accepted that these Cenozoic magmatic rocks are closely related to the India-Asia continental collision (Chung et al. 2003; Ji et al. 2009, 2012; Lee et al. 2009, 2012; Mo et al. 2007, 2008; Wen et al. 2008; Zhu et al. 2011, 2013, 2015, 2017).

The Paleocene-Eocene volcanic rocks consist mainly of Linzizong volcanic succession (LVS) or Linzizong Formation, locally also called the Meisu, Bugasi, Yulinshan, Bangba and Xiongba Formations in various localities in the Gangdese magmatic belt. The LVS are lithologically intermediate-rhyolitic lavas and pyroclastic rocks with interlayers of sedimentary rocks, and is widely considered to be formed during the India-Asia collision and provide constraints on the timing and processes associated with this collision.

The Gangdese magmatic belt (GMB) contains numerous granitoid batholiths, which are older in the north and younger towards south (e.g., Yin and Harrison 2000). The South Gangdese Batholith (shortly SGB) belt is dominated by 60–40 Ma composite plutons (Schärer et al. 1984; Copeland et al. 1987; Jiang et al. 1999; Mo et al. 2005), extends E-W for up to 2000 km, and accounts for ~45% of all the batholiths exposed in the Gangdese Belt. These granitoid intrusions are coeval with the LVS (Mo et al. 2005; Dong et al. 2005), and were previously interpreted as resulting from Tethyan seafloor subduction beneath the Andean-type Eurasian continental margin (Allègre et al. 1984; Dewey et al. 1988; Pearce and Houjun 1988).

3.1 The Linzizong Volcanic Rocks in the Gangdese

3.1.1 Distribution and Sequences of Linzizong Volcanic Succession (LVS)

The LVS rocks extend west-east for over 1200 km with varying widths along the southern Gangdese and are best developed in several basins such as the Linzhou, Cuoqin, Hoerba and Yare basins (Fig. 3.1). The geologic mapping at 1:250,000

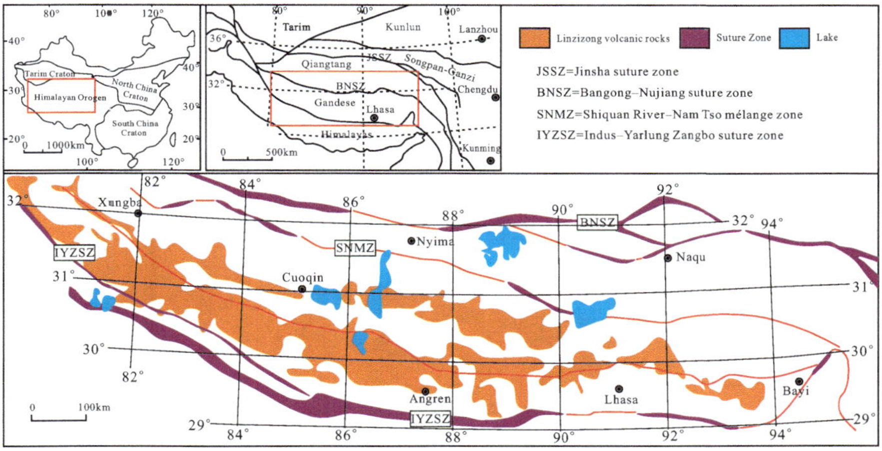

Fig. 3.1 The distribution of the Linzizong volcanic successions in Gangdese (Zhu et al. 2015)

scale shows extensive LVS distribution with a total outcrop area in excess of 39,600 km^2 in the Gangdese Belt (Pan et al. 2004).

Many studies have been carried out in the eastern segment of the Gangdese belt (e.g., Liu 1993; Dong 2002; Zhou et al. 2004; Dong et al. 2005, 2021; Mo et al. 2009; Zhu et al. 2015). The Linzhou volcanic basin in the eastern Gangdese near Lhasa was first studied by late Prof. Li Pu in 1960s and represents a type locality of the LVS with best outcrops and most complete stratigraphy (Liu 1993; Dong 2002; Mo et al. 2003). On the basis of detailed mapping and volcanology studies, the whole LVS sequence can be divided into three formations upwards into the lower Dianzhong, middle Nianbo and upper Pana formations (Fig. 3.2) with continuous between-formation relation contacts and a clear unconformity over the highly deformed Cretaceous meta-sedimentary rocks (Figs. 3.2 and 3.3) (e.g., Liu 1993; Dong 2002; Zhou et al. 2004; Dong et al. 2005, 2021; Zhu et al. 2015).

The LVS sequence and the interbedded sedimentary rocks have been well characterized through detailed studies in the Linzhou basin about 30 km north of the Lhasa city (Mo et al. 2003, 2006; Dong et al. 2005, 2021). In this basin, the LVS sequence has a total thickness of ~5200 m with the three stratigraphic formations well exposed. The lower Dianzhong Formation (LDF), distributed in the southern part of the basin, unconformably overlies the conglomerate of the late Cretaceous Shexing Formation and consists mainly of pyroclastic rocks, andesites and dacites (Fig. 3.3). The middle Nianbo Formation (MNF), distributed in the central part of the basin, comprises mainly of rhyolitic tuff and minor andesites and pyroclastic rocks along with the majority of the sedimentary rocks. The upper Pana formation (UPF) consists mainly of felsic ignimbrites and rhyolites, cropping out in the northern part of the basin with well developed columnar joints in some localities.

The LDF is composed of andesitic lava flow, dacitic and pyroclastic tuffs with a thickness of ~2360 m, which can be subdivided into three sections. Above the unconformity is a >800 m section dominated by rhyolitic pyroclastic tuff, overlain

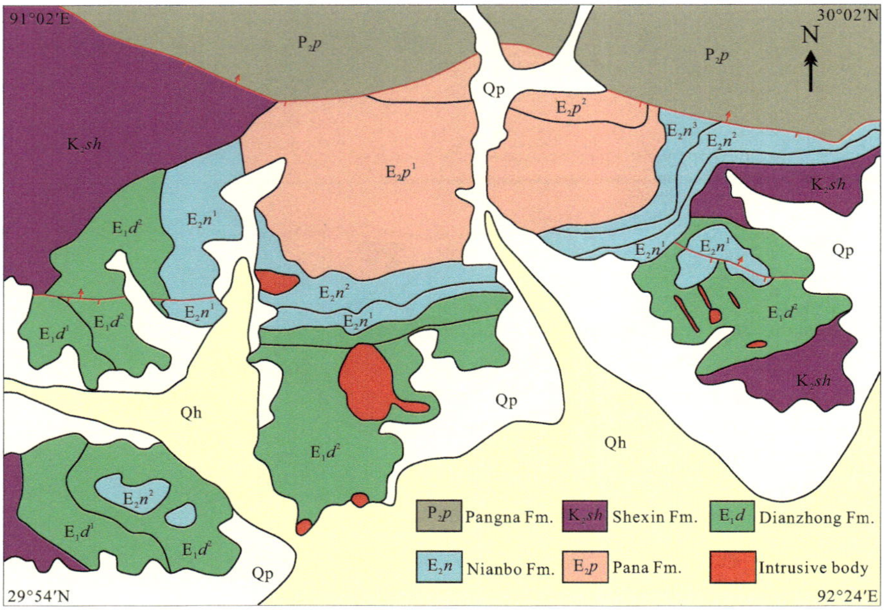

Fig. 3.2 Simplified geological map of the Linzhou Basin showing the distribution of the LVS. (After Dong et al. 2005)

by >1100 m interbedded andesite and andesitic pyroclastic rocks and a ~440 m section of dacite lavas, dacitic tuffs and tuffaceous mudstones (Hu et al. 2007).

The MNF has a total thickness of ~730 m, consists of conglomerate and interbedded fine-grained arkosic sandstone and siltstone at the bottom, meters to tens of meters-thick rhyolitic ignimbrites interbedded with yellow lime–mudstone, siltstone, sandstone and volcanic breccias above, and a ~280 m thick ash flow at the top (Fig. 3.3). Dike swarms are prevalent in this section.

The UPF overlies in disconformity on the MNF above an erosional surface with conglomerate layer at the base. The UPF is gray and reddish in color, has a total thickness of ~2180 m composed of pyroclastic rocks interbedded locally with sandstone, siltstone, mudstone and rhyolite and ash flows with conglomerates at the base in places.

Many studies have focused on the middle to western segments of the Gangdese in the last two decades thanks to the geological mapping program since 2000 (e.g., Wang 2007; Lee et al. 2012; Wang et al. 2015; Liu et al. 2018; Fu et al. 2014). The Coqin volcanic basin in the middle Gangdese has LDF of >1600 m thick (Bao 2014), comprising andesitic and rhyolitic pyroclastic rocks, MNF of ~4270 m thick, comprising rhyolitic pyroclastic rocks, and UPF of >6600 m thick, comprising rhyolitic rocks with diverse lithologies. The Horba volcanic basin (Lee et al. 2012; Liu et al. 2018) has thick MNF and UPF with no LDF observed. The Yare volcanic basin in the western Gangdese mainly develops MNF and UPF with ~588 m thick LDF (Fig. 3.4) (Wang 2007; Wang et al. 2015).

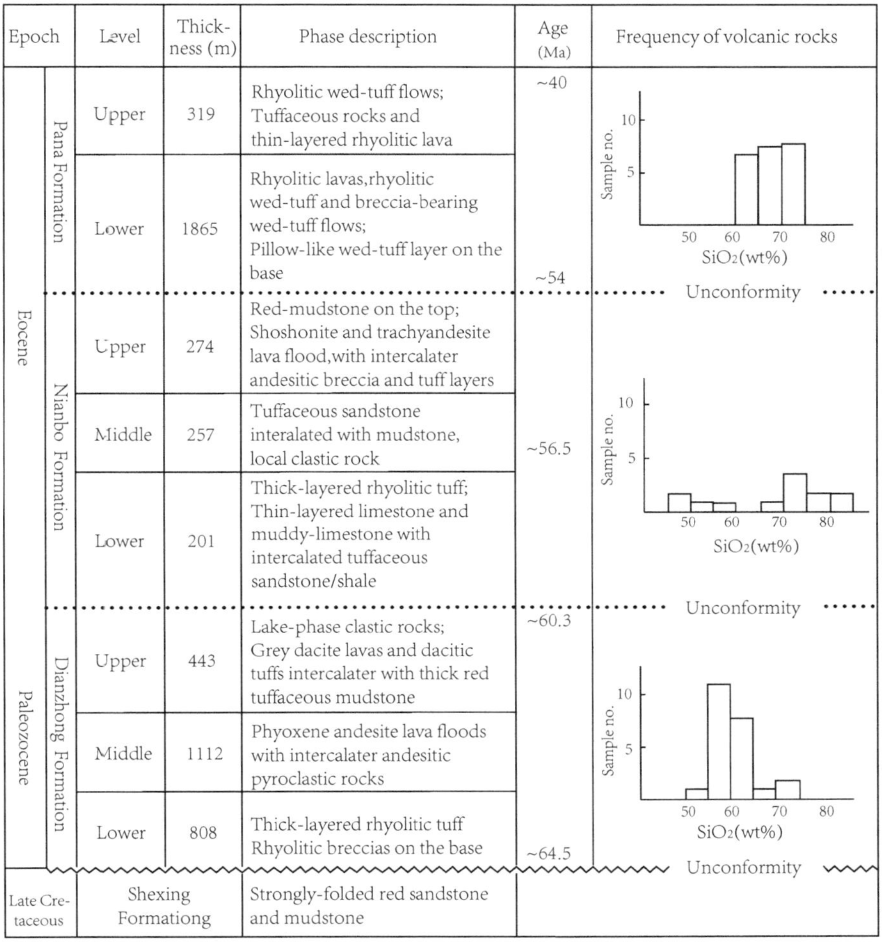

Fig. 3.3 Major lithology units and dominant phases of the Linzizong volcanic succession (LVS) in the Linzhou basin, southern Tibet. (After Mo et al. 2009)

The volcanic rocks in the Shiquanhe area at the western end of the Gandese, covering an area of ~1000 km², consisting locally of purple basalt, grey andesite, and lesser pyroclastic rocks such as andesitic ignimbrite, andesitic breccias and stuffs with zircon U–Pb age of 67.1–70.8 Ma (Cao et al. 2020). The volcanic rocks of andesitic, basaltic and minor rhyolitic compositions well correspond to the LDF; the sedimentary clastic rocks and andesitic-rhyolitic rock association are equivalent to the MNF; and the andesitic rocks with the overlaying rhyolitic rocks are equivalent to the UPF, which is quite similar to the volcanic rocks in the Linzhou basin in the Eastern Gangdese Belt.

The well-defined regional unconformity between the strongly deformed Cretaceous strata and its overlying undeformed LVS volcanic rocks of subaerial

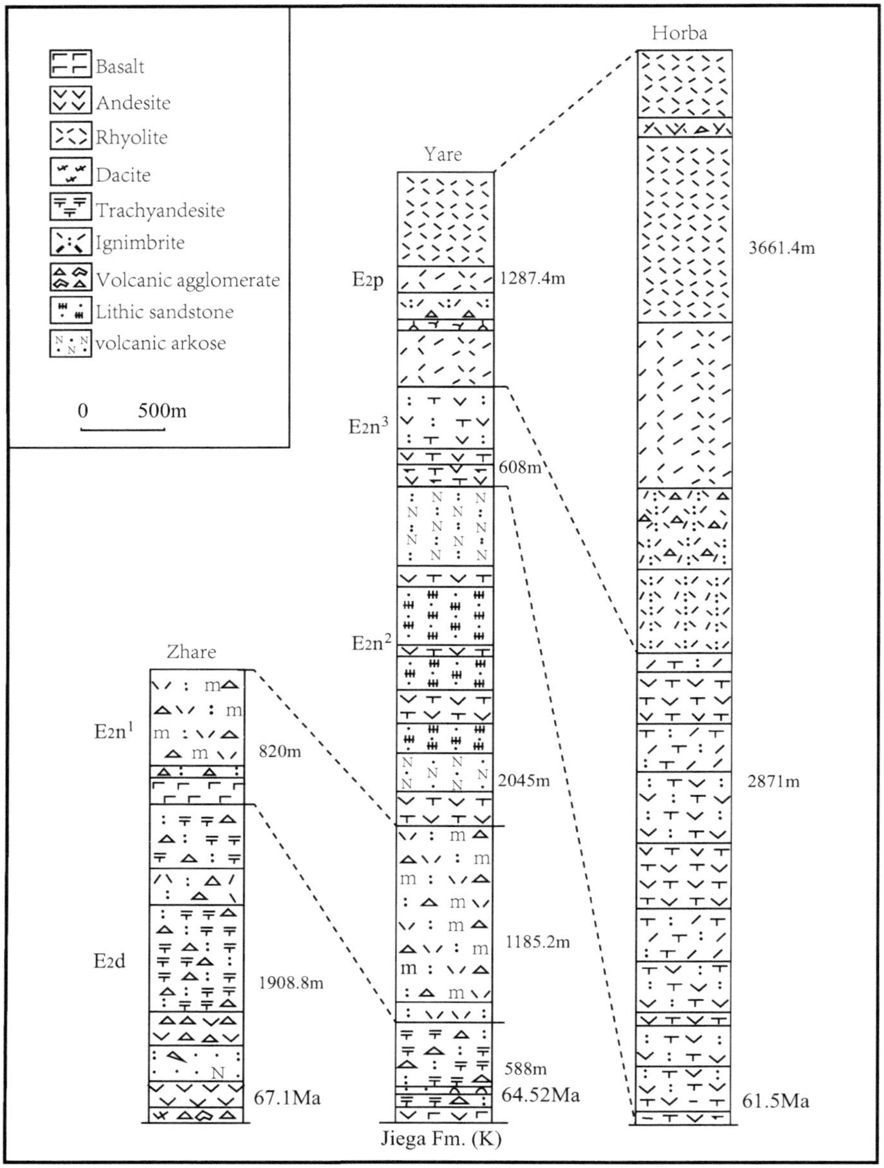

67.1Ma (MSWD=1.8) :Zircon U-Pb dating by Cao et al., 2020
64.52Ma (MSWD=1.32): Zr U-Pb dating by Wang, 2011
61.5Ma (MSWD=1.4): Rb-Sr dating by Liu Anlin, 2020

Fig. 3.4 Stratigraphic column of the LVS in middle to western segments of the Gangdese Belt eruption are prominent phenomenon in the Gangdese Belt (Allègre et al. 1984; Coulon et al. 1986; Willems et al. 1996; Dong 2002; Mo et al. 2003, 2006, 2007). This unconformity, together with the LVS throughout the entire Gangdese Belt, suggests a major tectonic event, which is best understood as representing the onset of the India–Asia collision at ≥65 Ma (Zhou et al. 2004; Dong 2002; Mo et al. 2002, 2003, 2006, 2007).

Therefore, the LVS volcanic rocks of andesitic to rhyolitic compositions offers a prime opportunity to test the hypothesis highlighted in the previous section. The LVS rocks are syncollisional, emplaced during the India–Asia collision from a "soft" touch at ~65 Ma to a "hard" clash at ~45–40 Ma (e.g., Yin and Harrison 2000; Flower et al. 2001; Mo et al. 2002, 2003, 2006, 2007), thus representing confirmation of the continental collision process with time in response to the collision.

3.1.2 Ages of the Linzizong Volcanic Successions

Many age data on the LVS volcanic rocks have been reported since 1990's, especially in the last two decades. So far, a detailed chronology study on the LVS is limited to the Linzhou Basin near Lhasa (Dong 2002; Mo et al. 2003; Zhou et al. 2004; Zhu et al. 2015), where the LVS rocks have been subdivided into the LDF, MNF and UPF three formations based on eruption hiatuses (Dong 2002).

Previously reported ages for the LVS rocks vary in a narrow range of 49.2–56.2 Ma (Jin and Zhou 1978; Maluski et al. 1982; Xu et al. 1985; Coulon et al. 1986), although the stratigraphic positions of the dated samples are unknown. In the Linzhou basin of eastern Gangdese, some representative samples of andesite and rhyolite place the LVS eruptions in the Paleocene to Early Eocene with Ar–Ar age of 63.5–43.9 Ma (Dong 2002; Zhou et al. 2004) and Zircon U–Pb isotopic age of 60.1–52.6 Ma (Zhu et al. 2015; Liu et al. 2018), suggesting a continuous volcanic activity of ~20 Myrs.

On the basis of detailed mapping and sampling of the three formations, we obtained systematic bulk-rock $^{40}Ar–^{39}Ar$ age dates on samples from the Linzhou Basin to be ~64.4 Ma for the lower section of the LDF, ~60.6 Ma for the upper section of the LDF, ~54 Ma for basal samples of the MNF, ~48.7 Ma for basal samples and 43.9 Ma for upper samples of the UPF (Dong 2002; Zhou et al. 2004). Of all these ages, 64.4–60.6 Ma for the LDF, ~54 Ma for the MNF and ~48.73–43.9 Ma for the UPF (Fig. 3.5), indicating that the LVS volcanic eruption took place from Paleocene to middle of Eocene (65–43 Ma).

The $^{40}Ar–^{39}Ar$ data of 64.4–60.6 Ma for the LDF, ~54 Ma for the MNF, 48.7–43.9 Ma for the UPF have been confirmed by recent zircon U–Pb dating (Fig. 3.5) (Lee et al. 2009; Liu et al. 2019; Liu et al. 2020). Lee et al. (2009) studied the LVS rocks in the Linzhou basin and yielded ages from 59.9 to 43.2 Ma from a large area (Lee et al. 2009). Chen et al. (2014) collected samples from the base of the LDF, base and top of the UPF in the Linzhou basin, and provided the $^{40}Ar–^{39}Ar$ ages of 64.1, 51.6 and 46.4 Ma, respectively. Recent SIMS zircon U–Pb age data

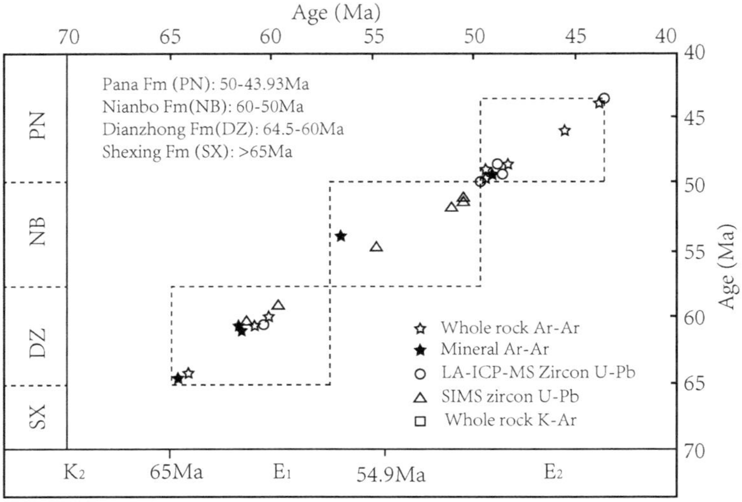

Fig. 3.5 The Ar–Ar dating on LVS volcanic rocks in the Linzhou basin. The age data are from the literature (Dong 2002; Zhou et al. 2004; Lee et al. 2009; Zhu et al. 2015; Liu et al. 2019; Liu et al. 2020)

indicate that the LVS volcanism in the Linzhou basin was active in the timeframe of 60.1–52.3 Ma, i.e., 60.1–58.3 Ma for the LDF, 55.4–52.6 Ma for the MNF, and 52.6–52.3 Ma for the UPF, respectively (Zhu et al. 2015). Some volcanic rocks of the UPF from the other basins give slightly younger zircon U–Pb ages than in the Linzhou volcanic basin, eastern Gangdese. Liu et al. (2018) obtained zircon U–Pb age of 49–50 for the UPF rhyolitic rocks east of the Linzhou basin (Liu et al. 2018; Liu et al. 2020). The SIMS and LA-ICPMS zircon U–Pb age data on dacite samples from Pangduo indicate that the UPF rocks erupted between 50.3 and 48.8 Ma (Liang et al. 2010).

The zircon U–Pb age data indicate that the LDF formed at 63.4 Ma near Geer (Cao et al. 2020) and at 62.2, 64.5 Ma in eastern Geer (Wang 2007), the UPF formed at 64.58 ± 0.66, 63.88 ± 0.59 Ma in Qure of Coqin (Ding et al. 2017) and the UPF were formed at 48.9 ± 0.8 Ma in Zhunuo (Liang et al. 2010) and at 49.8 ± 0.9 Ma in Sangsang (Xie et al. 2011) in the central segment of the southern Gangdese (Fig. 3.6). Wang et al. (2015) reported SIMS Zircon U–Pb age of 50.99 Ma for andesite near Shiquang river and 55.5 Ma for rhyolite in Gurong (Wang et al. 2015). Some basal andesites of the LVS from Cuoqin in middle Gangdese and Ngari in western Gangdese gave bulk-rock K–Ar ages of 58.6 Ma and 60.68 Ma (Guo et al. 1991), respectively. A recent 1:250,000 scale geologic mapping in the Gangdese Belt (Pan et al. 2004) has characterized the LVS rocks in the entire Gangdese Belt and agreed in general with the stratigraphic column in the Linzhou Basin. It would suggest a somewhat diachronous onset of the LVS volcanism along the Gangdese Belt, while recognizing factors such as spatial variation of individual lava flows and sampling uncertainties in topographically complex terrenes.

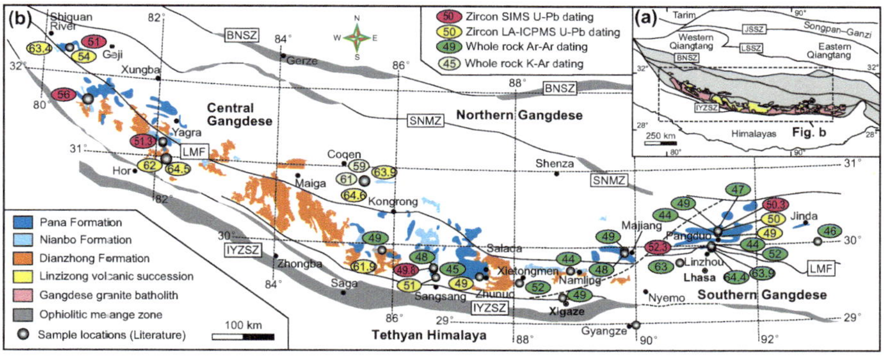

Fig. 3.6 Simplified tectonic units of the Lhasa Terrane (Zhu et al. 2011) showing spatial and temporal distributions of the LVS volcanic rocks. The data are from the literature (Liu et al. 2018; Wang et al. 2015)

Hence, the LVS volcanism occurred during the time interval of 64–44 Ma, e.g., 64–60 Ma for the LDF, 60–53 Ma for the MNF, and 53–44 Ma for the UPF, respectively.

3.1.3 Geochemical Characteristics of the Volcanic Rocks

Major and trace element data on selected samples (Table 3.1) (Dong 2002; Mo et al. 2003) confirmed the previous notion (Pearce and Houjun 1988) that the LVS volcanic rocks are mostly calc-alkaline although some are more alkalic, and can be further classified into several rock types. These rocks have varying SiO_2 content (SiO_2 = 47–79 wt.%), and consist mainly of trachybasalt, basaltic andesite, andesite, dacite and rhyolite in terms of total alkalis (K_2O + Na_2O) vs. SiO_2 (Fig. 3.7). The volcanic rocks of the LDF are compositionally intermediate. The volcanic rocks of the MNF and UPF show bimodal characteristics (Zhu et al. 2015). Most volcanic rocks of the LDF and MNF belong to sub-alkaline series, while the UPF volcanic rocks are of alkaline affinity. Most of the LDF and MNF volcanic rocks are largely median-K to high-K calc-alkaline (Fig. 3.8), but the UPF volcanic rocks are shoshonitic with varying high K_2O content (Fig. 3.8). There is a general increasing K_2O/Na_2O trend from the early LDF to the MNF and to the UPF (Fig. 3.9). Most LDF volcanic rocks are meta-aluminous, while MNF and UPF volcanic rocks range from metaluminous to peraluminous (Fig. 3.10).

The LVS rocks show enriched REE patterns, but the extent of enrichment is variable between rock suites (Fig. 3.11). All rocks from the LDF have a similar REE concentration level and relatively flat REE pattern, e.g., $\sum$REE = 103.64–155.13, $(La/Yb)_N$ = 5.2–7.8, with a slightly negative Eu anomaly (Eu/Eu* = 0.68–0.89), indicating limited or insignificant plagioclase crystallization during magma differentiation. As samples of the MNF show similar REE concentration levels and

Table 3.1 Major and trace element data on selected samples of LVS volcanic rocks

Sample	**Lower Dianzhong Formation (LDF)**								
	D-2	**D-15**	**BD-145**	**BD-151**	**BD-160**	**BD-123**	**BD-126**	**BD137-2**	**LZ9913**
Major element (wt.%)									
SiO_2	59.16	58.05	61.57	57.46	58.77	56.97	69.95	55.83	55.21
TiO_2	0.70	0.68	0.62	0.64	0.71	0.77	0.34	0.81	0.86
Al_2O_3	16.02	15.92	15.88	18.04	17.11	16.07	14.01	18.64	18.06
TFe_2O_3	6.21	6.75	5.46	5.28	7.42	6.73	3.13	6.63	8.3
MnO	0.15	0.15	0.11	0.11	0.1	0.1	0.08	0.1	0.14
MgO	3.66	2.53	1.77	1.91	1.77	2.3	1.04	2.1	4.01
CaO	3.67	6.49	5.62	7.92	3.13	5.35	1.77	7.07	7.29
Na_2O	4.11	2.73	3.03	2.61	5.63	4.76	4.48	3.95	2.98
K_2O	2.39	1.88	2.35	1.2	2.49	1.05	3.12	1.15	1.47
P_2O_5	0.19	0.17	0.17	0.18	0.22	0.2	0.07	0.25	0.22
LOI	3.62	4.22	3.48	4.56	3.07	5.72	1.93	3.32	1.3
Total	99.88	99.57	100.06	99.91	100.42	100.02	99.92	99.86	99.84
Trace element (ppm)									
Sc	15.3	15.2	13.3	12.6	11.8	18	8.43	18	18.1
Rb	41.4	51.5	63.1	21.8	71.3	34.5	92.7	23.2	36.8
Sr	450	386	351	436	566	377	249	656	477
Y	23.3	23.3	22.1	23.3	22.7	21	27.8	29	22.7
Zr	136	118	129	156	134	183	181	182	149
Nb	7.04	5.39	7.29	6.14	5.88	6.7	10.2	7.26	4.92
Cs	0.71	1.53	1.61	2.36	5.55	1.84	1.24	3.13	0.66
Ba	965	354	428	390	392	267	557	595	313
La	21.7	19.8	22.3	20	19.9	21	32.5	23.7	18.2
Ce	44.9	39.7	45.3	41.4	38.7	41	62.8	50.8	45.4

Pr	5.47	4.73	5.3	5.17	4.93	5.06	7.24	6.45	5.23
Nd	22	19.1	21	21.8	21	20.6	27.4	27.3	22.5
Sm	4.58	4.25	4.44	4.78	4.61	4.43	5.48	6.06	4.39
Eu	1.23	1.1	1.12	1.24	1.22	1.22	1.22	1.64	1.28
Gd	4.57	4.3	4.58	4.82	4.65	4.33	5.44	5.92	4.36
Tb	0.66	0.66	0.68	0.71	0.7	0.65	0.8	0.89	0.72
Dy	3.91	3.94	3.94	4.17	4.02	3.73	4.8	5.04	4.14
Ho	0.8	0.82	0.8	0.85	0.82	0.76	0.98	1.01	0.82
Er	2.15	2.17	2.15	2.24	2.11	1.98	2.63	2.63	2.37
Tm	0.32	0.33	0.32	0.33	0.31	0.29	0.4	0.39	0.33
Yb	2.39	2.39	2.34	2.41	2.31	2.11	2.99	2.8	2.5
Lu	0.36	0.35	0.35	0.36	0.34	0.32	0.45	0.41	0.37
Hf	3.5	3.14	3.54	4.17	3.57	4.27	4.96	4.65	3.92
Ta	0.47	0.4	0.56	0.44	0.42	0.46	0.76	0.41	0.51
Th	9.17	9.32	11.1	9.1	7.65	8.4	15.8	7.22	4.92
U	1.88	1.85	2.32	1.99	1.42	1.34	2.63	1.31	1.26
REE	115.04	103.64	114.62	110.28	105.62	107.48	155.13	135.04	112.61
$(La/Yb)_N$	6.51	5.94	6.84	5.95	6.18	7.14	7.80	6.07	5.22
Eu*	0.82	0.79	0.76	0.79	0.81	0.85	0.68	0.84	0.89
Sample	**Middle Nianbo Formation (MNF)**								
	LZ993	**LZ991**	**N-9**	**BD-65**	**LZ9921**	**BD-55**	**L060**	**LZ994**	
Major element (wt.%)									
SiO_2	68.13	73.9	51.57	49.69	74.69	74.01	77.64	71.56	
TiO_2	0.47	0.13	0.81	0.92	0.13	0.12	0.12	0.46	
Al_2O_3	15.31	14.28	17.04	17.08	12.03	11.78	11.42	12.4	
TFe_2O_3	3.25	1.68	8.53	9.46	1.58	1.47	1.498	3.53	
MnO	0.04	0.02	0.1	0.17	0.03	0.09	0.04	0.12	

(continued)

Table 3.1 (continued)

MgO	1.62	1.05	3.46	3.27	0.45	0.43	0.21	0.53	
CaO	1.76	0.95	7.89	8.67	2.28	2.35	1.6	2.34	
Na_2O	0.95	0.55	3.9	2.86	2.31	2.41	3.46	3.81	
K_2O	4.18	4	1.19	2.6	2.52	3.9	1.72	2.41	
P_2O_5	0.1	0.03	0.43	0.44	0.03	0.03	0.03	0.09	
LOI	4.5	3.54	5.01	5.03	3.68	3.09	2.12	3.3	
Total	100.31	100.13	99.93	100.19	99.73	99.68	99.86	100.55	
Trace element (ppm)									
Sc	9.29	7.41	12.1	16.3	6.74	6.35		5.37	
Rb	144	132	24.6	53.2	81.4	105	81	59.9	
Sr	87.2	86.1	1327	1021	76.5	132	178	183	
Y	19.3	21.5	24.8	39.3	24.3	29.5	35.5	17.8	
Zr	160	113	162	149	95.6	114	112	117	
Nb	10.35	9.52	10.2	10.8	7.56	8.56	11	7.48	
Cs	19.46	7.65	1.38	1.97	3.33	1.85		4.9	
Ba	115	503	558	1023	363	581	279	240	
La	32.8	36.9	47.1	48.2	33.7	31.9	40.9	28.6	
Ce	64.8	74.8	86.9	87.9	64.7	63	88.8	60.5	
Pr	7.84	9.21	10.6	11.1	8.12	7.24	10.9	7.31	
Nd	28.2	33.3	42.6	45.9	29.6	27.7	34.7	27.4	
Sm	5.25	6.43	8.23	9.29	5.67	5.71	7.58	5.11	
Eu	0.73	0.6	2.07	2.33	0.58	0.72	1.09	0.97	
Gd	4.15	5.1	7.85	9.02	4.58	5.96	6.95	4.15	
Tb	0.61	0.75	0.95	1.11	0.68	0.9	1.12	0.62	
Dy	4.03	4.86	4.85	5.69	4.31	5.23	5.88	3.97	
Ho	0.83	0.97	0.86	1.01	0.89	1.08	1.27	0.79	

Er	2.22	2.64	2.22	2.54	2.43	2.89	3.24	2.17	
Tm	0.3	0.37	0.31	0.36	0.34	0.43	0.52	0.31	
Yb	2.17	2.59	2.23	2.49	2.33	3.23	3.08	2.23	
Lu	0.31	0.39	0.32	0.36	0.34	0.48	0.49	0.33	
Hf	5.95	4.97	3.88	3.69	3.66	3.99	3.22	4.19	
Ta	1.13	1.07	0.46	0.46	0.9	0.71	0.71	0.75	
Th	20.2	22.5	15.4	13.9	16.5	19.1	13	14	
U	3.86	3.43	2.36	2.08	3.66	3.24	2.85	2.7	
REE	154.24	178.91	217.09	227.3	158.27	156.47	206.52	144.46	
$(La/Yb)_N$	10.8	10.2	15.2	13.9	10.4	7.1	9.5	9.2	
Eu*	0.48	0.32	0.79	0.78	0.35	0.38	0.46	0.64	
Sample	**Upper Pana Formation (UPF)**								
	LZ9917	**LZ9916**	**BD-103**	**BD-77**	**P-1**	**BD-106**	**BD-114**	**12PD01-1**	**12PD05-1**
Major element (wt.%)									
SiO_2	75.58	79.64	67.93	68.98	74.06	66.8	75.09	47.35	51.61
TiO_2	0.3	0.31	0.36	0.25	0.17	0.36	0.24	0.91	0.92
Al_2O_3	11.21	9.75	13.56	18.09	13.33	14.14	12.55	17.12	16.49
TFe_2O_3	2.6	2.39	2.99	0.72	1.33	3.19	1.47	10.00	9.00
MnO	0.03	0.02	0.1	0.01	0.08	0.11	0.08	0.18	0.24
MgO	0.48	0.23	0.84	0.27	0.28	0.7	0.34	5.45	4.21
CaO	0.59	0.39	2.78	0.28	1.06	3.09	0.46	9.36	5.48
Na_2O	1.71	2.3	2.83	4.15	3.33	2.45	3.23	3.03	3.40
K_2O	5.55	3.68	4.6	5.26	5.43	4.62	5.23	2.76	2.44
P_2O_5	0.14	0.16	0.14	0.06	0.04	0.15	0.05	0.58	0.38
LOI	1.77	1.12	3.39	1.68	0.8	3.93	1.42	3.02	7.04
Total	99.96	99.99	99.52	99.75	99.91	99.54	100.16	99.75	101.21
Trace element (ppm)									

(continued)

Table 3.1 (continued)

Sc	4.2	2.48	5.66	3.28	2.5	6.63	5.15	15.9	7.30
Rb	163	88.7	123	123	198	151	167	184	155
Sr	164	206	307	335	235	203	102	21.0	14.0
Y	17.7	13.7	21.4	28.9	17.8	21.6	26.7	28.2	22.8
Zr	113	101	150	234	142	176	213	43.2	25.4
Nb	8.67	7.88	12	19.4	11.9	13	22.4	38.8	0.0
Cs	13	8.16	4.34	2.61	2.31	6.09	2.15	80.3	219
Ba	813	1304	745	919	691	770	310	19.5	18.8
La	38.3	28.2	40.3	62.9	36.6	39.8	31.8	92.3	104
Ce	60.7	44	72.8	105	65.3	72.2	105	1305	526
Pr	8.24	6.21	8.04	12.4	6.72	7.94	6.67	36.5	27.0
Nd	29.1	23	28.5	42.5	22.9	28.6	23.2	169	137
Sm	5.32	4.27	5.51	6.74	4.1	5.41	4.71	13.4	7.60
Eu	1.1	0.96	1.21	1.48	0.75	1.15	0.59	0.99	8.69
Gd	4.02	3.24	5.61	6.92	4.22	5.35	5.3	868	424
Tb	0.57	0.46	0.73	0.88	0.55	0.72	0.78	51.7	39.0
Dy	3.47	2.51	4	4.82	3.11	4	4.87	102	76.9
Ho	0.6	0.49	0.76	0.95	0.63	0.79	1.05	12.7	9.31
Er	1.71	1.29	2.08	2.63	1.78	2.16	2.95	50.4	35.7
Tm	0.23	0.2	0.31	0.4	0.28	0.32	0.47	10.6	7.38
Yb	1.62	1.28	2.25	2.92	2.17	2.37	3.38	2.77	1.99
Lu	0.24	0.19	0.35	0.46	0.34	0.37	0.5	9.24	6.46
Hf	3.08	2.55	3.93	5.77	4.01	4.51	6.08	1.29	0.92
Ta	0.74	0.54	0.89	1.05	1.08	0.95	1.77	6.97	5.04
Th	21	13.2	27.7	24	35.3	30	51.3	1.29	0.95
U	4.23	5.48	6.23	6.13	6.76	5.48	8.52	3.10	2.42

REE	155.22	116.3	172.45	251	149.45	171.18	191.27	0.42	0.34
$(La/Yb)_N$	17.0	15.8	12.8	15.5	12.1	12.0	6.7	2.58	2.17
Eu*	0.73	0.79	0.67	0.66	0.55	0.65	0.36	0.37	0.31
	12PD05-2	**12PD05-3**	**12PD05-4**	**12PD02-1**	**12PD03-1**	**12PD04-1**	**12PD06-1**	**12PD07-1**	**12PD08-1**
Major element (wt.%)									
SiO_2	52.85	51.92	53.11	66.15	66.51	66.97	69.29	63.66	65.04
TiO_2	0.94	0.92	0.92	0.59	0.58	0.60	0.51	0.60	0.59
Al_2O_3	17.36	16.93	17.23	16.92	16.69	17.08	15.69	18.37	17.23
TFe_2O_3	8.23	8.07	8.10	2.82	2.43	2.54	1.95	2.94	2.69
MnO	0.16	0.15	0.15	0.08	0.06	0.07	0.05	0.09	0.05
MgO	4.34	4.15	4.10	0.84	0.60	0.66	0.37	0.60	0.67
CaO	6.43	6.74	6.74	2.05	1.83	2.27	1.44	2.65	2.51
Na_2O	2.53	2.61	2.24	3.56	3.62	3.66	3.18	3.91	3.46
K_2O	4.68	4.18	4.32	6.08	6.20	6.24	6.53	5.89	6.03
P_2O_5	0.40	0.39	0.39	0.13	0.12	0.13	0.09	0.19	0.14
LOI	3.26	3.14	2.70	1.98	1.59	0.71	0.40	1.08	1.26
Total	101.18	99.17	100.00	101.20	100.24	100.93	99.49	99.96	99.67
Trace element (ppm)									
Sc	15.8	14.5	15.6	4.30	5.50	5.60	10.7	6.30	5.80
Rb	151	157	133	338	359	352	410	340	327
Sr	1008	940	963	482	496	475	249	587	547
Y	27.3	26.8	27.2	29.6	35.4	29.9	32.0	28.9	28.1
Zr	142	140	143	404	403	408	447	420	371
Nb	7.80	7.80	8.00	19.1	19.8	19.6	23.1	16.5	18.6
Cs	7.47	10.17	3.54	6.44	8.71	9.88	13.3	10.5	9.00
Ba	770	672	693	768	731	858	414	818	937
La	38.3	37.7	38.9	60.7	80.9	60.2	66.9	59.3	59.5

(continued)

Table 3.1 (continued)

Ce	77.1	75.0	78.4	114	122	120	134	115	116
Pr	9.19	9.22	9.26	13.1	16.6	13.3	14.6	13.0	12.8
Nd	36.3	36.0	36.9	45.2	56.5	45.5	48.8	45.4	43.9
Sm	7.54	7.33	7.60	7.96	9.67	8.14	8.66	8.19	7.77
Eu	2.02	1.94	1.99	1.57	1.83	1.62	1.18	1.97	1.74
Gd	6.28	6.33	6.37	6.18	7.43	6.30	6.54	6.32	6.00
Tb	0.92	0.88	0.92	0.89	1.05	0.90	0.95	0.89	0.86
Dy	4.87	4.81	4.95	5.00	5.70	5.11	5.37	4.92	4.77
Ho	0.94	0.95	0.94	0.99	1.14	1.01	1.06	0.97	0.95
Er	2.48	2.44	2.49	2.68	3.00	2.76	2.93	2.60	2.56
Tm	0.35	0.35	0.36	0.40	0.45	0.42	0.46	0.39	0.39
Yb	2.17	2.24	2.26	2.69	2.85	2.72	2.97	2.50	2.55
Lu	0.33	0.33	0.33	0.40	0.43	0.41	0.44	0.38	0.38
Hf	3.51	3.56	3.54	8.81	9.06	9.06	10.1	8.93	8.44
Ta	0.53	0.53	0.53	1.50	1.54	1.52	1.83	1.21	1.42
Th	9.76	10.1	9.84	41.4	40.8	40.2	49.8	33.0	37.1
U	2.42	2.07	2.04	7.02	5.42	5.19	7.51	4.98	5.34
REE	188.802	185.5324	191.6364	262.1644	309.8546	268.1394	294.8089	261.9324	260.42
$(La/Yb)_N$	12.7	12.1	12.3	16.2	20.4	15.9	16.2	17.0	16.7
Eu*	0.90	0.87	0.88	0.69	0.66	0.69	0.48	0.84	0.78

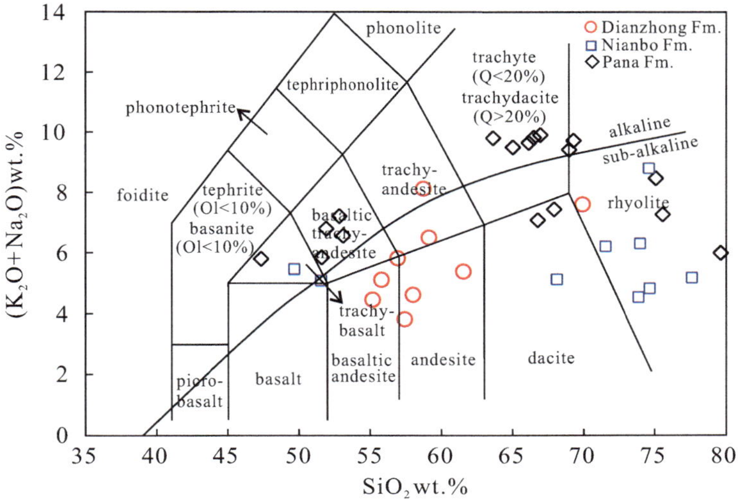

Fig. 3.7 Total alkalis vs. silica (TAS) diagram for volcanic rock classification

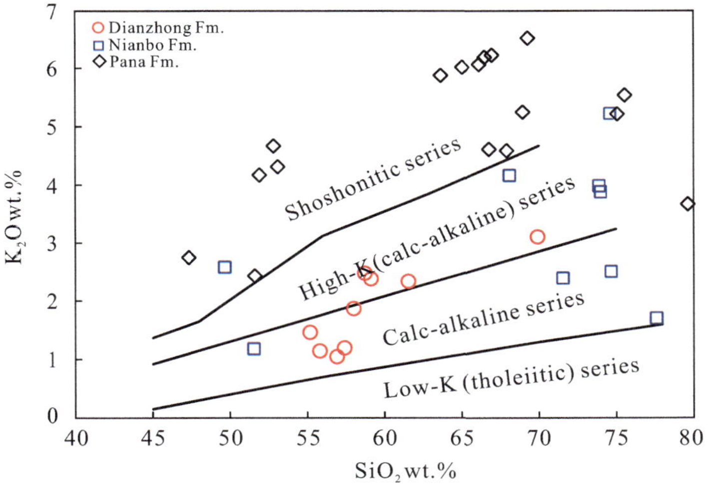

Fig. 3.8 SiO_2 vs. K_2O diagram for rock series classification

patterns, e.g., ∑REE = 144.46–217.09, $(La/Yb)_N$ = 7.1–15.2 with high SiO_2 samples having a stronger negative Eu anomaly (Eu/Eu* = 0.32–0.64) indicating obvious plagioclase crystallization during magma differentiation or plagioclase as residual phase in the melting source region. The shoshonitic suites of the UPF show the highest LREE concentrations and relatively steep REE patterns, e.g., ∑REE = 116.30–309.85, $(La/Yb)_N$ = 6.7–20.4. The negative Eu anomaly decreases from silicic to mafic compositions.

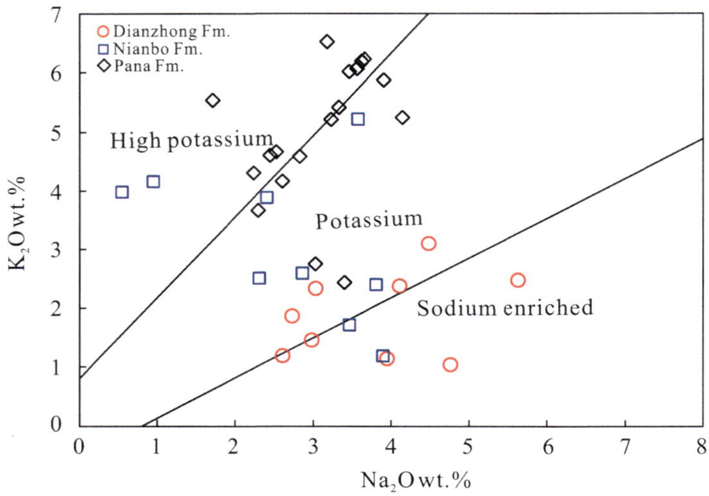

Fig. 3.9 Na_2O vs. K_2O diagram for rock type classification

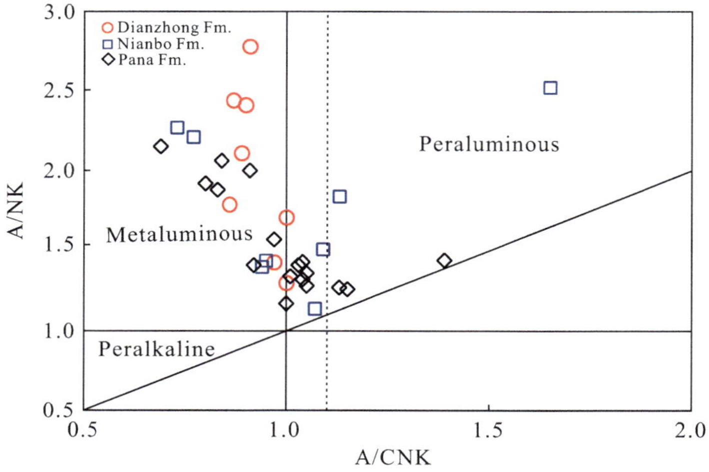

Fig. 3.10 A/NK vs. A/CNK diagram for the LVS volcanic rocks

On primitive mantle-normalized incompatible element diagrams, all the LVS volcanic rocks show significant enrichments in large ion lithophile elements (LILEs; e.g., Rb and Ba) and LREEs, but depletion of high field strength element (HFSE; e.g., Nb and Ta) (Fig. 3.11). These characteristics are similar to those of arc lavas from modern subduction zones, which is apparently consistent with magmas being of subduction origin. Another explanation is that the Nb and Ta negative anomalies, especially, of the felsic rocks, may result from source rock melting with ilmenite or

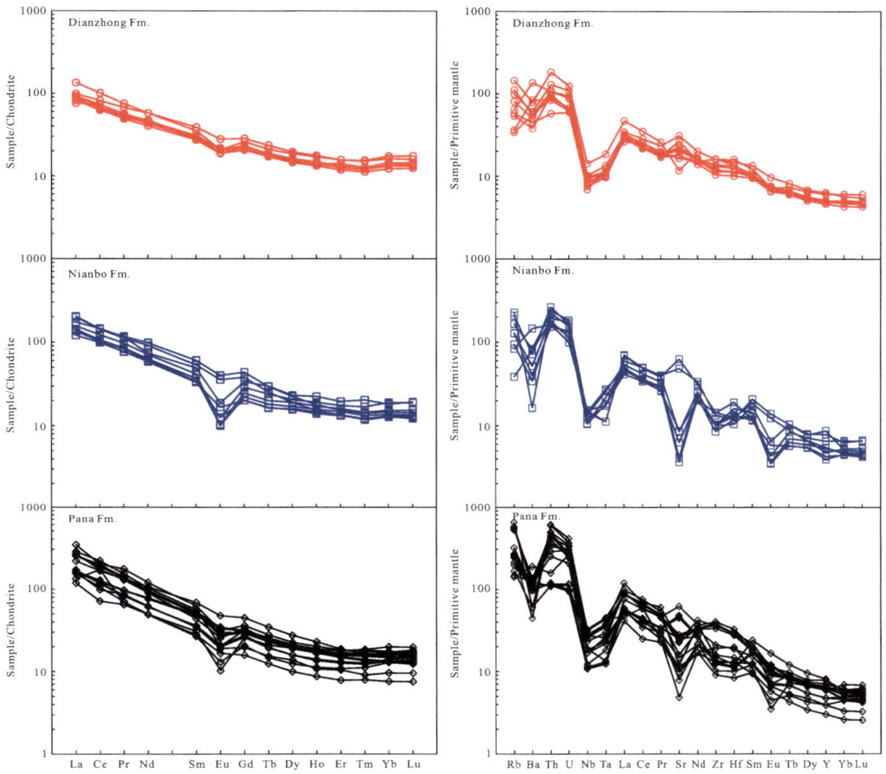

Fig. 3.11 Chondrite-normalized REE and primitive-mantle-normalized trace element patterns for the LVS volcanic rocks

rutile a residual phase, resembling the petrogenesis of the Quaternary high potassium volcanic rocks in northern Tibet. The mafic LVS volcanic rocks have relatively high Sr content with a positive Sr anomaly, while the felsic LVS rocks show varying extent of Sr depletion, pointing to the significance of plagioclase as a residual phase holding back Sr.

Most LDF samples display relatively depleted Nd isotope ($\varepsilon_{Nd}(t) = -0.42$ to 0.45) and intermediate Sr isotope [($^{86}Sr/^{87}Sr$)i = 0.7050–0.7062] compositions with one sample showing highly depleted Nd and Sr isotopic characteristics, e.g., $\varepsilon_{Nd}(t) = 9.45$ and ($^{86}Sr/^{87}Sr$)i = 0.7046. The MNF samples have slightly higher $\varepsilon_{Nd}(t)$ and ($^{86}Sr/^{87}Sr$)i values than those of the LDF samples (Fig. 3.12). In comparison, the UPF rocks have highly variable Nd and Sr isotope compositions, which can be divided into two groups; one with depleted Sr and Nd isotope compositions[$\varepsilon_{Nd}(t) = -2.91$ to 5.64 and ($^{86}Sr/^{87}Sr$)i = 0.7049–0.7055], and the other showing high ($^{86}Sr/^{87}Sr$)i (0.7061–0.7069) and low $\varepsilon_{Nd}(t)$ (−1.73 to −5.87) values.

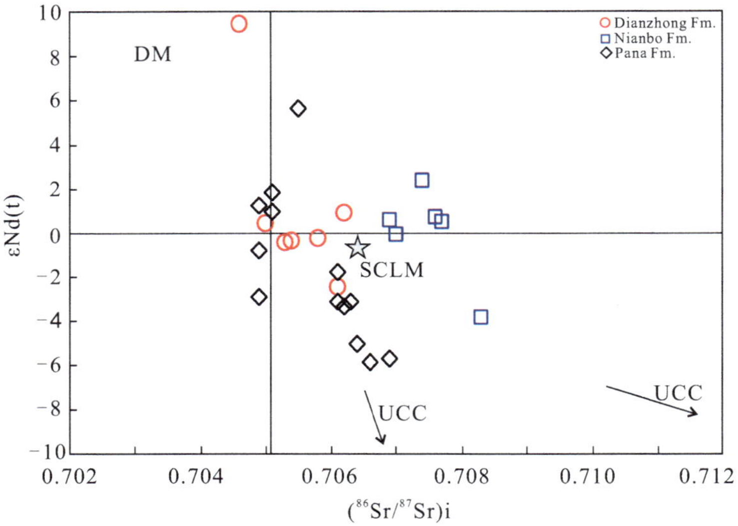

Fig. 3.12 $\varepsilon_{Nd}(t)$ vs. $(^{87}Sr/^{86}Sr)i$ for the LVS volcanic rocks

3.1.4 The Petrogenesis of the Linzizong Volcanic Succession and Associated Geodynamics

The LVS volcanic rocks of varying chemical compositions are indicative of their complex genesis, which might involve at least depleted mantle, metasomatized lithospheric mantle, lower continental crust and upper continental crust materials. Most of the LDF volcanic rocks have intermediate SiO_2 and MgO contents, and show a slightly negative Eu anomaly, and depleted Sr and Nd isotope compositions, indicating that they were likely derived from the hybrid of depleted asthenospheric mantle and metasomatized lithospheric mantle due to dehydration of the subducted seafloor slab, followed by a fractional crystallization. The MNF volcanic rocks are mostly exhibiting high SiO_2 content and low MgO content, with a negative Eu anomaly, suggesting that they were derived from a crustal source with plagioclase present as a residual phase. However, the relatively high $\varepsilon_{Nd}(t)$ and $(^{86}Sr/^{87}Sr)i$ values of these volcanic rocks indicate a juvenile crustal source that is most probably underplated basalts derived from an asthenospheric mantle source contaminated with subducting/subducted sediments. The UPF volcanic rocks have varying compositions with elevated K_2O.

The rocks of mafic compositions with relatively depleted Sr and Nd isotopes are mostly likely derived from partial melting of an enriched lithospheric mantle with asthenosphere input and lower crustal assimilation. The rocks of felsic compositions with relatively enriched Sr and Nd isotopes are likely derived from a hybrid crust source with both juvenile and ancient lower crust materials. Therefore, the petrogenesis of the LVS volcanic rocks from the LDF to the MNF and to the UPF resulted from melting of complex sources of mantle, juvenile crust and ancient mature crustal (terrestrial sediments) materials.

The andesitic rocks of the LDF show remarkable similarity (Fig. 3.11) in both abundances and systematics of incompatible elements to model bulk continental crust established using a comprehensive global approach (Rudnick and Gao 2003). Comparing all the major, minor and trace elements between the andesitic rock and continental crust (Rudnick and Gao 2003), the similarity between the two is striking with the exception of higher P and lower MgO, Ni and Cr in the andesitic units. Some key incompatible trace element ratios are even almost similar each other. The similar rare earth element (REE) abundances and La/Sm, Sm/Nd and Sm/Yb ratios (Fig. 3.13) point to the essentially identical REE patterns of the two without the so-called "garnet signature". Both have similar negative Sr and Eu anomalies, which is consistent with amphibolite melting with plagioclase as a residual phase that stabilizes Sr and Eu relative to Pr, Nd, Sm, Gd and other otherwise similarly incompatible elements.

The depletion in highly compatible elements like Mg, Ni and Cr in the andesitic lithologies of Dianzhong formation is a typical feature for andesites. High magnesian andesites have been proposed for making continental crust because they have elevated Mg, Ni and Cr relative to the more common andesites (Kelemen 1995; Tatsumi 2006). Therefore we suggest that the low Mg, Ni and Cr in syncollisional andesites are complemented by mafic and ultramafic lithologies in the continental crust (e.g., basaltic rocks and ophiolites).

Partial melting of the ocean crust under amphibolite conditions of ≤~40 km not only explains the lack of "garnet signature", but also the characteristic "arc-like

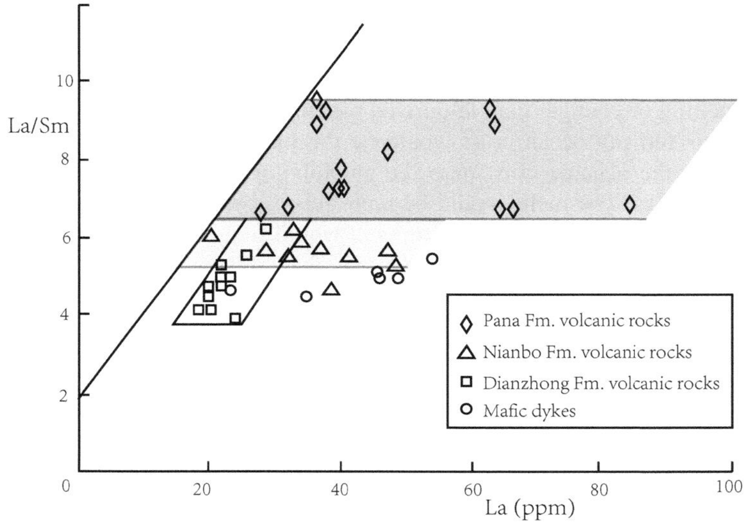

Fig. 3.13 La/Sm–La diagram illustrating partial melting trend and fractionational crystallization trend for the LVS volcanic rocks in the LVS in southern Tibet. (Modified after Mo et al. 2007)

signature" in the melt parental to the LVS volcanic rocks. The residual ocean crust after melt extraction will continue to enter the deep mantle along with the subjacent oceanic lithosphere.

3.1.5 Input of the Mantle Materials During the Indian–Asian Collision

The LVS in Gangdese is probably ideal for understanding the crustal thickening in response to the India–Asia collision (Mo et al. 2007). The lithological association of andesitic–dacitic–rhyolitic lavas of the LVS provided evidences for the link between the felsic and mafic rocks. Their identical trace element patterns (Fig. 3.11) suggest that they might share a parental magma with specific composition or derived from a common source. The felsic lithologies must have undergone complex evolution processes e.g., assimilation and fractional crystallization, or alternatively have derived by partial melting of mafic lithologies originated from mantle source. The latter is likely due to the difficulty in producing significant felsic melts by mantle melting and magma differentiation. The HFSEs such as Nb, Ta, Hf, P, Ti is depleted and LILEs like Rb, Ba, K is enriched. The signatures can also be produced by partial melting of amphibolite with accessory ilmenite. The residual amphibole and in particular ilmenite would hold HFSEs, causing the apparent depletion of these elements. The enrichment of Sr and the absence of Eu anomalies in the mafic rocks could be interpreted as melting result in plagioclase-absent source (Mo et al. 2007).

Most mafic dykes with Mg# of <0.62 are likely differentiated from the more primitive mantle-derived melts. Therefore, the felsic LVS rocks are unlikely the product of direct melting of mantle peridotite. The trace element similarity of the LVS rocks and the mafic dykes suggest that all the LVS rocks may have derived by partial melting of younger mantle-derived basaltic rocks (Mo et al. 2007). The subducted Neo-Tethyan ocean crust would be the ideal example. Thus, this process would bring the oceanic crust onto the amphibolite-solidus (Peacock 2003) at a depth of <60 km. The melts would be more felsic than peridotite-derived basaltic melts with the mantle isotopic signatures. It would suggest the mantle contribution to the mass of continental crust in response to the collision.

Plot of La vs. La/Sm (Fig. 3.13) provides further information on the genetic relation among LVS formations. The LDF andesitic rocks exhibit an invariable HREE, but a wide range of LREEs and LREE/HREE ratios, and a trend paralleling to the anticipated partial melting trend. This suggests that REEs of the andesite record both melting and crystallization processes. Hence, the LVS result likely from partial melting of a basaltic source with fractional crystallization. The evolved mafic dykes with Mg# of 0.53–0.62 with constant La/Sm could be explained by varying extents of fractional crystallization. They could be derived from metasomatized mantle peridotite or mantle peridotite with terrigenous sediment input (Mo et al. 2007).

To estimate the crustal/mantle contributions to the LVS volcanic rocks, we can use the Neodymium Crustal Index (NCI; Depaolo 1985; Depaolo et al. 1992):

$$NCI = \left[\varepsilon_{Nd}(rocks) - \varepsilon_{Nd}(M)\right] / \left[\varepsilon_{Nd}(C) - \varepsilon_{Nd}(M)\right]$$

where M and C refer to the mantle and the crust respectively.

This is a useful approach as it estimates mass contributions of crustal Nd in rocks experienced crustal assimilation during the evolution (Depaolo et al. 1992). The NCI for the LVS ranges from 0.16 to 0.47 with 0.30 on average and those for mafic dykes is 0.17. Obviously, the LVS and the associated dykes are dominated by up to 70–84% mantle contributions in terms of Nd isotopes (Mo et al. 2007). It is also inevitable that the crustal contributions to the LVS increased with advanced degrees of crustal level magma cooling and assimilation.

3.1.6 The Tectonic Significant of the Linzizong Volcanic Succession

It is reasonable to consider that the ~65–45 Ma LVS may reflect the active role of the last subducted bit of the Tethyan ocean crust during the collision (Mo et al. 2007, 2008; Guo et al. 2007). The lack of garnet signature in the LVS volcanic rocks place the constraint that the depth of amphibolite facies slab-melting is shallow at ~40 km, where garnet is not a stable phase in the subsolidus conditions, nor a peritectic phase during melting (Wyllie and Wolfe 1993; Vielzeuf and Schmidt 2001). Ilmenite as one of the primary hosts of Ti, Nb and Ta, is a common mineral in amphibolite (Niu and Lesher 1991). It is likely to be a residual phase during melting because of its refractory nature and to present a Nb–Ta–Ti depletion in the melts. All of the above, plus the Nd–Sr isotopic data (Fig. 3.12) with the involvement of recycled sediments (e.g., mature crustal materials), can account for the "arc-like signature" without actually requiring active subduction zones.

The LVS was considered as active continental margin magmatism prior to the India-Asia collision (Pearce and Houjun 1988) or syn-collision magmatism in response to the India-Asia collision (Mo et al. 2003, 2007, 2008). A comprehensive synthesis of geological, geochronological, and geochemical data reveals that the Gangdese magmatism migrates northward from 65 Ma, moves to the south at 64–48 Ma and flares-up at ca. 51 Ma (Zhu et al. 2015). The conspicuous unconformity between the Eogene LDF with the subjacent Mesozoic Shexing Formation represents the onset of the India–Asian collision, and the LDF volcanic rocks are products of syn-collision magmatism. The andesitic composition with obvious arc characteristics of these rocks suggests the presence of the subducting seafloor slab beneath the Gangdese. This explains a great deal of juvenile crustal formation at this time (60–53 Ma). Such melting is probably induced by subducting slab roll-back. The widespread and thick (>2000 m) UPF volcanic rocks with high alkalis, especially high K_2O content is best interpreted as a collision setting, in which the upper subducting seafloor crust melts and produce volumetrically significant magmas parental to the Gangdese batholith and the highly evolved UPF volcanic rocks.

The andesitic composition may have resulted from partial melting of the remaining part of the Tethyan ocean crust, from which the LVS inherited the mantle isotopic signatures (Mo et al. 2007). The syncollisional process of juvenile crust formation produces magmas from the ocean crust atop the subducting seafloor lithosphere within the mantle and convert the mafic arc assemblage to the andesitic continental crust required by the standard "island arc" model. Therefore, continental collision zones are ideal sites of juvenile crust formation and continental crust accretion.

3.2 The Cenozoic Collisional Granitoids in the Tibet Plateau

3.2.1 Spatial Distribution of the Cenozoic Collisional Granitoids in the Tibet

The Palaeocene-Eocene (66–41 Ma) granitoids that were considered as being collision related-mainly occur as the most intense stage of extensive magmatism in the Gangdese. The exposed areas of granitoid and volcanic rocks are roughly the same, each ~110,000 km^2 (Fig. 3.14), which form a giant magmatic belt stretching east-west for ~2500 km long and north-south for 100–300 km wide (Allégre et al. 1984; Yin and Harrison 2000; Wu et al. 2010). The eastern end of the belt is striking nearly north-south through the Yarlung-Tsangpo turn in connection with northern Myanmar, while to the west, the belt is connected to India, Ladakh, Pakistan and Kohistan.

The Gangdese magmatic belt can be roughly divided into three sub-belts, the northern, middle and southern belts. In general, rock types and ages differ between these belts, but there is no clear between-belt boundary with rock type age overlaps in places (e.g. Debon et al. 1986; Harris et al. 1988; Chung et al. 2003, 2005, 2009; Ding et al. 2003; Dong et al. 2005; Mo et al. 2005, 2007, 2009; Zhao et al. 2006; Ji et al. 2009, 2014, 2016; Wu et al. 2007, 2010). The early Cenozoic granitic rocks occur mainly in the south belt.

The Southern Belt is located in the south of the Longger-Gongbujiangda fault zone. The granitoids of the Gangdese belt have been widely studied in order to decipher the tectono-magmatic evolution of the southern Tibet (e.g., Chu et al. 2006; Dong et al. 2005, 2006, 2009, 2011, 2014; Ji et al. 2009, 2012, 2014; Mo et al. 2005, 2009; Zhang et al. 2010, 2013; Zhu et al. 2011, 2015). This belt, together with the widespread coeval intermediate-felsic volcanic rocks, accounts for more than 60% of the total area of the Gangdese magmatic belt. The collisional granitoids (65–45 Ma) are distributed throughout the belt, and mainly concentrated as batholiths in the middle-east and western segments of this belt.

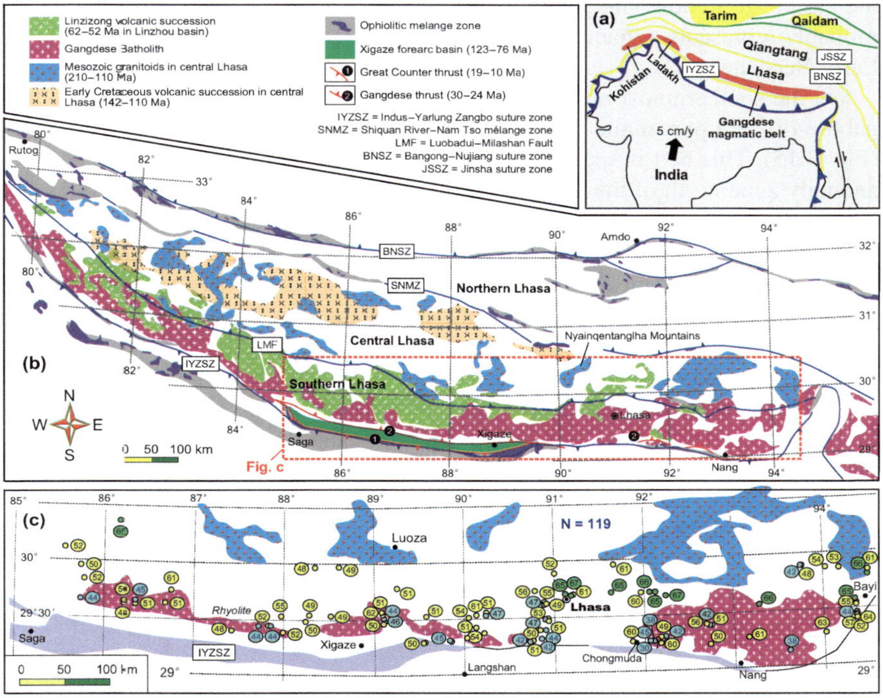

Fig. 3.14 Geological map of the Gangdese belt. (After Zhu et al. 2019)

The collisional granitoids in the Southern Belt, which have received much interest in recent years, mainly comprise granodiorites, granites and diorites (e.g. Dong et al. 2005; Mo et al. 2005, 2007, 2009). There are also minor olivine gabbros in the Dube pluton of the eastern Gangdese (Shui et al. 2021), and lenticular gabbros near Nimu of the middle-eastern Gangdese (Dong et al. 2005). The Qushui batholith is the most representative granite intrusion in the Xietongmen-Namulin-Nimu-Qushui area. The lithologies of the Qushui batholith are diverse, ranging from mafic to felsic rocks, including granodiorite, quartz diorite, quartz monzonite and monzogranite (Fig. 3.15). Mafic microgranular enclaves are abundant in these granites, indicating magma mixing process and interaction between mafic and felsic magmas (e.g. Dong et al. 2005; Mo et al. 2005).

The Namuru intrusion in the western Gangdese as one of the type intrusions consists mainly of biotite granite with gabbro inclusions and mafic micro-granular enclaves (shortly MMEs) (Fig. 3.16). Field investigation has found the transitional relationship between the gabbro inclusions and granite with abundant MMEs (Dong et al. 2011). Zircon U–Pb dating on 4 samples of both granite and gabbro inclusions gave the age of 46.11 ± 0.78, 45.47 ± 0.4, 46.7 ± 2.9 and 45.4 ± 1.4 Ma respectively, giving a tight magma crystallization age range of 45.4–46.7 Ma within analytical error (Dong et al. 2011). This magmatism is within the duration (40.0–52.5 Ma) of the eastern to middle Gangdese magmatism such as in the Quxu-Xigarze area. It probably represents a giant magmatic event throughout the south Gangdese in the Early Neocene.

On the southernmost margin of the Gangdese granitoid belt, a relatively small gabbro-diabase-pyroxenite belt is intermittently exposed from east to west (Fig. 3.15). This belt is spatially coincident with and parallel to the aeromagnetic anomaly zone north of the Yarlung-Zangbo ophiolite belt (Xiong et al. 2001), and these rocks are believed to be associated with or represent underplated mafic magmas. The zircon U–Pb dating on samples of gabbro, diabase, granite and MMEs in the Qushui-Xietongmen and Namuru areas indicate their emplacement at 45–53 Ma, suggesting that the magma underplating and magma mixing occurred simultaneously (Mo et al. 2005; Dong et al. 2005, 2011).

There are also ore-bearing porphyry intrusions in the collisional granites mentioned above, or in the Mesozoic-Cenozoic volcanic rocks. It is controlled by an E-W and N-S faults. The lithologies are dominated by monzogranite porphyry, quartz-monzonite porphyry and syenogranite porphyry, showing geochemical affinity of high-Sr/Y adakites or adakitic rocks (Chung et al. 2003; Hou et al. 2003, 2004). These porphyries have younger ages (~22–12 Ma) and host several giant Cu–Mo, Cu–Au, Cu–Fe deposits (Hou et al. 2003, 2004; Wang et al. 2015, 2018). It is believed that the ore-bearing porphyries were generated under a large-scale post-collisional extension background after the India-Asia collision (Hou et al. 2003, 2004; Wang et al. 2015, 2018).

It should be particularly noted that the Gangdese magmatic belt varies both along (E-W) and across (N-S) strike, such as outcrop ratios of granitoid intrusions over volcanic rocks, depth of denudation, age and geochemical characteristics of magmatic rocks. All these observations have shown that since the Mesozoic, especially in the Cenozoic, the Gangdese belt has been most magmatic active and also in terms of geothermal activities in the Tibet Plateau, providing favorable conditions for mineralization.

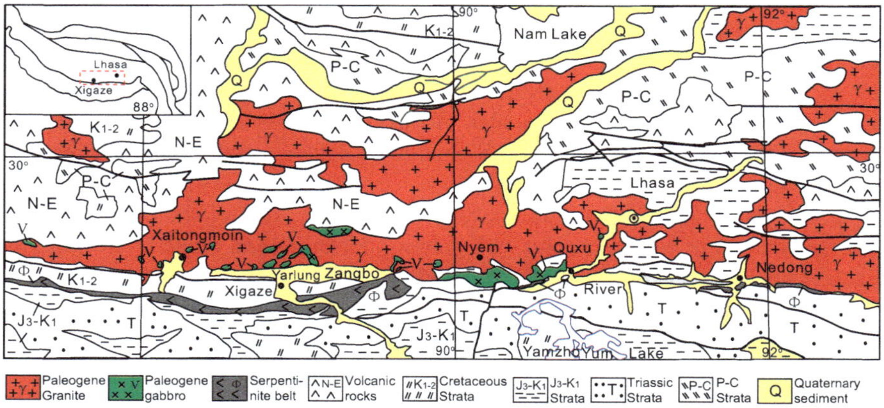

Fig. 3.15 Geological sketch of Qushui batholith. (After Mo 2005)

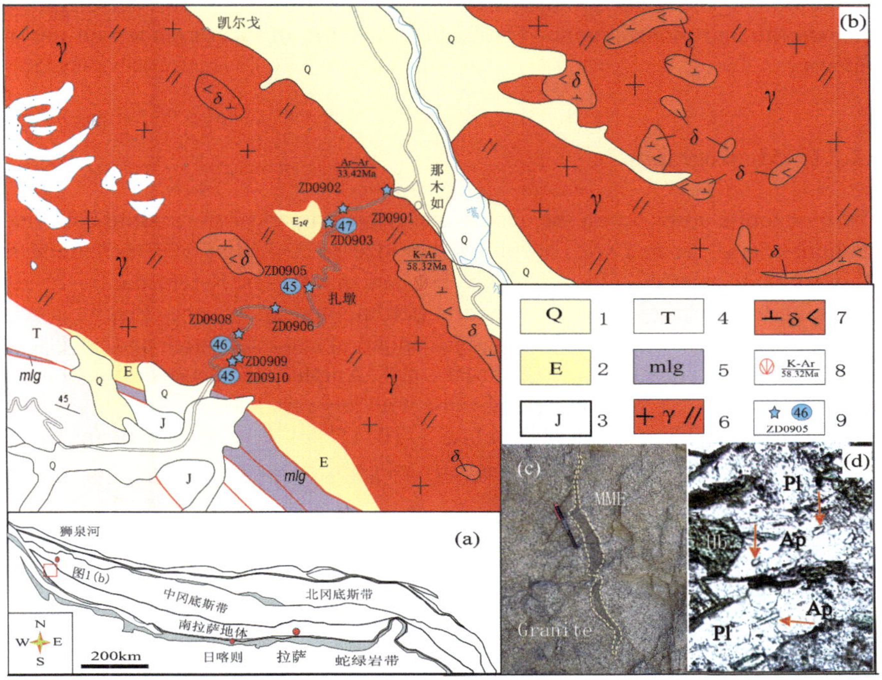

Fig. 3.16 The sketch of Cenozoic intrusions in western Gangdese. (**a**) Regional tectonic framework in southern Tibet; (**b**) geologic map of the western Gangdese; (**c**) MMEs and its host granitoid; (**d**) photomicrograph showing apatite needles in the MME. 1-Quaternary; 2-Paleogene; 3-Jurassic; 4-Triassic; 5-ophollite belt; 6-granite; 7-gabbrorocks; 8-isotopic dating; 9-sample locations

3.2.2 Petrological and Mineralogical Characteristics

According to their petrological and geochemical characteristics, the granitoids in the Northern Belt can be further divided into eastern section, middle section and western section. In the western and middle sections, the granitoids are spatially cogenetic with the coeval calc-alkaline volcanic rocks (Guo et al. 1991). They have low REE concentrations with a weak Eu anomaly and low initial Sr isotopic ratios, indicating that they belong to I-type granites (Chappell and White 1992; Whalen et al. 1987). By contrast, the coeval volcanic rocks and MMEs are rare in the eastern section of the Northern Belt. The granites are mainly S-type and show a significant negative Eu anomaly and high initial Sr isotopic ratios, which is related to the southward subduction of the Nujiang Ocean seafloor and the subsequent continental collision or arc-continent collision in PaleoTethys. The granitoids in the Middle Belt have high SiO_2 (>70 wt.%) and $Na_2O + K_2O$ contents, reflecting an anatexis origin caused by the collision of continents in PaleoTethys after the formation of the plateau.

The lithologies of the Southern Belt are dominated by medium-coarse-grained granodiorite and monzogranite. The main lithologies of each granite belt are as follows:

3.2.2.1 Monzogranite

The monzonite crops out in the central and northern part of the granite belt. It is medium-coarse-grained with varying grain size of 2–4 mm. The monzonite is mainly composed of K-feldspar, plagioclase and quartz. The K-feldspar is ~25–30 vol.%, and is mostly subhedral-anhedral perthite and microcline. Plagioclase is subhedral, and is ~15–20 vol.%. Quartz is anhedral and interstitial, and is ~20–25 vol.%. Biotite is euhedral, and is ~5–10 vol.%. Amphibole is euhedral, and is <5 vol.%. Some of these rocks are porphyritic with large (0.5–1 cm, up to 3 cm) phenocrysts of K-feldspar and plagioclase (~10–15 vol.%, and locally up to 30 vol.%). The accessory minerals are mostly magnetite, apatite, and zircon.

3.2.2.2 Granodiorite

Granodiorite is coarse-grained or porphyritic. The rock is mainly composed of K-feldspar (~20 vol.%), plagioclase (~35–55 vol.%) and quartz (~20–25 vol.%) with minor amphibole (~5 vol.%) and biotite (~5–10 vol.%). The K-feldspar is mainly microcline. The plagioclase is subhedral and shows zoning varying composition (An_{26-45}). Accessory minerals include zircon, apatite, magnetite and sphene. The phenocrysts are coarse (~1 cm and up to 3 cm) K-feldspars (~5–10 vol.%).

3.2.2.3 Quartz Diorite

Quartz diorite is medium-grained (~1–2 mm), mainly composed of plagioclase (55–60 vol.%), K-feldspar (~15 vol.%), quartz (10–15 vol.%), amphibole (~10 vol.%) and biotite (~5 vol.%). The plagioclase is subhedral with varying composition (andesine and labradorite, an average of An_{40}). The K-feldspar is anhedral and is mainly microcline. The quartz is anhedral and interstitial. The amphibole is subhedral with pyroxene relicts present in the cores of some crystals, and is usually altered to chlorite and calcite. The biotite is euhedral. The accessory minerals are sphene, apatite, magnetite, and zircon.

3.2.2.4 Tonalite

The tonalite is medium-grained (~0.5–1.5 mm) with the mineral assemblage of plagioclase, quartz, amphibole and biotite. The plagioclase is subhedral with relatively uniform composition (An_{25-30}). The quartz is anhedral and interstitial with varying abundance of 25–30 vol.%. The K-feldspar is anhedral and mainly microcline (~5 vol.%). The amphibole is subhedral (~10 vol.%). The biotite is euhedral (~10–15 vol.%). The accessory minerals are mainly zircon and magnetite.

3.2.2.5 Diorite

Diorite is mainly exposed in the southern part of the granite belt. It is medium to fine-grained (~1–1.5 mm), mainly composed of plagioclase (55–60 vol.%), amphibole (30–35 vol.%), biotite (~5 vol.%) and a small amount of pyroxene and quartz. The plagioclase is euhedral-subhedral with zoned composition (An_{35-60}). The amphibole is subhedral, shows sieve-texture with worm-like feldspar and quartz inclusions in some crystals. The biotite is euhedral, and the accessory minerals are magnetite, apatite and zircon. The apatite sometimes is acicular.

3.2.2.6 Gabbroic Diorite

The gabbro diorite is mainly exposed in the southern margin in a small area. It is medium-grained (1–2 mm) and massive. The mineral assemblage is mainly plagioclase, amphibole, pyroxene and biotite. The plagioclase (50–55 vol.%) is subhedral with zoning and twining and varying compositions (An_{36-79}). There are two types of plagioclase. One has large size with varying composition (An_{55-70}) and cloudy surfaces due to kaolinization and carbonate alteration. The other type of plagioclase has small size with less calcic composition ($An_{<45}$). The amphibole (~15–20 vol.%) is subhedral. The pyroxene (~10–15 vol.%) is subhedral-euhedral and mainly hypersthene. The pyroxene commonly rimmed with amphibole. The biotite (~5 vol.%) is euhedral. The accessory minerals are magnetite, sphene and zircon.

3.2.2.7 Gabbro

Gabbro is mainly exposed on the southern margin of the granite belt, and is distributed intermittently along the Yarlung-Zangbo suture. It is mainly composed of plagioclase (50–55 vol.%) varying composition (An_{55-91}), pyroxene (~25 vol.%), amphibole (~5–15 vol.%), biotite (~5 vol.%) and olivine (0–25 vol.%, Shui et al. 2021). The plagioclase is subhedral and auhedral with clean surfaces, zoning and twinning. The pyroxene is euhedral, commonly rimmed with amphibole. The amphibole is subhedral. The biotite is auhedral. The accessory minerals are magnetite (~5 vol.%), zircon and rutile. Magnetite is usually fine-grained and occur as inclusions in plagioclase and pyroxene or between mineral particles. The zircon is granular, with a length of 100–200 μm and a length-to-width ratio of 3:1. Its crystal surface is clear, and the crystal ridges are well developed.

3.2.2.8 Hornblendite

Hornblendite is subhedral and, medium and coarse grained (~2–3 mm), dominated by hornblende (>95 vol.%) with minor pyroxene (<5 vol.%) accessory magnetite and apatite. The rock is mainly found along the southern margin of the Gangdese belt in a small area, cropping out together with gabbro and diorite. It occurs locally in the form of enclaves in the gabbros.

3.2.3 Geochemical Characteristics

The gabbros have low SiO_2 (43–52 wt.%) and $Na_2O + K_2O$ (4.02–4.04 wt.%) contents, relatively high MgO (2.74–13.92 wt.%), Cr (1.2–374 ppm) and Ni (22–51 ppm) concentrations. On a total alkalis $Na_2O + K_2O$ versus SiO_2 (TAS) diagram (Middlemost 1994), they are classified as gabbro, monzogabbro, foid-gabbro, gabbroic diorite (Fig. 3.17a; e.g. Wang et al. 2019). They are alkalic to alkali calcic (Fig. 3.17b; Frost et al. 2001) and mostly plot in the field of shoshonitic rock series (Fig. 3.17c; Rickwood 1989).

In chondrite-normalized REE diagram (Fig. 3.18a) and primitive-mantle-normalized trace-element diagram (Fig. 3.18b), the gabbros are characterized by LILE enrichments, HFSE depletions, moderate REE fractionation with $La_N/Yb_N = 4.0–15.6$, and weak negative Eu anomalies with $Eu/Eu^* = 0.8–1.0$ (e.g. Wang et al. 2019).

The granitoids show large compositional variation with varying SiO_2 (59–80 wt.%), $Na_2O + K_2O$ (3.24–9.78 wt.%), MgO (0.08–3.05 wt.%) and CaO (0.12–5.32 wt.%). They plot in the fields of diorite, quartz monzonite, granodiorite and granite (Fig. 3.17a). The granitoids are alkali-calcic and calcic-alkali and calcic (Fig. 3.17b), most belonging to high-K calc-alkaline rock series (Fig. 3.17c) and metaluminous to peraluminous rock series (A/CNK = 0.87–1.60) (Fig. 3.17d). The granitoids are characterized by steep chondrite-normalized REE patterns with negative Eu anomalies, enriched in large ion lithophile elements (LILEs) and depleted in high field strength elements (HFSEs). By comparison with the gabbros and MMEs, the granitoids exhibit higher contents of HFSEs (Nb, Ta, Zr, and Hf) and LILEs (Rb, Ba, and K; Fig. 3.18c), and significant enrichment of HREEs and depletion of Eu, Ba Sr, P and Ti (Fig. 3.18d).

The MME display lower SiO_2 (47.5–57.7 wt.%), $Na_2O + K_2O$ (4.83–7.16 wt.%) and higher MgO (2.62–8.11 wt.%). They plot in the diorite and monzogabbro, monzodiorite and monzonite fields (Fig. 3.17a). The MMEs are alkali-calcic and calcic-alkali (Fig. 3.17b) and belong to high-K calc-alkaline to shoshonitic rock series (Fig. 3.17c) and metaluminous rock series (A/CNK = 0.71–1.01) (Fig. 3.17d). The MMEs display steep chondrite-normalized REE patterns with weak negative Eu

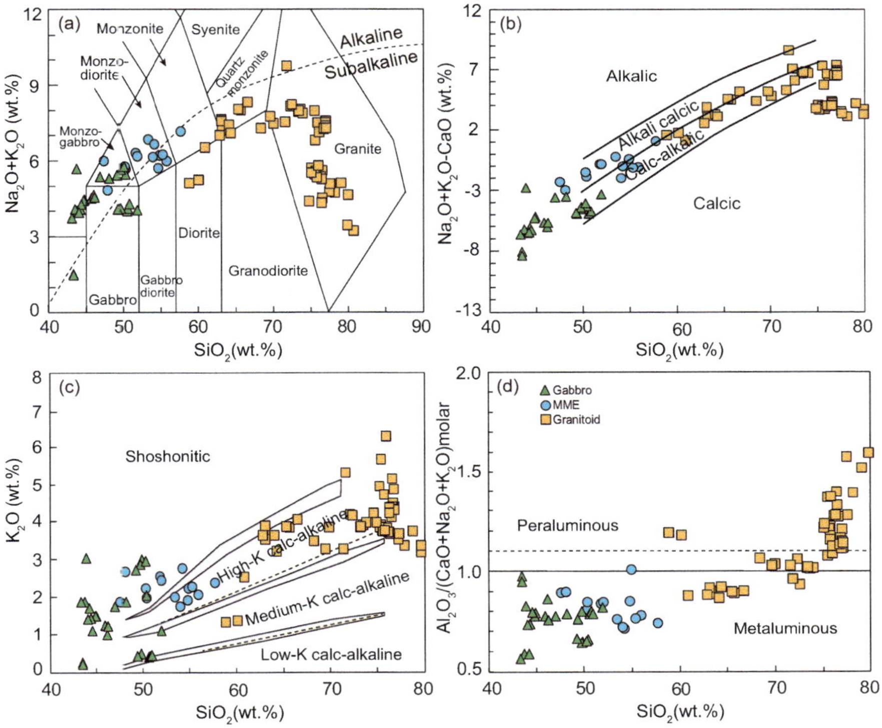

Fig. 3.17 Diagrams of major elements of the granitoid. (**a**) Total alkalis ($Na_2O + K_2O$) vs. SiO_2 (Middlemost 1994), (**b**) $Na_2O + K_2O$–CaO vs. SiO_2 (Frost et al. 2001), (**c**) K_2O vs. SiO_2 (Rickwood 1989), and (**d**) $Al_2O_3/(CaO + Na_2O + K_2O)$ molar (A/CNK) vs. SiO_2. (The data are from Duan et al. 2014, 2015; Dong et al. 2006; Ji et al. 2012; Ma et al. 2015; Meng et al. 2019; Mo et al. 2005; Shui et al. 2021; Wang et al. 2012, 2019; Wen et al. 2021)

anomalies. They are characterized by enriched large ion lithophile elements (LILEs) and depleted high field strength elements (HFSEs) (Fig. 3.18e, f). Compared with the granitoids, they exhibit lower REE concentrations and weaker negative Eu anomalies (Figs. 3.18e).

3.2.4 Petrogenesis of the Collisional Granitoids

The petrogenesis of the Gangdese collisional granitoids is one of the hot topics mainly focused on the granitoids, MMEs and gabbros.

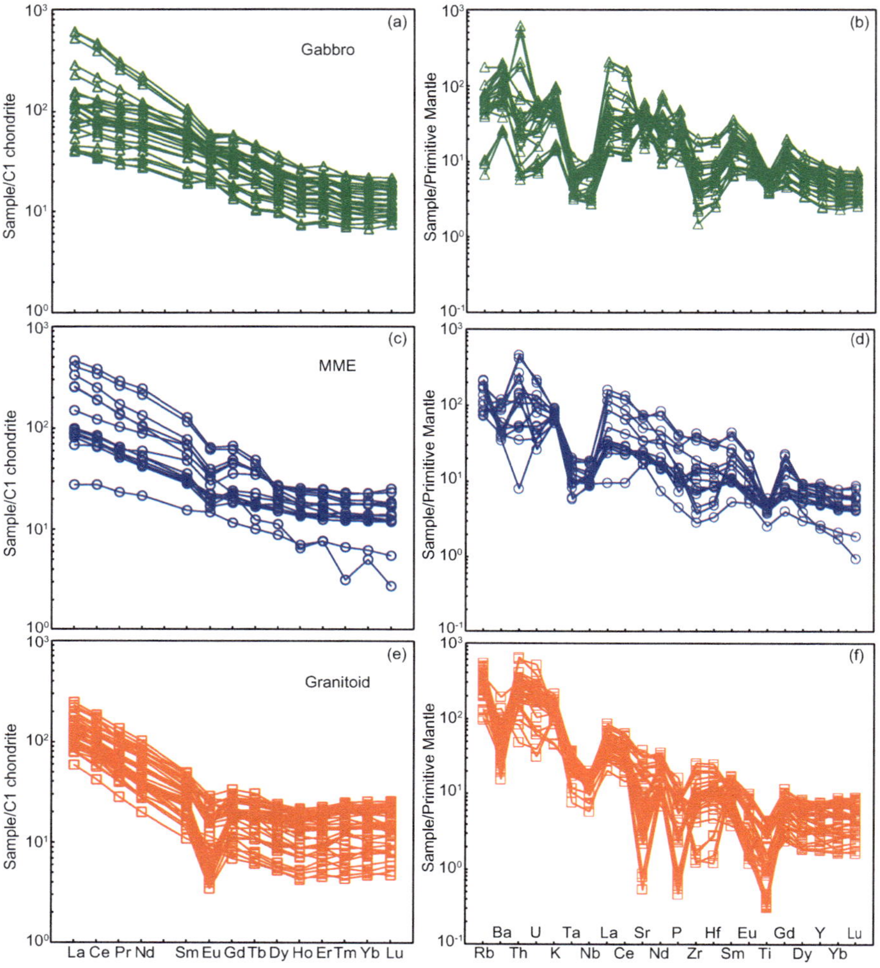

Fig. 3.18 Chondrite-normalized REE patterns (**a**, **c**) and primitive mantle-normalized trace element spider diagrams (**b**, **d**). Chondrite values and data for OIB, E-MORB and N-MORB are from Sun and McDonough (1989), and primitive mantle values are from McDonough and Sun (1995)

3.2.4.1 Granitoids

Regarding the petrogenesis of collisional granitoids in the Gangdese belt, the various lithologies and geochemical and Sr–Nd isotopic composition indicates that mantle-derived mafic magma is mixed with ancient crustal melts (Dong et al. 2006). The granitoids have varying chemical compositions of SiO_2 (59–80 wt.%), $Na_2O + K_2O$ (3.24–9.78 wt.%) and MgO (0.08–3.05 wt.%). They are characterized by enrichment of large ion lithophile elements (LILEs) and depletion in high field strength elements (HFSEs). Such a wide range of compositions imply that magma mixing is a possible mechanism for the generation of the collisional granitoids (Annen et al. 2006). Since granitic rock suites usually show diverse composition, which mainly depends on the source characteristics (e.g. Clemens and Stevens 2012). Zircon Hf isotope studies reveal that they display similar ε_{Hf} values to those of the mid-Paleocene volcanic rocks (Mo et al. 2009; Zhou 2017) and the highly evolved granites (Ji et al. 2012). The most evolved granites display similar zircon Hf and whole-rock Sr–Nd isotopic compositions to those of crust derived granites, indicating a cogenetic origin from the juvenile crust (e.g. Wen et al. 2021; Zhou et al. 2017).

In addition, some of the rocks contain abundant feldspar phenocrysts and the geochemical compositions show a liner relationship between the major elements and some trace elements (e.g. Dong et al. 2006; Wen et al. 2021). Combining these features with our petrological and geochemical observations, we consider that the collisional granitoids were derived from the partial melting of crustal materials or fractional crystallization of a mafic to intermediate magma.

3.2.4.2 Gabbros

The gabbros occur as small massive intrusions in the Gangdese belt. They are characterized by low SiO_2 (43–52 wt.%) and Al_2O_3 contents, high MgO (2.74–13.92 wt.%), Cr (1.2–374 ppm) and Ni (22–51 ppm), negative Eu anomalies and enrichments in LREEs (e.g., Shui et al. 2021; Wang et al. 2019), which are distinct from those of the lower crustal melts (Rapp and Watson 1995; Sisson et al. 2005). These features indicate that they probably crystallized from mafic melts. Furthermore, they display significantly higher Nb/U ratios (up to 18) than the bulk continental crust (6.15; Rudnick and Gao 2003), which is inconsistent with the results of crustal contamination.

Geochemical and isotopic studies reveal that the gabbro units show consistent Sr–Nd isotope composition, (e.g. Shui et al. 2021; Wang et al. 2019). They have low Nb, Ta concentrations and Nb/Ta ratios (Wang et al. 2019). The Sr–Nd isotope compositions suggest the gabbro source experienced varying degrees of enrichment (e.g., Shui et al. 2021; Wang et al. 2019). Therefore, they were likely generated from aqueous asthenospheric mantle.

3.2.4.3 The MMEs

There are abundant mafic micro-granular enclaves (MMEs, also known as mafic enclaves or microgranular enclaves; Didier 1973) in the collisional granitoids in the southern Gangdese belt. The MMEs are mainly distributed in granitoid batholith, especially abundant in the granodiorite, less so in monzogranite, extremely rare in gabbro-diorite and leucogranite. They occur generally as generally round, elliptical, lenticular, striped or irregular flame-like in shape with aspect ratios ranging from 1 to 8 (Fig. 3.19) and have sharp or diffuse contacts with their host rocks (e.g. Dong

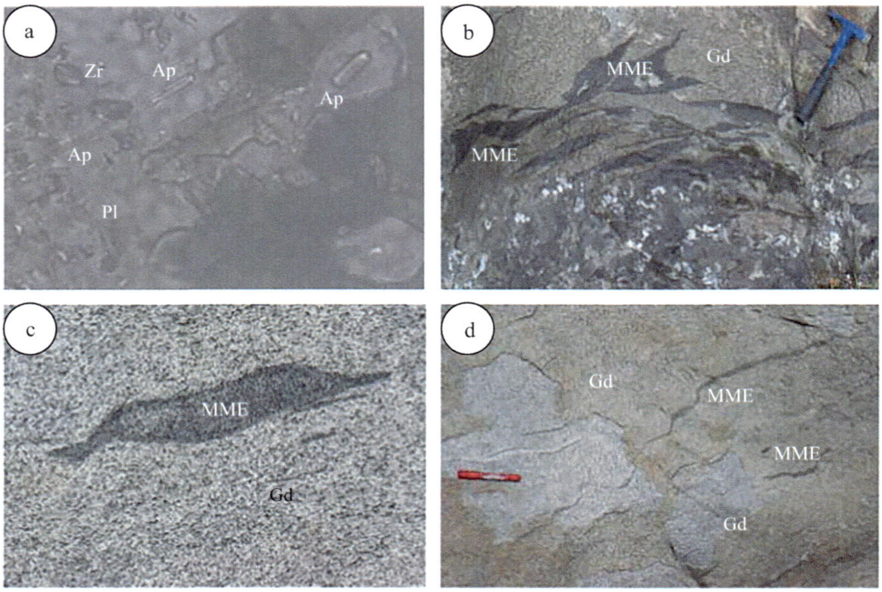

Fig. 3.19 Outcrop photos of gabbro, MMEs and the granotoid host rocks in the Gangdese belt. (After Dong et al. 2006)

et al. 2006, 2008, 2011; Mo et al. 2009; Wen et al. 2021) with varying size of 10s of cm to meters, commonly in the range of 20–40 cm in diameter (Mo et al. 2009). The MMEs account generally around 5 vol.% of their host granitoid rocks, locally to 15 vol.%. From north to south of the Gangdese belt, the size and content of MMEs in their host granitoids gradually increase. In the southern margin of the Gangdese belt near the Yarlung-Zangbo suture, a large number of such MMEs are closely distributed. The occurrence of fine-grained margins and acicular apatite in the MMEs indicate a rapid cooling process (Dong et al. 2006; Mo et al. 2005, 2009). Plagioclase in the MMEs is subhedral, have intermediate to slightly calcic composition (An_{55-65}), and constitue ~50% of the rock. Mafic minerals are mainly hornblende (20–25%), biotite (about 10%) and some pygoxene, while felsic mineral are quartz and K-feldspar.

The origin of the MME in granitoids has long been controversial. Popular explanations include wall-rock xenoliths (Maas et al. 1997), refractory or residual materials of the source rocks (Chappell and White 1991; Chappell et al. 1987), autoliths or early cumulates (Collins et al. 2006; Dahlquist 2002; Donaire et al. 2005; Xu et al. 2021), and mixing and mingling of mafic and felsic magmas (Barbarin 2005; Clemens and Stevens 2012; Dong et al. 2021). The MMEs in the Gangdese belt show typical igneous textures (Dong et al. 2006; Mo et al. 2005, 2009; Wen et al. 2021), which are different from the restites and xenoliths that usually show metamorphic structures (Collins 1998). Besides, the MMEs are fine grained and show no signature of cumulate.

The MMEs in the Gangdese belt are characterized by the following items:

1. The MMEs are a mafic and fine-grained and their volume gradually increases with the approach of the Yarlung-Zangbo suture.
2. The MMEs are calc-alkaline rocks, and the characteristics of major elements also show a close relationship with regional granitic and mafic magmatism.
3. The geochemical compositions of the MMEs are transitional between gabbros and granitoids, indicating a magma mixing trend of mafic and felsic magmas.
4. The disequilibrium textures of pyroxenes mantled by amphibole, biotite, plagioclase in the MMEs are generally accepted as evidence of magma mixing (Vernon 1984; Grogan and Reavy 2002; Foley et al. 2002; Wen et al. 2021).
5. The geochemical calculations suggest that the compositions of the MMEs can be the result of magma mixing (Dong et al. 2006, 2009, 2011; Mo et al. 2005, 2009).

Therefore, the MMEs are likely the products of magma mixing between felsic magma represented by their host granites and mafic magma represented by the mantle derived melts parental to gabbros in the Gangdese belt.

Magma mixing processes were employed in describing the petrogenesis of the Quxu pluton, and the proportions of mafic and felsic end-members involved in the mixing event were quantitatively estimated using major elemental compositions (Dong et al. 2005, 2006, 2008; Mo et al. 2005, 2007, 2009). The remaining issue is that the two proposed end-members cannot be recognized by Nd–Sr–Pb isotopes in terms of the most mafic portion having the highest Nd and the least Sr isotopic ratios (Mo et al. 2009). For instance, the $\varepsilon_{Nd}(t)$ values changed from 2.7 to 8.5, unrelated to their SiO_2 contents (47–56 wt.%) (Dong et al. 2008). This is explained as one of the geochemical features for supporting a magma mixing origin of granite and related rocks, indicating that the magma has been chemically homogenized during the mixing processes (Dong et al. 2006, 2008; Mo et al. 2005, 2009).

The Hf isotope studies of these rocks (Fig. 3.20) show no correlations between the $\varepsilon_{Hf}(t)$ values and their SiO_2 contents. For example, the most mafic sample has the lowest mean $\varepsilon_{Hf}(t)$ values of 8.8 among all the samples, even lower than that of the granodiorite with value of 9.1 (Dong et al. 2005). In this point, the zircon Hf isotopic results are consistent with the Nd–Sr data (Mo et al. 2009). This was supported by Ji et al. (2009), who pointed out that no significant Hf isotopic difference existed between the mantle-derived gabbro and the crustal-derived granite (Ji et al. 2009), further indicating that these zircons were crystallized from different kinds of host magmas.

However, it should be noted that there are large variations in zircon $\varepsilon_{Hf}(t)$ values, up to 5.4-ε units within a single rock and up to 15-ε units between zircons within the six contemporaneous samples of ~50 Ma (Mo et al. 2009). This heterogeneity requires an open system with more than 2 end-member mixtures of the magma source region instead of partial melting or fractional crystallization. In other words, the large variations in zircon $\varepsilon_{Hf}(t)$ values observed in the Quxu pluton are indicative of magma mixing.

There is no zircon with crustal signature ($\varepsilon_{Hf}(t) < 0$) with an age range of ~65–47 Ma (Fig. 3.20), suggesting that the source materials of these rocks are of

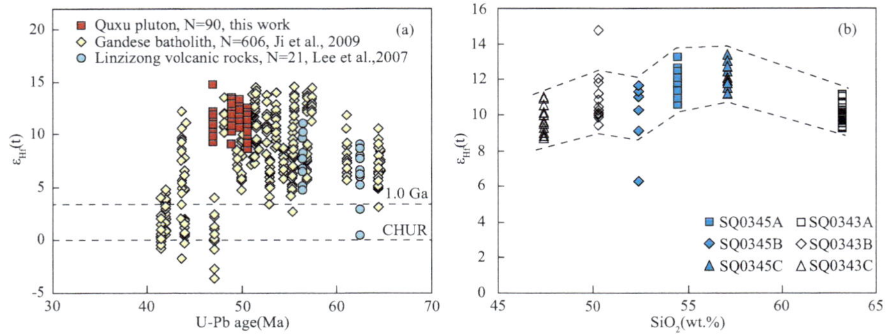

Fig. 3.20 Plots of $\varepsilon_{Hf}(t)$ vs. ages (**a**) and whole-rock SiO_2 content (**b**) of the Quxu gabbro-granitoid complex in the southern Gangdese Belt. (After Mo et al. 2009)

mantle origin in terms of Hf isotope composition with rather limited crustal input. But at ~47 Ma, the crustal components with negative $\varepsilon_{Hf}(t)$ values appear to be evident. The emplacement age of the diorite significantly predates by about 10 Ma the onset of the Indian continental crust beneath the Gangdese batholith (Chung et al. 2005). Therefore, it is shown that the crustal materials involved in the magma mixing of the Gangdese batholith may have come from the Lhasa terrane itself rather than the Indian continent.

All these features suggest that magma mixing took place between granite and mafic magma during the Neocene (45.4–46.7 Ma), forming the vast granitic and gabbro rocks as an important magmatic event in the western Gangdese. Therefore, it is inferred on the basis of igneous rocks that the India-Asia collision is acting simultaneously in both eastern and western Gangdese in the Eocene, resulting in basaltic magma underplating below and then magma mixing along the entire Gangdese belt produced the Gangdese giant magmatic belt.

3.2.5 The Tectonic Implications to the Continental Collision

The Paleocene-Eocene magmatic rocks in the Gangdese are closely related signifying the incorporation of Indian continental crustal materials into the source of Gangdese magmatic rocks (Ji et al. 2009). The southward migration phenomenon strengthens the general notion for slab rollback during the India–Asia continental collision and thus steepening of the Neotethyan subduction occurred before the collision starting at ~60–55 Ma (Fig. 3.21a). The subsequent intensification of magmatic flare-up event at ~50 Ma, is related to the break-off of the steepened Neotethyan oceanic slab in the early stage of the India–Asia collision (Fig. 3.21b). Such a break-off would result in significant thermal perturbation to cause partial melting of various sources including the ascending asthenosphere and the overlying continental crust (Lee et al. 2009). Magmas generated under this condition are typical of heterogeneous compositions (Davies and von Blanckenburg 1995; Coulon et al. 2002).

Isotope data show that the magma genesis of the south Gangdese basement was involved during the continental collision despite the mantle source differences beneath the eastern and middle central Gangdese (Guo et al. 2011; Ji et al. 2017; Zhang et al. 2010). The zircon Hf isotopic data of the Quxu pluton from Mianjiang and Niedang gave the average $\varepsilon_{Hf}(t)$ values from 10 to 12.2 among the samples of gabbros, gabbroic MMEs, and granodiorites. The variation within one sample can be as high as 5.4 units of $\varepsilon_{Hf}(t)$. They have a young Hf model age (230–340 Ma). These depleted mantle-like high and positive $\varepsilon_{Hf}(t)$ values suggest that the Quxu granitoid complex, as one of the type Gangdese batholiths, is generated from either the mantle wedge or a pre-existing mafic lower crust beneath the collision zone instead of the granitoids originated from the anatexis of lower continental crust of prior sedimentary protolith. The Hf isotope compositions of the central Gangdese batholith and related volcanic rocks support that the mantle materials were added to the crust via the partial melting of the subducted remaining part of the Tethyan Ocean crust (Lee et al. 2007; Mo et al. 2008, 2009; Niu et al. 2013) or the recycling of the pre-existing oceanic arc terrane (with positive $\varepsilon_{Nd}(t)$ values) (Ji et al. 2009). Therefore, such a significant feature of high and positive $\varepsilon_{Hf}(t)$ values and a young Hf model age in a large spatial frame in the Gangdese batholith suggests a volumetrically significant mantle-to-crust mass transfer process through magmatism in response to the India-Asia continental collision in the Cenozoic (Ji et al. 2009).

The Paleogene felsic magmatism along the south Gangdese Belt was considered as resulting from partial melting of the final moment of the Tethyan ocean crust in response to the India–Asia continental collision (Mo et al. 2008). The interpretation concerns key issues of how and where the mantle mass contributed to continental crust. The slab break-off model can be used to account for both the formation of the collisional magmatic event (Fig. 3.21) (e.g., Chung et al. 2005; Lee et al. 2009, 2012; Wen et al. 2008; Zhu et al. 2011, 2015; Ma et al. 2014). With continental collision, the tensile stresses lead to the separation of the oceanic slab from the continental lithosphere, resulting in slab break-off (e.g., Davies and von Blanckenburg 1995; von Blanckenburg and Davis 1995). The slab break-off related to the India-Asia collision probably took place in the Early Eocene (ca. 51–46 Ma), and can place important constraints on associated collision dynamics (Hunen and Allen 2011; Chung et al. 1998).

It is the slab break-off that started with a narrow slab window between the continent and the subducted oceanic slab and resulted in a linear magmatic belt (Davies and von Blanckenburg 1995). The upwelling of asthenosphere can lead to melting and trigger the formation of a variety of magmas, especially tholeiitic basaltic magma (e.g., von Blanckenburg and Davis 1995; Xu et al. 2008). The Paleocene–Eocene Gangdese magmatic intrusions and coeval Linzizong volcanic rocks along the southern Gangdese (Fig. 3.21) represent an intensive magmatism confined to a narrow zone on the southern Gangdese belt (e.g., Lee et al. 2009, 2012; Mo et al. 2007; Wen et al. 2008; Ma et al. 2014), which would have contributed to the Cenozoic "flare-up" of mafic to felsic magmatic event in the southern Gangdese Belt (e.g., Chung et al. 2005; Lee et al. 2009; Mo et al. 2007; Ma et al. 2014).

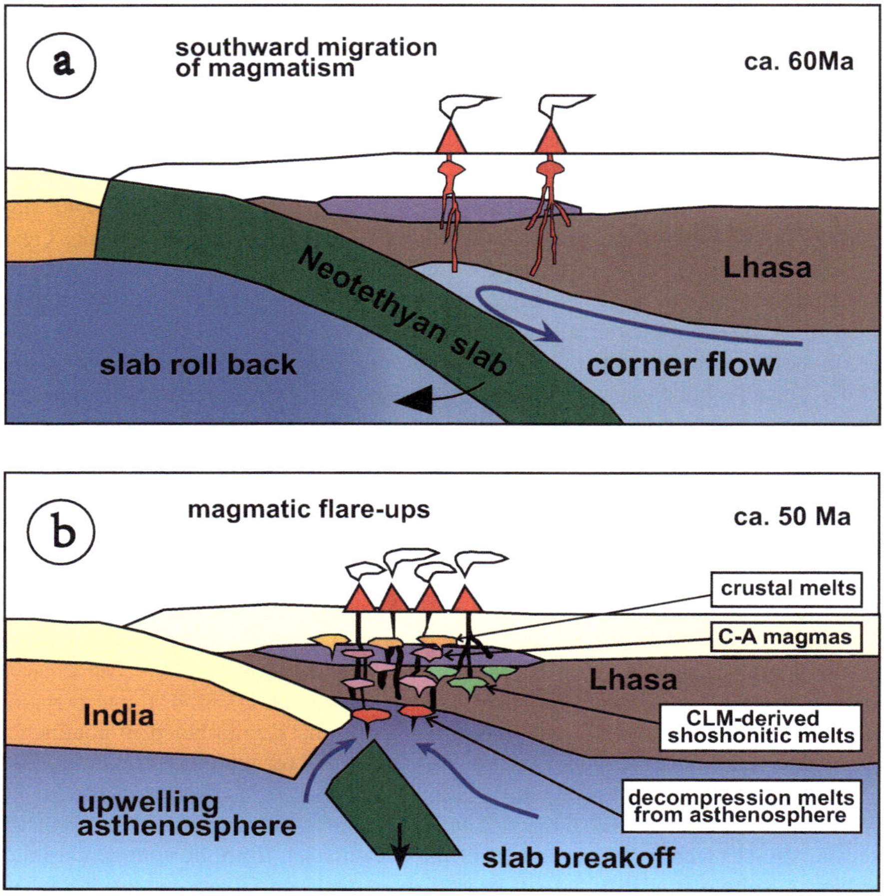

Fig. 3.21 Schematic models for magmatism during the India-Asia collision in southern Tibet. (**a**) Rollback and (**b**) breakoff of the Neotethyan slab. *C–A* calc–alkaline, *CLM* continental lithospheric mantle

In summary, the collisional granitoids were likely produced by partial melting of the mafic lower crust, which was triggered by the upwelling of asthenosphere through the slab window during the break-off of the Indian slab.

References

Allégre CJ, Courtillot V, Tapponnier P, Hirn A, Mattauer M, Coulon C, Jaeger JJ, Achache J, Schärer U, Marcoux J, Burg JP, Girardeau J, Armijo R, Gariépy C, Göpel C, Li TD, Xiao XC, Chang CF, Li GQ, Lin BY, Teng JW, Wang NW, Chen GM, Han TL, Wang XB, Deng WM, Sheng HB, Cao YG, Zhou J, Qiu HG, Bao PS, Wang SC, Wang BX, Zhou YX, Xu RH (1984) Structure and evolution of the Himalaya-Tibet Orogenic belt. Nature 307:17–22

Annen C, Blundy JD, Sparks RSJ (2006) The genesis of intermediate and silicic magmas in deep crustal hot zones. J Petrol 47:505–539

Bao C (2014) Genesis and geochemistry characteristics of Dianzhong formation volcanic rocks of Linzizong Group in the Cuoqin area, Tibet, Master degree thesis, p 62

Barbarin B (2005) Mafic magmatic enclaves and mafic rocks associated with some granitoids of the central Sierra Nevada batholith, California: nature, origin, and relations with the hosts. Lithos 80:155–177

Cao Y, Kang ZQ, Xu JF et al (2020) Geochronology, geochemistry and geological significance of volcanic rocks of the Dianzhong Formation, Shiquanhe Area, Western Lhasa Block. Earth Sci 45(5):1573–1592 (in Chinese with English abstract)

Chappell BW, White AJR (1991) Restite enclaves and the restite model. In: Dider J, Barbarin B (eds) Enclaves and granite petrology, developments in petrology. Elsevier, Amsterdam, pp 375–381

Chappell BW, White AJR (1992) I- and S-type granites in the Lachlan Fold Belt. Trans R Soc Edinb Earth Sci 83:1–26

Chappell BW, White A, Wyborn D (1987) The importance of residual source material (restite) in granite petrogenesis. J Petrol 28:1111–1138

Chen J, Huang B, Yi Z et al (2014) Paleomagnetic and $^{40}Ar/^{39}Ar$ geochronological results from the Linzizong Group, Linzhou Basin, Lhasa Terrane, Tibet: implications to Paleogene paleolatitude and onset of the India–Asia collision. J Asian Earth Sci 96(15):162–177

Chung S-L, Lo C-H, Lee T-Y, Zhang Y-Q, Xie Y-W, Li X-H, Wang K-L, Wang P-L (1998) Dischronous uplift of the Tibetan plateau starting from 40 My ago. Nature 349:769–773

Chung S-L, Liu D-Y, Ji J-Q, Chu M-F, Lee H-Y, Wen D-J, Lo C-H, Lee T-Y, Yian Q, Zhang Q (2003) Adakites from continental collision zone: melting of thicken lower-crust beneath southern Tibet. Geology 31:1021–1024

Chung SL, Chu M, Zhang Y, Xie Y, Lo CH, Lee TY, Lan CY, Li X, Zhang YQ, Wang Y (2005) Tibetan tectonic evolution inferred from spatial and temporal variations in post-collisional magmatism. Earth Sci Rev 68:173–196

Chung SL, Chu MF, Ji JQ, O'Reilly SY, Pearson NJ, Liu DY, Lee TY, Lo CH (2009) The nature and timing of crustal thickening in Southern Tibet: geochemical and zircon Hf isotopic constraints from postcollisional adakites. Tectonophysics 477:36–48

Chu MF, Chung SL, Song B, Liu DY, O'Reilly SY, Pearson NJ, Ji JQ, Wen DJ (2006). Zircon U–Pb and Hf isotope constraints on the Mesozoic tectonics and crustal evolution of Southern Tibet. Geology 34:745–748

Clemens JD, Stevens G (2012) What controls chemical variation in granitic magmas? Lithos 134–135:317–329

Collins WJ (1998) Evaluation of petrogenetic models for Lachlan Fold Belt granitoids: implications for crustal architecture and tectonic models. Aust J Earth Sci 45:483–500

Collins WJ, Wiebe RA, Healy B, Richards SW (2006) Replenishment, crystal accumulation and floor aggradation in the megacrystic Kameruka Suite, Australia. J Petrol 47:2073–2104

Copeland P, Harrison TM, Kidd WSF, Xu RH, Zhang YQ (1987) Rapid Early Miocene acceleration of uplift in the Gangdese Belt, Xizang (southern Tibet), and its bearing on accommodation mechanisms of the India-Asia collision. Earth Planet Sci Lett 86(2):240–252

Coulon C, Maluski H, Bollinger C, Wang S (1986) Mesozoic and Cenozoic volcanic rocks from central and southern Tibet: 39Ar–40Ar dating, petrological characteristics and geodynamical significance. Earth Planet Sci Lett 79:281–302

Coulon C, Megartsi M, Fourcade S, Maury RC, Bellon H, Louni-Hacini A, Cotton J, Coutelle A, Hermitte D (2002) Post-collisional transition from calc-alkaline to alkaline volcanism during the Neogene in Oranie (Algeria): magmatic expression of a slab breakoff. Lithos 62(3):87–110

Dahlquist JA (2002) Mafic microgranular enclaves: early segregation from metaluminous magma (Sierra de Chepes) Pampean Ranges, NW Argentina. J South Am Earth Sci 15:643–655

Davies JH, Von Blanckenburg F (1995) Slab breakoff: A model of lithosphere detachment and its test in the magmatism and deformation of collisional orogens. Earth Planet Sci Lett 129(1):85–102

Debon F, LeFort P, Sheppard SM, Sonet J (1986) The four plutonic belts of the Transhimalaya–Himalaya: a chemical, mineralogical, isotopic, and chronological synthesis along a Tibet–Nepal section. J Petrol 27:219–250
DePaolo DJ (1985) Isotopic studies of processes in mafic magma chambers: I. The Kiglapait Intrusion, Labrador. J Petrol 26(4):925–951
DePaolo DJ, Perry FV, Baldridge WS (1992) Crustal versus Mantle Sources of Granitic Magmas: A Twoparameter model based on Nd isotopic studies. Earth and Environmental Science Transactions of the Royal Society of Edinburgh 83(1/2):439–446
Dewey JF, Shackleton RM, Chang C, Sun Y (1988) The tectonic evolution of the Tibetan plateau. Philosophical Transactions of the Royal Society, London A327:379–413
Didier J (1973) Granites and their enclaves. Elsevier, Amsterdam
Ding L, Paul K, Zhong D, Deng W (2003) Cenozoic volcanism in Tibet; evidence for a transition from oceanic to continental subduction. J Petrol 44(10):1833–1865
Ding F, Xu Z, Liu S, Li Y, Gao J, Fan Y, Li Q (2017) LA-ICP-MS zircon U-Pb ages, petrochemical characteristics and petrogenesis of the volcanic rocks from the Paleocene-Eocene Nianbo formation in Chima area, Coqen County, Xizang (Tibet). Geol Rev 63(4):1102–1116
Donaire T, Pascual E, Pin C, Duthou J (2005) Microgranular enclaves as evidence of rapid cooling in granitoid rocks: the case of the Los Pedroches granodiorite, Iberian Massif, Spain. Contrib Mineral Petrol 149:247–265
Dong GC (2002) Linzizong volcanic rocks and implications for probing India-Eurasia collision process in Linzhou Volcanic Basin, Tibet, PhD thesis. China University of Geosciences, 150 pp. (in Chinese)
Dong GC, Mo XX, Zhao ZD, Wang LL, Chen T (2005) Geochronologic constraints by SHRIMP II zircon U-Pb dating on magma underplating in the Gangdese Belt following India-Eurasia collision. Acta Geol Sin 79:787–794 (in Chinese with English abstract)
Dong GC, Mo XX, Zhao ZD, Zhu DC, Wang LL, Chen T, Li B (2006) Magma mixing in middle part of Gangdese magma belt: evidences from Granitoid complex. Acta Petrol Sin 22:835–844 (in Chinese with English abstract)
Dong GC, Mo X, Zhao ZD, Zhu DC, Song YT, Wang L (2008) Gabbros in southern gangdese belt of Tibet and the implication for crust-mantle interaction. Acta Petrol Sin 24(2):203–210
Dong GC, Mo XX, Zhao ZD, Zhu DC, Song YT, Wang L (2009) Gabbros from southern Gangdese: implication for mass exchange between mantle and crust. Acta Petrol Sin 24:203–210
Dong GC, Mo XX, Zhao ZD, Zhu DC, Xie XF, Dong ML (2011) The Neocene magmatism from Namuru intrusion in western Gangdese, Tibet and its tectonic significance. Acta Petrologica Sinica 27(7):1983–1992 (In Chinese with English abstract)
Dong X, Zhang ZM, Liu F, He ZY, Lin YH (2014) Late Paleozoic intrusive rocks from the southeastern Lhasa terrane, Tibetan plateau, and their late Mesozoic metamorphism and tectonic implications. Lithos 198–199:249–262
Dong P, Dong G, Santosh M, Mo X, Wang P, Li H, Ketchaya YB (2021) Multiple sources and magmatic evolution of the late Triassic Daocheng batholith in the Yidun Terrane: implications for the Paleo-Tethys Ocean evolution in the eastern Tibetan plateau. Geol Soc Am Bull 134:1660–1680. https://doi.org/10.1130/B35970.1
Duan ZM, Li GM, Li YX, Wang BD, Zhang H, Huang Y, Guo L, Duan YY (2014) Geochronology and geochemical characteristics of ore-bearing porphyry in longgen lead-zinc deposit of middle-Gangdese metallogenic belt, Tibet. Mineral Deposits 33:625–638 (in Chinese with English abstract)
Duan ZM, Li GM, Wang BD, Li YX, Huang Y, Guo L, Duan YY (2015) Geochronology and its geological significance of the ore-bearing porphyry in Chagele lead-zinc deposit in middle-Gangdese metallogenic belt, Tibet. J Jilin Univ 45:1667–1690 (in Chinese with English abstract)
Flower M, Yu XH, Mo XX, Su SG, Zhao X (2001) Volcanism and tectonic setting of the Cenozoic kamafugite from western Qinling, Gansu province. China Acta Petrologica Sinica 17(3):1000–0569
Foley SF, Tiepolo M, Vannucci R (2002) Growth of early continental crust controlled by melting of amphibolite in subduction zones. Nature 417:837–840

Frost CD, Bell JM, Frost BR, Chamberlain KR (2001) Crustal growth by magmatic underplating: isotopic evidence from the northern Sherman batholith. Geology 29:515–518

Fu WC, Kang ZQ, Pan HB (2014) Geochemistry, zircon U-Pb age and implications of the Linzizong group volcanic rocks in Shiquan River area, Tibet. Geol Bull China 33(6):850–859 (in Chinese with English abstract)

Grogan SE, Reavy RJ (2002) Disequilibrium textures in the Leinster Granite Complex, SE Ireland: evidence for acid–acid magma mixing. Mineral Mag 66:929–939

Guo TY, Liang DY, Zhang YZ, Zhao CH (1991) Geology of Ngarli, Tibet (Xizang). China University of Geosciences Press, Wuhan, pp 1–464

Guo Z, Wilson M, Liu J (2007) Post-collisional adakites in south Tibet: Products of partial melting of subduction-modified lower crust. Lithos 96:205–224

Guo L, Zhang HF, Harris N, Pan FB, Xu WC (2011) Origin and evolution of multi-stage felsic melts in eastern Gangdese belt: constraints from U–Pb zircon dating and Hf isotopic composition, Lithos127:54–67

Harris NBW, Xu R, Lewis CL, Jin CW (1988) Plutonic rocks of the 1985 Tibet Geotraverse, Lhasa to Golmud. Philosophical Transactions of the Royal Society, London A327:145–168

Hou ZQ, Ma HW, Zaw K, Zhang YQ, Wang MJ, Wang Z, Pan GT, Tang RL (2003) The Himalayan Yulong porphyry copper belt: Product of large-scale strike-slip faulting in eastern Tibet. Econ Geol 98(1):125–145

Hou ZQ, Gao YF, Qu XM, Rui ZY, Mo X, Zhao ZD (2004) Origin of adakitic intrusives generated during mid-Miocene east-west extension in southern Tibet. Earth Planet Sci Lett 220:139–155

Hu X-W, Ma R-Z, Tao X-F et al (2007) Lithogeochemical characteristics and tectonic setting of volcanic rocks of Dianzhong Formation in the Coqen area, Tibet, China. J Chengdu Univ Technol Sci Technol Ed 34(1):15–27 (in Chinese with English abstract)

Ji WQ, Wu FY, Chung SL et al (2009) Zircon U-Pb geochronology and Hf isotopic constraints on on petrogenesis of the Gangdese batolith, southern Tibet. Chem Geol 262:229–245

Ji WQ, Wu FY, Liu CZ, Chung SL (2012) Early Eocene crustal thickening in southern Tibet: New age and geochemical constraints from the Gangdese batholith. J Asian Earth Sci 53:82–95

Jiang W, Mo X, Zhao C, Guo TY, Zhang SQ (1999) Geochemical characteristics of granites and their micromafic enclaves in Central Gangdese in the Tibetan Plateau. Acta Petrol Sin 15(1):89–97

Jin CW, Zhou YS (1978) Magmatic belt in the Himalaya and Gangdese arc range and its genetic model. Geological Sciences 4:297–312. (in Chinese with English abstract)

Kelemen PB (1995) Genesis of high Mg-Number andesites and the continental-crust. Contributions to Mineralogy and Petrology 120(1):1–19

Lee HY, Chung SL, Wang YB, Zhu DC, Yang JH, Song B, Liu DY, Wu FY (2007) Age, petrogenesis and geological significance ofthe Linzizong volcanic successions in the Linzhou basin, southern Tibet: Evidence from zircon U-Pb dates and Hf isotopes. Acta Petrol Sin 23(2):493–500

Lee HY, Chun SL, Lo CH, Ji JQ, Lee TY, Qian Q, Zhang Q (2009) Eocene Neotethyan slab breakoff in southern Tibet inferred from the Linzizong volcanic record. Tectonophysics 477(1-2):20–35

Lee HY, Chung SL, Ji J et al (2012) Geochemical and Sr–Nd isotopic constraints on the genesis of the Cenozoic Linzizong volcanic successions, southern Tibet. J Asian Earth Sci 53:96–114

Liang YP, Zhu J, Ci Q et al (2010) Zircon U-Pb ages and geochemistry of volcanic rock from Linzizong Group in Zhunuo area in middle Gangdise belt, Tibet Plateau. Earth Sci 35(2):211–223

Liu HF (1993) Division of Linzizong volcanic rock system and belong to time in Lhasa area. Tibet Geology 2:59–69. (in Chinese with English abstract)

Liu AL, Wang Q, Zhu DC et al (2018) Origin of the ca. 50 Ma Linzizong shoshonitic volcanic rocks in the eastern Gangdese arc, southern Tibet. Lithos 304:374–387

Liu X, Su BX, Patrick Xiao Y, Chen C, Uysal I, Jing JJ, Zhang PF, Chu Y, Lin W, Sakyi PA (2019) Initial subduction of Neo-Tethyan ocean: geochemical records in chromite and mineral inclusions in the Pozantı-Karsantı ophiolite, southern Turkey. Ore Geol Rev 110:102926.

Liu YM, Dai JG, Wang CS, Li H.A, Wang Q, Zhang LL (2020) Provenance and tectonic setting of upper triassic turbidites in the eastern Tethyan Himalaya: implications for early-stage evolution of the Neo–Tethys. Earth Sci Rev 200:103030. https://doi.org/10.1016/j.earscirev.2019.103030

Maas R, Nicholls IA, Legg C (1997) Igneous and metamorphic encalves in the S-type Deddick Granodiorite, Lachlan Fold Belt, SE Australia: petrographic, geochemical and Nd–Sr isotopic evidence for crustal melting and magma mixing. J Petrol 38:815–841
Ma L, Wang BD, Jiang ZQ, Wang Q, Li ZX, Wyman DA, Zhao SR, Yang JH, Gou GN, Guo HF, (2014) Petrogenesis of the early eocene adakitic rocks in the napuri area, southern Lhasa: Partial melting of thickened lower crust during slab break-off and implications for crustal thickening in southern Tibet. Lithos 196:321–338
Ma L, Wang Q, Wyman DA, Jiang ZQ, Wu FY, Li XH, Yang JH, Gou GN, Guo HF (2015) Late Cretaceous back-arc extension and arc system evolution in the Gangdese area, southern Tibet: geochronological, petrological, and Sr-Nd-Hf-O isotopic evidence from Dagze diabases. J Geophys Res Solid Earth 120(9):6159–6181
Maluski H, Proust F, Xiao XC (1982) 39Ar/40Ar dating of the trans-Himalayan calc-alkaline magmatism of southern Tibet. Nature 298:152–156
McDonough WF, Sun SS (1995) The composition of the Earth. Chem Geol 120:223–253
Meng Y, Xiong F, Xu Z, Ma X (2019) Petrogenesis of late Cretaceous mafic enclaves and their host granites in the Nyemo region of southern Tibet: implications for the tectonic-magmatic evolution of the Central Gangdese Belt. J Asian Earth Sci 176:27–41
Middlemost EAK (1994) Naming materials in the magma/igneous rock system. Earth Sci Rev 37:215–224
Mo X, Zhao Z, Zhou S, Dong GC (2002) Evidence for timing of the initiation of India-Asia collision from Igneous rocks in Tibet. AGU Fall Meeting Abstracts 83(47):F1003, S62B–1201
Mo X, Dong GC, Zhao ZD, Zhou S, Wang LL, Qiu RZ, Zhang FQ (2005) Temporal- spatial distribution and Characteristics of granites in the Gangdese belt, Tibet and implications for crustal growth and evolution. Geol J China Univ 11(3):281–290
Mo X, Zhao Z, Deng J et al (2003) Response of volcanism to the India-Asia collision. Earth Sci Front 10(3):135–148
Mo X, Zhao Z, Deng J, Flower M, Yu X, Luo Z, Li Y, Zhou S, Dong G, Zhu D, Wang L (2006) Petrology and geochemistry of post-collisional volcanic rocks from the Tibetan plateau: implications for lithosphere heterogeneity and collisioninduced asthenospheric mantle flow. In: Dilek Y, Pavlides S (eds) Postcollisional tectonics and magmatism in the Mediterranean region and Asia. Geological Society of America, Special Paper 409, pp 507–530
Mo X, Hou Z, Niu Y et al (2007) Mantle contributions to crustal thickening during continental collision: evidence from Cenozoic igneous rocks in southern Tibet. Lithos 96(1–2):225–242
Mo X, Niu Y, Dong G et al (2008) Contribution of syncollisional felsic magmatism to continental crust growth: a case study of the Paleogene Linzizong volcanic succession in southern Tibet. Chem Geol 250(1–4):49–67
Mo XX, Dong GC, Zhao ZD, Zhu DC, Zhou S, Niu YL (2009) Mantle input to the crust in southern Gangdese, Tibet, during the Cenozoic: zircon Hf isotopic evidence. J Earth Sci 20:241–249
Niu Y, Lesher CM (1991) Hydrothermal alteration of mafic metavolcanic rocks and genesis of Fe–Zn–Cu sulfide deposits, Stone Hill district. Alabama Econ Geol 86:983–1001
Niu YL, Zhao ZD, Zhu DC, Mo X (2013) Continental collision zones are primary sites for net continental crust growth – A testable hypothesis. Earth Sci Rev 127:96–110
Pan GT, Ding J, Yao DS, Wang LQ, Luo JN, Yan YJ, Yong YY, Zheng JK, Liang XZ, Qin DH, Jiang XS, Wang QH, Li RS, Geng QR, Liao ZL, Zhu DC (2004) Geological Map of the Tibetan Plateau and Adjacent Areas (1:1500 000) and the Manual. Chengdu Map Publishing House, Chengdu 133pp
Peacock SM (2003) Thermal structure and metamorphic evolution of subducting slabs, in inside the subduction factory. Geophys Monogr 238:7–22
Pearce JA, Houjun M (1988) Volcanic rocks of the 1985 Tibet geotraverse: Lhasa to Golmud. Philos Trans R Soc Lond A Math Phys Sci 327(1594):169–201
Rapp RP, Watson EB (1995) Dehydration melting of metabasalt at 8-32-Kbar—implications for continental growth and crust–mantle recycling. J Petrol 36:891–931

Rickwood PC (1989) Boundary lines within petrologic diagrams which use oxides of major and minor elements. Lithos 22:247–263
Rudnick RL, Gao S (2003) Composition of the continental crust. In: Holland HD, Turekian KK (eds) The crust: treatise on geochemistry, vol 3. Elsevier-Pergamon, Oxford, pp 1–64
Schärer U, Xu RH, Allègre CJ (1984). U–Pb geochronology of Gangdese (Transhimalaya) plutonism in the Lhasa-Xigaze region, Tibet Earth Planet Sci Lett 69, 311–320.
Shui XF, Klemd R, He ZY, Mao JW, Zhao YY (2021) Geochronology and petrogenesis of eocene gabbros and granitic rocks of the eastern gangdese belt, southern Tibet: implications for the timing of India-Asia collision. Gondwana Res 97:145–157
Sisson TW, Ratajeski K, Hankins WB, Glazner AF (2005) Voluminous granitic magmas from common basaltic sources. Contrib Mineral Petrol 148:635–661
Sun, S.S., McDonough, W., 1989. Chemical and isotopic systematics of oceanic basalts: implications for mantle composition and processes. In Magmatism in the ocean basins (eds AD Saunders and MJ Norry), pp. 313–345. Geological Society of London, Special Publication No. 42
Tatsumi Y (2006) High-Mg andesites in the Setouchi volcanic belt, Southwestern Japan: analogy to Archean magmatism and continental crust formation? Annu Rev Earth Planet Sci 34:467–499
van Hunen J, Allen MB (2011) Continental collision and slab break-off: a comparison of 3-D numerical models with observations. Earth Planet Sci Lett 302:27–37
Vernon RH (1984) Microgranitoid enclaves in granites-globules of hybrid magma quenched in a plutonic environment. Nature 309:438–439
Vielzeuf D, Schmidt MW (2001) Melting relations in hydrous systems revisited: application to metapelites, metagreywackes and metabasalts. Contrib Mineral Petrol 141:251–267
von Blanckenburg F, Davies JH (1995) Slab Breakoff: A Model for Syncollisional Magmatism and Tectonics in the Alps. Tectonics 14(1):120–131
Wang Q (2007) Geochemistry and zircon U-Pb chronology of Linzizong Group volcanic rocks in western Gangdese, Tibet, Master thesis. China University of Geosciences, Beijing, 81 pp. (in Chinese)
Wang Q, Chung SL, Li XH, Wyman DA, Li ZX, Sun WD, Qiu HN, Liu YS, Zhu YT (2012) Crustal melting and flow beneath Northern Tibet: evidence from Mid-Miocene to quaternary strongly peraluminous rhyolites in the Southern Kunlun Range. J Petrol 53(12):2523–2566
Wang R, Richards JP, Hou ZQ, An F, Creaser RA (2015) Zircon U-Pb age and Sr-Nd Hf-O isotope geochemistry of the Paleocene–Eocene igneous rocks in western Gangdese: evidence for the timing of Neo-Tethyan slab break off. Lithos 224–225:179–194
Wang Q, Zhu DC, Liu AL, Cawood PA, Liu SA, Xia Y, Chen Y, Wang H, Zhang LL, Zhao ZD (2018) Survival of the lhasa terrane during its collision with Asia due to crust-mantle coupling revealed by ca.114Ma intrusive rocks in western Tibet. Lithos 304–307, 200–210
Wang YF, Zeng L, Gao J, Zhao L, Gao LE, Shang Z (2019) Along-arc variations in isotope and trace element compositions of Paleogene gabbroic rocks in the Gangdese batholith, southern Tibet. Lithos 324–325:877–892
Wen DR, Liu DY, Chung SL, Chu MF, Ji JQ, Zhang Q, Song B, Lee TY, Yeh MW, Lo CH (2008) Zircon SHRIMP U-Pb ages of the Gangdese batholith and implications for Neotethyan subduction in southern Tibet. Chem Geol 252(3–4):191–201
Wen DJ, Hu XM, Qiu JS, Yu JH, Wang RQ, He ZY, Li YF (2021) Petrogenesis of early Eocene granites and associated mafic enclaves in the Gangdese batholith, Tibet: implications for net crustal growth in collision zones. Lithos 394–395:106170
Whalen JB, Currie KL, Chappell BW (1987) A-type granites: geochemical characteristics, discrimination and petrogenesis. Contrib Mineral Petrol 95:407–419
Willems H, Zhou Z, Zhang B, Gräfe KU (1996) Stratigraphy of the upper Cretaceous and lower tertiary strata in the Tethyan Himalayas of Tibet (Tingri area, China). Geol Rundsch 85:723–754
Wu FY, Clift PD, Yang JH (2007) Zircon Hf isotopic constraints on the sources of the Indus Molasse, Ladakh Himalaya, India Tectonics 26. https://doi.org/10.1029/2006TC002051TC2014

Wu, FY, Ji, WQ, Liu CZ, Chung SL (2010) Detrital zircon U–Pb and Hf isotopic data from the Xigaze fore-arc basin: constraints on Transhimalayan magmatic evolution in southern Tibet. Chem Geol 271, 13–25

Wyllie PJ, Wolfe MB (1993) Dehydration-melting of amphibolite: sorting out the solidus. In: Prichard HM, Alabaster T, Harris NBW, Near CR (Eds.), Magmatic Processes and plate tectonics. Geol Soc Spec Publ 76:405–416

Xiong SQ, Zhou FH, Yao ZX et al (2001) Aeromagnetic Survey of Central and Western Tibetan Plateau. Geological Publishing House, Beijing, pp 75–87

Xie KJ, Zeng LS, Liu J et al (2011) Timing and geochemistry of the Linzizong Group volcanic rocks in Sangsang area, Ngamring County, southern Tibet. Geol Bull China 30(9):1339–1352

Xu RH, Schärer U, Allègre CJ (1985) Magmatism and metamorphism in the Lhasa block (Tibet): a geochronological study. J Geol 93:41–57

Xu YG, Lan JB, Yang QJ, Huang XL, Qiu HN (2008) Eocene break-off of the Neo-Tethyan slab as inferred from intraplate-type mafic dykes in the Gaoligong orogenic belt, eastern Tibet. Chem Geol 255:439–453

Xu W, Zhu DC, Wang Q, Weinberg RF, Wang R, Li SM, Zhao ZD (2021) Mafic microgranular enclaves formed by gas-driven filter pressing during rapid cooling: an example from the Gangdese Batholith in southern Tibet. J Petrol 61:11–12

Yin A, Harrison TM (2000) Geologic evolution of the Himalayan-Tibetan Orogen. Annu Rev Earth Planet Sci 28(1):211–280

Zhang HF, Harris N, Guo L, Xu WC (2010) The significance of Cenozoic magmatism from the western margin of the eastern syntaxis, southeast Tibet. Contrib Mineral Petrol 160:83–98

Zhang KX, Wang GC, Hong HL, Xu YD, Wang A, Cao K, Luo MS, Ji JL, Xiao GQ, Lin X (2013) The study of the Cenozoic uplift in the Tibetan Plateau: A review. Geological Bulletin of China 32(1):1–18

Zhao ZD, Mo X, Nomade P, Renne PR, Zhou S, Dong GC, Wang LL, Zhu DC, Liao ZL (2006) Post-collisional ultrapotassic rocks in the Lhasa Block, the Tibetan Plateau: spatial and temporal distribution and its implications. Acta Petrol Sin 22(4):787–794

Zhou JS (2017) Ore genesis of the Narusongduo lead–zinc deposit, Tibet and the implications for regional metallogeny, PhD thesis. Peking University, Beijing, 133 pp. (in Chinese with English abstract)

Zhou S, Mo X, Dong G et al (2004) ^{40}Ar-^{39}Ar geochronology of Cenozoic Linzizong volcanic rocks from Linzhou Basin, Tibet, China, and their geological implications. Chin Sci Bull 49(18):1970–1979

Zhou J, Yang Z, Hou Z, Liu Y, Zhao X, Zhang X, Zhao M, Ma W (2017) The geochemical evolution of syncollisional magmatism and the implications for significant magmatic-hydrothermal lead–zinc mineralization (Gangdese, Tibet). Lithos 288-289:143–155

Zhu DC, Zhao ZD, Niu YL, Mo X, Chung SL, Hou ZQ, Wang LQ, Wu FY (2011) The Lhasa Terrane: record of a microcontinent and its histories of drift and growth. Earth Planet Sci Lett 301(1/2):241–255

Zhu DC, Zhao ZD, Niu YL, Dilek Y, Hou ZQ, Mo X (2013) The origin and pre-Cenozoic evolution of the Tibetan Plateau. Gondwana Res 23:1429–1454

Zhu DC, Wang Q, Zhao ZD et al (2015) Magmatic record of India-Asia collision. Sci Rep 5(1):1–9

Zhu DC, Wang Q, Zhao ZD, Chung SL, Cawood PA, Niu YL, Liu SA, Wu FY, Mo X (2015) Magmatic record of India-Asia collision. Sci Rep 5:14289. https://doi.org/10.1038/srep14289

Zhu DC, Wang Q, Cawood PA, Zhao ZD, Mo XX (2017) Raising the Gangdese Mountains in Southern Tibet. J Geophys Res Solid Earth 122(1):214–223

Zhu DC, Wang Q, Chung SL, Cawood P, Zhao ZD (2019) Gangdese magmatism in southern Tibet and India–Asia convergence since 120 Ma. Geol Soc Lond Spec Publ 483:583–604

Chapter 4
Cenozoic Post-collisional Igneous Rocks in the Tibetan Plateau

Abstract The post-collisional igneous rocks (31–9 Ma) are widespread in the central and southern Lhasa subterranes and extend ~1500 km between 80°E and 91°E. They are associated with the N-S-striking normal faults, N-S elongated lakes, and Neogene basins or grabens. They mainly consist of three magmatic suites, namely, ultrapotassic volcanic rocks (UPVRs), potassic volcanic rocks (PVRs), and adakitic intrusions. The petrography, geochemistry, and petrogenesis are summarized in this chapter. The UPVRs were generated by the partial melting from a phlogopite-bearing peridotite upper mantle source region. The PVRs is proposed to be the combined results of the input of the UPVRs, partial melting of amphibole-bearing eclogite-facies lower crust, and fractional crystallization dominated by feldspar and clinopyroxene. The coeval adakitic rocks in the Lhasa terrane display high Sr/Y and La/Yb, low Y features, suggesting the source region is most probably thickened lower crust (garnet-bearing, hydrous amphibolite, or amphibole eclogite).

Keywords Post-collisional · Ultrapotassic volcanic rocks · Potassic volcanic rocks · Adakitic intrusions · Tibetan plateau

4.1 Spatial-Temporal Distributions in the Lhasa Terrane

The Lhasa Terrane, located between the Neo-Tethyan Bangong-Nujiang suture (BNS) to the north and the Yarlung Zangbo suture (IYZS) to the south, is a huge tectonomagmatic belt with extensive exposure of the Late Triassic to Cenozoic igneous rocks in southern Tibet (Zhu et al. 2011; Fig. 4.1). The Lhasa terrane can be divided into the northern, central and southern Lhasa subterranes, separated by the Shiquanhe-Nam Tso mélange zone (SNMZ) and the Luobadui-Milashan fault (LMF), respectively. The southern Lhasa subterrane is dominated by the Gangdese batholith (mainly 65–41 Ma, Ji et al. 2009), syn-collisional Linzizong volcanic

Prof. Zhidan Zhao has fully contributed to this chapter.

X. Mo et al., *Magmatism and Crustal Evolution of the Tibetan Plateau*, Modern Approaches in Solid Earth Sciences 25,
https://doi.org/10.1007/978-981-97-2109-2_4

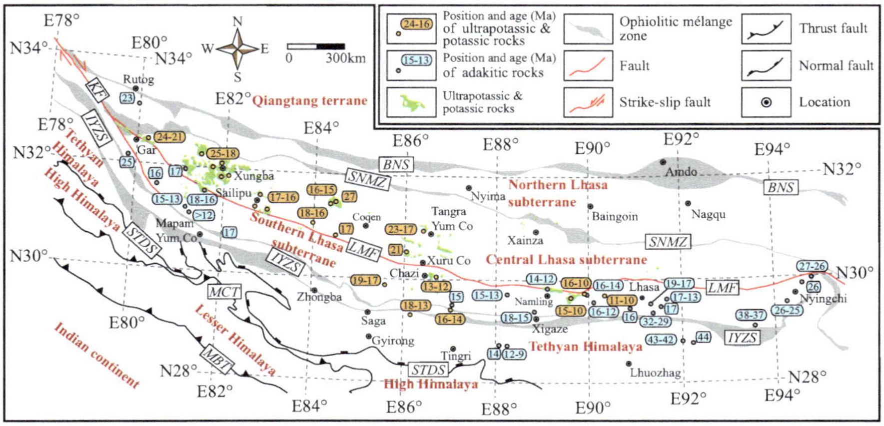

Fig. 4.1 Simplified geological map of southern Tibet showing the main tectonic units (modified from Zhu et al. 2011) and spatial-temporal distribution of the Cenozoic post-collisional magmatic rocks (modified from Liu et al. 2014b). The post-collisional potassic-ultrapotassic rocks in the Lhasa terrane include Shiquanhe (24–21 Ma, Williams et al. 2004), Xungba-Bangba-Yare (25–18 Ma, Miller et al. 1999; Chen et al. 2011; Liu et al. 2011), Sailipu (17–16 Ma, Sun et al. 2008), Zabuye-Bugasi (27–15 Ma, Nomade et al. 2004), Maiga-Gongmutang (18–16 Ma, Ding et al. 2006; Zhao et al. 2006), Tangra Yum Co-Konglong Xiang (23–17 Ma, Ding et al. 2003; Chen et al. 2010), and Xuru Co (13–12 Ma, Ding et al. 2003; Zhao et al. 2006), Daggyai Tso-Pabbai Zong (19–13 Ma, Williams et al. 2001, 2004), Xuena-Xigaze (18–14 Ma, Chan et al. 2009; Liu et al. 2013), Namling-Wuyu (15–10 Ma, Zhao et al. 2001; Spicer et al. 2003; Williams et al. 2004), Majiang (16–10 Ma, Coulon et al. 1986), and Yangying (11–10 Ma, Nomade et al. 2004). Data of adakitic rocks in the southern Lhasa subterrane are from Guo et al. (2007), Chung et al. (2009), and Guan et al. (2012). *BNS* Bangong-Nujiang suture zone, *SNMZ* Shiquan river-Nam Tso mélange zone, *LMF* Luobadui-Milashan fault, *KF* Karakorum fault, *IYZS* Indus-Yarlung Zangbo suture zone

successions (65–45 Ma, Mo et al. 2008), and the post-collisional igneous rocks (mainly 31–9 Ma, Liu et al. 2017). The post-collisional igneous rocks are widespread in the central and southern Lhasa subterranes and extend ~1500 km between 80°E and 91°E, roughly parallel to the IYZS (Fig. 4.1, Zhao et al. 2009). They are largely associated with the N-S-striking normal faults, N-S elongated lakes, and Neogene basins or grabens (Zhao et al. 2009). They predominantly consist of three magmatic suites, namely, ultrapotassic volcanic rocks (UPVRs), potassic volcanic rocks (PVRs), and adakitic intrusions (Chung et al. 2005; Liu et al. 2017).

The potassic and ultrapotassic volcanic rocks are scattered irregularly along the IYZS and LMF in southern Tibet (Fig. 4.1). The ultrapotassic volcanic rocks (UPVRs) mainly outcrop west of 87°E throughout the central Lhasa subterrane and have been recognized in the Shiquanhe, Bangba, Xungba, Jarga, Zabuye salt lake, Maiga, Wenbu, Yiqian, Chazi, and Pabbai Zong areas (Fig. 4.1, Zhao et al. 2009; Liu et al. 2015). They have zircon U–Pb ages varying from 24 to 8 Ma, getting progressively younger towards east (Zhao et al. 2009; Liu et al. 2014a). Comparatively, the potassic volcanic rocks (PVRs) have a wider spatial distribution and predominantly crop out in the central and southern Lhasa subterranes from Shiquanhe to Lhasa City (Fig. 4.1). Their Ar–Ar dating, together with zircon U–Pb dating, yield the ages from 22 to 10 Ma (Miller et al. 1999; Liu et al. 2014a), nearly coeval with the UPVRs. Particularly, a remarkable feature for the early Miocene PVRs in southwestern Tibet is their bimodal distribution with overlying UPVRs (Liu et al. 2014b). In contrast, adakitic rocks are concentrated along a narrow E-W trending belt in the southern Lhasa subterrane, more abundant to the east of 87°E. They crop out in the Shiquanhe, Gegar, Guoqin, Dajia Co, Xigaze, Nanmuling, Majiang, Yangying, Wuyu, Nanmu, Jiama and Linzhi areas (Fig. 4.1). These rocks intruded the Gangdese batholith and Linzizong volcanic succession (Liu et al. 2017). The majority of adakitic rocks give zircon U–Pb ages of 31–9 Ma (e.g., Hou et al. 2004; Liu et al. 2014b; Hu et al. 2017), and the rest of them were emplaced as early as the Eocene (51–38 Ma, Guan et al. 2012; Ji et al. 2012). Importantly, almost all the reported porphyry Cu (±Mo) deposits are associated with these adakitic intrusions (e.g., Hou et al. 2013; Huang et al. 2019).

4.2 Petrography

In the Lhasa terrane, the Cenozoic post-collisional igneous rocks occur as either small-volume plugs or dike swarms, intruding the Gangdese batholith, Linzizong volcanic successions, and underlying sedimentary formations (Williams et al. 2001). In general, the petrography of the UPVRs and PVRs is similar to the orogenic lamproites from central Italy, SE Spain (Conticelli et al. 2009), and the East European Alpine belt (Altherr et al. 2004). The UPVRs, emplaced as lava flows, are predominantly trachyte, trachyandesite, basaltic trachyandesite, phonolite, and tephriphonolite (Zhao et al. 2009). The majority of the UPVRs are porphyritic with up to 47 vol.% phenocrysts (Fig. 4.2a, b). The mafic suites generally contain <20 vol.% phenocrysts of olivine, clinopyroxene, phlogopite and sanidine with the groundmass having an assemblage of olivine, clinopyroxene, phlogopite, sanidine plus apatite, Fe–Ti oxides and glass. The felsic members, on the other hand, are highly phyric with up to 47 vol.% phenocrysts of clinopyroxene, sanidine and phlogopite with no or minor olivine. Besides, nepheline and leucite are common in the groundmass.

The PVRs occur as small-volume lava flows and minor plugs or dykes (Zhang et al. 2017). The PVRs are mostly trachyte with minor trachydacite and trachyandesite. They are typically porphyritic with a phenocryst assemblage of sanidine, plagioclase, clinopyroxene and minor phlogopite (Fig. 4.2c, d, Liu et al. 2017). Phlogopite phenocrysts are commonly zoned and oxidized (Miller et al. 1999). The groundmass is dominated by sanidine, plagioclase, clinopyroxene, phlogopite, minor apatite plus Fe–Ti oxides and glass.

The adakitic rocks occur as small bodies of granitic porphyry, dykes, plugs and minor lava flows with a typical porphyritic texture (Liu et al. 2017). In the eastern Lhasa terrane, the adakitic rocks occur as small intrusive bodies of granitic porphyry, monzonitic granitic porphyry, and quartz monzonitic porphyry (Chen et al. 2011). The main phenocryst phases (~10 modal%) are sanidine, plagioclase, biotite and minor clinopyroxene. The groundmass minerals include sanidine, plagioclase, biotite, quartz plus accessory zircon and apatite (Fig. 4.2e, f). In the western Lhasa terrane, the adakitic rocks occur as both intrusions and lava flows with cryptocrystalline groundmass. The adakitic rocks in the western Lhasa terrane share similar petrographic characteristics to those in the eastern segment.

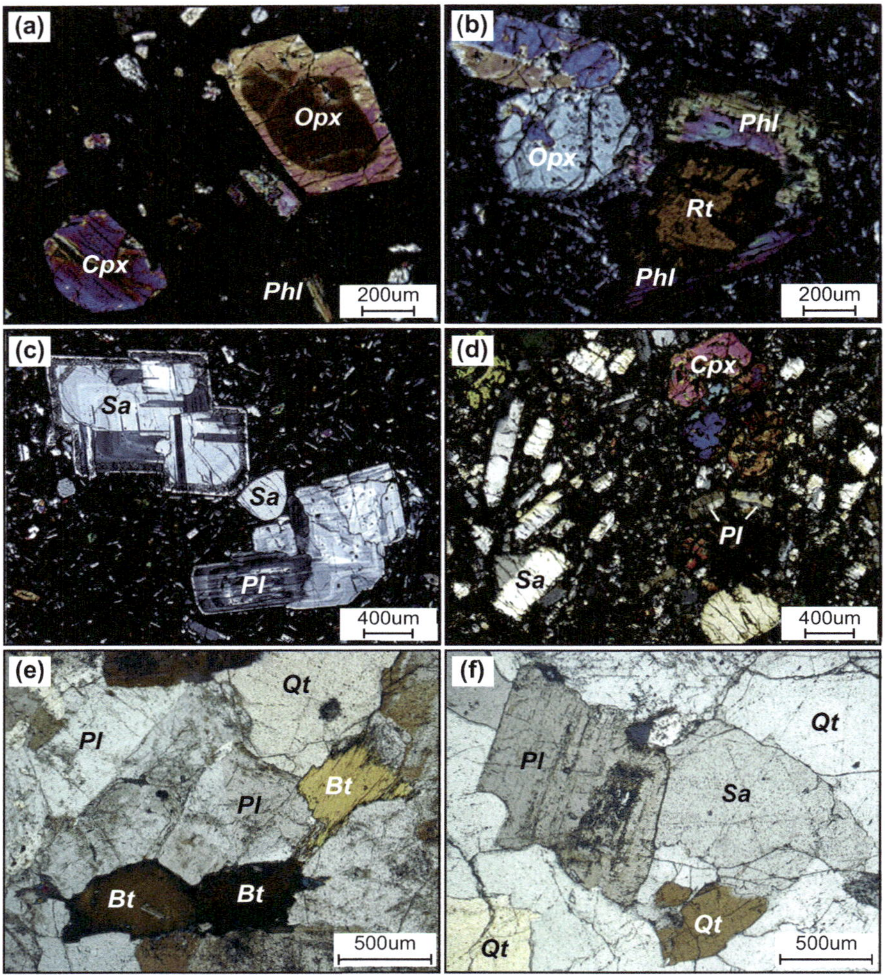

Fig. 4.2 Representative photomicrographs for post-collisional ultrapotassic volcanic rocks (**a**, **b**), potassic volcanic rocks (**c**, **d**) and adakitic rocks (**e**, **f**) in the Lhasa terrane. Abbreviations: *Bt* biotite, *Cpx* clinopyroxene, *Opx* orthopyroxene, *Phl* phlogopite, *Pl* plagioclase, *Qt* quartz, *Rt* rutile, *Sa* sandine. (Modified from Liu et al. 2014b, 2017)

4.3 Geochemistry

4.3.1 Ultrapotassic Volcanic Rocks in the Lhasa Terrane

The group of ultrapotassic volcanic rocks (UPVRs) in the Lhasa terrane are classified using the geochemical criteria of Foley et al. (1987) with high K_2O (>3 wt.%), MgO (>3 wt.%) and K_2O/Na_2O (>2). They have a broad range of SiO_2 (46.44–69.58 wt.%), MgO (mostly >3 wt.%), CaO (1.26–10.66 wt.%), K_2O (4.70–11.84 wt.%) and K_2O/Na_2O ratios (1.69–14.47). The LOI (loss on ignition) is relatively low, reflecting the contribution of phlogopite in these UPVRs rather than alteration, which generally decreases with increasing MgO. Most UPVRs are trachyandesite and trachyte, accompanied by minor basaltic tranchyandesite, tephriphonolite and phonolite (Fig. 4.3a). The high K_2O contents, together with high K_2O/Na_2O ratios, show the ultrapotassic-dominant affinity. Besides, most oxides show linear trends against MgO (Zhao et al. 2009) with SiO_2, Al_2O_3, Na_2O and K_2O inversely and Fe_2O_3, TiO_2 and CaO positively correlating with MgO. Additionally, the UPVRs have significantly varying high Rb (102–2518 ppm), Cs (1.80–348 ppm), Sr (331–3476 ppm), Ba (1256–6436 ppm) and Th (14–336 ppm), which are independent of MgO contents. They are strongly enriched in light rare earth elements (LREEs; La: 115 to 1168 times chondrite, Sun and McDonough 1989) with mostly significant Eu negative anomalies (δEu = 0.37–1.05). In the primitive mantle-normalized spider diagram (Fig. 4.4), the UPVRs show remarkable features of highly enriched large ion lithophile elements (LILEs; e.g., Rb, Pb and Sr) and negative anomalies in high field-strength elements (HFSEs; e.g., Nb, Ta, Ti and P).

The UPVRs show highly radiogenic Sr ($^{87}Sr/^{86}Sr$ = 0.710631–0.739246), unradiogenic Nd ($\varepsilon_{Nd}(0) = -18.63$ to -7.60) and fairly radiogenic Pb isotopic signatures ($^{206}Pb/^{204}Pb$ = 18.33–19.35; $^{207}Pb/^{204}Pb$ = 15.65–15.89; $^{208}Pb/^{204}Pb$ = 39.08–40.25), revealing that the ultrapotassic melts are sourced from an EM II-type enriched upper mantle (Figs. 4.5 and 4.6). Their Sr isotopic compositions are distinguishable from the hosted gabbroic (0.709763, Liu et al. 2015) and granitic (0.713443, Liu et al. 2015) xenoliths. The varying elemental and isotopic systematics of the UPVRs point to the presence of extensive crustal contamination in their mantle sources. Also, it implies that the mantle source of the UPVRs was replenished by old continental crust materials (T_{DM} = 1.3–2.1 Ga, Zhao et al. 2009). The same cases have been found in Italy (e.g., Foley et al. 1987; Casalini et al. 2019), Roman province (e.g., Conticelli et al. 2015), United States (Davis et al. 1996) and other localities. In addition, these UPVRs display a wide range of Re (0.04–0.416 ppb) and Os (0.004–0.698 ppb), and give more radiogenic Os isotopic compositions ($^{187}Os/^{188}Os$ = 0.1315–0.7197, Fig. 4.7) than the primitive upper mantle ($^{187}Os/^{188}Os$ = 0.1296, Meisel et al. 2001) and the hosted peridotite xenoliths ($^{187}Os/^{188}Os$ = 0.1246, Liu et al. 2015). Besides, a positive correlation can be observed between 1/Os and $^{187}Os/^{188}Os$ (Liu et al. 2015).

Whole-rock $\delta^{18}O$ values of the UPVRs vary from 6.3‰ to 12.3‰, while those of phlogopite and zircon range from 6.3‰ to 12.3‰ (Zhao et al. 2009), 5.5‰ to 8.5‰ (Tian et al. 2017) respectively. These oxygen isotopic compositions are irrelevant to LOI, and exhibit lower $\delta^{18}O$ values than those of Himalayan basement (10.9–13.3‰; Cheng and Guo 2017). The above means that the oxygen isotopic compositions of the UPRVs are inherited from the mantle source, not post-crystallization hydration or low-temperature alteration (e.g., Benito et al. 1999). In contrast to mantle peridotite ($\delta^{26}Mg = -0.25 \pm 0.04$‰, Teng et al. 2010), the UPVRs present varying lighter Mg isotopic compositions, ranging from −0.22‰ to −0.46‰ in $\delta^{26}Mg$ values. Their Mg isotopic compositions are significantly different from the hosted gabbroic ($\delta^{26}Mg = -0.12 \pm 0.06$‰, Liu et al. 2015) and granitic ($\delta^{26}Mg = -0.49 \pm 0.05$‰, Liu et al. 2015) xenoliths. In addition, $\delta^{7}Li$ values of the UPVRs vary from −4.9‰ to 3.2‰, with the corresponding Li contents of 12.7–64.9 ppm (Tian et al. 2020).

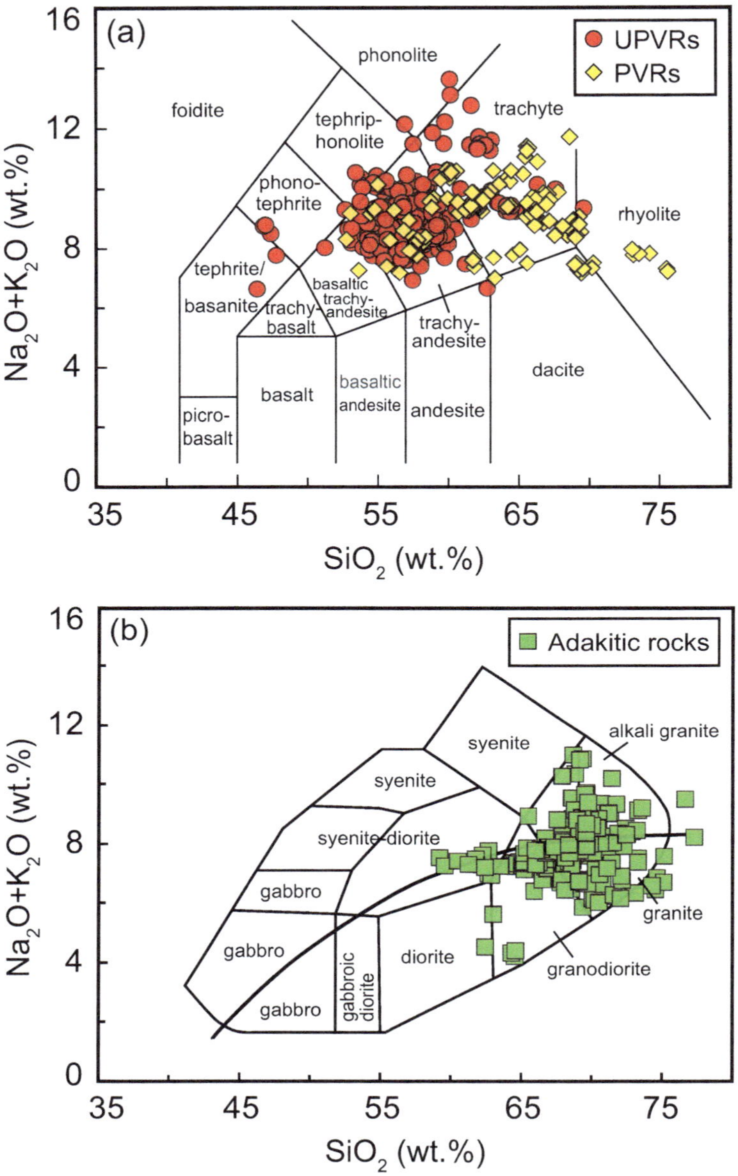

Fig. 4.3 Total alkalis vs. silica plot of ultrapotassic-potassic and potassic volcanic rocks (**a**) and the adakitic intrusive rocks (**b**) in the Lhasa terrane. The classification boundaries are from Le Maitre (1989), respectively. Data sources in the Lhasa terrane: the UPVRs and PVRs are from Miller et al. (1999), Ding et al. (2003), Nomade et al. (2004), Williams et al. (2004), Zhao et al. (2009), Guo et al. (2013, 2015), Liu et al. (2014b, 2015, 2017), Cheng and Guo (2017), Zhang et al. (2017), Tian et al. (2017, 2020), Guo and Wilson (2019), and the adakitic rocks are after Qu et al. (2004), Hou et al. (2004, 2013), Guo et al. (2007), Gao et al. (2010), Zheng et al. (2012), Yang et al. (2015), Hu et al. (2017), Li et al. (2017), Liu et al. (2017), Huang et al. (2019)

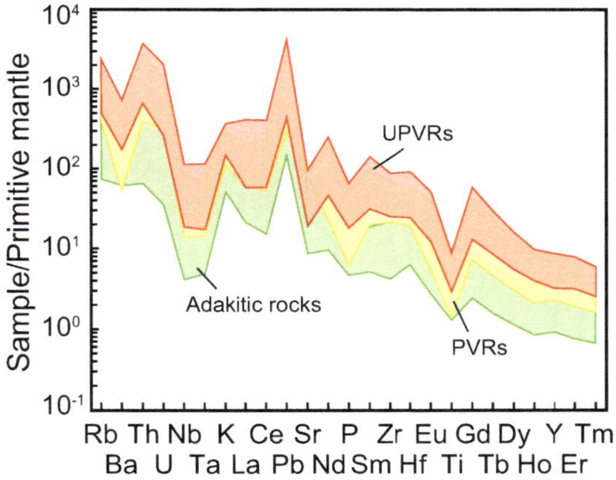

Fig. 4.4 Primitive mantle-normalized (Sun and McDonough 1989) trace-element patterns for the post-collisional igneous rocks in the Lhasa terrane. The rock types are as in Fig. 4.3

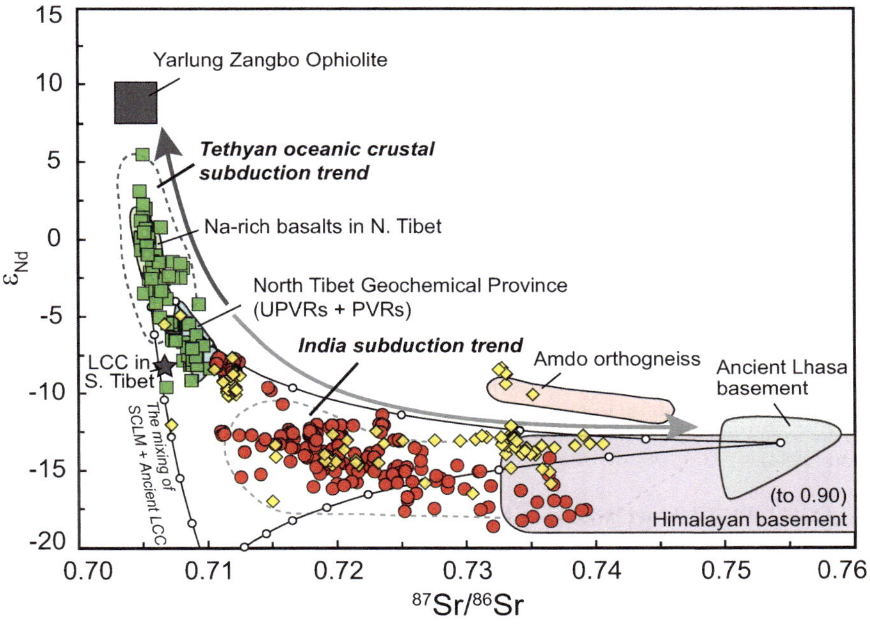

Fig. 4.5 Sr–Nd isotopic compositions of the post-collision igneous rocks in the Lhasa terrane (modified from Zhao et al. 2009; Liu et al. 2017). The data sources of the igneous rocks in the Lhasa terrane are as those in Fig. 4.3

4.3.2 *Potassic Volcanic Rocks in the Lhasa Terrane*

Potassic volcanic rocks (PVRs) in the Lhasa terrane are mainly trachyte and trachydacite, and only a few of them are trachyandesite and rhyolite (Fig. 4.3a). Compared with the UPVRs, they generally have varying higher SiO_2 (52.76–75.56 wt.%),

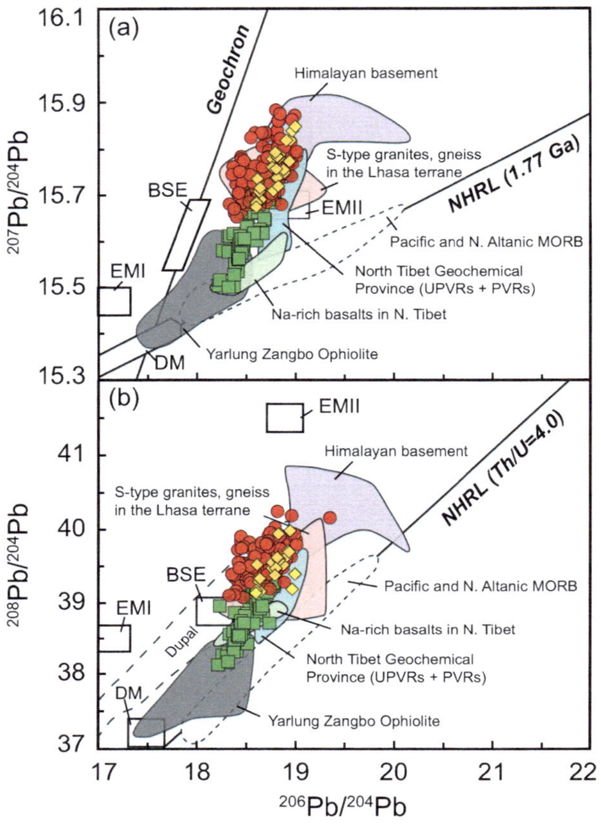

Fig. 4.6 $^{207}Pb/^{204}Pb$ (**a**) and $^{208}Pb/^{204}Pb$ (**b**) vs. $^{206}Pb/^{204}Pb$ of the post-collisional igneous rocks in the Lhasa terrane (Modified from Zhao et al. 2009). BSE, mantle endmembers (DM, EM I, and EM II) are from Zindler and Hart (1986). Northern Hemisphere Reference Line (NHRL) is based on Hart (1984)

Al_2O_3 (11.34–17.34 wt.%), lower MgO (0.25–11.79 wt.%), Fe_2O_3 (1.43–7.56 wt.%), K_2O (2.90–9.29 wt.%) and K_2O/Na_2O (0.71–6.67), exhibiting the shoshonitic-dominant affinity. In general, the less evolved PVRs (SiO_2 <60%) have relatively low K_2O/Na_2O ratios. And Fe_2O_3, CaO and TiO_2 correlate positively with MgO. Relative to the PVRs in the Qiangtang terrane, these in the Lhasa terrane show a much wider range in MgO contents, and give lower Fe_2O_3 and TiO_2 at a given MgO content (Williams et al. 2004). Besides, the PVRs display varying lower Rb (66.70–1359 ppm), Cs (4.17–121 ppm), Sr (32–1895 ppm), Ba (378–4294 ppm) and Th (16–253 ppm) than the aforementioned coeval UPVRs, showing enrichments in LREEs (La: 100–497 times chondrite) and negative anomalies in Eu (δEu = 0.45–0.89). In the primitive mantle-normalized spider diagram (Fig. 4.4), the PVRs show significant enrichment in LILEs (e.g., Rb, Th and U) and negative anomalies of HFSEs (e.g., Nb, Ta and Ti).

The PVRs have radiogenic Sr ($^{87}Sr/^{86}Sr$ = 0.706694–0.740593) and unradiogenic Nd ($\varepsilon_{Nd}(0)$ = −16.99 to −4.27) isotopic compositions relative to Bulk Silicate Earth

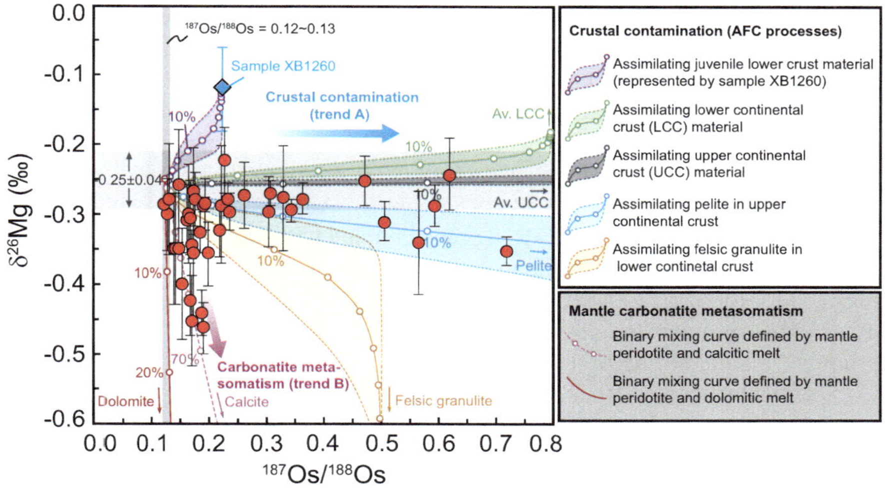

Fig. 4.7 Mg and Os isotopic compositions of the ultrapotassic volcanic rocks in the Lhasa terrane (modified from Liu et al. 2015). The UPVRs data are from Schaefer et al. (2000) and Liu et al. (2015)

(BSE) with a slightly wider range than those of the UPVRs (Fig. 4.5). At a given $^{206}Pb/^{204}Pb$ (18.59–18.99), $^{207}Pb/^{204}Pb$ (15.68–15.85) and $^{208}Pb/^{204}Pb$ (39.14–39.99) in the PVRs (Fig. 4.6), partly overlapping with those of the UPVRs, are higher than the Northern Hemisphere Reference Line (NHRL; Hart 1984). Compared with the North Tibet Geochemical Province (Figs. 4.5 and 4.6), the PVRs in the Lhasa terrane, together with UPVRs, exhibit higher $^{87}Sr/^{86}Sr$, $^{206}Pb/^{206}Pb$, but lower $^{143}Nd/^{144}Nd$. Similar to the UPVRs, the elemental and isotopic characteristics of the PVRs also point to the involvement of crustal material.

The limited whole-rock oxygen isotopic compositions of the PVRs are distinguishable from those of the UPVRs. They vary from 9.4‰ to 13.7‰ in $\delta^{18}O$ values (Liu et al. 2014b), matching well with the $\delta^{18}O$ ranges of the hosted comagmatic zircons (5.5–11.3‰, Tian et al. 2017; Zhang et al. 2017). Besides, their oxygen isotopic compositions are close to those of granitoids derived from the Lhasa basement ($\delta^{18}O$ = 8.9–13.3‰, Liu et al. 2014b). Furthermore, the PVRs have similar Li isotopic compositions ($\delta^{7}Li$ = −4.9‰ to 3.2‰) to those of the UPVRs, which are irrelevant to MgO. The corresponding Li concentrations range from 15.3–46.1 ppm.

4.3.3 Adakitic Rocks in the Lhasa Terrane

The adakitic rocks in the Lhasa terrane have a wide range of SiO_2 (59.23–78.21 wt.%), Fe_2O_3 (0.72–8.17 wt.%), MgO (0.51–5.36 wt.%), CaO (0.35–5.51 wt.%), and K_2O (1.07–8.65 wt.%), equivalent to granite, granodiorite and quartz monzonite (Fig. 4.3b). They display higher SiO_2, lower MgO, K_2O, TiO_2, K_2O/Na_2O than the UPVRs and PVRs. Particularly, the adakitic rocks of ore-bearing group are dominated by shoshonitic and high-K calc-alkaline rocks with minor calc-alkaline examples, whereas the adakitic rocks of ore-barren group are mainly of high-K calc-alkaline and calc-alkaline affinities (Hou et al. 2004, 2013). Besides, the adakitic rocks have the lowest incompatible element concentrations (e.g., Th, REEs) among the UPVRs and PVRs (e.g., Liu et al. 2014b). They display varying concentrations of Rb (25.3–494 ppm), Cs (1.49–37.0 ppm), Sr (47.3–1175 ppm), Ba (87–1870 ppm), and Th (2.58–48.82 ppm). And their low values of Y (2.90–19.70 ppm) and Yb (0.27–1.78 ppm), coupled with high Sr/Y (16.1–209) and La/Yb (7.46–122), indicate the geochemical affinities with collision-type adakitic rocks (e.g., Hou et al. 2004). They present the enrichment in LREEs (La: 17.81–265 times chondrite) and the absence of Eu anomalies (average δEu = 0.92). In the primitive mantle-normalized spider diagram (Fig. 4.4), the adakitic rocks, with LILE enrichment (e.g., Rb, Pb, and Sr) and negative HFSE (e.g., Nb, Ti, and Ta) anomalies, have similar trace-element patterns to the UPVRs and PVRs.

These adakitic rocks in the Lhasa terrane display radiogenic Sr ($^{87}Sr/^{86}Sr$ = 0.704807–0.710192) and unradiogenic Nd ($\varepsilon_{Nd}(0) = -9.58$ to 5.50) isotopic compositions relative to BSE (Fig. 4.5). At a given $^{206}Pb/^{204}Pb$ (18.21–18.78), $^{207}Pb/^{204}Pb$ (15.50–15.74) and $^{208}Pb/^{204}Pb$ (38.13–39.31) of the adakitic rocks (Fig. 4.6) also plot above the NHRL line. Their Sr–Nd–Pb isotopic compositions are close to those reported for Na-rich basalts in north Tibet (Figs. 4.5 and 4.6). Compared with the UPVRs and PVRs in the Lhasa terrane, they give significantly lower $^{87}Sr/^{86}Sr$, $^{206}Pb/^{206}Pb$, but higher $^{143}Nd/^{144}Nd$.

4.4 Petrogenesis

4.4.1 Ultrapotassic Volcanic Rocks in the Lhasa Terrane

Ultrapotassic volcanic rocks in the Lhasa terrane plot on a strongly convex-upward O-Mg isotopic mixing hyperboles between the lithospheric mantle and ancient Lhasa basement (Fig. 4.8a). This process also accounts for the negative correlations between $^{87}Sr/^{86}Sr$ and MgO (Liu et al. 2014b). Meanwhile, the Sr–Nd–Pb isotopic data of the UPVRs show a clear mixing trend between the North Tibet Geochemical Province and the Himalayan continental crust reservoirs (Figs. 4.5 and 4.6). Coupled with partly varying radiogenic Os isotopic compositions, it implies the extensive crustal contamination with high-$^{187}Os/^{188}Os$ crustal material in the UPVRs, matching well with the observations of hosted crustal xenolith and Paleozoic-Proterozoic zircon xenocrysts (Liu et al. 2014a). Previously, the positive La/Yb-Dy/Yb correlation in the UPVRs was explained by the presence of residual garnet in the source region (e.g., Miller et al. 1999; Ding et al. 2003). It seems to be a paradox with the observations of the hosted mantle xenoliths, because these xenoliths, proposed as the source fragments of the UPVRs (Zhao et al. 2008), are present at depths of 50–65 km that is significantly shallower than the depth of the spinel-garnet transition zone (Liu et al. 2011). Additionally, several lines of evidence support lower crust contribution: (1) the close association of temporal-spatial distributions between the UPVRs and adakitic rocks (Fig. 4.1, Chen et al. 2011, 2012); (2) the observations of gabbro and granulite xenoliths hosted in the UPVRs (e.g., Miller et al. 1999; Chan et al. 2009); (3) the positive trend between Sr–O isotopic ratios is toward the lower crust with low $^{87}Sr/^{86}Sr$ and $\delta^{18}O$ ratios (Liu et al. 2014b). Also, the Mesozoic-Cenozoic magmatic zircons with positive $\varepsilon_{Hf}(t)$ have been identified in these UPVRs, further supporting the contributions from the juvenile crust produced during episodic convergent events (e.g., Sun et al. 2008; Liu et al. 2011). Compared with depleted mantle (DM), the primitive UPVRs with less crustal contamination display low Hf/Sm, $^{187}Os/^{188}Os$, $\delta^{26}Mg$, and high CaO/Al_2O_3, $^{87}Sr/^{86}Sr$, well-plotted along mantle peridotite-carbonate trends (Fig. 4.7). Meanwhile, their Hf/Sm ratios are positively related to $\delta^{26}Mg$, whereas a negative correlation is observed between $^{87}Sr/^{86}Sr$ and $\delta^{26}Mg$. Therefore, the primitive UPVRs, coupled with low $^{187}Os/^{188}Os$ and $\delta^{26}Mg$, plausibly reflect the recycled carbonate-bearing sediments into the mantle sources (Liu et al. 2015). Besides, the varying light Li isotopic compositions were explained by the interaction with subducted slab-derived fluids/melts (Tian et al. 2020). Heretofore, it has reached a consensus by the partial melting modeling results that the UPVRs in the Lhasa terrane were generally originated from a phlogopite-bearing peridotite source (e.g., Ding et al. 2003; Zhao et al. 2009; Liu et al. 2014b).

4.4.2 *Potassic Volcanic Rocks in the Lhasa Terrane*

In the Lhasa terrane, a close correlation between the PVRs and UPVRs is suggested by the similarities in temporal-spatial distributions (Fig. 4.1), incompatible element patterns (Fig. 4.4), and Sr–Nd–Pb–O isotopic compositions (Figs. 4.5 and 4.6). Hence, the PVRs in the Lhasa terrane were proposed to have genetic links with the coeval UPVRs (e.g., Chen et al. 2012; Liu et al. 2017). The La/Yb–Sm/Yb positive correlations defined by the PVRs can be well-explained by the involvement of the UPVRs, further supported by the variably low $(La/Sm)_N$ ("N" denotes normalization by chondrite) and elevated Y concentrations in high-K_2O/Na_2O PVRs (Liu et al. 2017). However, mixing with the UPVRs fails to account for high K_2O/Na_2O shown by several high-Si PVRs, which should be ascribed to magma sources and magmatic processes. The positive relationships between varying Rb/Sr and Sr/CaO in the PVRs reveal clinopyroxene, sanidine, and plagioclase as the dominant liquidus phases in the magmatic differential process (Liu et al. 2017). Additionally, several lines of evidence pointing to the involvement of the Lhasa terrane crust in the PVRs: (1) the varying negative co-magmatic zircon Hf isotopic compositions are comparable with magmatic zircons from the Paleozoic-Mesozoic granitoids in the Lhasa terrane (e.g., Wu et al. 2014); (2) the PVRs display similar Pb isotopic data with S-type granite and gneiss in the Lhasa terrane (average $^{207}Pb/^{204}Pb = 15.727 \pm 0.011$, Liu et al. 2017), but give lower Pb isotopic ratios than those of the Tethyan Himalaya gneiss (average $^{207}Pb/^{204}Pb = 15.824 \pm 0.020$; Vidal and Le Fort 1982; Gariépy et al. 1985); (3) Sr–Nd isotopic compositions of the PVRs can be reproduced in terms of binary mixing between lithospheric mantle-crust components (Liu et al. 2017); (4) the elevated $\delta^{18}O$ and $^{87}Sr/^{86}Sr$ ratios are negatively proportional to MgO, plotting in the crustal contamination trend (Fig. 4.8a). Collectively, their negative zircon $\varepsilon_{Hf}(t)$ values and whole-rock Sr–Nd–Pb–O isotopic signatures plausibly reflect the nature of the Lhasa terrane crust. Comparable to the UPVRs, the PVRs exhibit varying light Li isotopic compositions, suggesting the interaction with fluids/melts in the mantle source (Tian et al. 2020). As a consequence, the origin of the PVRs is proposed to be the combined results of the input of the UPVRs, partial melting of amphibole-bearing eclogite-facies lower crust, and fractional crystallization dominated by feldspar and clinopyroxene.

4.4.3 Adakitic Rocks in the Lhasa Terrane

Compared with ultrapotassic-potassic volcanic rocks, the coeval adakitic rocks in the Lhasa terrane display lower incompatible elements (e.g., Ba, Sr, Fig. 4.4) and distinct Sr–Nd–Pb isotopes, ruling out an origin by fractional crystallization from the UPVRs (Liu et al. 2014b). As to these adakitic rocks, the varying Sr/Y and Eu anomalies can

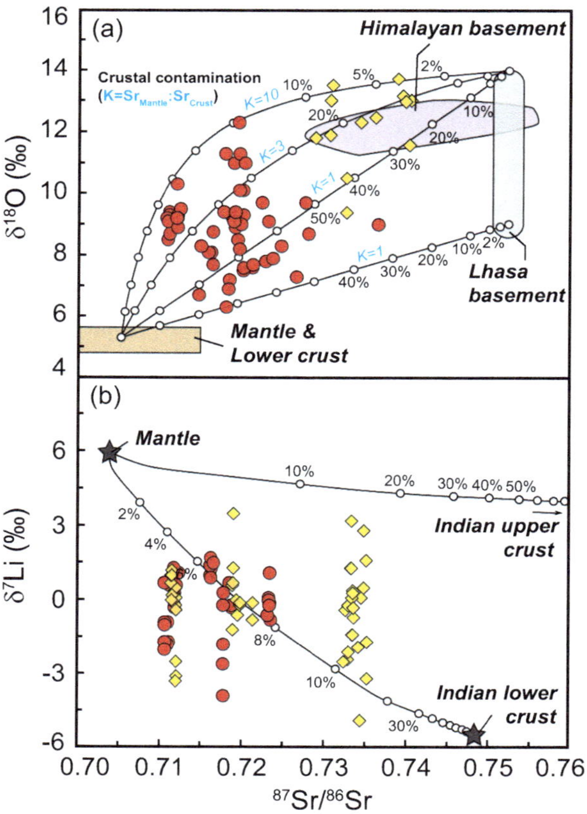

Fig. 4.8 $\delta^{18}O$ (**a**) and $\delta^{7}Li$ (**b**) vs. $^{87}Sr/^{86}Sr$ of the post-collisional ultrapotassic-potassic volcanic rocks in the Lhasa terrane (modified from Liu et al. 2015; Tian et al. 2020). The sources of the igneous rocks: Sr–O isotopic compositions (Miller et al. 1999; Zhao et al. 2009; Liu et al. 2014b; Cheng and Guo 2017), and Li–Sr isotopic data (Tian et al. 2017, 2020)

be reconciled with the crystallization of plagioclase. In the Lhasa terrane, the isotopic compositions of the juvenile crust accreted during the formation of the Gangdese batholith are reasonably speculated by the Linzizong mafic volcanic rocks (Mo et al. 2007, 52–41 Ma), the Dazi basalts (Gao et al. 2008, 40–38 Ma) and the Quxu gabbros (e.g., Dong et al. 2008, 52–41 Ma). The Sr–Nd–Pb isotopic compositions of the adakitic rocks partly overlap with the aforementioned basalts, suggesting a genetic link with a juvenile lower crust source (Zhang et al. 2014). Nevertheless, such a proposal cannot explain the entire adakitic rocks, especially those with higher $^{87}Sr/^{86}Sr$, $^{206}Pb/^{204}Pb$, and lower $^{143}Nd/^{144}Nd$ ratios, which require an additional component with enriched Sr–Nd compositions. Furthermore, this enriched additional component was proposed to be the Paleozoic-Neoproterozoic basement of the Lhasa terrane (e.g., Liu et al. 2014b, 2017). The enriched trace-element patterns, coupled with high Sr/Y and La/Yb, suggest the existence of garnet as a residual phase in the source region (e.g., Defant and Drummond 1990). The elements of Nb and Ti show strong depletion in these adakitic rocks, indicating the presence of rutile and amphibole as residual minerals. Thus, the source of the adakitic rocks is most probably garnet-bearing, hydrous amphibolite or amphibole eclogite (Hou et al. 2004). Finally, it was proposed that the adakitic rocks in the Lhasa terrane were derived from the thickened lower crust and then modified by crustal contamination with ancient Lhasa basement (e.g., Liu et al. 2014b; Zhang et al. 2014).

4.5 Plausible Geodynamic Mechanism

The ultrapotassic volcanic rocks in the central Lhasa subterrane display varying lower Hf/Sm and $\delta^{26}Mg$ values, whereas those UPVRs in the southern Lhasa subterrane appear to have mantle-like Hf/Sm and $\delta^{26}Mg$. Such northward geochemical variations were explained by Liu et al. (2015) to represent a clear pattern of mantle re-fertilization during the northward subduction of the Indus-Yarlung Zangbo Neo-Tethyan oceanic slab. Because slab metamorphic dehydration would be gradually replaced by a melting regime at greater depth (Schmidt and Poli 2003), the decreasing $\delta^{26}Mg$ and Hf/Sm from south to north plausibly reflect the enhanced mantle carbonation as the Neo-Tethyan oceanic slab goes deeper. This interpretation is consistent with the result of the thermodynamic calculation (Connolly 2005), pointing out that a significant amount of marine carbonates may have survived from the decarbonation and dissolution in sub-arc region and been transferred into deep earth.

In this regard, a two-stage geodynamic model was proposed to illustrate the mechanism of mantle metasomatism and the generation of the post-collisional igneous rocks in the Lhasa terrane (Zhao et al. 2009; Liu et al. 2015, 2017). With the slab devolatilization processes gradually changed from metamorphic dehydration to melting regime at greater depths, partial melts of carbonate-bearing sediments stemmed from downgoing oceanic slab to flux into the overlying mantle wedge, causing clinopyroxene-rich mantle assemblage (e.g. wehrlite, Yang et al. 2012) with isotopically light Mg (Fig. 4.9a).

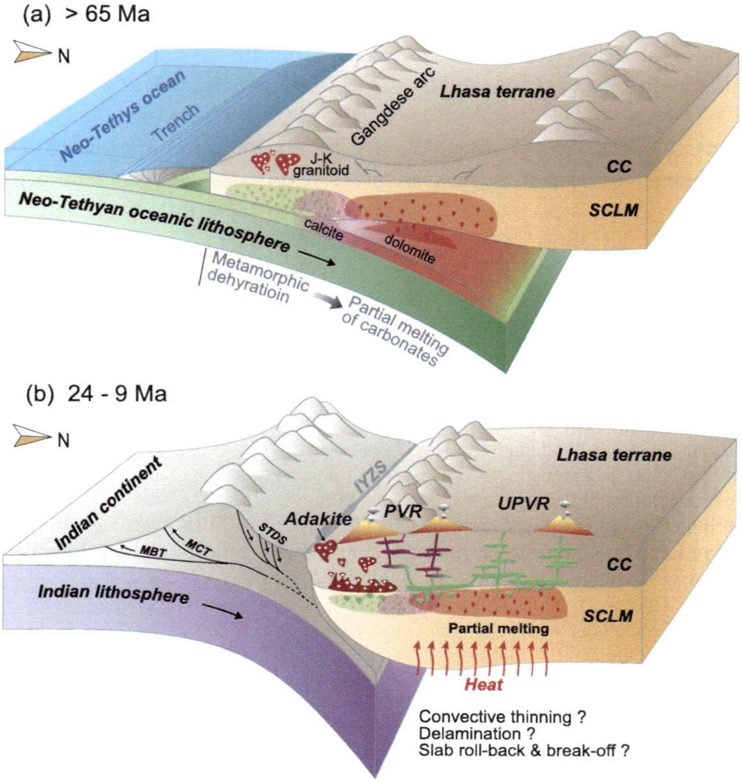

Fig. 4.9 A two-stage geodynamic model (modified from Liu et al. 2015, 2017) consisting of (**a**) the mechanism of mantle metasomatism during the northward subduction of the Neo-Tethyan seafloor, and (**b**) the generation of the Cenozoic post-collisional igneous rocks (UPVRs, PVRs, and adakitic rocks) in response to the continued convergence. *CC* continental crust, *MBT* main boundary thrust, *MCT* main central thrust, *STDS* southern Tibetan detachment system

Secondly, the isotopic characteristics of the mantle-derived volcanic rocks in north Tibet has not changed over the last 46 million years, implying the North Tibet Geochemical Province has not been disturbed by continuously subducted Indian crust materials. Thus, the subducted Indian plate only reach as far northward as the Bangong-Nujiang suture zone, consistent with the geophysical investigations (e.g., Tilmann et al. 2003; Priestley et al. 2008) and the He isotopic pattern in Tibet (Hoke et al. 2000). As evidenced by the uncomformable relationship between highly folded Cretaceous formation and the Linzizong volcanic successions in southern Tibet (e.g., Yin and Harrison 2000; Mo et al. 2007, 2008), the subduction of Indian plate beneath the Lhasa terrane would have absorbed >400 km of shortening and thickened the crust by tectonic underplating, finally causing elevation of the Tibetan Plateau. Under this tectonic background, thermal perturbation induced by deep geodynamic processes (e.g. convective thinning and/or delamination of the thickened lithospheric mantle, roll-back, and break-off of Indian slab) finally gives rise to the partial melting of metasomatized mantle sources and thickened lower crust, generating the widespread Cenozoic post-collisional magmatism in the Lhasa terrane (Fig. 4.9b). As evidenced by hosted crustal xenoliths, zircon xenocrysts with pre-eruptive older ages, highly variable Sr–Nd–Pb–O isotopic data, and radiogenic Os isotopic composition, these post-collisional magmas more or less underwent crustal contamination during their ascent through the thickened crust in the Lhasa terrane.

Acknowledgements Zhidan Zhao thanks his students Dong Liu, Hangshan Lei, Chenguang Sun for their assistance of this chapter, and Guochen Dong, Zengqian Hou, Yaoling Niu, Di-Cheng Zhu, and Xuanxue Mo for their long-term help during joint research over 20 years. This work and the on-going research have been supported by Ministry of Science and Technology of China (grants: 2019QZKK0702, 2002CB211702, 2011CB403102, 2016YFC0600304), National Natural Science Foundation of China (grants: 42121002, 42073035, 40473020, 40873023, 40830317, 41273044, 91755207), and the 111 project (B18048).

References

Altherr R, Meyer HP, Holl A, Volker F, Alibert C, McCulloch MT, Majer V (2004) Geochemical and Sr-Nd-Pb isotopic characteristics of late Cenozoic leucite lamproites from the east European Alpine belt (Macedonia and Yugoslavia). Contrib Mineral Petrol 147:58–73

Benito R, López-Ruiz J, Cebriá JM, Hertogen J, Doblas M, Oyarzun R, Demaiffe D (1999) Sr and O isotope constraints on source and crustal contamination in the high-K calc-alkaline and shoshonitic Neogene volcanic rocks of SE Spain. Lithos 46:773–802

Casalini M, Avanzinelli R, Tommasini S, Elliott T, Conticelli S (2019) Ce/Mo and molybdenum isotope systematics in subduction-related orogenic potassic magmas of Central-Southern Italy. Geochem Geophys Geosyst 20:2753–2768

Chan GHN, Waters DJ, Searle MP, Aitchison JC, Horstwood MSA, Growley Q, Lo CH, Chan JSL (2009) Probing the basement of southern Tibet: evidence from crustal xenoliths entrained in a Miocene ultrapotassic dyke. J Geol Soc Lond 166:45–52

Chen JL, Xu JF, Wang BD, Kang ZQ, Jie L (2010) Origin of Cenozoic alkaline potassic volcanic rocks at KonglongXiang, Lhasa terrane, Tibetan plateau: products of partial melting of a mafic lower-crustal source? Chem Geol 273:286–299

Chen JL, Xu JF, Zhao WX, Dong YH, Wang BD, Kang ZQ (2011) Geochemical variations in Miocene adakitic rocks from the western and eastern Lhasa terrane: implications for lower crustal flow beneath the southern Tibetan plateau. Lithos 125:928–939

Chen JL, Zhao WX, Xu JF, Wang BD, Kang ZQ (2012) Geochemistry of Miocene trachytes in Bugasi, Lhasa block, Tibetan plateau: mixing products between mantle- and crust-derived melts? Gondwana Res 21:112–122

Cheng Z, Guo Z (2017) Post-collisional ultrapotassic rocks and mantle xenoliths in the Sailipu volcanic field of Lhasa terrane, south Tibet: petrological and geochemical constraints on mantle source and geodynamic setting. Gondwana Res 46:17–42

Chung SL, Chu MF, Zhang Y, Xie Y, Lo CH, Lee TY, Lan CY, Li X, Zhang Q, Wang Y (2005) Tibetan tectonic evolution inferred from spatial and temporal variations in post-collisional magmatism. Earth Sci Rev 68:173–196

Chung SL, Chu MF, Ji J, O'Reilly SY, Pearson NJ, Liu D, Lee TY, Lo CH (2009) The nature and timing of crustal thickening in southern Tibet: geochemical and zircon Hf isotopic constraints from postcollisional adakites. Tectonophysics 477:36–48

Connolly JAD (2005) Computation of phase equilibria by linear programming: a tool for geodynamic modeling and its application to subduction zone decarbonation. Earth Planet Sci Lett 236:524–525

Conticelli S, Guarnieri L, Farinelli A, Mattei M, Avanzinelli R, Bianchini G, Boari E, Tommasini S, Tiepolo M, Prelević D, Venturelli G (2009) Trace elements and Sr–Nd–Pb isotopes of K-rich, shoshonitic, and calc-alkaline magmatism of the western Mediterranean region: genesis of ultrapotassic to calc-alkaline magmatic associations in a post-collisional geodynamic setting. Lithos 107:68–92

Conticelli S, Avanzinelli R, Ammannati E, Casalini M (2015) The role of carbon from recycled sediments in the origin of ultrapotassic igneous rocks in the Central Mediterranean. Lithos 232:174–196

Coulon C, Maluski H, Bollinger C, Wang S (1986) Mesozoic and Cenozoic volcanic rocks from central and southern Tibet: ^{39}Ar-^{40}Ar dating, petrological characteristics and geodynamical significance. Earth Planet Sci Lett 79:281–302

Davis LL, Smith D, McDowell FW, Walker NW, Borg LE (1996) Eocene potassic magmatism at two buttes, Colorado, with implications for Cenozoic tectonics and magma generation in the western United States. Geol Soc Am Bull 108:1567–1579

Defant MJ, Drummond MS (1990) Derivation of some modern arc magmas by melting of young subducted lithosphere. Nature 34:662–665

Ding L, Kapp P, Zhong D, Deng W (2003) Cenozoic volcanism in Tibet: evidence for a transition from oceanic to continental subduction. J Petrol 44:1833–1865

Ding L, Yue Y, Cai F, Xu X, Zhang Q, Lai Q (2006) $^{40}Ar/^{39}Ar$ geochronology, geochemical and Sr-Nd-O isotopic characteristics of the high-Mg ultrapotassic rocks in Lhasa block of Tibet: implications in the onset time and depth of NS-striking rift system. Acta Geol Sin 80:1252–1261

Dong G, Mo X, Zhao Z, Zhu D, Song Y, Wang L (2008) Gabbros from southern Gangdese: implication for mass exchange between mantle and crust. Acta Petrol Sin 24:203–210 (in Chinese with English abstract)

Foley SF, Venturelli G, Green DH, Toscani L (1987) The ultrapotassic rocks: characteristics, classification, and constraints for petrogenetic models. Earth Sci Rev 24:81–134

Gao Y, Wei R, Hou Z, Tian S, Zhao R (2008) Eocene high-MgO volcanism in southern Tibet: new constraints for mantle source characteristics and deep processes. Lithos 105:63–72

Gao Y, Yang Z, Santosh M, Hou Z, Wei R, Tian S (2010) Adakitic rocks from slab melt-modified mantle sources in the continental collision zone of southern Tibet. Lithos 119:651–663

Gariépy C, Allègre CJ, Xu RH (1985) The Pb-isotope geochemistry of granitoids from the Himalaya-Tibet collision zone: implications for crustal evolution. Earth Planet Sci Lett 74:220–234

Guan Q, Zhu DC, Zhao ZD, Dong GC, Zhang LL, Li XW, Liu M, Mo XX, Liu YS, Yuan HL (2012) Crustal thickening prior to 38 Ma in southern Tibet: evidence from lower crust-derived adakitic magmatism in the Gangdese batholith. Gondwana Res 21:88–99

Guo Z, Wilson M (2019) Late Oligocene–early Miocene transformation of postcollisional magmatism in Tibet. Geology 47:776–780

Guo Z, Wilson M, Liu J (2007) Post-collisional adakites in south Tibet: products of partial melting of subduction-modified lower crust. Lithos 96:205–224

Guo Z, Wilson M, Zhang M, Cheng Z, Zhang L (2013) Post-collisional, K-rich mafic magmatism in south Tibet: constraints on Indian slab-to-wedge transport processes and plateau uplift. Contrib Mineral Petrol 165:1311–1340

Guo Z, Wilson M, Zhang M, Cheng Z, Zhang L (2015) Post-collisional ultrapotassic mafic magmatism in south Tibet: products of partial melting of pyroxenite in the mantle wedge induced by roll-back and delamination of the subducted Indian continental lithosphere slab. J Petrol 56:1365–1406

Hart SR (1984) A large-scale isotope anomaly in the southern hemisphere mantle. Nature 309:753–787

Hoke L, Lamb S, Hilton DR, Poreda RJ (2000) Southern limit of mantle-derived geothermal helium emissions in Tibet: implications for lithospheric structure. Earth Planet Sci Lett 180:297–308

Hou ZQ, Gao YF, Qu XM, Rui ZY, Mo XX (2004) Origin of adakitic intrusives generated during mid-Miocene east-west extension in southern Tibet. Earth Planet Sci Lett 220:139–155

Hou Z, Zheng Y, Yang Z, Rui Z, Zhao Z, Jiang S, Qu X, Sun Q (2013) Contribution of mantle components within juvenile lower-crust to collisional zone porphyry Cu systems in Tibet. Mineral Deposita 48:173–192
Hu Y, Liu J, Ling M, Liu Y, Ding X, Liu D, Sun W (2017) Constraints on the origin of adakites and porphyry Cu-Mo mineralization in Chongjiang, southern Gangdese, the Tibetan plateau. Lithos 292-293:424–436
Huang ML, Bi XW, Hu RZ, Gao JF, Xu LL, Zhu JJ, Shang LB (2019) Geochemistry, in-situ Sr-Nd-Hf-O isotopes, and mineralogical constraints on origin and magmatic-hydrothermal evolution of the Yulong porphyry Cu-Mo deposit, eastern Tibet. Gondwana Res 76:98–114
Ji WQ, Wu FY, Chung SL, Li JX, Liu CZ (2009) Zircon U–Pb geochronology and Hf isotopic constraints on petrogenesis of the Gangdese batholith, southern Tibet. Chem Geol 262:229–245
Ji WQ, Wu FY, Liu CZ, Chung SL (2012) Early Eocene crustal thickening in southern Tibet: new age and geochemical constraints from the Gangdese batholith. J Asian Earth Sci 53:82–95
Le Maitre RW (1989) A classification of igneous rock and glossary of terms. Blackwell Science Publication, Oxford
Li Y, Li X, Wang C, Wei Y, Chen X, He J, Xu M, Hou Y (2017) Miocene adakitic intrusions in the Zhongba terrane: implications for the origin and geochemical variations of post-collisional adakitic rocks in southern Tibet. Gondwana Res 41:65–76
Liu D, Zhao ZD, Zhu DC, Wang Q, Sui QL, Liu YS, Hu ZC, Mo XX (2011) The petrogenesis of postcollisional potassic-ultrapotassic rocks in Xungba basin, western Lhasa terrane: constraints from zircon U-Pb geochronology and geochemistry. Acta Petrol Sin 27:2045–2059
Liu D, Zhao Z, Zhu D, Niu Y, Liu S, Wang Q, Liu Y, Hu Z (2013) Zircon records of Miocene ultrapotassic rocks from southern Lhasa subterrane, Tibetan plateau. Acta Petrol Sin 29:3703–3715 (in Chinese with English abstract)
Liu D, Zhao Z, Zhu DC, Niu Y, Harrison TM (2014a) Zircon xenocrysts in Tibetan ultrapotassic magmas: imaging the deep crust through time. Geology 43:43–46
Liu D, Zhao Z, Zhu DC, Niu Y, DePaolo DJ, Harrison TM, Mo X, Dong G, Zhou S, Sun C, Zhang Z, Liu J (2014b) Postcollisional potassic and ultrapotassic rocks in southern Tibet: mantle and crustal origins in response to India–Asia collision and convergence. Geochim Cosmochim Acta 143:207–231
Liu D, Zhao Z, Zhu DC, Niu Y, Widom E, Teng FZ, DePaolo DJ, Ke S, Xu JF, Wang Q, Mo X (2015) Identifying mantle carbonatite metasomatism through Os–Sr–Mg isotopes in Tibetan ultrapotassic rocks. Earth Planet Sci Lett 430:458–469
Liu D, Zhao Z, DePaolo DJ, Zhu DC, Meng FY, Shi Q, Wang Q (2017) Potassic volcanic rocks and adakitic intrusions in southern Tibet: insights into mantle–crust interaction and mass transfer from Indian plate. Lithos 268-271:48–64
Meisel T, Walker RJ, Irving AJ, Lorand JP (2001) Osmium isotopic compositions of mantle xenoliths: a global perspective. Geochim Cosmochim Acta 65:1311–1323
Miller C, Schuster R, Klötzli U, Frank W, Purtscheller F (1999) Post-collisional potassic and ultrapotassic magmatism in SW Tibet: geochemical and Sr–Nd–Pb–O isotopic constraints for mantle source characteristics and petrogenesis. J Petrol 40:1399–1424
Mo X, Hou Z, Niu Y, Dong G, Qu X, Zhao Z, Yang Z (2007) Mantle contributions to crustal thickening during continental collision: evidence from Cenozoic igneous rocks in southern Tibet. Lithos 96:225–242
Mo X, Niu Y, Dong G, Zhao Z, Hou Z, Zhou S, Ke S (2008) Contribution of syncollisional felsic magmatism to continental crust growth: a case study of the Paleogene Linzizong volcanic succession in southern Tibet. Chem Geol 250:49–67
Nomade S, Renne PR, Mo X, Zhao Z, Zhou S (2004) Miocene volcanism in the Lhasa block, Tibet: spatial trends and geodynamic implications. Earth Planet Sci Lett 221:227–243
Priestley K, Jackson J, McKenzie D (2008) Lithospheric structure and deep earthquakes beneath India, the Himalaya and southern Tibet. Geophys J Int 172:345–362
Qu X, Hou Z, Li Y (2004) Melt components derived from a subducted slab in late orogenic ore-bearing porphyries in the Gangdese copper belt, southern Tibetan plateau. Lithos 74:131–148

Schaefer BF, Turner SP, Rogers NW, Hawkesworth CJ, Williams HM, Pearson DG, Nowell GM (2000) Re-Os isotope characteristics of postorogenic lavas: implications for the nature of young lithospheric mantle and its contribution to basaltic magmas. Geology 28:563–566
Schmidt MW, Poli S (2003) Generation of mobile components during subduction of oceanic crust. Treatise Geochem 3:567–591
Spicer RA, Harris NBW, Widdowson M, Herman AB, Guo S, Valdes PJ, Wolfe JA, Kelley SP (2003) Constant elevation of southern Tibet over the past 15 million years. Nature 421:622–624
Sun SS, McDonough WF (1989) Chemical and isotopic systematics of oceanic basalts: implications for mantle composition and processes. Geol Soc Lond Special Publ 42:313–345
Sun CG, Zhao ZD, Mo XX, Zhu DC, Dong GC, Zhou S, Chen HH, Xie LW, Yang YH, Sun JF, Yu F (2008) Enriched mantle source and petrogenesis of Sailipu ultrapotassic rocks in southwestern Tibetan plateau: constraints from zircon U-Pb geochronology and Hf isotopic compositions. Acta Petrol Sin 24:249–264 (in Chinese with English abstract)
Teng FZ, Li WY, Ke S, Marty B, Dauphas N, Huang S, Wu FY, Pourmand A (2010) Magnesium isotopic composition of the Earth and chondrites. Geochim Cosmochim Acta 74:4150–4166
Tian SH, Yang ZS, Hou ZQ, Mo XX, Hu WJ, Zhao Y, Zhao XY (2017) Subduction of the Indian lower crust beneath southern Tibet revealed by the post-collisional potassic and ultrapotassic rocks in SW Tibet. Gondwana Res 41:29–50
Tian SH, Hou ZQ, Mo XX, Tian Y, Zhao Y, Hou K, Yang Z, Hu W, Li X, Zhang Y (2020) Lithium isotopic evidence for subduction of the Indian lower crust beneath southern Tibet. Gondwana Res 77:168–183
Tilmann F, Ni J, INDEPTH III Seismic Team (2003) Seismic imaging of the downwelling Indian lithosphere beneath Central Tibet. Science 300:1424–1427
Vidal P, Le Fort ACP (1982) Geochemical investigations of the origin of the Manaslu leucogranite (Himalaya, Nepal). Geochim Cosmochim Acta 46:2279–2292
Williams H, Turner S, Kelley S, Harris N (2001) Age and composition of dikes in southern Tibet: new constraints on the timing of east-west extension and its relationship to postcollisional volcanism. Geology 29:339–342
Williams HM, Turner SP, Pearce JA, Kelley SP, Harris NBW (2004) Nature of the source regions for post-collisional, potassic magmatism in southern and northern Tibet from geochemical variations and inverse trace element modeling. J Petrol 45:555–607
Wu FY, Ji WQ, Wang JG, Liu CZ, Chung SL, Clift PD (2014) Zircon U–Pb and Hf isotopic constraints on the onset time of India-Asia collision. Am J Sci 314:548–579
Yang W, Teng FZ, Zhang HF, Li SG (2012) Magnesium isotopic systematics of continental basalts from the North China craton: implications for tracing subducted carbonate in the mantle. Chem Geol 328:185–194
Yang ZM, Lu YJ, Hou ZQ, Chang ZS (2015) High-Mg diorite from Qulong in southern Tibet: implications for the genesis of adakite-like intrusions and associated porphyry Cu deposits in collisional orogens. J Petrol 56:227–254
Yin A, Harrison TM (2000) Geologic evolution of the Himalayan–Tibetan orogen. In: Jeanloz R, Albee AL, Burke KC (eds) Annu Rev Earth Planet Sci 28:211–280
Zhang LY, Ducea MN, Ding L, Pullen A, Kapp P, Hoffman D (2014) Southern Tibetan Oligocene–Miocene adakites: a record of Indian slab tearing. Lithos 210-211:209–223
Zhang L, Guo Z, Zhang M, Cheng Z, Sun Y (2017) Post-collisional potassic magmatism in the eastern Lhasa terrane, South Tibet: products of partial melting of mélanges in a continental subduction channel. Gondwana Res 41:9–28
Zhao Z, Mo X, Zhang S, Guo T, Zhou S, Dong G, Wang Y (2001) Post-collisional magmatism in Wuyu basin, central Tibet: evidence for recycling of subducted Tethyan oceanic crust. Sci China Ser D Earth Sci 44:27–34
Zhao ZD, Mo XX, Nomade S, Renne PR, Zhou S, Dong GC, Wang LL, Zhu DC, Liao ZL (2006) Post-collisional ultrapotassic rocks in Lhasa block, Tibetan plateau: spatial and temporal distribution and its' implications. Acta Petrol Sin 22:787–794 (in Chinese with English abstract)
Zhao ZD, Mo XX, Sun CG, Zhu DC, Niu YL, Dong GC, Zhou S, Dong X, Liu YS (2008) Mantle xenoliths in southern Tibet: geochemistry and constraints for the nature of the mantle. Acta Petrol Sin 24:193–202 (in Chinese with English abstract)

Zhao Z, Mo X, Dilek Y, Niu Y, DePaolo DJ, Robinson P, Zhu D, Sun C, Dong G, Zhou S, Luo Z, Hou Z (2009) Geochemical and Sr–Nd–Pb–O isotopic compositions of the post-collisional ultrapotassic magmatism in SW Tibet: petrogenesis and implications for India intra-continental subduction beneath southern Tibet. Lithos 113:190–212

Zheng YC, Hou ZQ, Li QY, Sun QZ, Liang W, Fu Q, Li W, Huang KX (2012) Origin of Late Oligocene adakitic intrusives in the southeastern Lhasa terrane: evidence from in situ zircon U–Pb dating, Hf–O isotopes, and whole-rock geochemistry. Lithos 148:296–311

Zhu DC, Zhao ZD, Niu Y, Mo XX, Chung SL, Hou ZQ, Wang LQ, Wu FY (2011) The Lhasa terrane: record of a microcontinent and its histories of drift and growth. Earth Planet Sci Lett 301:241–255

Zindler A, Hart SR (1986) Chemical geodynamics. Annu Rev Earth Planet Sci 14:493–571

Chapter 5
Tectonic Evolution from Pangea to the Neo-Tethys, and to the Tibetan Plateau

Abstract This chapter discussed the tectonic evolution of the Pangea supercontinent to the Neo-Tethys, and to the formation of the Tibetan Plateau, in the frame of plate tectonic concept. This work infer that Pangea opened eastward in the Permian-Triassic and formed a triangular bay known as the Tethys. Before the Triassic, the Tethys ocean was pressed and subsequently closed between Angaraland and Gondwanaland, and subsequently caused the growth of the Alps-Himalaya orogeny belt. This study emphasized three tectonic facies model of lithospheric root evolution in continental-continental collision orogeny in the Tibetan Plateau and adjacent areas, including Pamir-type active lithospheric cold root, Nyainqentanglha-type detached orogenic lithospheric root, and Qiangtang-type hot orogenic lithospheric root. These three type tectonic facies, with the evidences from magmatism, deep geophysical data, geochronology, etc., have significant power in explaining the tectonic evolution of the continental lithosphere in the Tibetan Plateau.

Keywords Pangea supercontinent · Neo-Tethys · Tectonic facies model · Evolution · The Tibetan Plateau

5.1 The Breakup of Pangea

5.1.1 Pangea Supercontinent

The initial timing of the Tethys formation is so far still under debate. Following the theory of plate tectonics, we infer that Pangea opened eastward in the Permian-Triassic and formed a triangular bay known as the Tethys. Before the Triassic, the Tethys was squeezed and subsequently closed between Angaraland in the north and Gondwanaland in the south which induced the growth of the Alps-Himalaya orogeny belt (Sengor 1988). According to the monography "Solid Earth Science and

Prof. Jinfu Deng has fully contributed to this chapter.

X. Mo et al., *Magmatism and Crustal Evolution of the Tibetan Plateau*, Modern Approaches in Solid Earth Sciences 25,
https://doi.org/10.1007/978-981-97-2109-2_5

Society" (NRC (National Research Council) 1993), Gondwanaland was formed at ~600 Ma and subsequently collided with Laurasia at ~290 Ma. The collage of Gondwanaland and Laurasia constituted Pangea which initially broke up at ~200 Ma. Dilek and Furnes (2011) proposed that the breakup of Pangea generated a series of Tethyan ophiolites. The life span of Pangea is relatively short. Before the formation of Pangea, Gondwanaland had already existed at ~600 Ma, but Laurasia had not yet assembled from several isolated continental blocks at least during much of the Paleozoic (Pan 1994). Noteworthily, the breakup of Pangea was largely due to the rifting and breakup of Gondwanaland (NRC (National Research Council) 1993). Thus, the global pattern of continental masses constituted by the development of Gondwanaland and the breakup of Laurasia in the early Paleozoic (before Pangea) is in sharp contrast with the continental mass distribution created by the breakup of Gondwana and the growth of the Eurasian continent since the Mesozoic (after Pangea) (Deng et al. 1996). Pangea is significantly a turning point in the pattern of global sea-land distribution and the re-organization of global plates. Therefore, it is reasonable to regard the breakup of Pangea as the beginning of the Tethys development. Furthermore, the paleomagnetism investigation revealed that the superchron of "reversed" polarity of the Earth's magnetic field occurred at 320–245 Ma (Irvine 1989) or 323–248 Ma (Larson 1991), roughly coeval with Pangea. This means that the formation of Pangea is a response and manifestation of major adjustment and transformation of the Earth's interior energy. From either the perspective of plate re-organization and sea-land distribution in response to shallow Earth processes or the possibiolity of deep Earth's interior energy variation, the breakup of Pangea as the beginning of the Tethys development has its global tectonic significance. In other words, the formation and development of the Tethys is not just a regional tectonic event, but is a rather important part of global events. Thus, it is important to discuss the Mesozoic-Cenozoic tectonics and magmatism of the Tibetan Plateau in the context of global tectonics.

5.1.2 Continental Flood Basalt (CFB) at the Permian-Triassic Interval

At the Permian-Triassic interval, continental flood basalt (CFB) massively and temporarily erupted at the eastern margin of Pangea, coinciding with the end of Permian reversed superchron (Irvine 1989), marking a sudden transition in the energy state of the Earth's interior. As Pangea is a response and manifestation of reversed superchron of the Earth's magnetism, the end of the reversed superchron marked by CFB should be regarded as the initial rifting of Pangea and also the beginning of the Tethys development, which is a starting point for re-organizing the sea-land distribution. Ren et al. (1999) considered that the large-scale eruption of Permian basalts in southwestern China and Southeast Asia is a precursor to the opening of the Tethys.

The distribution of these Permian CFB is shown in Fig. 5.1. At the eastern margin of Pangea, the CFB is mainly distributed in the following three regions: Tunguska

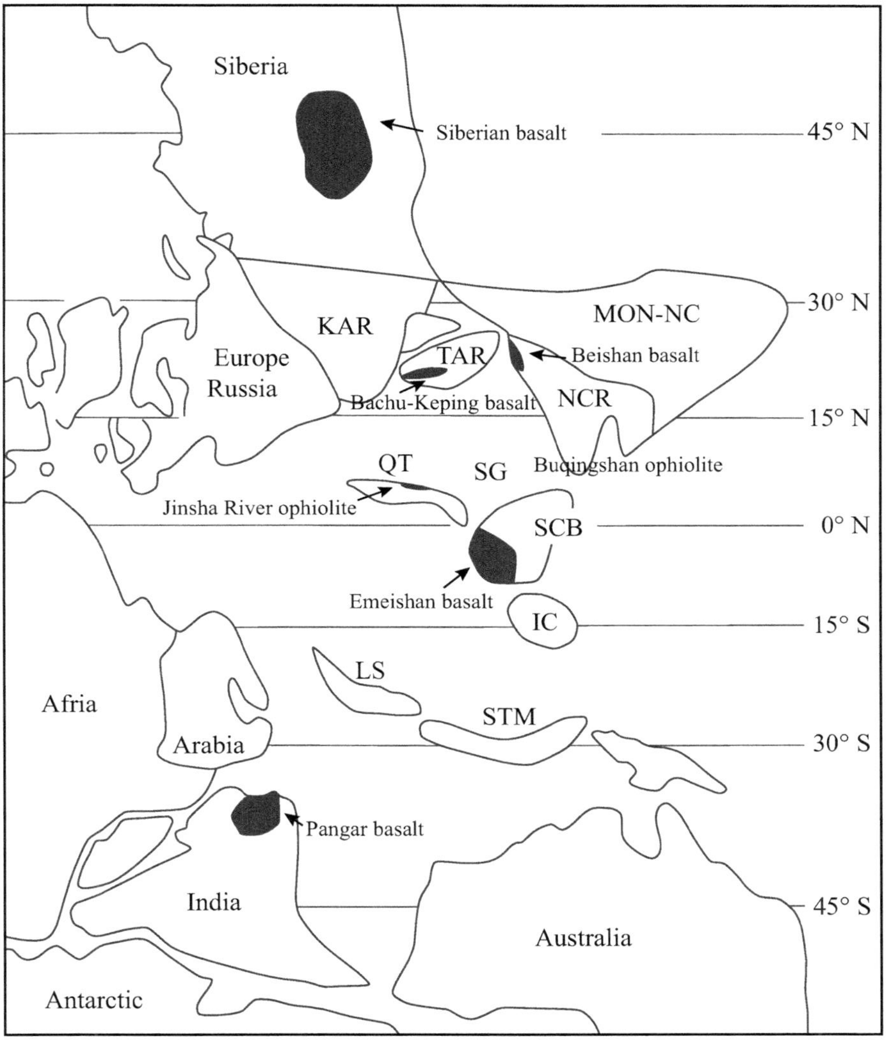

Fig. 5.1 Permian-Triassic Eastern Asian paleocontinent reconstruction and the distribution of continental flood basalts (after Cao 1999)

(Siberian continent) in the north, and Pangar (Indian continent) in the south, Emeishan (Yangtze Craton) between the former two regions (Chung et al. 1998). Additionally, the Permian CFB was also probably scattered in: Tarim continent (e.g., Bachu-Keping, Jinta, A'nyemaqen-Buqingshan, and Mianlue; Cao 1999), the Middle Tethys Himalayas belt (e.g., Selong Group, Zhu 2003), Southeast Asia (Song Da rift zone in Vietnam and Nan-Ultaradit OIB-type ophiolite in Thailand; Chung et al. 1998), southwestern China (Jinsha River and Litang ophiolites, Mo et al. 1993; and Youjiang region, Ma 1992). The CFB erupted a large volume in a short time (ca. 250 Ma). For instance, the CFB in Siberia is ca. 4×10^6 km^2 with a volume of >2×10^6 km^3 (Fedorenko and Czamanske 1997); the area of Emeishan basalts reaches 2.5×10^5 km^2 with a thickness of several hundred meters to 5 km (mostly 1–2 km, Chung et al. 1998); the area of Pangar basalts exceeds 10^5 km^2 in >2 km thickness (after Zhu 2003).

Except for massive basalts, the CFB also developed a series of high-Mg volcanic rocks, including picrite, meimechite and komatiite. The compositions of representative CFBs are given in Table 5.1. In Table 5.1, except for the volcanic rocks in the Selong Group that are basalts, others are mainly high-Mg and picritic volcanic rocks. Using the new IUGS classification (Le Bas et al. 1986), high-Mg and picritic rocks are categorized by (1) Picrite: MgO > 12 wt.%, 30 wt.% < SiO_2 < 52 wt.%, $Na_2O + K_2O < 3$ wt.%; (2) Komatite ($TiO_2 < 1$ wt.%) and meimechite ($TiO_2 > 1$ wt.%): MgO > 18 wt.%, 30 wt.% < SiO_2 < 52 wt.%, $Na_2O + K_2O < 2$ wt.%; (3) Boninite: MgO > 8 wt.%, SiO_2 > 52 wt.%, TiO_2 < 0.5 wt.%.

According to MgO and Ni contents (Table 5.1), these CFB are typical products of mantle plumes as shown in Fig. 5.2. With the same MgO content, the plume-derived basaltic melts tend to have higher Ni content, indicating a higher melting temperature (Condie 2001). The potential temperature of the asthenosphere was considered to be ~1280 °C, while that of a mantle plume in deeper depth could reach 1500–1600 °C (Condie 2001; Wyllie 1988). Based on the experimental study, Wyllie (1988) proposed that the temperature at the axis (or tail) of Hawaii mantle plume needed to reach ≥1500 °C for picritic and high-Mg basaltic melt generation. As investigated by Condie (2001), high-Mg picrites are commonly observed in the lower part of CFB. And they are produced by partial melting of the plume axis (or tail) at high temperatures, which are sourced from the hot materials at the core-mantle boundary. As given in Table 5.1, picritic melts (MgO = 17.82 wt.%) can be generated at 3.0 GPa (approximately 100 km), 1525 °C. Compared with the experimental melts, picritic melts (MgO = 15–27 wt.%) in the P_2–T_1 CFB at the eastern margin of Pangea should be generated at ≥1525 °C. Thus, these magmas can be considered as the melts extracted from the axis of the mantle plume.

Consequently, the presence of the aforementioned high-Mg and picritic melts suggests that the development of the CFB is closely associated with a thermal anomaly event in the lower mantle. Therefore, the formation of the Tethys has an important deep geodynamic significance.

5.1.3 The Tectonic Framework of the Breakup of the Eastern Margin of Pangea

5.1.3.1 Global Biogeographic Provinces in the Permian

In the early Triassic, the biological activity in Pangea reached its peak, causing the worldwide migration of terrestrial reptiles during the Permian-Triassic (Liu and Quan 1996). Wang (1981) pointed out that the biogeographic provinces in the

Table 5.1 Major element compositions of representative continental flood basalts

Region	Siberia			Emeishan		Heishuihe (Vietnam)		Selong group	Experimental melt
Rock type	Picrite	Komatiite	Meimechite	Picrite	Meimechite	Picrite	Komatiite	Basalt	Picrite
SiO_2	48.58	43.41	41.64	47.24	44.36	46.23	45.71	49.49	46.77
TiO_2	1.54	0.44	3.29	2.4	1.26	1.44	0.61	1.89	0.55
Al_2O_3	10.26	13.91	4.15	9.59	6.53	10.59	10.46	16.4	12.87
TFe_2O_3	–	–	–	13.16	12.33	–	12.03	1.54	–
TFeO	13.37	13.79	14.71	–	–	12.05	–	10.69	9.81
MnO	0.18	0.21	0.22	0.21	0.21	0.21	0.15	0.16	0.29
MgO	15.85	19.27	24.96	14.18	14.18	15.45	21.47	11.62	17.82
CaO	8.37	7.4	9.15	9.8	9.8	12.62	8.58	6.12	10.63
Na_2O	1.5	1.39	0.3	1.35	1.35	0.82	0.86	1.76	0.87
K_2O	0.22	0.12	1.14	1.15	1.15	0.35	0.08	0.16	0.09
P_2O_5	0.13	0.06	0.44	0.31	0.31	0.24	0.05	0.18	–
Ni	698	619	950	–	–	–	–	271	–
Source	1	1	2	3	3	4	5	6	7

[a] (1) The major element oxides are in wt.% recalculated to 100% on an anhydrous basis. Total Fe is expressed as either TFe_2O_3 or TFeO; (2) Rock types are determined using the classification method by Le Bas et al. (1986). (3) Data sources: 1—Fedorenko et al. (1996); 2—Fedorenko and Czamanske (1997); 3—Song et al. (2001); 4—Balykin et al. (2001); 5—Yem et al. (2002); 6—Zhu (2003); 7—The experimental melt quoted from Hirose et al. (1993), which is melted from KLB-1 lherzolite at 3.0 GPa, 1525 °C

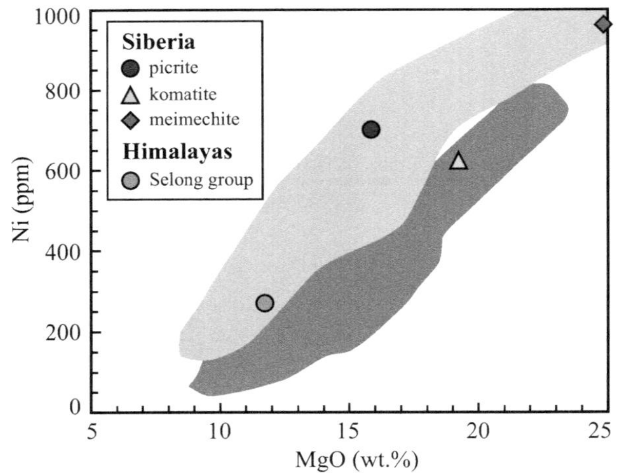

Fig. 5.2 Ni-MgO correlation of P_2–T_1 continental flood basalts at the eastern margin of Pangea (after Condie 2001)

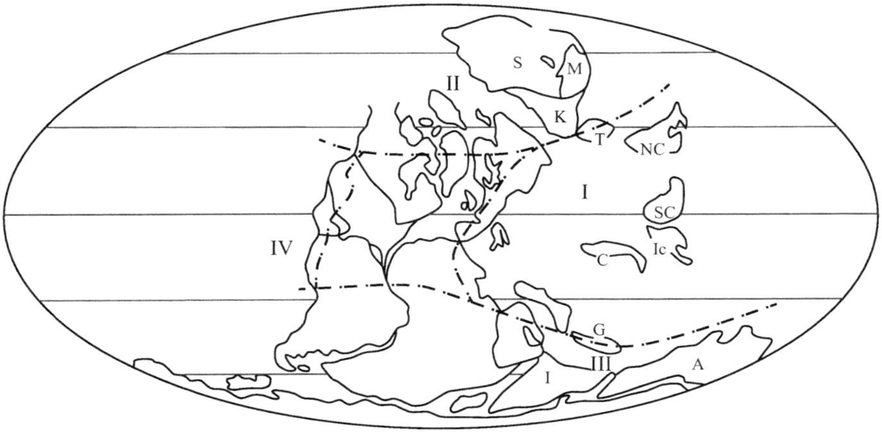

Fig. 5.3 Global biogeographic provinces in the Permian (Yin 1988). Four biogeographic provinces: I—Tethys province; II—Boreal province; III—Gondwana province; IV—Western America province; Continental plates or blocks: S—Siberia; M—Mongolia; K—Kazakhstan; T—Tarim; NC—North China; SC—South China; IC—Indochina; C—Qiangtang; G—Gangdese; I—India; A—Australia

Permian are significant. In this time, the zonal pattern of the global paleoclimate was apparent. As shown in Fig. 5.3, there are three biogeographic provinces: Boreal province to the north, Gondwana province to the south, and Tethys province in the centre. The Boreal province was located in a cool and warm climate zone at a relatively high north latitude, predominantly consisting of clastic rocks intercalated with marine strata and containing Angaran flora. The Tethys province was situated in tropical-subtropical zones around the equator at relatively low latitudes. In this province, massive carbonate rocks were deposited with the widespread development of reefs and complex corals, which were also accompanied by Cathaysian flora, coal, gypsum salt among other sedimentary deposits. The Himalayan orogenic belt in the Gondwana province was located at a higher south latitude, developing a suite of moraine and glacial-marine deposits of late Carboniferous to early Permian age (Yin 1988). Additionally, the biogeographic provinces could also be divided into three continental block groups (Chen 1994; Pan 1994). The northern part was called Laurasia, characterized by the presence of Angaran flora, while the central part belonged to Cathaysian block group, marked by the growth of Cathaysian or Gigantopteris floras. Glacial-marine deposits, together with Glossopteris flora, were widely developed in the southern part which was named Gondwanaland. As shown in Fig. 5.3, in China, (1) the boundary between the Boreal province and the Tethys province (Cathaysian block group) is along the belt of Aibi lake-Juyan Sea-Xilamulun River (Yin 1988). Because Siberian continent and Sino-Korean continent were almost assembled in the Permian, the Cathaysian floras were mixed with the Angaran floras in Gansu (China), Xinjiang (China), etc. (2) the boundary between the Tethys and the Gondwana provinces is along the Yarlung-Zangbo River. However, Gondwana and Tethys floras were alternatively developed in the transition zone between the Yarlung-Zangbo River in the south and Longmuco-Jinsha river in the north, including Gangdese, Nyainqentanglha, Karakorum, Tanggula, Hoh Xil, and Hengduan Mount. Meanwhile, mixed Gondwana-Cathaysian floras were observed in Xiagangjiang area, Tibet (Yin 1988; Zhong 1998). All this suggests that the southernmost segment of the Cathaysian block group was nearly attached to the northernmost part of Gondwanaland. The climate zone of each biogeographic province was variable in geographic position. In the earliest Permian, the Gondwana province in the south and the Boreal province in the north were both positioned in a cool and warm climate zone or a colder climate zone. Then, as the warm-hot climate zone expanded in the late Early Permian, the margins of the Gondwana and Boreal provinces respectively shifted to the southern and northern warm temperate zone, where massive Tethyan paleontology migrated to. In the late Permian, the Boreal province including Tarim, Beishan and Qilian moved northward and then attached to Siberian continent. Meanwhile, the climate zone would shift from tropics or subtropics to a warm temperate zone. Consequently, the biogeographic provinces are obliquely intersected with the climate zones (Yin 1988).

The mixed biotas at the margins of the biogeographic provinces, together with the aforementioned wide distributions of the terrestrial reptiles in Pangea, reveal that Pangea already existed as a unified collage block. In addition, on the basis of the obliquely intersecting relationship between the biogeographic province and the climate zone, a horn shape expanding eastward in the Tethys biogeographic province (Fig. 5.3), and the distribution of mantle plume as implied by CFB (Fig. 5.1), the warming characteristics reflected by Tethys biota may be not only affected by the latitude but also related to the plume-induced heating. Because a heating process is required prior to the formation of a mantle plume, which needs further study.

5.1.3.2 The Breakup of Pangea and the Tethyan Growth

Irvine (1989) considered the rifting of Scoresbyland in the middle Triassic as marking the opening of the Tethys, but the exact opening time remains under debate and need further research. Besides, the meaning of the Tethyan opening refers to the formation of the initial oceanic crust or the rifting with the absence of oceanic crust, which has not been determined. The discussion in this chapter focuses on two points: (1) To discuss the breakup of Laurasia, Cathaysia, and Gondwana continents, we have considered the splitting of the eastern margin of Pangea, the aforementioned biogeographic provinces, and the three largest CFB provinces corresponding to the biogeographic provinces. Although the Mongol-Okhotsk ocean, formed by the splitting at the southern and eastern margins of Siberia, is not part of the Tethys, it is likely to be the product of the splitting at the eastern margin of Pangea. (2) Due to the scarcity of research, especially on the geochronology of ocean crust formation, the discussion is qualitative with intracontinental or continental rifting considered as the splitting event at the eastern margin of Pangea.

1. The breakup of the northern margin of eastern Gondwanaland. The presence of CFB in Pangar plausibly indicates the initial splitting at the northern margin of eastern Gondwanaland. As proposed by Sengor (1988), (a) in the northern Gondwanaland, the rifting associated with the formation of Cimmerian continent was induced by regional normal faults and massive Permian basic extrusive rocks, including the Pangar trap in Kashmir. (b) Prior to the closure of the Paleo-Tethys, the formation of back-arc basin underwent two stages of extension. In the early stage, a small oceanic basin opened along the Pindos-Budava-Karakaya suture in the timeframe of late Permian to early Triassic. The northeastern part of this small basin was closed in the late Triassic, while that in the Pindos-Budava segment survived until the early Cenozoic. In the late stage, a multi-forked ocean was formed, constituting the northern branch of the Neo-Tethys in the Mediterranean. (c) Cimmerian continent was split from northern Gondwanaland in the late Permian-Triassic, and collided with Eurasia continent in the late Triassic-to-middle Jurassic, thus forming Cimmerian orogenic belt. Subsequently, the Neo-Tethys was opened, and then initially closed in the late Cretaceous until the formation of Alpine-Himalayan orogenic belt in the Eocene. As formed by the closure of the Paleo-Tethys, Cimmerian orogenic belt included Balkans-Carpathian, North Turkey, Caucasus, and northern Iran-Afghanistan. And this orogenic belt, together with a series of blocks on its southern side (e.g., South Turkey, Iran, Afghanistan), constituted the Cimmerian continent. Meanwhile, the Cimmerian orogenic belt was parallel to the Neo-Tethys and Alpine orogenic belt represented by Zagros in the south (Chen 1994).

   There are still several different views on the opening of the Tethys. As above, Sengor (1988) considered that the opening of the Neo-Tethys was closely related and simultaneous to the closure of the Paleo-Tethys. In addition, the Bangong Co-Nujiang suture was regarded as Cimmerian suture, belonging to the Paleo-Tethys, while Indus-Yarlung Zangbo suture was classified in the Neo-Tethys

(Opening time: the late Permian-Triassic? Closing time: Eocene). Chang et al. (1982) suggested that the Carboniferous-Permian strata on both sides of Yarlung Zangbo suture have significant similarities in petrology and animal fossils. After the Triassic, the Tethys Himalayas can be divided into two distinct parallel lithofacies belts. Its southern belt is dominated by marine shelf deposits under a stable environment. A flysch belt, namely South Tibet, dividing range flysch belt, stretches between the southern Tethys Himalayas belt and Yarlung Zangbo ophiolite belt. The early-middle Triassic fossils were discovered in the flysch belt which was covered by a suite of fault trough-type thick flysch intercalated with intermediate-mafic volcanic rocks and rare fossils. Besides, the breakup of the platform and the presence of passive margin were characterized by Permian shallow-marine carbonates and Triassic-Cretaceous semideep-sea and deep-sea facies turbidites intercalated with basic volcanic rocks and ophiolite suites, etc. The lithofacies changes in the Permian-Triassic should be considered as marking the initial time of the Neo-Tethys evolution. Constrained by stratigraphy and sedimentary facies, Xiao and Li (2000) pointed out that Yarlung Zangbo suture was opened in the late Triassic and formed a small ocean basin in the Triassic. Recently, using sedimentary facies and petrology and geochemistry of volcanic rocks, Zhu (2003) proposed: (1) The Permian volcanic rocks did not show the affinities of oceanic crust and as such cannot represent the existence of Yarlung Zangbo ocean basin; (2) The early-middle Triassic basalt was MORB-type basalts, suggesting the Yarlung Zangbo ocean basin was opened at the end of the Permian-the early Triassic. And the opened basin was still small in the early-middle Triassic; (3) In the middle-late Triassic, the ocean basin underwent a rapid extension event, and was fully opened with a certain scale; (4) The second rapid extension was recorded by the basalts in the upper-middle Sangxiu group (J_3–K_1). And the ocean basin reached its maximum width. Nevertheless, based on the investigation of ophiolite (Xiao and Li 2000), the middle segment of the ocean basin was developed at 160–80 Ma, while the western and southern segments evolved at 210–90 Ma. The Rikangba pluton in the western segment of the Yarlung Zangbo ophiolite belt had a ^{40}Ar–^{39}Ar age of 180 Ma, while the Xiugugabu gabbro in the western segment, the Angren gabbro in the central segment, and the Luobusa gabbro in the eastern segment dated at 173 Ma, 166 Ma, and 177 ± 23 Ma respectively using inter-mineral Sm-Nd isochronous method (Zhou 2002). Consequently, the geochronologic results supported that the ocean basin was simultaneously opened in J_1–J_2.

Based on the above, a rough framework can be established, which might be: (1) P_2–T_1, the CFB in Pangar represents the initial rifting at the northern margin of Gondwanaland, south of Pangea; (2) T, it is predominantly characterized by thinned continental margin, regional volcanic-type passive margin, even the presence of initial oceanic crust; (3) J–K, the typical oceanic crust was widely formed.

2. The breakup of the Cathaysian block. The Emeishan CFB at the western margin of the Yangtze Craton indicated the initial splitting of Cathaysian blocks. As proposed by Liu and Quan (1996), (a) The lithofacies in the Maokou period (the

late Early Permian) were significantly differentiated. In Central Hunan and Lower Yangtze regions, the siliceous and muddy deposits in Dangchong and Gufeng Formationss had a few benthic organisms and enrichment of planktonic ammonites and radiolarians, suggesting a stagnant hydrostatic environment under a hypoxic condition; (b) In the late Maokou period, South China plate underwent generally enhanced tectonic differentiation. Meanwhile, the upwelling of mantle plume at the western margin of Yangtze Craton induced the eruption of massive basalts and the Emei Taphrogenesis. Besides, radiolarian micrites and siliceous rocks were observed in the middle-upper part of the Upper Permian Wujiaping Formation, pointing to a deep-sea and low-energy sedimentary environment, while the limestones in the middle-late Changxing period partially changed into deep-sea siliceous rocks (Dalong Formation); (c) Jinsha River-Tengtiaohe suture was situated at the southwestern margin of South China plate, and several blocks (e.g., Qiangtang, Qambo, Simao) appeared to its west, representing these small block groups were rifted from the margin of South China plate. As investigated by Zhong (1998), Carboniferous continental rift-related volcanic rocks in the Laochang mining area along the Changning-Menglian belt were re-identified as late Permian oceanic-type volcanic rocks. Besides, the late Permian shallow-water volcanic rocks were also recognised, and its tuffaceous clastic rocks were deposited on the paleokarst surface of Carboniferous carbonates. Thus, the late Permian oceanic-type volcanic rocks in the Laochang mining area were probably originally related to the Emeishan CFB, which might be the fingerprint of the breakup at the western margin of the Yangtze Craton. Additionally, Cao (1999) finished a distribution map of Permian-Triassic basalts, ophiolites, and siliceous rocks in the Yangtze Craton and its adjacent areas (Fig. 5.4). As shown in Figs. 5.1 and 5.4, except for the Emeishan basalt, the nearly contemporaneous basalts also erupted in the Cathaysian block, such as Bachu-Keping region, Jinta area. Combined with the presence of deep-water siliceous rocks, All this suggests that the CFB is associated with large-scale lithospheric thinning, marking the formation of the Tethys.

3. The breakup of the southern and eastern margins of the Siberian platform. The breakup event was probably induced by the activity of Tunguska trap. Based on paleomagnetism research, North China, South China, and Mongolia blocks had been assembled in the late Jurassic (J_3) but were still far away from the Siberian platform (Fig. 5.5). As suggested by the aforementioned biogeographic provinces (Fig. 5.3), the Siberian platform was assembled to Mongolia block in the Permian. Hence, it can be inferred that the initial formation of Mongolian-Okhotsk tectonic belt is probably related to the eruption of CFB in Siberia. And it means Mongolian block was rifted from the margin of Siberian platform. Parfenov and Natalin (1986) pointed out that in Northeast Asia, several small blocks including Okhotsk, Omolonsk, Bureirsk, Khankaisk, and Okhotoworsk, were likely split from the margin of Siberian platform in the early Mesozoic and separated by Mongolian-Okhotsk ocean and Anyuisk ocean which were evidenced by the presence of Mesozoic ophiolites. The closure of the former ocean

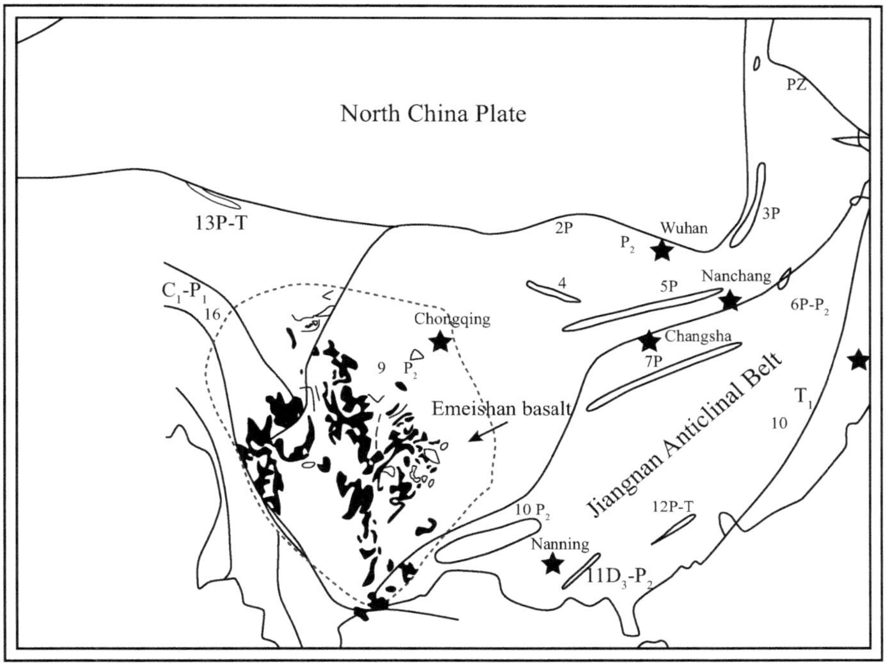

Fig. 5.4 The concise distributions of Permian-Triassic basalts, ophiolite, and siliceous rocks in Yangtze Craton and its adjacent areas (Cao 1999)

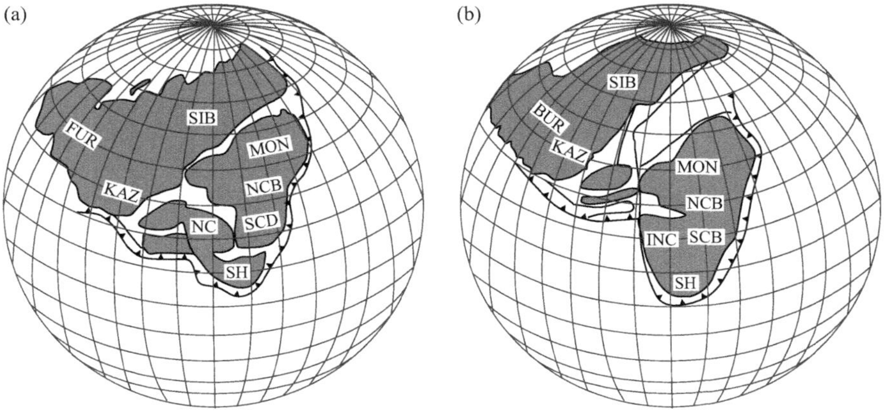

Fig. 5.5 Paleogeographic reconstruction of Siberia and its adjacent Asian blocks at (a) 125–95 Ma (the early Cretaceous) and (b)155–135 Ma (the late Jurassic) (modified from Vander et al. 1999). SIB—Siberia; EUP—Europe; Kaz—Kazakhstan; Mon—Mongolia; NCB—North China block; SCB—South China block; INC—Indochina; SH—Shan Tai block

occurred in the late Jurassic-early Cretaceous and subsequently formed Mongolian-Okhotsk orogenic belt, while the latter was closed in the early Cretaceous followed by the formation of Verkhoyansk-Chukchi orogenic belt, which also formed the suture zone between Siberian and North American (Alaska) platforms.

As above, this section focuses on discussing the breakup of the eastern margin of Pangea, which is associated with the Tethys growth. And the splitting of Pangea predominantly occurred in Gondwanaland whose breakup resulted in the opening of the Atlantic and Indian oceans as discussed in the last section of this chapter.

5.2 Ophiolite of the Tethyan Supra-Subduction Zone (SSZ)

5.2.1 Introduction

Previous studies suggest that ophiolites are remnants of ancient oceanic crust and lithosphere generated at seafloor spreading centers (or ocean ridges). However, as proposed by Miyashiro (1973, 1975), typical Troodos ophiolite originated in an island arc setting rather than a mid-ocean ridge setting (Zhang et al. 2001). Therefore, the tectonic setting of ophiolite generation is still debated. Subsequent studies have shown that ophiolite suites can be lithospheric remnants developed in various tectonic settings (e.g., Xiao 1995; Wang et al. 1987; Zhong 1998; Zhang et al. 2001). Currently, ophiolites can be predominantly divided into MORS-type (mid-ocean ridge spreading) and SSZ-type (supra-subduction zone) ophiolites (Pearce et al. 1984; Zhang et al. 2001). Malpas and Robinson (2000) insisted that ophiolites in the geological records belonged to the SSZ type, while others are the MORS type.

A suite of ophiolites generally has the following types of volcanic rocks, MORB (mid-ocean ridge basalt), IAT (island arc tholeiite), and boninite, including andesite, rhyodacite and alkaline basalt. Malpas and Robinson (2000) pointed out that each type of volcanic rocks represent a different initial component in SSZ (e.g., island arc, forearc, back-arc basin), and old oceanic crust could also be involved in several ophiolites. Meanwhile, the presence of alkaline basalts reveals the involvement of seamounts in the ophiolite suite. The mantle peridotite in ophiolite mainly consists of refractory harzburgite and minor lherzolite. In general, SSZ-type ophiolite is significantly characterized by the development of a boninite-harzburgite suite. Till now, the boninites are mainly found in forearc basins whose generation requires a refractory mantle source enriched by subducted slab-dehydrated fluids (Pearce et al. 1984; Kushiro 1990; Falloon and Danyushevsky 2000). Additionally, the emplacement of ophiolite distributed along continental margins is not necessarily confirmed as closed ocean basin (Malpas and Robinson 2000). And ophiolite is the lithospheric remnant overlying the SSZ and accretes to continental margin prior to the closure of ocean. Therefore, the sutures among blocks can develop a series of ophiolite belts whose emplacements are established prior to the closure of ocean. As investigated by Pearce et al. (1984), several characteristics of SSZ-type ophiolite are as follows: (a) SSZ-type ophiolite has the similar geochemical characteristic to island arcs; (b) mantle sequence of SSZ-type ophiolites is geochemically depleted; (c) podiform chromite deposits are developed in its cumulate rocks. Because the crystallization of plagioclase is after that of pyroxene, the proportion of wehrlite is higher relative to troctolite; (d) most SSZ-type ophiolites are formed during the initial stage of subduction before the development of volcanic arc. Boninite is the first batch of magmas in response to intra-oceanic subduction, which is derived by partial melting of hydrated oceanic lithosphere in a "mantle wedge". As the subduction continues, the magma shifts to island-arc tholeiite, plausibly ascribed to the mantle source changing into hydrated asthenosphere in "mantle wedges". Meanwhile, other SSZ-type ophiolites are formed in the early stages of back-arc extension following the splitting of a pre-existing arc. Nevertheless, the most popular mechanism of generating SSZ-type ophiolite appears to be pre-arc spreading rather than back-arc spreading. For distinguishing "pre-arc" and "forearc", several researchers focus on "time sequence", and use "pre-arc" to describe the formation of boninite, while "forearc" emphasizing "spatial position" is used to illustrate that boninite is commonly developed in forearc basin. And the use of "back-arc" also emphasizes "spatial position".

Recently, Dilek and Furnes (2011) published a widely accepted classification of ophiolite: (a) subduction-unrelated ophiolite, subdivided into continental margin-type, mantle plume-type, and mid-ocean ridge-type; (b) subduction-related ophiolite, including SSZ-type and volcanic arcs.

5.2.2 *Mantle Peridotite in the Tethyan Ophiolite Suite*

Mantle peridotites in the Tethyan ophiolite suite are distributed along the belts of Garze-Litang, Jinsha River, Bangongco-Nujiang, Yarlung Zangbo, predominantly consisting of harzburgite, dunite, and lherzolite (Mo et al. 1993; Wang et al. 1987; Qiu 2002; Qiu et al. 2005; Yang et al. 2012; Xu et al. 2011; Bao et al. 2015). And the compositions of major elements are shown in Fig. 5.6 (Qiu et al. 2005).

SSZ-type harzburgite ophiolite (HOT). Olivines in these mantle peridotites are all forsterite rich (Fo: 90–94, Wang et al. 1987), similar to those in depleted harzburgite. It means that these peridotites are the refractory residual after basaltic melts are extracted, and constitute the main body of oceanic lithospheric mantle. However, most meta-peridotites have high Th concentrations (Wang et al. 1987, $Th_N \geq 10$, the subscript "N" denotes primitive mantle normalization), the enrichment of LREEs (Qiu et al. 2005), and variable initial $^{87}Sr/^{86}Sr$ values (0.70802–0.71862, Qiu 2002; Qiu et al. 2005), suggesting that the depleted oceanic lithospheric mantle was enriched by fluids which were closely related to mantle wedge. Thus, these meta-peridotites have the characteristics of SSZ-type ophiolites (Fig. 5.7).

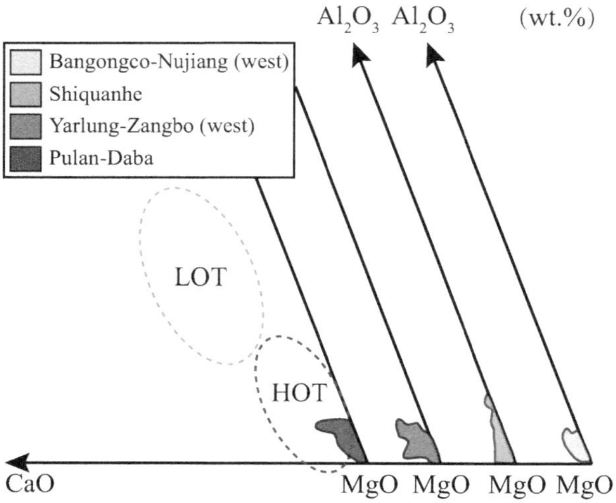

Fig. 5.6 Al_2O_3-CaO-MgO correlations of mantle peridotites (after Qiu et al. 2005)

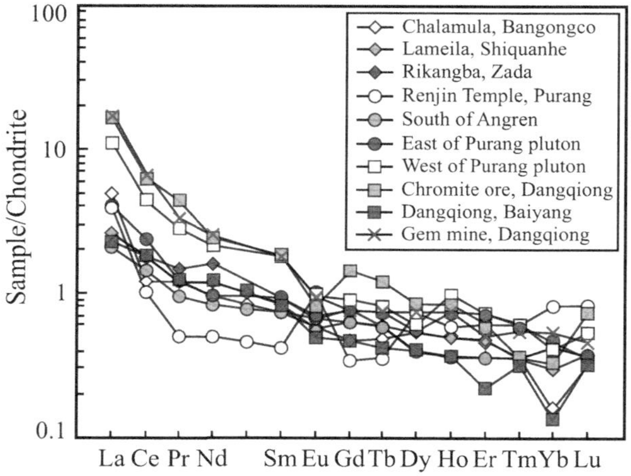

Fig. 5.7 Chondrite-normalized REE patterns of harzburgite and dunite in ophiolites in the western Tibetan Plateau (modified from Qiu et al. 2005)

5.2.3 Volcanic Rocks, Sheeted Dyke Swarms and Cumulates in the Tethyan Ophiolite Suites

1. Cumulate. The crystallization of plagioclase is after that of clinopyroxene in most cumulates, thus generating wehrlite, while few cumulates experience the crystallization of plagioclase before clinopyroxene, forming troctolite (Wang et al. 1987; Mo et al. 1993; Zhang et al. 2001). Therefore, the cumulates have the characteristics of SSZ-type ophiolite.

2. Boninite. Boninites are commonly developed in volcanic lavas and sheeted dyke swarms which were observed in Garze-Litang belt (Mo et al. 1993), Bangong Co-Nujiang belt (Zhang et al. 2001; Wang et al. 1987; Qiu 2002), and Yarlung Zangbo River belt (Wang et al. 1987; Zhang et al. 2001; Qiu 2002). These boninites have MgO < 16 wt.%, mostly TiO_2 < 0.5 wt.%, similar to the boninitic melts produced by partial melting of hydrous lherzolite (Kushiro 1990), but distinct from the boninitic melts generated by partial melting of harzburgite (MgO up to 19–24 wt.%)
3. The trace-element characteristics of volcanic rocks (e.g., boninite) and sheeted dyke swarms. Pearce et al. (1984) distinguished SSZ-type and MOBS-type mafic ophiolites using Cr-Y correlations. As shown in Fig. 5.8, these two types of ophiolites are separated by Y concentration at 20 ppm, and MOBS-type ophiolite has relatively higher Y concentrations. In SSZ-type ophiolite, Line B and A show the affinities of island arc tholeiite and boninite, respectively. The Cr and Y concentrations of volcanic rocks and sheeted dyke swarms are listed in Table 5.2 and shown in Fig. 5.8. Consequently, volcanic rocks and sheeted dyke swarms in the Tethyan ophiolites are dominated by SSZ-type ophiolites with minor MORS-type ophiolites. And in these SSZ-type ophiolites, Line B-type and Line A-type magmatic rocks are both equally developed.

Additionally, on N-MORB-normalized trace-element spider diagrams (Fig. 5.9; Wang et al. 1987; Mo et al. 1998), Th concentrations of the aforementioned volcanic rocks and sheeted dyke swarms are significantly higher than that of N-MORB, which are considered as one of the most important fingerprints to track the involvement of slab-dehydrated fluids into "mantle wedge"-derived melts (Pearce et al. 1984). Thus, it further supports that the Tethyan ophiolite suites predominantly consist of SSZ-type ophiolites, rather than MORS-type ophiolites.

5.2.4 Discussion

5.2.4.1 SSZ-Type and MORS-Type Ophiolites

Recently, a consensus was reached that the Tethyan ophiolites are dominated by SSZ-type ophiolites, rather than MORB-type ophiolites in large ocean basins (e.g., Wang et al. 1987; Xiao 1995; Xiao and Li 2000; Zhong 1998; Zhang and Zhou 2001; Mo et al. 1993, 1998; Zhang et al. 2001; Pearce et al. 1984; Flower et al. 2000; Malpas and Robinson 2000). Pearce et al. (1984) summarized the characteristics of the Tethyan ophiolite suites and further put forward that Jurassic ophiolites in the western Mediterranean (Calabria, Corsica, Northern Apennines, and the Alps in France and Austria) were almost MORS-type ophiolites in small ocean basins, similar to those in Red Sea. To the east, Cretaceous ophiolites along the belt of Cyprus-Turkey-Iran-Oman are almost SSZ-type ophiolites. Thus, it indicates that the Tethyan ophiolites shifted from MORS-type ophiolites in small ocean basins to SSZ-type ophiolites over time. Besides, Cretaceous ophiolites in the eastern Tethys (Masirah island, Eastern Iran, Yarlung Zangbo suture) are all MORS-type

Table 5.2 Cr and Y concentrations of representative volcanic lavas and sheeted dyke swarms

Garze-Litang	Y	19.74	16.85	27.84	17.84	14.50	29.55	19.51	26.21	15.66	14.90	15.80	23.15	29.75					Mo et al. (1993)
	Cr																		
Mid-Jinsha River	Y	15.50	20.90	17.60	21.60	18.60	29.30	15.83	35.37										Mo et al. (1993)
	Cr	74.03	289.14	355.06	214.92		205.0	224.1	85.63										
Bangong Co-Nujiang	Y	14	17	28	10	12	39	31	16	23	23	10	11	10	9	10	7	17	Wang et al. (1987)
	Cr	330	280	110	120	100	20	20	81	250	420	1000	690	480	440	740	890	545	
	Y	16	11	12	27	25	24	27	26	18	31	24	23	31	23	18	6	6	
	Cr	18	100	1094	1108	1102	972	1485	677	1204	821	1013	807	32	84	230	200	220	
	Y	6	16	13	8	23	24	17	13	20	15	20	18	11	23	26	10		
	Cr	34	675	787	376	513	472	294	405	385	200	164	157	445	355	212	320		
Yarlung Zangbo	Y	17	27	19	21	30	28	21	25	27	32	25							Wang et al. (1987)
	Cr	329	120	267	93	123	345	317	300	220	317	260							
	Y	15	11	10	13	14	8	16	10	5	9	5	16	18		25	23	22	Zhang et al. (2001)
	Cr	887	455	335	583	681	577	577	469	968	418	364	1020	118	128	577	182	55	
	Y	17	10	54	24														
	Cr	546	328	83	618														

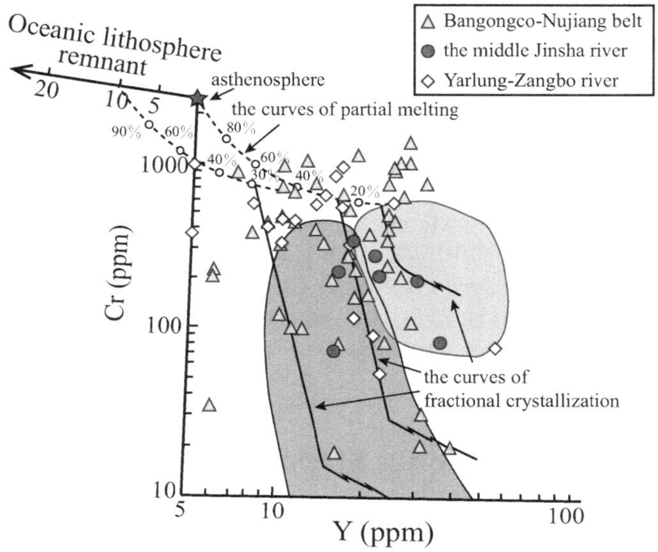

Fig. 5.8 The Cr-Y plot finger-printing SSZ-type and MORS-type mafic magmatic rocks (after Pearce et al. 1984)

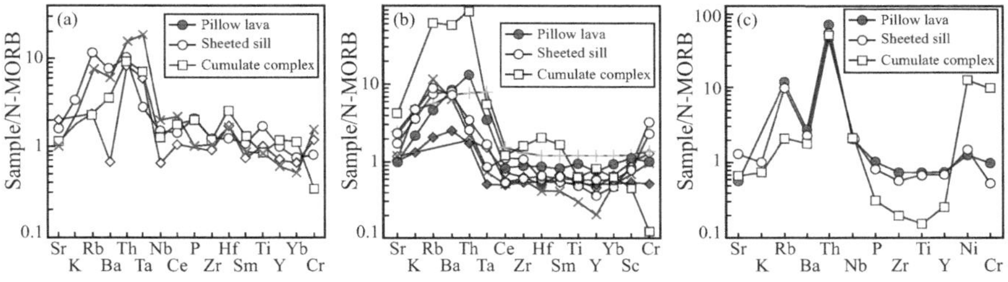

Fig. 5.9 Primitive mantle-normalized spider diagrams of SSZ-type and MORS-type mafic magmatic rocks (modified from Wang et al. 1987; Mo et al. 1998)

ophiolites. Nonetheless, as discussed above, the Tethyan ophiolites in Tibet (e.g., Yarlung Zangbo suture) and Sanjiang region simultaneously developed a series of SSZ-type and MORS-type ophiolites with SSZ-type dominant (Figs. 5.8 and 5.9). Recently, Dilek and Furnes (2011) emphasized that the ophiolites in Xigaza and Zedong were respectively the representative SSZ-type and volcanic arc ophiolites. The works in the doctoral dissertation by Li (2013) demonstrated in detail that three suites of ophiolites in Xigaza (Xiapu, Qunrang, Bailang) were SSZ-type (FA-FB) ophiolites using the classification scheme of ophiolite by Dilek and Furnes (2011). Moreover, 273 sets of ophiolite data were collected, including 234 sets of literature data in the Yarlung Zangbo suture and 39 sets from the works of Li (2013) in Xigaza. Using Th/Yb-Nb/Yb and Nb × 2-Zr/4-Y diagrams, it shows that in these ophiolites, SSZ-type ophiolites account for 74.3%, while the rest are subduction-irrelevant ophiolite (MORS + OIB, 31.7%).

5.2.4.2 The Initial Time of the Tethys Opening: A Case Study in Yarlung Zangbo Suture

In the past, it was commonly thought that the Yarlung Zangbo ocean was opened in the Triassic. However, ophiolite developed before the Triassic has been discovered in recent years. For instance, gabbros in Bailang subduction-accretionary complex had a zircon $^{206}Pb/^{238}U$ age of 273 ± 6 Ma (Mcdermid et al. 2002); layered gabbros in Luobusa dated at 352–353 Ma using plagioclase and clinopyroxene ^{40}Ar–^{39}Ar methods (Mo et al. 2008); OIB-type alkali gabbros in Najiu, the western Yarlung Zangbo suture had a SIMS zircon U-Pb age of 363.7 ± 1.7 Ma (1δ), belonging to the Upper Devonian (D_3, Dai et al. 2011). Consequently, the Yarlung Zangbo ocean had been already opened in the late Devonian.

5.3 Gangdese Magmatic Arc System

5.3.1 Introduction

It is known that the Gangdese belt developed a giant granite batholith and widely exposed Mesozoic-Cenozoic volcanic rocks, bounded by Bangong Co-Nujiang suture in the north and Indus River-Yarlung Zangbo suture in the south. Besides, this belt is a giant magmatic belt with a length of ~2500 km from east to west and a width of 100–300 km from north to south. It is noteworthy that Shiquanhe-Zhongzan-Jiali ophiolite mélange suture zone is distributed in the Gangdese belt (Mo 2013; Mo et al. 2020).

5.3.2 Major Characteristics of Volcanic and Intrusive Rocks in the Gangdese Magmatic Arc System

5.3.2.1 The Volcanic-Dominated Division of Tectonic-Magmatic Unites (Pan et al. 2006)

As summarized by Pan et al. (2006), there are three important characteristics of the Gangdese magmatic arc: (1) the Gangdese belt is not a simple block, massif or terrane. However, it takes Longer-Nyainqentanglha as the main axis and has experienced six periods of arc accretion processes, i.e., Carboniferous-Permian, the early-middle Triassic, the late Triassic, the early-middle Jurassic, the late Jurassic-early Cretaceous, and the late Cretaceous-Eocene; (2) The magmatic evolution of the Gangdese belt is probably constrained by the subduction of Banggong Co-Nujiang Tethyan ocean in the north and Yarlung Zangbo ocean in the south, rather than single subduction of Banggong Co-Nujiang Tethyan ocean or Yarlung Zangbo ocean; (3) Accretionary arc defines that the magmatic arc is formed in the late stage rooted in the previous accretionary complex. And it is further proposed that the magmatic arcs in Anglonggangri, Dongqia Co, and Sangri area are all accretionary arcs.

5.3.2.2 Intrusive Petrotectonic Assemblage and the Division of Tectonic Units (Deng et al. 2015, 2017a, b)

The tectonic-magmatic units in the simplified Gangdese intrusive petrotectonic sub-province (VII-3, Є-N_1) are shown in Fig. 5.11, whose upper-level unit is Sanjiang Tethys tectonic-magmatic province (VII) in southwestern China. Gangdese intrusive petrotectonic sub-province is bounded by Bangong-Nujiang sub-province οφ (VII-2, οφ) in the north and Yarlung Zangbo sub-province οφ (VIII(οφ), T-K) in the south. The sub-province is called Gangdese-Baoshan petrotectonic sub-province in the original map. However, due to focusing on the geology of Gangdese, the name "Baoshan" is deleted. The notes behind the codes and petrotectonic assemblages represent the ages of formation (Fig. 5.10).

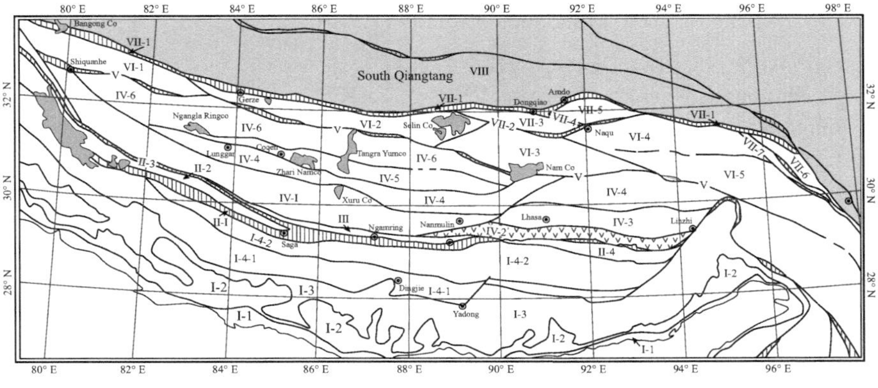

Fig. 5.10 Distribution of tectonic-magmatic belts in Gangdese volcanic arc system. I—Himalayan belt; II-1—Yarlung Zangbo suture; II-2—Zhongba block; II-3—Yarlung Zangbo (Gongzhu Co-Angren-Milin) main suture; II-4—Langjiexue accretionary wedge thrust belt; III—Xigaze forearc basin; IV—Gangdese-Nyainqentanglha composite magmatic arc; IV-1—Southern Gangdese magmatic arc (K_2–E_2); IV-2—Sangri volcanic arc (J_3–K_1); IV-3 Yeba volcanic arc (J_1–J_2); IV-4—Lunggar-Nyainqentanglha composite volcanic arc (P–T_3); IV-5—Coqen-Duowa composite back-arc foreland basin; IV-6—Zenong volcanic arc (J_3–K_1); V—Shiquanhe-Lhaguo Tso-Asuo-Yongzhu-Nam Co-Jiali-Bomi arc-arc collisional zone (i.e., Slainajsp zone); VI—Anglonggangri-Bange-Boshula Range magmatic arc; VI-1—Anglonggangri magmatic arc; VI-2 Bocangzangbo-Selin Co forearc shelf facies zone (K_1); VI-3—Bange-Bengco volcanic arc (K_1–K_2); VI-4—Naqu forearc basin; VI-5—Boshula Range magmatic arc (T_3, J–K); VII—Bangong Co-Nujiang suture; VII-1—Ritu-Gerze belt—Mugagangri belt—Anduo-Suoxian belt—Dingqing-Bitu belt; VII-2—Guiya arc-Jiaoweng arc-arc collisional zone; VII-3—Dongqia Co accretionary arc (K_2); VII-4—Dongqiao-Yilashan arc-arc collisional zone; VII-5—Nierong relic arc; VII-6—Jiayuqiao relic arc; VII-7—Luolong-Basu arc-arc collisional zone; VIII—Southern Qiangtang and Zuogong foreland basin

1. Composite intrusive rocks in the northern Gangdese (VII-3-1, J–E_2)

   (a) Ritu-North Gaer intrusive rocks (VII-3-1-1, K–E_2). Ritu TTG assemblage (K_2-E_2) is distributed in the northern part (VII-3-1-1-1), while North Gaer GG assemblage (K) is situated in the southern part (VII-3-1-1-2). Their spatial distribution indicates the subduction polarity that the ocean of Bangong οφ (VII-2, οφ) subducted southwestward.
   (b) Yanhu-Chaka intrusive rocks (VII-3-1-2, K). Its northern segment (VII-3-1-2-1) is occupied by Yanhu TTG assemblage (K), while the southern segment (VII-3-1-2-2) is dominated by Chaka GG assemblage (K), revealing the ocean of Bangong οφ (VII-2, οφ) subducted southwestward.
   (c) Bange-Naqu intrusive rocks (VII-3-1-3, K–N_1). Bange TTG (K–N_1) and Naqu-Bomi GG (K–E) assemblages are distributed in the southern (VII-3-1-3-1) and northern (VII-3-1-3-2) parts respectively, suggesting that the ocean represented by Shiquanhe-Shenzha-Jiali οφ (VII-3-2, T_3–K) subducted northeastward.
   (d) Nierong-Bomi intrusive rocks (VII-3-1-4, J). Nierong-Basu TTG (J) and Naqu-Bomi GG (J) assemblages are situated in the northeastern (VII-3-1-4-1) and southwestern segments respectively, indicating the ocean of Bangong-Nujiang οφ (VII-2) subducted southwestward.

2. Shiquanhe-Shenzha-Jiali οφ (VII-3-2, T_3–K)
   Shiquanhe-Shenzha-Jiali οφ (VII-3-2, T_3–K) is a composite arc or tectonic-magmatic zone οφ.
3. Composite intrusive rocks in the southern Gangdese (VII-3-3, Є–N_1)

   (a) Coqen-Xuru Co intrusive rocks (VII-3-3-1, J–E_2). Coqen TTG (J–K) and Palong Co-Xuru Co GG (J–E_2) assemblages are respectively allocated in the northern (VII-3-3-1-1) and southern (VII-3-3-1-2) segments, showing the ocean represented by Shiquanhe-Shenzha οφ (VII-3-2, T_3–K) subducted southwestward.
   (b) Nanmulin-Linzhi-Chayu intrusive rocks (VII-3-3-2, K–N_1). Nanmulin-Linzhi-Modog TTG (K–N_1) and Dangxiong-Linzhi-Chayu GG (K–N_1) assemblages are respectively distributed in the southern (VII-3-3-2-1) and northern (VII-3-3-2-2) segments, suggesting the northern direction of the oceanic subduction represented by Yarlung Zangbo οφ (VIII, T–K)
   (c) Three relic arcs. Northwest Milin (VII-3-3-3, Є) and West Chayu (VII-3-3-4, Є) relict arcs are dominated by TTG assemblages, while North Nanmulin relict arc (VII-3-3-9, T_3) predominantly consists of GG (T_3) assemblage.

5.3.3 Discussion

The formation of the Gangdese magmatic arc is not only restricted by the southward subduction of the ocean represented by Bangong-Nujiang οφ and the northward subduction of the ocean recorded by Yarlung Zangbo οφ, but also related to the subduction of the ocean represented by Shiquanhe-Shenzha-Jiali oφ, as suggested by compositional polarity recorded by spatial distributions of intrusive petrotectonic assemblages. Besides, if there is no oceanic subduction of Shiquanhe-Shenzha-Jiali oφ and its subducted slab into the deep mantle, the magmatic arcs in the northern and southern Gangdese could not form a unified Gangdese magmatic arc system in arc accretion process.

As pointed out by Pan et al. (2006), it is not reasonable that the Gangdese belt was a single block, massif, or terrane. Since the Carboniferous-Permian, the Gangdese belt had undergone six periods of arc accretion process, corresponding arc-continent and continent-continent collisions, and finally formed the composite orogenic belt in the late Cenozoic. Meanwhile, another arc accretion event was recorded by Cambrian-Ordovician intrusive relic arcs (Fig. 5.11), implying seven periods of arc accretion process in forming the Gangdese belt. Considering the magmatic arc system, fossil arc massifs (arc massif) by Stern and Scholl (2010) would be better to describe the significance of the tectonic environment. Thus, the Gangdese arc system can be called Gangdese arc massif.

The concept of accretionary arc proposed by Pan et al. (2006) is important. It is pointed out that the basement covered by the early Cretaceous Anglonggangri

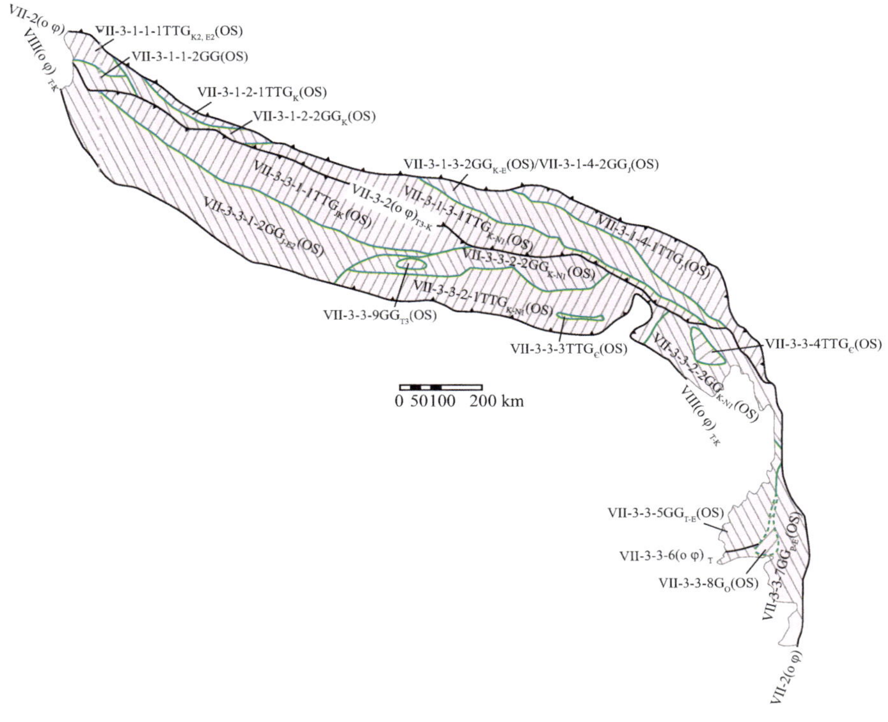

Fig. 5.11 Tectonic-magmatic units in the simplified Gangdese intrusive petrotectonic subprovince, and the distribution framework of oφ subprovinces at the northern and southern boundaries (modified from Deng et al. 2015, 2017a, b)

volcanic arc is a Jurassic accretionary complex, while the early-middle Jurassic accretionary complex formed by the subduction of Yarlung Zangbo ocean is the basement beneath the late Jurassic-early Cretaceous Sangri volcanic arc. As recognized on accretionary arcs, (a) before the formation of an accretionary magmatic arc, the seafloor subduction already occurs; (b) the seafloor subduction associated with the formation of accretionary magmatic arc was regressive, which is important in tracking the evolution of seafloor subduction; (c) if the temperature is high enough in magma chambers for arc magma genesis, it may induce melting of peraluminous accretionary complex and generate strongly peraluminous granitoids. Therefore, strongly peraluminous granitoids could also be observed in the petrotectonic assemblage of magmatic arcs. Different from those of continental collision-related origin, strongly peraluminous granitoids in magmatic arc coexist with TTG; (d) furthermore, once strongly peraluminous granitoids are identified in a magmatic arc, this type of magmatic arc can be determined as accretionary arc. Then, it can thus be inferred that its underlying basement is plausibly to be the accretionary complex generated by the last sefloor subduction event, further providing a useful fingerprint for identifying concealed subduction-related accretionary complex beneath the magmatic arc system. And this is of great significance for further understanding the generation and evolution of magmatic arc system. For instance, in the eastern Gandese arc system, strongly peraluminous granitoids coexisting with TTG are distributed in three belts, from north to south namely Bange-Boshulaling belt, Coqen-Shenzha belt, and the southern Gangdese belt (Liao et al. 2006, 2007; Mo 2009). Hence, it can be inferred that concealed subduction-related accretionary complexes are probably below these granitoid belts in magmatic arc settings with accretionary arc affinities, rather than in settings of intracontinental subduction or continental-continental collision. Furthermore, except for Shiquanhe-Shenzha-Jiali οφ belt, the Gangdese magmatic arc system probably developed other concealed Tethyan οφ belts covered by accretionary magmatic arcs, although these belts are not exposed. Therefore, this discussion further supports the rationality and importance of naming Gangdese magmatic arc system as Gangdese arc massif.

5.4 The Tibetan Plateau

5.4.1 The Closure of the Tethys

The closure of the Tethys is still highly debated and has been reviewed in detail by Mo (2009), Mo et al. (2020), and Hou et al. (2008). Hou et al. (2008) put forward a three-stage model of continental collision orogeny, i.e., the main collision stage (65–41 Ma), the late collision stage (40–26 Ma), and the post-collision stage (25–0 Ma). This issue is discussed from various aspects in other chapters. In this monograph, we provide several new investigations for discussion.

5.4.1.1 Discovery of Pelagic Radiolarian Chert

As investigated by Wang (2013), since the late Paleogene radiolaria in Jiangzi and Saga were discovered in 2000, it was also found in Zhongba with an age of 59–56.5 Ma. Meanwhile, a suite of contemporaneous basalt developed in the subduction-related accretionary complex was dated at 59.1 Ma using zircon SHRIMP U-Pb method. Therefore, the age of 59 Ma can be considered as the initial time that the pelagic radiolarian chert layer left from the mid-ocean ridge, i.e., the first key interface in OPS sequence (Ocean Plate Stratigraphy, defined by Isozaki 1996). However, in the OPS sequence, the stratigraphic boundary between bathyal siliceous mudstone and oceanic trench turbidite is still unknown, which means the residence time of pelagic sediment is unclear in the ocean. And the residence time of pelagic sediment reached 2.5 Myrs. Besides, it is also unknown on the relative timing of seafloor arrival at trenches and subduction into the mantle. Nevertheless, the initial time of oceanic crust subduction is expected to be later than 56 Ma.

5.4.1.2 The Latest Time of Magmatic Arc Formation

According to the intrusive petrotectonic assemblage, the Gangdese magmatic arc system was developed until the Miocene (N_1). As investigated by Mo (2009), adakite in the arc system was generated at 25–12 Ma. If the age of 25 Ma (E_3^2) is considered as the initial time of the last-stage magmatic arc formation, it differs by 35 Myrs from the youngest oceanic crust (59 Ma). Defant and Drummond (1990) pointed out that adakite was produced by partial melting of young oceanic crust (<25 Ma). Gutscher et al. (2000) listed ten examples of worldwide flat subductions, and adakitic magmas had been found in eight of them. Although the age of subducted oceanic crusts could reach 43–45 Ma, thermal simulation results support that subducted oceanic crust (43–45 Ma) could generate adakitic melts when it is heated by the hot upwelling asthenosphere. Besides, the simulation further shows that as long as the front end of the subducted slab slides at the depth of 70–80 km in ~5 Ma, there should be enough time for the subducted ocean crust to heat up to 700–800 °C, inducing local melting with the input of slab-dehydrated fluids. And this process is irrelevant to the dip angle of plate subduction.

As above, there is a 34 Myrs difference between the youngest ocean crust (59 Ma) and the earliest adakite in the last-stage magmatic arc (25 Ma), which means the age of the subducted ocean crust would be 34 Ma. As for the age of ocean crust, the results of 34 Ma in this work are between <25 Ma proposed by Defant and Drummond (1990) and 43–45 Ma investigated by Gutscher et al. (2000). Consequently, it implies that the subducted ocean crust with an age of 34 Ma can generate adakitic melts as long as it subducts and slides through the hot asthenosphere in ~2.5 Ma.

5.4.1.3 The Latest Period of Marine Strata Development

In southern Tibet, the marine strata with the highest altitude were developed in the early Bartanian (~40 Ma, Li 2004; Mo 2009; Mo et al. 2020). The latest research (Wang et al. 2002; Aitchison et al. 2007; Wang 2013) reported that near Qumiba in the north of Mount Qomolangma, the uppermost marine sediments in the Himalayan succession were NP20 nannofossil assemblage at the end of Eocene (Priabonian, ~35 Ma). In the Jurassic-Paleogene, marine units with a thickness of 5 km were deposited in Sulaiman and Mount Kirta, the western Pamir syntaxis (Pan et al. 1990; Deng et al. 2001). The regression of Tarim basin was later than that of the main body of the Tibetan Plateau. This regression event mainly occurred at the end of the Oligocene, and finally ended until the Miocene (Xiao and Li 2000; Deng et al. 2001). Thus, it seems that the time of continental-continental collision in Pamir is later than that of the main body of the Tibetan Plateau, which is ascribed to the tectonic period on the growth of orogenic lithospheric root discussed as follows.

Integrated with the generation of the Eocene-Miocene (E_3–N_1) biotite-two-mica-muscovite granite assemblages in Lhagoi Kangri-High Himalaya passive continental margin, it can be inferred by the aforementioned geological records that the early stage of Mediterranean-type continental-continental collision occurred in the Eocene-Miocene (E_3–N_1). In the middle Miocene, as the development of TTG ended, the residual ocean completely closed and thereafter shifted to the late stage of continental-continental collision. Until now, the Tibetan Plateau is still shortening horizontally and not in the post-orogenic stage (Deng et al. 2015, 2017a, b).

5.4.2 The Formation of Mountain Root and the Uplift of the Tibetan Plateau

Mountain root is formed by horizontal compression and vertical thickening during the orogeny of continental-continental collision. And the formation of mountain root would induce the continuous uplift of the plateau.

Deng et al. (1996) put forward that aluminosilicate continental crust between the surface and Moho discontinuity of mountain base was defined as mountain body. The mountain body of the Tibetan Plateau-Himalayas is shown in Fig. 5.12. In Fig. 5.12, the topography is drawn based on the elevation of mountain surface, basin surface, and main mountain peak at the margin of the plateau (Table 5.3). For the sake of clarity, the vertical direction in Fig. 5.12 is not shown on a scale. It can be seen in Fig. 5.12 that the flat and slightly concave mountain surface between Kunlun mountain and the Himalayas corresponds to the basin surface. The giant mountain root and body have an average composition of diorite and granitoid, respectively. In general, this giant mountain body is characterized by relatively homogeneous low density, providing an important basis for the overall uplift of the plateau. Besides, the Himalayas at the margin of the plateau has a relatively small mountain root with

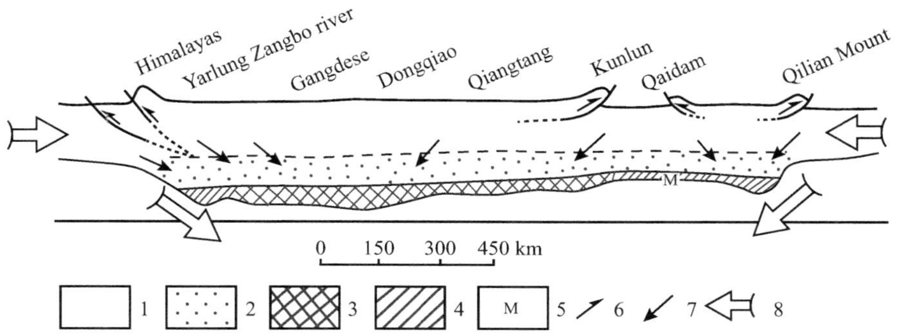

Fig. 5.12 Cartoon showing the mountain body contour of Qinghai-Tibet-Himalaya orogenic belts (after Deng et al. 1996). 1—Upper, middle, lower crust, granitic composition on average; 2—Thicken lower crust, high-pressure granulite facies, granitic composition on average; 3—Lithospheric root zone, dioritic eclogite; 4—Lithospheric root zone, granitic eclogite; 5—Moho discontinuity; 6—Thrust direction of thrust fault; 7—Direction of material flow during continental crust thickening; 8—Forces of horizontal extrusion and downward convergence

a granitic average composition and relatively lower density, similar to Qilian mountain. It reveals that the Himalayas plausibly has a faster uplift rate than the interior of the plateau. Taking Dongqiao as the central axis belt (e.g., Bangong Co-Nujiang belt), the southern and northern mountain bodies are asymmetric. The northern part has high density and small thickness, while the southern part is characterized by low density and large thickness. These differences are probably ascribed to different orogenic processes. However, the areas of their mountain bodies are nearly the same, implying the same volumes. It seems that the horizontal compressional forces from the north and south directions are roughly equal in the orogenic processes. Finally, the shape of the mountain bodies shows that the mountain roots beneath the Himalayas and Qilian mountains are staggered and asymmetric with the overlying mountain peaks on the surface. It generally appears as a fan-shaped mountain body as the mountain roots are connected with the peaks. And the fan-shaped mountain body implies that: (1) The Himalayan orogenic belt is still in the process of horizontal compression orogeny, consistent with the above discussion that this belt is in the stage of continued continental convergence at present; (2) Geodynamics of the Himalayan belt comes from the deep downward convergence, which is related to the downward pull of the underlying dense lithospheric root (see the discussion below); (3) The pull would inevitably lead to bidirectional horizontal compression. Namely, the formation of the mountain root supports the dynamic model of bidirectional horizontal compression from Indian continent in the south and Asian continent in the north (e.g., Deng et al. 1996), rather than that of single northward horizontal extrusion from Indian continent (e.g., Tapponnier et al. 1986); (4) Because the surface is a kind of "free-interface" when the lower crust material plastically flows and cumulates downward to form a giant mountain root, the shallow brittle crust would thrust upward and to the periphery, resulting in the formation of fan-shaped mountain body.

Table 5.3 Height of the plateau surface and mountain peak of the Tibetan Plateau (after Deng et al. 1996)

	South Bangong-Nujiang as the central axis North						
Mountain surface	Northern foot of Himalayas, 5500 m	Gangdese, 5200–5300 m		Hoh Xil, 5300 m	Front of Kunlun Mount, 5500 m	Qilian Mount, 4000 m	
Basin surface	Northern slope of Himalayas, 5000 m	Southern foot of Gangdese, 4400–4600 m	Northern foot of Gandese, 4800–4900 m	Bangong Co-Dongqiao, 4400–4600 m	Qiangtang, 4700–4800 m	Southern foot of Kunlun Mount, 5000 m	Qaidam, 2600–3100 m
Adjacent mountain peaks	Mountain peak of Himalayas, >8000 m			Peak of Kunlun Mount, >7000 m		Peak of Qilian mount, >6000 m	

5.4.3 Tectonic Facies Model of Lithospheric Root Evolution in Continental-Continental Collision Orogeny

5.4.3.1 Tectonic Facies Model of Lithospheric Evolution of the Tibetan Plateau

Integrated with geophysical investigations and geological explanations in continental-continental collision process, Mo and Deng (1999) put forward a tectonic facies model of lithospheric evolution in the Tibetan Plateau, shown in Fig. 5.13.

5.4.3.2 Pamir-Type Active Lithospheric Cold Root

Teng et al. (1984) proposed that in Hindu Kush and Pamir, a large number of earthquakes were concentrated in the area at 70°–72°E with 36°–37°N, forming a "V"-type dense seismic active zone at ~300 km depth (Fig. 5.14). This zone is

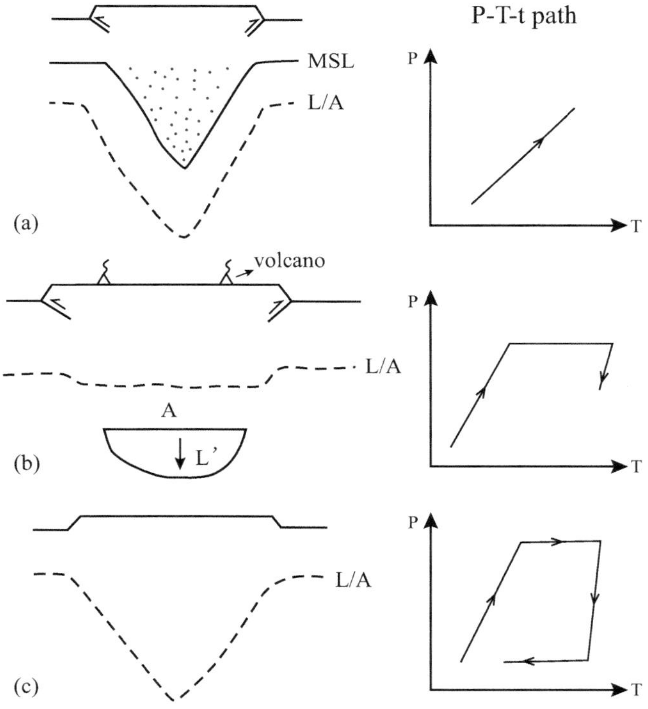

Fig. 5.13 Tectonic facies model of lithospheric evolution in the Tibetan Plateau (after Deng et al. 2001). MSL—mechanic strong lithospheric boundary (Black dot represents intermediate-focus earthquake); L/A—thermal lithosphere; L—detached lithosphere; A—asthenosphere; (**a**)—the early-stage facies: Parmir-type lithospheric cold root; (**b**)—the medium-stage facies: Nyainqentanglha-type detached lithosphere; (**c**)—Qiangtang-type hot lithosphere

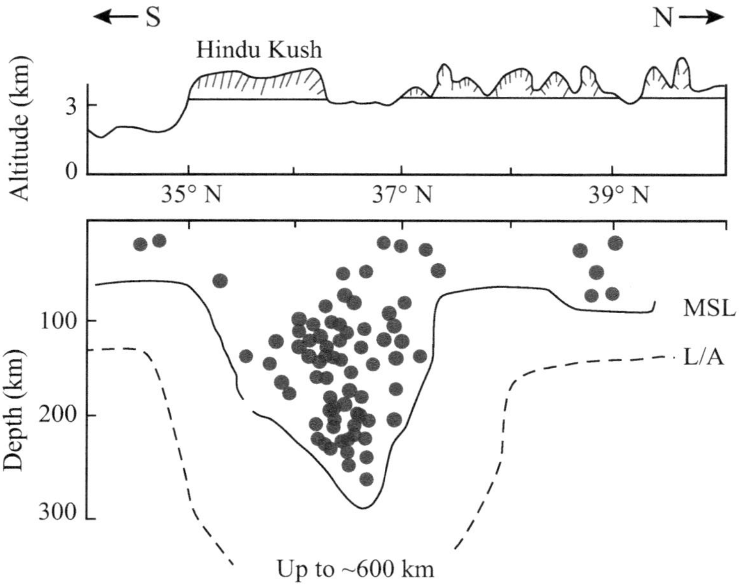

Fig. 5.14 The active lithospheric cold root in Pamir at ~71°E longitude (after Deng et al. 2001). The black dot represents an intermediate-focus earthquake (after Teng et al. 1984). MSL—lower boundary of mechanic strong lithosphere outlined by intermediate-focus earthquakes; L/A—lower boundary of the thermal lithosphere

called "active lithospheric cold root" (Mo and Deng 1999; Deng et al. 2001), while the brittle (rigid) lithosphere where the seismic activities are concentrated is named mechanical lithosphere (Anderson 1995). According to the thermal definition, the term "Lithosphere" is defined as the outer shell of the Earth where there is a temperature gradient. Generally, mechanical lithosphere is about one-half thickness of thermal lithosphere. And mechanical lithosphere up to 300 km in depth and thermal lithosphere up to 600 km in depth are outlined in Fig. 5.14 (Deng et al. 2001). According to Teng et al. (1984), there are several lithospheric cold roots in the Himalayan orogenic belt similar to that in Pamir: (1) The focal plane is also distributed in a "V" shape with a maximum depth of 200 km beneath Indus-Yarlung Zangbo oφ belt, but its intermediate-focus earthquakes are far less frequent than those in Pamir; (2) There is a lithospheric cold root up to 200 km in depth beneath Burma arc. And its intensity of seismic activity is lower than that in Pamir but much stronger than that beneath Yarlung-Zangbo oφ belt; (3) There are also intermediate-focus earthquakes with depths of 71–100 km and 101–200 km near the Altun mountain in the northern Tibetan Plateau. It indicates that these active lithospheric cold roots were formed recently and are still active at present, which are considered to be the early-stage tectonic facies of lithospheric roots beneath continent-continent collision orogenic belt in the Tibetan Plateau (Mo and Deng 1999; Deng et al. 2001).

5.4.3.3 Nyainqentanglha-Type Detached Orogenic Lithospheric Root

As suggested in the Yadong-Gelmud geological section (Wu et al. 1989) and the investigation of two-dimensional magnetotelluric sounding (edited by Xiao and Li 2000), the lithosphere thickness of Gangdese is the thinnest with a depth of ~120 km. Meanwhile, the results of three-dimensional P-S wave seismic travel time inversion show that there is a low-velocity zone in the depth of 100–250 km beneath Nyainqentanglha. Consequently, it reveals that the depth of the asthenosphere beneath these regions is the shallowest with a depth of 100–120 km, which is called "Nyainqentanglha-type orogenic lithospheric root". And this lithospheric root is classified as the medium-stage tectonic facies of lithospheric root beneath continent-continent collision orogenic belt in the Tibetan Plateau (Mo and Deng 1999; Deng et al. 2001).

5.4.3.4 Qiangtang-Type Hot Orogenic Lithospheric Root

According to the Cenozoic volcanic rocks in Qiangtang-Hoh Xil-Bayankala, the asthenosphere beneath this region is relatively shallow. Nevertheless, the current depth of the asthenosphere is ~200 km indicated by Yadong-Gelmud geological section (Wu et al. 1989) and the investigations of three-dimensional magnetotelluric sounding (Wu et al. 1989; Lu et al. 1996). And the different understandings above can be explained by the following geological observations. The volcanic rocks in this area erupted in the late Eocene-Miocene (~39–12 Ma), and the main eruption period is in the Oligocene (~29–27 Ma) (Deng et al. 2001). In this period, the asthenosphere was relatively shallow with a depth of 80–100 km (Deng et al. 1996). After the Miocene, the asthenosphere became cool and was transformed into the lithosphere. So far, ~100 km-thick asthenosphere has been transformed into the lithosphere so that the current lithosphere has a high potential temperature. In the seismic activity (Teng et al. 1984), earthquake in this area was predominantly developed at shallow depth ($\leq$30 km) and significantly decreased with deep, agreeing well with the above speculation. Compared with Pamir-type active lithospheric cold root, the lithospheric root in this area can reach 200 km in depth and has a high potential temperature, which is called "Qiangtang-type hot orogenic lithospheric root" (Mo and Deng 1999; Deng et al. 2001).

Acknowledgement Many thanks to Qinghui Xiao and Yanfang Feng for valuable discussions, and to Cui Liu, Yongjun Di and Hangsheng Lei for figure-drawing, typewriting and text correction for the chapter. This work has been supported by China Geological Survey, Ministry of Natural Resources (Grant Number: DD20190370).

References

Aitchison JC, Ali JR, Davis AM (2007) When and where did India and Asia collide? J Geophys Res 112:B05423. https://doi.org/10.1029/2006JB004706

Anderson DL (1995) Lithosphere, asthenosphere, and perisphere. Rev Geophys 33(1):125–149. https://doi.org/10.1029/94RG02785
Balykin PA, Polyakov GV, Shelepaeu PA et al (2001) Compositions of primary melts of Permian-Triassic and Triassic-Jurassic Utramafic-mafic complexes of North Vietnam. Dokey Earth Sci 378(4):444–448
Bao PS, Su L, Wang J et al (2015) Yarlung Zangbo ophiolite. Geological Publishing House, Beijing, pp 1–267. (In Chinese, with English Abstract)
Cao YQ (1999) Magmatism, tectonic evolution, deep processes and metallogenic. China University of Geosciences, Beijing, pp 1–95. (In Chinese, with English Abstract)
Chang CF, Pan YS, Zhen XL et al (1982) Tectonics of the Tibetan plateau. Science Press, Beijing, pp 1–91. (In Chinese, with English Abstract)
Chen ZL (1994) Tethyan geology for 100 years. Tethyan. Geology 18:1–22. (In Chinese, with English Abstract)
Chung S L, John B M, Wu G et al (1998) The Emeishan flood basalt in SW China: a mantle plume initiation model and its connection with continental breakup and mass extinction at the Permian-Triassic boundary. In: Flower MFJ et al (eds) Mantle dynamics and plate interactions in East Asia. American Geophysical Union, Washington DC, (27), pp 47–58
Condie KC (2001) Mantle plumes and their record in earth history. Cambridge University Press, pp 1–306
Dai J, Wang C, Hébert R et al (2011) Late Devonian OIB alkaline gabbro in the Yarlung Zangbo suture zone: remnants of the paleo-Tethys? Gondwana Res 19(1):232–243
Defant MJ, Drummond MS (1990) Derivation of some modern arc magmas by melting of young subducted lithosphere. Nature 347(6294):662–665
Deng JF, Zhao HL, Mo XX et al (1996) Continental roots-plume tectonics—the key to continental dynamics in China. Geological Publishing House, Beijing, pp 1–110. (In Chinese, with English Abstract)
Deng JF, Mo XX, Luo ZH et al (2001) Heterogeneity of lithosphere in the Tibetan Plateau and its geodynamics science in China (Series D), 31:55–60. (In Chinese, with English Abstract)
Deng JF, Feng YF, Di YJ et al (2015) The intrusive spatial-temporal evolutional framework in the paleo-Asian tectonic domain. Geol Rev 61(6):1211–1224. (In Chinese, with English Abstract)
Deng JF, Feng YF, Di YJ et al (2017a) Geotectonic map of intrusive rocks of China (1: 2500000). Geological Publishing House, Beijing, pp 1–583 (In Chinese, with English Abstract)
Deng JF, Feng YF, Di YJ et al (2017b) Instructions of geotectonic map of intrusive rocks of China (1: 2500000). Geological Publishing House, Beijing, pp 1–81 (In Chinese, with English Abstract)
Dilek Y, Furnes H (2011) Ophiolite genesis and global tectonics: geochemical and tectonic fingerprinting of ancient oceanic lithosphere. Geol Soc Am Bull 123(3–4):387–411
Falloon TJ, Danyushevsky LV (2000) Melting of refractory mantle at 1.5, 2 and 2.5 GPa under anhydrous and H2O-undersaturated conditions: implications for the petrogenesis of high-Ca Boninites and the influence of subduction components on mantle melting. J Petrol 41(2):257–283
Fedorenko V, Czamanske G (1997) Results of new field and geochemical studies of the volcanic and intrusive rocks of the Maymecha-Kotuy area, Siberian flood-Basalt Province, Russia. Int Geol Rev 39(6):479–531
Fedorenko VA, Lightfoot TPC, Naldert AJ et al (1996) Petrogenesis of the flood-basalt sequence at Norilsk, north Central Siberia. Int Geol Rev 38(2):99–135
Flower MFJ, Russo RM, Widom E et al (2000) The signification of boninite in Tethyan subduction-accretion complexes. IGCP430 workshop I, Corasna, Romania, pp 20–21
Gutscher MA, Maury R, Eissen JP et al (2000) Can slab melting be caused by flat subduction? Geology 28(6):535–538
Hirose K, Kushiro I et al (1993) Partial melting of dry peridotites at high pressures: determination of compositions of melts segregated from peridotite using aggregates of diamond. Earth Planet Sci Lett 114(4):477–489

Hou ZQ, Wang EQ, Mo XX et al (2008) Collisional orogeny and metallogeny in the Tibetan plateau. Geological Publishing House. (In Chinese, with English Abstract)
Irvine TN (1989) A global convection framework: concepts of symmetry, stratification and system in the Earth's dynamic structure. Econ Geol 84(8):2059–2114
Isozaki Y (1996) Anatomy and genesis of a subduction related orogen: a new view of geotectonic subdivision and evolution of the Japanese Islands. Island Arc 5(3):289–320
Kushiro I (1990) Partial melting of mantle wedge and evolution of Island arc crust. J Geophys Res 95(951):15929–15939
Larson RL (1991) Geological consequences of superplumes. Geology 19(10):963–966
Le Bas MJ, Le Maitre RW, Streckeisen A et al (1986) IUGS subcommission on the systematics of igneous rocks, a chemical classification of volcanic rocks based on the Total alkali-silica diagram. J Petrol 27(3):745–750
Li GB (2004) Paleogene micropaleontology and basin evolution in southern Tibet. China University of Geosciences, Beijing. (In Chinese, with English Abstract)
Li WX (2013) Geochemical characteristics and tectonic environments of the Yarlung Zangbo ophiolite in Xigaze, Tibet. Doctoral dissertation of China University of Geosciences Beijing, pp 1–109. (In Chinese, with English Abstract)
Liao ZL, Mo XX, Pan GT et al (2006) On peraluminous granites in Tibet, China. Geol Bull China 25(7):812–821. (In Chinese, with English Abstract)
Liao ZL, Mo XX, Pan GT et al (2007) Spatial and temporal distribution of peraluminous granites in Tibet and their tectonic significance. J Asian Earth Sci 29:378–389
Liu BP, Quan QQ (1996) Historical geology, 3rd edn. Geological Publishing House, Beijing, pp 1–277. (In Chinese, with English Abstract)
Lu QT, Jiang M, Ma KY et al (1996) Three-dimensional travel time inversion and deep structure in southern Tibetan plateau. Acta Seismol Sin 18(4):451–459. (In Chinese, with English Abstract)
Ma WP (1992) Analysis of regional tectonics: theoretical studies and plate tectonics in China. Geological Publishing House, Beijing, pp 1–308. (In Chinese, with English Abstract)
Malpas JG, Robinson PT (2000) Oceanic lithosphere 4. The origin and evolution of oceanic lithosphere: magmatic processes at oceanic spreading centres. Geosci Can 27(3):131–146
Mcdermid I, Aitchison J, Davis A et al (2002) The Zedong terrane: a late Jurassic intra-oceanic magmatic arc within the Yarlung-Tsangpo suture zone, southeastern Tibet. Chem Geol 187(3):267–277
Miyashiro A (1973) The Troodos ophiolitic complex was probably formed in an Island arc. Earth Planet Sci Lett 19(2):218–224
Miyashiro A (1975) Classification, characteristics, and origin of ophiolites. J Geodyn 83:249–281
Mo XX (2009) Cenozoic collisional-postcollistional volcanic rocks in the Tibetan Plateau. Geological Publishing House (In Chinese, with English Abstract)
Mo XX (2013) Map of tectonic-magmatic rocks and instructions in the Tibetan plateau and its adjacent areas. Geological Publishing House (In Chinese, with English Abstract)
Mo XX, Deng JF (1999) Three types of lithospheric structure in the Tibetan plateau. 14th HKT workshop Abst. Terra nostra, Z: 99–100
Mo XX, Lu FX, Shen SY et al (1993) Volcanism and metallogeny in Sanjiang Tethys. Geological Publishing House, Beijing, pp 1–267. (In Chinese, with English Abstract)
Mo XX, Shen SW, Zhu QW et al (1998) Volcanic rocks-ophiolite and metallogeny in mid-southern segment of the Sanjiang Tethys. Geological Publishing House, Beijing, pp 1–128. (In Chinese, with English Abstract)
Mo XX, Zhao ZD, Zhou S (2008) Complexity in ages and tectonic settings of Yarlung Zangbo Ophiolite Belt
Mo XX, Zhu DC, Hou ZQ et al (2020) Mesozoic-Cenozoic tectonic-magmatism in the Tibetan plateau. Guangdong Liberation Publishing House, Guangzhou, pp 1–206. (In Chinese, with English Abstract)
NRC (National Research Council) (1993) Solid-Earth Sciences and Society. National Academy Press, Washington D. C

Pan GT (1994) An evolution of Tethys in global ocean-continent transformation. Tethyan Geol 18:23–40. (In Chinese, with English Abstract)
Pan GT, Wang PS, Xu RY et al (1990) Cenozoic tectonic evolution in the Tibetan plateau. Geological Publishing House, pp 1–190 (In Chinese, with English Abstract)
Pan GT, Mo XX, Hou ZQ et al (2006) Spatial-temporal framework of the Gangdese Orogenic Belt and its evolution. Acta Petrol Sin 22(3):521–533. (In Chinese, with English Abstract)
Parfenov LM, Natalin BA (1986) Mesozoic tectonic evolution of northeastern Asia. Tectonophysics 127(3):291–304
Pearce J A, Lippard S J, Roberts S, 1984. Characteristics and tectonic significance of supra-subduction zone ophiolites, In Kokelaar BP, Howells MF (eds) Marginal basin geology, Blackwell Science Publishers, Ames, pp 77–94
Qiu RZ (2002) Igneous rocks and tectonic evolution of the neo-Tethyan in the western Tibetan plateau. China University of Geosciences, Beijing, pp 1–99. (In Chinese, with English Abstract)
Qiu RZ, Deng JF, Zhou S et al (2005) Ophiolite types in western Qinghai-Tibetan plateau: evidences from petrology and geochemistry. Earth science frontiers. China University of Geosciences, Beijing, 12(2), pp 277–291. (In Chinese, with English Abstract)
Ren JS, Wang ZX, Chen BW et al (1999) A global view of China tectonics: a brief instruction to the tectonic map of China and its adjacent areas. Geological Publishing House, Beijing, pp 1–50. (In Chinese, with English Abstract)
Sengor AMC (1988) The story of Tethys: how many wives did Okeanos have? Episodes 8(1):3–12
Song XY, Zhou MF, Hou ZQ et al (2001) Geochemical constraints on the mantle source of the upper Permian Emeishan continental flood basalts, southwestern China. Int Geol Rev 43(3):213–225
Stern RJ, Scholl DW (2010) Yin and yang of continental crust creation and destruction by plate tectonic processes. Int Geol Rev 52(1):1–31
Tapponnier P, Peltzer G, Armijo R (1986) On the mechanics of the collision between India and Asia. Geological Society of London Special Publications 19(1):113–157
Teng JW, Sun KZ, Wei SY et al (1984) The characteristics of seismic activity in the Tibetan Plateau and its marginal region of China. In: Abstracts of Institute of Geophysics, Chinese Academy of Sciences (In Chinese, with English Abstract)
Vander VR, Spakman W, Bijwaard H (1999) Mesozoic subducted slabs under Siberia. Nature 397(6716):246–249
Wang HZ (1981) Geotectonic units of China from the view-point of Mobilism. Earth Sci 1:42–66. (In Chinese, with English Abstract)
Wang LQ (2013) Geological map and its instructions in the Tibetan plateau and its adjacent areas. Geological Publishing House. (In Chinese, with English Abstract)
Wang XB, Bao PS, Deng WM et al (1987) Ophiolite in Tibet: tectonic evolution of Himalayan lithosphere. Geological Publishing House, Beijing, pp 1–336. (In Chinese, with English Abstract)
Wang C, Li X, Hu X et al (2002) Latest marine horizon north of Qomolangma (Mt Everest): implications for closure of Tethys seaway and collision tectonics. Terra Nova 14(2):114–112
Wu GJ, Xiao XC, Li TD (1989) Yadong-Gelmud geological section in the Tibetan Platea. Acta Geol Sin 4:285–296. (In Chinese, with English Abstract)
Wyllie PJ (1988) Solidus curves, mantle plumes, and magma generation beneath Hawaii. J Geophys Res Atmos 93(B5):4171–4181
Xiao XC (1995) Discussion on the classification of ophiolites by spreading rate. Acta Petrol Sin 11:10–23. (In Chinese, with English Abstract)
Xiao XC, Li TD (2000) Tectonic evolution and uplift mechanism of the Tibetan plateau. Guangdong Science and Technology Press, Guangzhou, pp 1–313. (In Chinese, with English Abstract)
Xu XZ, Yang JS, Guo GL et al (2011) Lithological research on the Purang mantle peridotite in western Yarlung Zangbo suture zone in Tibet. Acta Petrol Sin 11:3179–3196. (In Chinese, with English Abstract)
Yang JS, Xu ZQ, Duan XD et al (2012) Discovery of a Jurassic SSZ ophiolite in the Myitkyina region of Myanmar. Acta Petrol Sin 6:16–36. (In Chinese, with English Abstract)
Yem NT, Tri TV, Hoa TT et al (2002) Field trip guide of the IGCP430 second workshop Halong, Vietnam Halong:[s.n.]: 6–9

Yin HF (1988) Paleobiogeography in China. China University of Geosciences, Wuhan, pp 1–329. (In Chinese, with English Abstract)
Zhang Q, Zhou GQ (2001) Ophiolite in China. Science Press, Beijing, pp 1–182. (In Chinese, with English Abstract)
Zhang GW, Zhang BR, Yuan XC et al (2001) Qilian orogenic belt and continental dynamics. Science Press, Beijing, pp 1–855. (In Chinese, with English Abstract)
Zhong DL (1998) Paleo-Tethyan orogenic belt in the western Chuan-Dian area. Science Press, Beijing, pp 1–217. (In Chinese, with English Abstract)
Zhou S (2002) Study on the geochronology of pivotal regions of Gangdese magmatic and Yarlung Zangbo ophiolite belts, Tibet. China University of Geosciences, Beijing, pp 1–69. (In Chinese, with English Abstract)
Zhu DC (2003) Volcanic rocks in the central belt of Tethyan Himalayas since the late Paleozoic and their implications. Chinese Academy of Geological Sciences, Beijing. (In Chinese, with English Abstract)

Chapter 6
India-Asia Collision and the Uplift of the Tibetan Plateau

Abstract The processes of the India-Asia continental collision, and the uplift and growth of the Tibetan Plateau are discussed in this chapter. The closure of the Neo-Tethyan ocean and the India-Asia collision are the prerequisite and basis for the formation of the Tibetan Plateau. Based on various lines of evidence, the initial India-Asia collision took place at 65 ± 5 Ma with the main phase of collision at ~45 ± 5 Ma, and then followed by post-collisional processes. The main phase of the collision is characterized by the emplacement of the volumetrically significant Gangdese batholith and widespread of high temperature volcanic eruptions with asthenospheric mantle material contribution. The India-Asia collision and the continued convergence gave long-term effects on the Plateau. The formation of the extremely thickened Tibetan crust is closely related to the India-Asia collision by means of combined contributions of asthenospheric mantle materials and tectonic compression. The collisional and post-collisional igneous rocks in the Tibetan Plateau show significant mantle contributions. It is proved that two types of continental crust coexist in the Tibetan Plateau, the juvenile crust and reworked mature crust. The crust of the northern and southern Lhasa subterranes is dominated by juvenile crust, while the crust of the central Lhasa subterrane is dominated by reworked mature crust. Syncollisional felsic magmatism may in fact account for much of the net contribution to continental crust growth. At least 1400 km of north-south shortening has been absorbed since the onset of the India-Asian collision. The surface uplift history of the Tibetan Plateau and Himalaya is among the most important topics in geosciences because of its effects on regional and global climate variation since the Cenozoic, monsoon intensity, regional vegetation ecological models, biodiversity and watershed models. In general, there are three general views on the pattern and mechanism of uplift and growth of the Plateau: (1) crustal and lithospheric shortening, thickening and convective instability caused uplift of the plateau and growth of the plateau from south to center and then to north; (2) the oblique collision between the Indian plate and the Eurasian plate caused successive uplift

Prof. Xuanxue Mo has fully contributed to this chapter.

X. Mo et al., *Magmatism and Crustal Evolution of the Tibetan Plateau*, Modern Approaches in Solid Earth Sciences 25,
https://doi.org/10.1007/978-981-97-2109-2_6

and growth from south to north; (3) the uplift of the Plateau started in pre-Eocene time from the Lhasa and Qiangtang terranes and expanded stepwise throughout the Neogene toward its present southern and northern margins by multiple driving forces. There have been six accelerated periods of uplift of the Tibetan Plateau since the Miocene, i.e., 25–20 Ma, 16–14 Ma, 10–7 Ma, 4.5–2.6 Ma, 1.6–1.0 Ma and 0.7 Ma, which well correspond to the key episodes of Asian monsoon and aridity evolution.

Keywords The India-Asia continental collision · The uplift and growth of the Tibetan Plateau · Timing and mechanism · Juvenile crust · Reworked mature crust · Crustal shortening · Mantle contribution · The pattern and mechanism of uplift and growth of the Plateau

6.1 The India-Asia Continental Collision Process

The closure of the Neo-Tethyan ocean and the India-Asia continental collision are ultimately responsible for the formation of the Tibetan Plateau. Two major events, i.e., the Bangong-Nujiang Ocean closure that led to the Qiangtang-Lhasa collision, and the Indus-Yalung Zangbo Ocean closure that led to the India-Asia collision and continued convergence, contributed to the transition of the tectonic constituents from an ocean-continental regime into a continental regime.

Most researchers believe that the continental collision was a complex and lengthy process rather than a short-lived event. The collision between two continents can be divided into three stages: initial collision, main collision and post collision. The initial collision refers to the time when the passive margin lithosphere reaches the trench and its continental crust becomes in contact with the active margin continental crust. At this point, the leading edge of the oceanic plate subduction may continue until the beginning of main collisional stage. Furthermore, due to the complexity and diversity of continental margins (including continental margin arcs, island arcs and marginal sea, etc.), the starting time of collision may vary from place to place, resulting in the diachronicity of collision. That is, the collision is a lengthy process from contact points to contact sutures, from surface to depth, and from edges to the amalgamation, including initial collision, main collision and post-collision stages. In the initial collision stage, records (e.g., magmatic record) may contain some information and signatures of volcanic arcs, because it is in the context of transition from the end of subduction to the beginning of collision, and the subduction of oceanic plate may continue. After the completion of the main collision or amalgamation, it turns into the post-collision stage. At early time of this stage, the newly assembled blocks may not have been welded together, and there may be relative movements between the blocks (e.g., extrusion, stretching or shearing, etc.).

Judging the timing and course of continental collision requires multi-disciplinary efforts and arguably convincing lines of evidence. For instance, some geological records may indicate the upper limit of collision time (i.e., an oldest age), some may indicate the lower limit of collision time (i.e., a youngest age), and some may indicate a certain stage of the collision process. Several consensus ones, for example, during the emplacement of ophiolites, metamorphic garnet amphibolite base may be formed, which can be used to define the time of the emplacement, but this is generally earlier than the initiation of collision; cessation of marine deposition and disappearance of residual epicontinental sea are later than the initial collision. Therefore, the highest marine horizon represents the youngest age of collision time; molasse deposits are the products of the uplift of the collisional suture zone or foreland thrust belt and represent the time of the uplift of orogenic belt, which is later than the end of marine deposition and the initial continental collision. Changes in the nature of magmatism usually lag behind some tectonic events (e.g., continental collision) or deep process (e.g., subducting slab break-off).

6.1.1 *Overview of the Timing of India-Asia Collision*

The timing of the India-Asia continental collision is key to the genuine understanding of the formation and evolution of the Tibetan Plateau, yet the onset of this collisional remains hotly debated at present, varying from >70 Ma to ~34 Ma, roughly in three age groups. (1) ~55 Ma (e.g., Achache et al. 1984; Allégre et al. 1984; Besse et al. 1984); (2) <50 Ma (e.g., Aitchison et al. 2007; Dewey et al. 1989; Rowley 1996, 1998; Searle et al. 1987; Wang 2017), and (3) >55 Ma (e.g., Jaeger et al. 1989; Yin and Harrison 2000; Wan et al. 2002a, b; Mo et al. 2002; Wang et al. 2003; Ding 2003; Ding et al. 2005, 2017, 2022; Huang et al. 2010; Zhu et al. 2004, 2015; Hu et al. 2017).

Yin and Harrison (2000) suggested that the initial India-Asia collision may have started as early as the late Cretaceous (~70 Ma) with stratigraphic evidence in support of this older timing. Jaeger et al. (1989) proposed that an early age of India-Asia collision was close to the K/T boundary according to the early terrestrial faunal resemblances between India and Asia, together with the lack of any significant endemism within the fauna, corresponding to the Deccan Traps volcanism in India centered on a main reversed chron (65.7 ± 2.0 Ma), which was related to the extrusion of the Réunion hotspot. Ding (2003), Ding et al. (2005, 2017) agreed on the ~65 Ma collision timing based on the foreland basin development. Wan et al. (2002a, b) proposed the collision timing of ~65 Ma according to the dramatic changes in sedimentary facies and microfauna content across the Cretaceous-Paleogene boundary. Wang et al. (2003) suggested a slightly younger collision timing of ~65–60 Ma at the end of Cretaceous to the beginning of Paleogene based on sedimentological studies of some typical areas at the northern margin of the Indian continent. Leech et al. (2005) used the age of the eclogite in the western Himalayas as indicating the collision time. Willems et al. (1996) interpreted the change in depositional patterns along the Indian continental passive margin between 70 and 64 Ma as indicating the timing of initial contact between India and Asia continents. On the other hand, Searle et al. (1987) inferred that India-Asia collision and the closure of Tethys occurred in Eocene (ca.40–50 Ma) based on then geological and geophysical data from southern Tibet. Najman et al. (2010, 2017), by comparing the youngest

Tethyan succession in the east (Tingri, Tibet) and west (Ladakh, India) of the orogen, found that the collision in Ladakh is constrained by arrival of Asian detritus on the Indian plate by 54 Ma, but the Tingri calcareous nannofossil species are compatible with an age corresponding to Zones NP11–12 (50.6–53.5 Ma). They thus conclude that the India-Asia collision occurred by 54 Ma in the west, with only extremely limited, if any diachroneity eastward. Garzanti et al. (1987) believed that the initial India-Asia collision in Ladakh in the NW Himalaya commenced close to the Paleocene/Eocene boundary (~56 Ma) according to a stratigraphic and petrographic analysis of the Cretaceous to Eocene Tibetan sedimentary succession. Rowley (1996) proposed a diachronous collision initiating in the late Ypresian (<52 Ma) in the west and progressing into and through the Lutetian (~48–41 Ma) in the east based on review of stratigraphic data, especially the study of thick submarine delta-fan complexes along-strike preserved in the Indus- Yarlung Zangbo suture. Aitchison et al. (2007) on the basis of new field evidence from Tibet and a reassessment of published data suggested that the India-Asia collision began much later, around the Eocene/Oligocene boundary (~34 Ma). Chung et al. (2005) believed that the soft phase of India-Asia collision occurred at ~60 Ma and the hard phase of the collision occurred at ~40 Ma.

For several decades, there have been continued studies on the timing of the India-Asia continental collision and the growth of the Tibetan plateau using different observations with different perspectives. It is time to review these efforts and offer a more credible view consistent all the observations.

6.1.2 *Our View on the Timing of India-Asia Collision and Supporting Evidences*

Based on comprehensive lines of evidence, we consider that the India-Asia continental collision has been a long process, starting ~65 ± 5 Ma, completing 45 ± 5 Ma and evolving into the post-collision stage with the convergence continuing to this day. Various lines of evidence are discussed below.

6.1.2.1 Magmatic Evidence

Magmas and associated igneous rocks are the results of the interactions between different spheres of the Earth and are important carriers of the exchange of mass and energy between the spheres, and can be taken as probes into and windows of processes of Earth's deep interiors while also recording tectonic events. Igneous rocks of the Tibetan Plateau contain abundant information about the Indo-Asia collision and the uplifting process of the plateau and are thus material evidence for multidisciplinary research on the formation and evolution of the Tibetan Plateau.

The India-Asia continental collision (Mo et al. 2002, 2003) produced syncollisional igneous rocks widely distributed in southern Gangdese magmatic belt, i.e., the Linzizong volcanic succession (~65–45 Ma), southern Gangdese granitoids (51–49 Ma, Peak at 50 Ma), and post-collisional igneous rocks (~40 Ma–Quaternary in the Tibetan plateau, 25 Ma–8 Ma in the Gangdese belt).

Linzizong Volcanic Succession (LVS)

The Linzizong volcanic rocks, with varying thickness up to 5000 m, extend east-west for more than 1000 km along the southern Gangdese belt, overlying the upper Cretaceous and older strata with angular unconformity (see Fig. 3.14, this book). The lithology, thickness and eruption time of the Linzizong volcanic rocks show along-strike variation. The Linzizong volcanic rocks are better developed and exposed in the Linzhou basin near Lhasa, covering an area of ~230 km^2, which is the classic place where the Linzizong volcanic series was named by Li Pu in 1953. It was divided into the lower Dianzhong Formation, middle Nianbo Formation and upper Pana Formation firstly by Xizang Regional Geology Survey Team (Regional Geology of Tibet Autonomous Region 1993). Many researchers have studied the Linzizong volcanic rocks. On the basis of previous work, Dong (2002) further established the stratigraphic framework of the Linzizong volcanic series in the Linzhou Basin through detailed mapping (see Fig. 3.2, this book).

The Stratigraphic Framework of the LVS in the Linzhou Basin

The LVS can be divided into three formations from bottom to top (see Fig. 3.3 in Chap. 3, this book):

The lower Dianzhong Formation (LDF, E_1d) is 2363 m thick and can be divided into three lithologic sections: the lower section outcropped only in the southwest of the basin, which is a set of rhyolitic pyroclastic rocks overlying the red mudstone of the upper Cretaceous Shexing Formation, with basal conglomerates. The middle section is most widely distributed in the basin and mainly consists of a set of andesitic lava and volcanoclastic rock series with some dacites. In many places, the strata directly unconformably overlay the upper Cretaceous Shexing Formation. The basal conglomerate is well developed with rounded red siltstone and sandy mudstone clasts of varying size from underlaying Shexing Formation. The upper section is composed of interbedded lacustrine red mudstone and pyroxene andesite, with fan-facies conglomerate layers at the top.

The middle Nianbo Formation (MNF, E_2n) is 732 m thick, distributed in the middle of the basin in para-unconformity or small angle unconformity above the Dianzhong Formation. In places it unconformably overlies on the upper Cretaceous purple siltstone. It is a volcanic-sedimentary rock series and contains the most sedimentary rocks in the Linzizong volcanic series, making up to ~70% of the thickness of the formation. It can also be divided into three lithologic sections: The lower section is composed of thin-bedded limestone marls with a small amount of rhyolite tuffs. The limestones contain freshwater fossils and have mud cracks on the bed plane. The bottom of the section is commonly conglomerate of varying thickness with clasts of andesite and andesitic tuff. At the top of this section there is a 7–8 m thick layer of grayish white rhyolitic glassy tuffs. The middle section is composed of interbedded tuffaceous sandstone, siltstone and mudstone with some tuffs, and sedimentary clastic rocks of channel facies locally developed. The lower part of the upper section consists of basaltic andesites and andesitic tuffs and volcanic breccia/agglomerates. The upper part of the section is composed of shoshonitic lavas and pyroclastic rocks. There is a 10 m- thick layer of fluvial-lacustrine facies purple mudstone at the top.

The upper Pana Formation (UPF, E_2p) is 2184 m thick and distributed in the north of the basin and overlies different layers of the Nianbo Formation with an eruptive unconformity. It can be divided into two lithologic sections: the lower section consists of rhyolitic ignimbrites (weld-tuffs) and lavas, with well-developed columnar joints. There is a 10-m layer with pillow structure at the bottom. The upper section is composed of rhyolitic ignimbrites intercalated with fluvial and lacustrine sedimentary clastic rocks locally with cross-cutting bedding and tabular bedding. The most upper part of the Pana Formation is truncated by the regional Luobadu-Gulu-Zuogang-Tanga thrust.

Dong (2002) mapped that the Linzhou volcanic basin showing that it is a well-preserved Cenozoic caldera system with various eruptive facies of volcanic rocks. The cryptovolcanic rocks and dikes exposed in the Linzhou basin include crypto-rhyolites, crypto-andesites, diabase dikes, diorite porphyrite dikes, shoshonite dikes, rhyolite porphyry dikes and fine-grained granitic dikes.

Petrological and Petrochemical Characteristics of the LVS Volcanic Rocks

The total-alkali vs. SiO_2 (TAS) scheme (Le Bas and Streckeisen 1991; Le Maitre et al. 2020) has been used for the classification and nomenclature of volcanic rocks. It should be noted that in the TAS diagram S1 (trachybasalt), S2 (basaltic trachyandite) and S3 (trachyandite) are divided into potassium series (Na_2Owt%–2 < K_2O wt%) and sodium series (Na_2Owt%–2 ≥ K_2Owt%), with different rock terms For instance, in the case of the potassic series, the rocks are called S1 (potassic trachyte basalt), S2 (shoshonite) and S3 (latite), respectively. At the same time, the classification scheme of Foley et al. (1987) was also used for the classification of potassic-ultrapotassic rocks, which can be divided into five types, namely: (1) ultrapotassic, with $K_2O > 3$ wt%, $MgO > 3$ wt%, and $K_2O/Na_2O > 2$; (2) potassic mafic-ultramafic, characterized by <52 wt% SiO_2 and K_2O/Na_2O 0.5–2, including kamafugite, carbonatite, absarokite, trachybasalt, some shoshonites, and lamprophyres; (3) potassic intermediate, with SiO_2 ranging from 52 to 63 wt% and K_2O/Na_2O from 0.5 to 2, including latite, potassic trachyte, and some shoshonites (the most common type in the Tibetan plateau); (4) potassic acidic, characterized by SiO_2 contents >63 wt% but K_2O/Na_2O between 0.5 and 2, often associated with type 3; and (5) sodic types, with $K_2O/Na_2O < 0.5$, relatively common in western Qiangtang and Maguan, the Sanjiang, but also appearing in regions dominated by potassic-ultrapotassic rocks.

The LVS volcanic rocks include andesite, basaltic andesite, basaltic trachyandesite, shoshonite, trachyandesite, latite, dacite, rhyolite and corresponding pyroclastic rocks, and the ratio of lava to pyroclastic rock is about 3:7.

The LDF volcanic rocks are mainly represented by andesitic lavas and pyroclastic rocks, with subordinary dacitic- rhyolitic rocks and a small amount of basaltic andesite, trachyandesite and pyroclastic rocks. The andesitic rocks generally contain about 5% pyroxene and almost no amphibole and biotite. The K_2O-Na_2O diagram shows that the volcanic rocks of the LDF are both potassic and sodic, and the ratio of K_2O/Na_2O is 0.51 on average. In the K_2O-SiO_2 diagram, the majority of the LDF volcanic rocks are medium potassic (CA). In terms of Al-saturation index, the LDF volcanic rocks are mainly metaluminous (A/CNK < 1). They were intruded by diabase and find-gained granitic dikes. The evolution trend of the LDF is characterized by andesite-dacite-rhyolite intruded by bimodal granitic + diabase dikes.

The MNF volcanic rocks are mainly rhyolitic- dacitic pyroclastic rocks in the lower and middle parts and shoshonites in the upper part of the Formation. They are intruded by shoshonitic dikes, diabase dikes and diorite porphyrite dikes. In the K_2O-Na_2O diagram, the volcanic rocks plot mainly in the field of potassium to high-potassium rocks with an average K_2O/Na_2O of 2.14 and 7.27. In the K_2O-SiO_2 diagram, >40% of volcanic rocks plot into the HKCA (high-K calc-alkaline) and SHO (shoshonite) fields. In terms of Al-saturation index, >50% of MNF rocks are strongly peraluminous (A/CNK > 1.1). These characteristics indicate that crustal materials have a significant influence on magma compositions, especially the appearance of shoshonites and high-K calc-alkaline igneous rocks is considered as typical of continental collisional orogenic belts (Deng et al. 1996, 2004a, b; Flower et al. 1998; Leigeois 1998).

The lithology of UPF volcanic rocks is relatively simple, mainly composed of high-K rhyolitic ignimbrite or strongly welded tuff with a few lava flows. Strongly welded tuffs are most developed in the UPF, although welded tuff is also present in the MNF. Very thick welded tuffs with a large amount of plastic glass and rock debris mark a spectacular scene of huge hot ash flows flowing over tens of thousands square kilometers. The average K_2O content of UPF volcanic rocks is 5.25%, up to 7.94%. The average K_2O/Na_2O ratio is 1.77, up to 3.48. In K_2O–Na_2O diagram, all the UPF samples plot in potassic and high-potassic fields. In K_2O–SiO_2 diagram, the majority of the samples plot in the SHO and HKCA fields. The strong peraluminous rocks account for >60% of the UPF. The UPF are locally intruded by shoshonitic dikes. Thus, the UPF volcanic rocks clearly show the tectonic environment of continental collision and crustal thickening.

REE Characteristics of the LVS Volcanic Rocks

All LVS volcanic rocks generally have LREE-enriched patterns. $\sum$REE in the LDF volcanic rocks vary in the range of 107–288 ppm with an average of 141 ppm. LREE/HREE are 5.6–7.4 and $(La/Yb)_N$ 7.31–15.13 with an average of 9.53 in average. There are weak to very weak negative Eu anomalies in REE patterns.

∑REE in the MNF volcanic rocks vary in the range of 97–404 ppm, with an average of 204 ppm. LREE/HREE are 6.7–10.1 and $(La/Yb)_N$ 9.9–21.1 with an average of 13.49. There are no negative Eu anomalies for shoshonite samples, but obvious negative Eu anomalies for rhyolitic samples (δEu 0.35–0.46).

∑REE in the UPF volcanic rocks vary in the range of 147–358 ppm, with an average of 209 ppm. LREE/HRE are 8.9–16.5, with an average of 10.8 in average, and $(La/Yb)_N$ 9.4–23.9 with an average of 7.05. In general, all the three formations have right-leaning parallel patterns, the UPF samples show steeper than those of the LDF and MNF samples with obvious negative Eu anomalies (δEu 0.34–0.73).

In conclusion, ∑REE and $(La/Yb)_N$ of the LVS volcanic rocks increases from bottom to top (i.e., from the LDM to MNF and to UPF) with the UPF rhyolite samples show obvious negative Eu anomalies.

Characteristics of Trace Elements in the LVS Volcanic Rocks

On primitive-normalized spidergrams, the LVS show negative anomalies of Nb, Ta and Ti, and positive anomalies of Rb, Ba, Th, U, especially Pb, resembling the average composition of the bulk continental crust (Rudnick and Fountain 1995; Rudnick and Gao 2003). The andesitic LDF are characterized by enrichments of Rb, Ba, Th, U, obvious Nb-Ta troughs and Ti valley with essentially no P and Sr anomaly. The MNF shows both an obvious Nb-Ta trough and Ti, Sr, P, Ba valleys. Compared with LDF and MNF, the UPF are more enriched in Th and U with progressively deeper Nb, Ta, Ti, P and Sr valleys. This trend is also accompanied by gradually steeper slopes on spidergrams and greater peak-valley contrast. In general, these spidergram patterns are very similar to typical post-collision high-K volcanic rocks in southern Tibet. The patterns of shoshonite in the MNF resemble characteristics of the LDF and the patterns of rhyolite in the MNF resemble those of the UPF.

In summary, from the andesitic LDF to the dacitic MNF and the rhyolitic UPF, SiO_2 increases while mafic components decrease, in parallel with increasing large ion lithophile elements such as K, Rb, Th and U (especially in the UPF) and greater negative anomalies of high field strength elements such as Nb, Ta and Ti.

Recently, Liu et al. (2022) reported a comprehensive dataset on LVS rocks from four locations with varying distance across-strike of the Indus-Yarlung suture zone in southern Tibet. They divided the LVS rocks into two groups: a calc-alkaline Group 1 (69–55 Ma), mainly including basaltic-andesitic varieties, and a shoshonitic Group 2 (52–50 Ma), predominantly consisting of silicic rocks with minor mafic compositions. They believed that Group 1 samples were likely derived from the fractional crystallization of primitive basaltic melts as a result of the partial melting of a metasomatized mantle wedge, and Group 2 silicic samples can be attributed to the involvement of ancient lower crust-derived melt from Tethyan Himalaya and central Lhasa subterrane basement. They concluded that the LVS were the magmatic response to the process from late subduction (Group 1) to initial India-Asia collision and slab breakoff (Group 2).

Sr-Nd Isotopic Characteristics of the LVS Volcanic Rocks

The Nd and Sr isotopic compositions of the LVS rocks vary in relatively limited ranges in terms of εNd(t) (+9.20 to −3.96) and initial $^{87}Sr/^{86}Sr$ (0.704690–0.708316). Most of the LVS rocks plot in a restricted field between the two mixing lines of 'Yarlung MORB–Lower Crust' and 'Yarlung MORB–Amdo Orthogneiss'. Herein the Yarlung MORB (Mahoney et al. 1998) is assumed to represent the Neo-Tethyan oceanic lithospheric mantle component whereas the Amdo Orthogneiss (Miller et al. 1999) represents typical upper crust material in Tibet, respectively. The lower crustal values of $^{87}Sr/^{86}Sr$ and $^{143}Nd/^{144}Nd$ are cited from Ben Othman et al. (1984). Within the data set, the LDF rocks have $(^{87}Sr/^{86}Sr)_i$ = 0.70460–0.70609, εNd (t) +9.20 to −3.96, mostly concentrated at +5.43 to −0.52. The MNF rocks have $(^{87}Sr/^{86}Sr)_i$ = 0.706871–0.708316, εNd (t) = +2.32 to −3.96. The UPF rocks have $(^{87}Sr/^{86}Sr)_i$ = 0.704957–0.705692, εNd (t) = −3.01 to +5.43. The mafic dykes intruding the LVS have εNd(t) = +3.29, $(^{87}Sr/^{86}Sr)_i$ = 0.704955. In εNd vs. $(^{87}Sr/^{86}Sr)_i$ diagram, the LVS rocks plot in the triangular region defined by the Yarlung Zangbo MORB-lower crust-Amdo gneiss.

The significance of these data can be elaborated as follows.

1. The LVS volcanic rocks have a significant asthenospheric mantle contribution. To estimate this contribution, we can apply the method of Neodymium Crustal Index (NCI; DePaolo 1985; DePaolo et al. 1992; Perry et al. 1993), i.e., mass contributions of crustal Nd in mantle derived magmas through crustal assimilation during its evolution. The NCI values for the entire LVS system range from 0.16 to 0.47 with an average of 0.30. That is, isotopically, the LVS rocks are dominated by mantle materials, equivalent to 84–53% with an average of 70% mass of mantle. Specifically, NCI (mantle mass %) values for mafic dykes, the LDF, MNF and UPF are 0.17 (83%), 0.31(69%)–0.34 (66%), 0.30 (70%)–0.47 (53%) and 0.16 (84%)–0.29 (70%), respectively. In simple clarity, in terms of Nd isotopes, the LVS and the associated dykes are dominated by mantle contributions of up to ~70–84%.
2. Nd–Sr isotope compositions of the LVS show no systematic change with increasing SiO_2 and evolving bulk-rock major and trace element systematics from LDF, MNF to UPF, but have some overlap. It suggests the likelihood that the more evolved MNF and UPF may in fact have derived from melting of the earlier andesitic LDF—the most primitive LVS end-member (or intrusive equivalent). It also suggests that the crustal contribution to the LVS petrogenesis may not be simple crustal level assimilation, but probably occurs in the melting region. Melting of recycled terrigenous sediments or subduction-eroded crustal materials is highly likely.
3. As mentioned above, dikes intruding all the three LVS formations of varying compositions. They occur as find-grained granitic and diabasic dikes in the LDF, as shoshonitic, diabasic and dioritic-porphyrite dikes in the MNF, and as rhyolitic-porphyritic veins and shoshonitic dikes in the UPF. A diabasic dike (Lz9932) in the LDF has Mg# value ($Mg/[Mg + Fe^{2+}]$) of 71.86, close to primary magmatic composition (with some cumulate mafic phases), while the Mg#

values for most mafic dikes are in the range of 53.40–62.45, representing variably evolved magma. There is no isotopic data for the most mafic dikes with the highest Mg# value, but the isotope data for a more evolved mafic dike with Mg# = 53.40 give a positive εNd value (+3.29), a low $(^{87}Sr/^{86}Sr)_i$ (0.7050), ∑REE 267 ppm, $(La/Yb)_N$ 15.69, δEu 0.83. A diabasic dike in the MNF has ∑REE 179.5 ppm, $(La/Yb)_N$ 10.27, δEu 0.78. In view of the importance of dikes, further research is needed to obtain more valuable information.

The LVS Volcanic Rocks Record the Process of India-Asia Continental Collision

1. The Linzizong volcanism consists of three cycles, i.e., the early Dianzhong cycle, the later Nianbo cycle and the latest Pana cycle, represented by the three formations, i.e., the LDF, MNB and UPF, respectively. There are between-formation unconformable contacts, but the scale and nature of these unconformities vary. While the unconformity between the LDF and the underlying strata is a regional unconformity, the LDF-MNF and MNF-UPN are local conformities or even eruptive unconformities. They are thus of different tectonic significance (see below).
2. Each cycle of the LVS started with felsic magmatic activity and ended with mafic-shoshonitic or bimodal magmatism. The Dianzhong cycle began with rhyolitic- dacitic magma, followed by eruption of andesitic magma and then ended with bimodal magmatic activity consisting of mafic and felsic dikes. The Nianbo cycle also began with rhyolite-dacitic magmatism and ended with shoshonitic and diabasic magmatism. The Pana cycle was dominated by rhyolitic magmatism, but it also ended with shoshonitic eruptions.

From the perspective of K_2O-SiO_2 and K_2O-Na_2O co-variations, the volcanic rocks formed in early cycle are dominated by medium K_2O calc-alkaline (CA), those in the middle cycle comprise the existence of high potassium calc-alkaline (HKCA), shoshonitic (SHO) and CA, and those in the late cycle are completely HKCA and SHO rocks. In terms of Al_2O_3 saturation, the rocks are mainly metaluminous in the early stage, and strongly peraluminous in the late stage.

3. The trace elements (including rare earth elements) of the early LDF rocks show characteristics of active continental arc volcanic rocks (i.e., Andean type). Such "arc" characteristics gradually declined towards characteristics of continental volcanic rocks with time. By the time of late Pana cycle, they were consistent with the typical continental volcanic rocks.
4. The Nd and Sr isotopic compositions of the LVS rocks give two types of common information: (1) All the three formations show significant asthenosphere mantle contributions, about 70–90% in terms of Nd crustal index (NCI; DePaolo 1985): NCI = [εNd (rock) − εNd (M)]/[εNd (C) − εNd (M)] = 0.1 − 0.3) (Mo et al. 2007, 2008; Niu et al. 2013; Zhu et al. 2019). (2) Isotopic compositions of the LVS rocks also reflect contributions of subducted oceanic crust, oceanic mantle and continental crust, suggesting a geodynamic environment of transition from Neo-Tethyan subduction regime to continental collision regime. In εNd–$(^{87}Sr/^{86}Sr)_i$ diagram, the LVS rocks plot in the triangular region defined by

Yarlung Zangbo MORB-lower crust-Amdo gneiss, in which the mafic and intermediate-mafic rocks are closer to the PREMA, while the felsic rocks partly overlap with the Miocene postcollisional potassic volcanic rocks in the eastern segment of southern Tibet.

5. The LDF volcanic rocks were formed in the transition period from the end of subduction to the beginning of collision, thus having both the characteristics of continental margin arc rocks and their own unique characteristics in comparison to normal subduction arc volcanic rocks (e.g., the Sangri Group and Yeba Formation in southern Gangdese). They are the records of the initial collision.
6. The late MNF and especially the UPF volcanic rocks are the products in response to the main collision. The first appearance of shoshonite in late MNF is an important indication of continental collision. The UPF volcanic rocks represent the peak period of main collision. The main collision stage is characterized by huge volume eruption, extremely high temperature and injection of asthenosphere materials. In a very short period of time, huge amounts of fiery hot magmas (melts and pyroclastics) erupted over tens of thousands square kilometers, formed ~2500 m thick of ignimbrites (weld-tuffs) and intruded Gangdese batholith in places. It is also proven by an increase in zircon saturation temperature at ca. 52 Ma documented by the rhyolitic rocks in the UPF, indicating the anomalously high heat input from the mantle (Zhu et al. 2015). As aforementioned, all LVS (including MNF and UPF) volcanic rocks have significant asthenosphere mantle contribution. Zhu et al. (2015) interpreted all these as resulting from slab rollback followed by slab breakoff, causing partial melting of differing magma source regions through increasing temperature of lithosphere and decreasing pressure of asthenosphere because of the slab breakoff.

Discussion on the Nature and Tectonic Setting of the LDF and Others

1. It is important to explain why the LDF volcanic rocks with active continental arc characteristics can be regarded as one of the principal records of India-Asia initiation collision (Niu et al. 2007, 2013; Mo et al. 2008). The LDF volcanic rocks were produced in the broad context of Tethyan ocean closing and India-Asia collision (see Niu et al. 2013; Fig. 6). Therefore, the LDF volcanic rocks formed in this tectonic setting must have both the characteristics of arc igneous rocks and their own unique characteristics. Upon India-Asia collision, the convergence still continued, but the rate of Tethyan crust underthrusting must be retarded. The latter facilitated thermal equilibrium of the crust with the overlying mantle/crust, making the underthrusting crust evolve along a higher T/P geotherm. This hydrated underthrusting crust will inevitably melt when intersecting the hydrous basaltic solidus in the amphibolite facies. This melting mechanism is defined as compression (vs. decompression) and equilibrium heating induced melting of H_2O-saturated basaltic rocks (Niu 2005). Sediments of the more felsic compositions (e.g., granitic or tonalitic) going along with the underthrusting Tethyan crust would melt under similar conditions. Thus, the remaining part of the Tethyan ocean crust took a high T/P evolution path upon collision and then

underwent hydrous melting under amphibolite facies conditions. This melting produced primary andesitic melts with mantle isotopic signatures, either erupted and fast solidified as andesites (i.e., the LDF) or intruded, slowly cooled and solidified as diorites/granodiorites (i.e., the SGB).

Melts produced from such unusually "warm slabs" in the amphibolite facies (1) would be andesitic or more silicic, depending on the extent of melting and amount of sediments melted; (2) would have mantle dominated isotopic signatures inherited from the ocean crust (MORB-like); (3) may show no garnet signature; (4) would have "arc-like signature", i.e., relative depletion in Nb, Ta and Ti because these high-field strength elements (HFSE) are compatible in ilmenite (or pseudobrookite), which is a common phase in amphibolite (Niu and Lesher 1991); (5) would have sub-chondritic Nb/Ta ratio because of amphibole controlled fractionation ($Kd_{Nb}{}^{amp/L} > Kd_{Ta}{}^{amp/L}$); and (6) would not have excess Sr (vs. adakitic and arc magmas) because of calcic plagioclase as a residual phase preferentially holds Sr and Eu relative to Nd, Sm, Gd and other similarly incompatible elements.

Indeed, the LVS volcanic rocks have all these properties. The scatter in the Nd–Sr isotope space is likely caused by "mixing" of mature crustal materials (e.g., recycled sediments, subduction-eroded materials, crustal assimilation) with the Tethyan ocean crust (altered MORB) with inherited isotopically depleted MORB mantle signature in the recent past. The LVS volcanic rocks would have 70–90% mantle contribution in terms of Nd crustal index (NCI; DePaolo 1985; Mo et al. 2007). Contribution of mature crustal materials is also apparent from trace elements such as the Rb, K and Pb spikes. In addition, the negative Nb, Ta and Ti anomalies are conspicuous, the negative Sr anomaly is weak, but discernable, and the lack of garnet signature (i.e., flat Dy–Yb pattern) is obvious. In contrast to the chondritic Nb/Ta ~ 17.57 (e.g., Sun and McDonough 1989), the LVS andesites have a mean sub-chondritic Nb/Ta ratio of ~13.

Thus, the concept of "amphibolite-facies" slab (plus terrigeneous sediments) melting could explain all the features of the LVS lithologies well. It is particularly straightforward to explain all the LDF. Given the similar Sr-Nd isotopic variability in the three formations with progressively stronger mantle signature in the less "contaminated" samples, it is plausible that both MNF and UPF may in fact have derived from melting of the earlier lithologies, i.e., the LDF (and intrusive equivalent).

2. Deng et al. (1996, 2004a, b), Deng (2021) also regarded the LVS volcanic rocks and the granites in southern Gangdese as the products of Neo-Tethys closure and India-Asia collision and proposed a model to illustrate the process of the collision. They suggested that while the Indian and Asian continents have collided at the surface, the Tethyan oceanic crust is still colliding with the Asian lithosphere at depth, and with the collision, the Indian continent was pushed down until the Indian continental lithosphere completely collided with the Asian continental lithosphere, thus completing the collision process of the two continents. In more detail, (1) abrupt change of sedimentary facies on the surface, evolution of stratigraphic paleontology and deformation of thrust nappe structure recorded the

beginning of the collision; (2) after the initial collision, oceanic crust subduction at depth was still going on and therefore the nature of the magmatism and igneous rocks on the overlying continent are still of oceanic crust subduction; (3) with the collision proceeding, the underthrusting continental crust dragged by the subducting oceanic crust begin to contact with the overlying continental lithosphere. As a result, the involved oceanic crust decreases, while the continental crust gradually increases; (4) when the two continental lithospheres fully contact, the high density of oceanic lithosphere will no longer drag continental lithosphere going downwards, the buoyancy of downward thrusting lithosphere will terminate the collision, and thrust nappe deformation will again appear on the surface. The thrust after the end of magmatic activity will mark the end of the continental collision processes.

3. Forty years ago, Gill (1981) pointed out that continents, island arcs, seamounts and anomalous oceanic crust such as the Ontong-Java Plateau were relatively unsubductable, and the appearance of such objects at a convergent plate boundary as part of underthrust plate would cause collision and terminate subduction. He further pointed out that andesitic volcanism sometimes continues during and after collision. For instance, an ARC-PCM (island arc-passive continental margin) collision along Luzon-Taiwan arc commenced in the late Miocene and was completed in the Pleistocene in Taiwan; an ACM-PCM (active continental margin- passive continental margin) collision was associated with active orogenic andesite volcanism in the Turkey-Iran region, where collision began in the late Cretaceous to early Tertiary. These observations indicate that it is not unexpected to see active continental arc characteristics in the LDF volcanic rocks of the LVS.

Radimetric Dating of the LVS Volcanic Series

Prior to 2002, there were only a few isotopic ages reported on the LVS volcanic rocks with large uncertainties (ranging from 88 to 40 Ma). Zhou (2002) and Dong (2002) carried out $^{40}Ar/^{39}Ar$ dating for different units of the LVS based on detailed geological mapping and sampling in the Linzhou volcano-sedimentary basin. Zhou et al. (2004) reported $^{40}Ar/^{39}Ar$ ages on 10 representative samples the LVS volcanic strata, including four samples from the LDF, one from the MNF, four from the UPF in the Linzhou Basin and one from the LDF in the Maqu area. The bottom-to-top ages for the LDF are 64.43 Ma ± 0.63 Ma, 61.54 Ma ± 0.60 Ma, 61.46 Ma ± 0.71 Ma, 60.63 Ma ± 0.35 Ma, respectively, on samples from the Linzhou Basin and 60.50 ± 1.6 Ma on the sample from Maqu. The age on the sample from the middle part of the MNF is 56.51 ± 0.34 Ma. The bottom-to-top ages for the UPF are 51.60 Ma ± 2.6 Ma, 48.73 Ma ± 0.49 Ma, 48.03 Ma ± 3.1 Ma and 43.93 Ma ± 1.3 Ma, respectively. The top of the UPF was cut by a regional thrust and therefore no samples could be collected. To sum up, the age system of LVS volcanic rocks is as follows: LDF 64.43 Ma–60 Ma, MNF 60–50 Ma and UPF 50 Ma–<43 Ma.

Zhu et al. (2015) undertook SIMS (secondary ion mass spectrometry) zircon U-Pb dating on 23 samples from the stratigraphic boundaries of each formation. The U-Pb zircon age data of two samples from the lowermost LDF (60.2 Ma ± 0.6

and 60.2 Ma ± 0.8 Ma, respectively) and of one sample from the uppermost LDF (58.3 ± 1.3 Ma) indicate that the LDF andesitic volcanism had been active during 60.2–58.3 Ma. One sample from the lowermost MNF (55.4 ± 0.5 Ma) and two samples from the upper MNF (52.6 Ma ± 0.4 and 52.7 Ma ± 1.9 Ma, respectively) constrain the MNF age to be 55.4–52.6 Ma. This is very compatible with the age data obtained from two marl samples from the lower MNF (54.4 ± 0.5 and 54.5 ± 0.7 Ma). Three samples from the lower, middle, and upper UPF give U-Pb zircon ages of 52.3 ± 0.5, 52.6± 0.4, and 52.3 ± 0.6 Ma, indistinguishable within analytical errors. These data bracket the duration of each formation of the LVS volcanic rocks from the Linzhou Basin: 60.2–58.3 Ma for the LDF, 55.4–52.6 Ma for the MNF, and 52.6–52.3 Ma for the UPF.

As mentioned above, the upper UPF is truncated by the regional Luobadu-Tanga thrust and thus has no age constrain on the latest eruption. However, we can estimate the age from other evidence. Postcollisional volcanic rocks in Tibet occur in three stages: ~45–25 Ma, 25–5 Ma, and 5–0 Ma. Postcollisional muscovite granites aged in 35–8 Ma occur in the Gangdese. In combination of radiometric ages for the UPF given, we infer that the age of the upmost UPF is likely to be 45 ± 5 Ma.

Furthermore, the oldest ages corresponding to the base of the LDF show east-to-west variation along the strike of the Gangdese as follows (Fig. 6.1): 64.4 Ma ($^{40}Ar/^{39}Ar$, Zhou et al. 2004) and 60.2 Ma (SIMS zircon U-Pb, Zhu et al. 2015) in the Linzhou basin, 60.5 Ma ($^{40}Ar/^{39}Ar$) and 68.9 Ma (SIMS zircon U-Pb) in Maqu; farther to the west, the ages are 63.5 Ma in Coqin (Rb-Sr, Hu et al. 2007), 61.5 Ma in Horba/Zhongba(zircon U-Pb, by Liu, 2018), 64.5 in Yare/Pulan (zircon U-Pb, Wang et al. 2007) and 67.1 Ma in Zhare/Ge'er (zircon U-Pb, Cao et al. 2020), respectively.

Gangdese Granitic Rocks

Granites and granitoids are widespread with about 12% areal coverage of Tibet 80% of which are distributed in the Gangdese magmatic belt with the outcrop area roughly equal to that of volcanic rocks, each about 110,000 km^2 (see Fig. 3.14, this book). The Quxu granitic batholith in the middle part of the southern Gangdese belt (SGB) are representative of collisional granitoids, which are distributed in Xaitongmoin–Namlin–Nyemo–and Quxu area (see Fig. 3.15, this book). The granitoids are mainly granodiorite, quartz diorite, quartz monzonite and monzogranite, containing abundant mafic microgranular enclaves (MMEs).

The granitic rocks of the batholith have the following compositional variations: SiO_2 54.49–63.86 wt. %, $Na_2O + K_2O$ 5.18–7.01 wt. %, $Na_2O > K_2O$, $\sum$REE 50.60–193.30 ppm, $(La/Yb)_N$ 3.09–33.75, LREE-enriched patterns with negative Eu anomalies, and belong to metaluminous-slightly peraluminous and calc-alkaline series. The spidergrams show enrichment of Th, U, K and Nb-Ta-Ti valleys. Their Zircon U-Pb ages are in the range of 49.3–64.7 Ma, mostly around 50 Ma (Dong et al. 2005, Mo et al. 2005a, b, Ji et al. 2009, the same below).

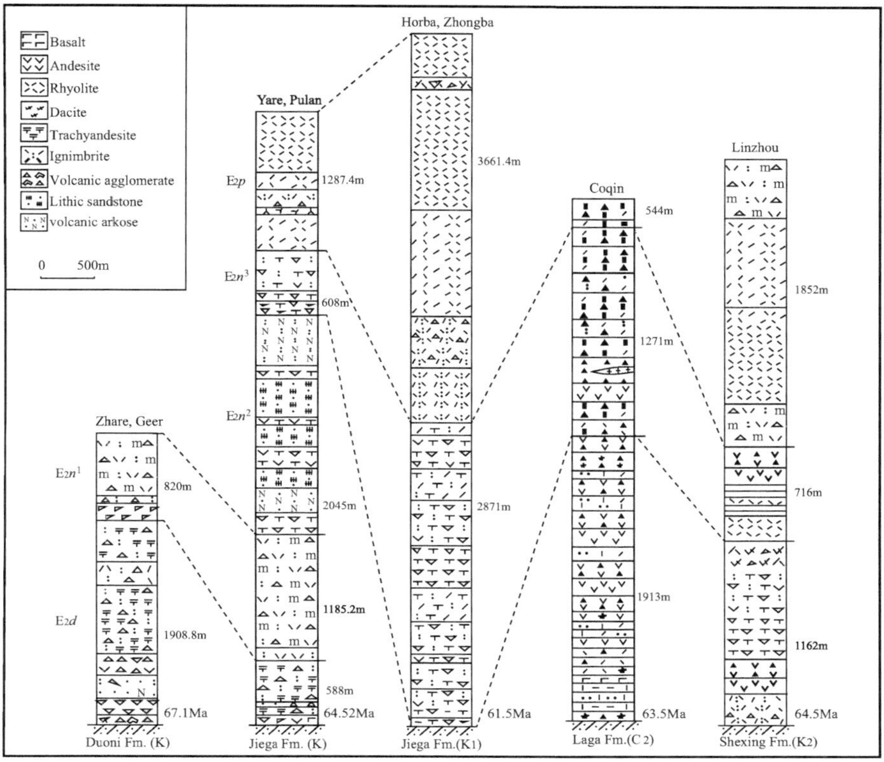

67.1Ma (MSWD=1.8) :Zircon U-Pb dating by Cao et al., 2020
64.52Ma (MSWD=1.32): Zr U-Pb dating by Wang, 2011
61.5Ma (MSWD=1.4) : Zr U-Pb dating by Liu Anlin, 2020
63.5Ma: Rb-Sr dating by Hu et al, 2007

Fig. 6.1 Stratigraphic correlation of the Linzizong Formation in five representative localities along the Gangdese magmatic belt in southern Tibet. The consistent basal radiometric age data (67–61 Ma) are given. Data from Zhou (2002), Dong (2002) and Zhou et al. (2004) for Linzhou, Hu et al. (2007) for Coqin, Liu (2020) for Horba, Zhongba, Wang et al. (2011a, b) for Yare, Pulan, and Cao et al. (2020) for Zhare, Ge'er

Conspicuously, an east-to-west trending gabbro-diabase-pyroxenite belt is exposed on the southernmost margin of the Gangdese granite belt, which is spatially consistent with aeromagnetic anomalies to the north of the Yarlung Zangbo ophiolite belt (Xiong et al. 2001). This mafic rock belt is dominated by gabbros and diabases (consisting of 50–55% plagioclase (An_{55-90}), 25 vol. % pyroxene, 5–15 vol. % hornblende, biotite 5 vol. %), and a few gabbroic diorites (mainly consisting of 50–55 vol. % plagioclase (An_{36-70}), 10–15 vol. % pyroxene, 15–20 vol. % hornblende and 5 vol. % biotite), pyroxenites and hornblendites (as small bodies or inclusions in gabbros and generally consisting of >95 vol. % hornblende and <5 vol. % pyroxene). These rocks show compositional variations: SiO_2 46.66–52.00 wt.%, TiO_2 0.64–1.56 wt.%, MgO 4.70–8.98 wt.%, $Na_2O + K_2O$ 2.76–4.88 wt.%, $\sum$REE 41.43–101.30 ppm, and LREE-enriched patterns with weak positive Eu anomalies.

Their spidergrams show incompatible element patterns similar to those of granitoids, but have more pronounced negative Nb-Ta-Ti anomalies, elevated K-spikes relative to other large ion lithophile elements. Their zircon U-Pb ages are 47.0–56.9 Ma.

There are also mafic dikes that intruded the granites with irregular edge contacts. According to the rheological theory (Sparks and Marshall 1986; Fernandez and Barbarin 1991), when these granites had not been completely cooled and consolidated but above 70%, their rheological properties would be significantly changed and closely approached solid state (pseudo-solid state), thus cracks could be generated to provide a space for dike injection. In this case, limited material and heat exchange take place between the dike and granite wall-rock, resulting in irregular contact edges.

MMEs of varying size (tens of meters to centimeters in diameter) are widespread and locally abundant mainly in granodiorite and also in monzogranite. They consist of plagioclase (An_{55-65}) ~ 50 vol. %, ferromagnesian minerals ~35 vol. % (mainly pyroxene/hornblende 20–25 vol. %, biotite ~10 vol. %), and minor quartz and potassium feldspar, commonly with doleritic structure. Bulk-rock analysis of 11 MME samples give the following major element compositional variations: SiO_2 47.4–53.5 wt. %, $Na_2O + K_2O$ 2.77–7.71 wt. %, with an average of 5.12 wt. %, TFeO, Na_2O, K_2O are 7.2–9.4 wt. %, 2.3–5.2 wt% and 0.2–2.9 wt%, respectively. Compared with gabbros, the MMEs have higher SiO_2, Na_2O and K_2O, but lower TFeO. Relative to granodiorite, the MMEs have lower SiO_2, Na_2O and K_2O, but higher TFeO. Overall, the bulk-rock chemical compositions of MMEs are between gabbro and granodiorite, but closer to gabbro. The REE patterns of MEEs show obvious negative Eu anomalies. The spidegrams show characteristics similar to those of both granitoids and gabbros. The MMEs give essentially identical zircon U-Pb ages within error (49.9–51.1 Ma).

All rock types mentioned above give εNd(T) ranging from +1.64 to +5.21 and zircon εHf(T) ranging from +6.3 to +14.7, showing obvious and significant asthenosphere mantle contribution in their origin.

In summary, Dong et al. (2005, 2008), Mo et al. (2005a, b), Ji et al. (2009) obtained zircon U-Pb ages of 47.0–64.7 Ma with a peak value of ~50 Ma, for the granitic and mafic rocks and MEEs in the the Quxu batholith in southern Gangdese belt. Zhu et al. (2015) suggested that a magmatic flare-up event at ~51 Ma for the Gangdese batholith, which is roughly consistent with the rapid eruption of the ~2 km-thick rhyolitic ignimbrite and lavas (52.5–52.3 Ma) represented by the UPF of the LVS volcanic rocks. Overall, collisional granites are exposed in the entire southern Gangdise belt, with the oldest age up to ~65 Ma, but the majority is about 50 Ma. This indicates that the most intense period of collisional activity occurred at ~50 Ma.

In addition, strongly peraluminous granites are also considered to be the products of the collision-postcollisional stage magmatism. Fifteen strongly peraluminous granite bodies have been found in the Gangdese belt, which are muscovite granites, often containing almandine garnet with the presence of tourmaline and andalusite in places (Liao et al. 2007). The cordierite-type of peraluminous granites, the characteristics of the Lachlan orogenic belt in Australia (Chappell and White 1992; Barbarin 1996; Sylvester 1998), are not found in Tibet, indicating that the Tibetan peraluminous granites were formed in the environment of relatively high pressure, low temperature and thick crust. The muscovite granites in the Gangdese belt show a large range of ages (from Jurassic to Neogene), which differs from the leucogranites in the Himalayas that concentrated in 20–10 Ma. They were formed roughly in four ages: (1) 196–189 Ma (muscovite K-Ar age, Wu 2003), such as the Ningzhong pluton to the south of Dangxiong, which may be related to the closure of the Paleotethys. (2) 145–74 Ma, distributed in the North and Central Gangdese, typical intrusions such as the Luoza muscovite granitic pluton in Namlin County (zircon U-Pb age 133.9 ± 0.9 Ma, Zhai et al. 2005), Bangor pluton (biotite K-Ar age 110–145 Ma, Liu and Wang 1994), Demla pluton (muscovite K-Ar age 79 Ma, Zhang et al. 1981), which may be related to the closure of Bangong-Nujiang Ocean and the Qiangtang-Lhasa collision. (3) 56–50 Ma, distributed in Chayu, Yambajing, Langxian and so on, corresponding to the period of India-Asia continental collision. (4) 35–8 Ma (peak 24–18 Ma), mainly distributed in Chayu, Sangri, Linzhi, Ritu, which may be related to post- India-Asia collision, corresponding to the main period of leucogranites in the Himalayas. Thus, it can be considered that the muscovite granites in 56–50 Ma are syncollisional and the muscovite granites in 35–8 Ma are post-collisional products.

6.1.2.2 Evidence from the Regional Uncomformity

As mentioned above, the LVS volcanic rocks unconformably overlie the upper Cretaceous Shexing Formation or older strata. This is a large regional unconformity extending for more than 1000 km in the southern Gangdese Belt. The rock units above and below the unconformity are entirely different in lithofacies, paleogeography, and intensity and style of deformation (Fig. 6.2). Above the unconformity, the LVS are subaerial volcanic rocks with near-horizontal occurrence, but the Shixing Formation or older strata below the unconformity are marine facies with strong folds, suggesting that the unconformity represents a major geological event. Obviously, no other geological events were larger and more important than the India-Asia continental collision during the late Cretaceous-Paleogene period around the Tibetan Plateau and the adjacent regions.

The latest time limit of this regional unconformity is best defined by the radiometric age of the basal LDF units of the LVS volcanic rocks. Zhou et al. (2004) obtained an $^{40}Ar/^{39}Ar$ age of 64.47 Ma on samples from the Linzhou Basin and 60.5 Ma on samples from the Maqu area. Zhu et al. (2015) obtained a zircon U-Pb age of 60.2 Ma on samples from the Linzhou Basin and 68.9 Ma on samples from the Maqu area. Further far to the west, the isotopic ages are 63.5 Ma in Coqin (using Rb-Sr method; Hu et al. 2007), 61.5 Ma in Horba/Zhongba (zircon U-Pb method; Liu et al. 2018), 64.5 in Yare/Pulan (zircon U-Pb method; Wang et al. 2007) and 67.1 Ma in Zhare/Ge'er (zircon U-Pb method; Cao et al. 2020), respectively. These

Fig. 6.2 The regional unconformity between the Linzizong Formation $(E_{1-2})_L$ and the highly deformed underlaying upper Cretaceous Shexing Formation (K_{2sh}) at Maqu, near Lhasa. The strata above and below the unconformity are entirely different in lithofacies, paleogeography, and intensity and style of deformation. This is a large regional unconformity extending for more than 1000 km in the southern Gangdese Belt

ages can be considered to represent the time of initiation of India-Asia continental collision (see Fig. 6.1).

As noted above, the LDF of the LVS volcanic rocks were formed in the transition period from the end of subduction to the beginning of collision. Therefore, they have both the characteristics of continental margin arc rocks and their own unique characteristics in comparison to normal subduction arc volcanic rocks (e.g. the Sangri Group and Yeba Formation in the southern Gangdese Belt). The volcanic rocks of MNF and especially UPF are the products in response to the main collision. The first appearance of shoshonite in upper MNF is an important indication of continental collision. The UPF volcanic rocks show the characteristics of volcanic rocks during the climax of the main collision.

The rock units below the unconformity are the upper Cretaceous Shexing Formation in the Linzhou Basin and the middle-eastern segment of the south Gangdese. Zhu et al. (2019) determined that the detrital zircon age of the upper Cretaceous Shexing Formation is 85.1 Ma and the age of the volcanic rocks at the bottom of the LDF is 68.9 Ma in the Maqu area and believed that the strong fold deformation of the upper Cretaceous Shexing Formation occurred between 85 and 69 Ma. Xing et al. (2020) studied the petrology, geochemistry and chronology of sandstones and basalt interlayers in the upmost Shexing Formation in the Linzhou Basin, and concluded that most of the detrital materials of the Shexing Formation came from the Central Lhasa block to the north, and a few came from the Gangdese batholith to the south. The age of the youngest detrital zircons in the sandstones indicates that the Shexing Formation was deposited later than 98 Ma and the eruption of basalts of the Shexing Formation was later than 110 Ma, related to the subduction of the Neo-Tethys. The formation time of the Shexing Formation (later than 98 Ma) and the overlying LVS volcanic rocks (with a bottom age of 65–60 Ma), implying an about 35 Myr structural discontinuity and representing the uplifting and denudation process before the continental collision. The the Shexing Formation was obviously folded, in sharp contrast to the sub-horizontal LVS volcanic rocks. Wang (2017) believed that the age of tuff intercalation in the upper Shexing Formation is 80 Ma, and even younger at its top. The folding of the Shexing Formation and older strata indicates a compression-uplift event along the Gangdese belt, which is related to the end of marine environment and the crustal deformation and therefore suggested that the end of magmatism may be corresponding to the earliest deformation. Therefore, it seems that the formation time of the unconformity are 72–65 Ma and the deformation and erosion had lasted for ~7 Myrs. Subsequently, the volcanic activity of the LVS steadily continued from early Paleocene to late Eocene (about 65–45 Ma) and the Gangdese magmatic belt began to be deformed again by the end of Oligocene (ca. 25 Ma).

6.1.2.3 Sedimentary and Stratigraphic Paleontological Evidence

Sedimentary and biostratigraphic records are also very important for constraining the timing of India–Asia collision. Gaetani and Garzanti (1991) show that an abrupt change from marine to terrestrial deposition occurred at the end of the early Eocene in the Zanskar region of northeastern India (52 Ma). Le Fort (1996) and Rowley (1996) quoted it as representing the initial age of the India-Asia collision. However, Beck et al. (1995) thought that it only constrains a lower limit (i.e., the youngest age) on the onset of the collision and showed that accretionary-prism and trench rocks along the southern edge of Asia in northwestern Pakistan were thrust over the passive continental margin of India after 66 Ma but before 55 Ma. These facts enhanced Yin and Harrison (2000)'s interpretation that this relationship suggested that the elimination of the oceanic lithosphere between India and Asia must have occurred before 55 Ma, because a significant portion of the Indian passive continental margin could have been underthrust beneath Asia before 55 Ma, and this implied that 55 Ma is a minimum age for the initial India-Asia collision in northwestern Pakistan.

Willems et al. (1996) found that a drastic change in sedimentary facies and depositional patterns occurred in the Middle Maastrichtian (70 Ma) in the continuous Upper Cretaceous to Lower Tertiary Indian passive continental margin marine strata near Tingri in south-central Tibet. The Middle Maastrichtian rocks are characterized by a sudden transition from siliciclastic turbidite sediments to marly sandstone. In the Lower Paleocene rocks, directly above the Maastrichtian strata (66–64 Ma), redeposition of shallow-water clastic rocks are found. They interpreted this change in depositional patterns along the Indian continental passive margin between 70 and 64 Ma as indicating the initial contact between India and Asia.

Wang et al. (2003) divided the sedimentary records corresponding to the initial stage of continental collision into two types: collisional delta and residual shallow sea/bay deposition and conducted a detailed study on collisional fan-delta deposition in the southern margin of the Asian continent. A set of marine deposits, named Cuojiangding Group, developed in Saga and Zhongba areas to the north of Yarlung Zangbo Suture zone in western Tibet from the end of Cretaceous to the early -middle Paleogene. It is a collisional delta developed with the subduction of the NeoTethyan Ocean in the Cretaceous/Paleogene boundary (Wesott 1980) based on the Xigaze Group submarine fan in the Xigaze forearc basin (Ensele et al. 1994; Wang et al. 1999). The Cuojiangding Group is dominated by middle-coarse-grained terrigenous clastic rocks, with more shale and bioclastic limestone in the middle and lower part of the group, containing three lithostratigraphic units. The lower member (Qubeiya Formation) is composed of gray bioclastic limestone, siltstone and purplish red calc silty shale, which is rich in benthic macroforaminifera. It mainly formed in Maastrichtian stage at the end of Cretaceous (Wan et al. 2001) in a finger-like cross mixing environment of offshore shelf limestone and pre-delta sandy shale. In the middle part (Quxia Formation) is composed of medium-coarse-grained lithic sandstone conglomerate with interlayers of purple-red shale, containing plant and ostrea fossils. Within which gravels mainly came from the basement dominated by metamorphic and granitic rocks and conglomerates are probably the fillings of river channel. The first appearance of granite boulders imply that the basement had been eroded because of the uplift of the Gangdese in the front of Asian continent during the collision stage. Shale and clastic rocks with more than three upward tapering depositional cycles suggest that they formed in a sedimentary environment of braided river delta front in the Danian stage of Paleocene (Wan et al. 2001). The lower part of the upper member (Galazi Formation) is dominated by sheeted calcareous coarse-grained sandstone intercalated with bioclastic limestone; the middle bioshoal biolimestone is rich in foraminifera fossils; the upper part is gray-green siderite nodule-bearing shale with nummulite limestone. They belong to the Selandian stage, Thanetian stage and Ypresian stage of Paleocene, respectively (Wan et al. 2001). The Gangdese group overlying the Cuojiangding Group is called the Mela conglomerate, which is characterized by variegated and purplish red conglomerates, containing complex and diverse of gravels, such as granite, schist, intermediate to basic igneous rock, quartzite, sandstone, limestone, siliceous rock and so on. The lithologic profile structure indicates that the sedimentary environment is the underwater alluvial fan in delta plain stage deposited in the middle Eocene (Wan et al. 2001).

Wang et al. (2003) suggested that the Zhongba Fan Delta was collisional type deposits and the products of the early stage of the India-Asia continental collision, based on the unusual fan delta features shown in the sedimentary records of Chuojiangding Group and Gangdese Group mentioned above. The shelf is very irregular on the coast where continents collided with mountain arcs. Considering the age of the Quxia Formation, the India-Asia continental collision there likely started from the end of Cretaceous to the beginning of Paleogene, which is equivalent to the Late Maastrichtian- Danian period (65–60 Ma).

They also studied the sedimentary system in the northern margin of the Indian continent, such as the upper Cretaceous-Paleogene sedimentary system in the Gamba-Tingri Basin. They believed that this sedimentary system should contain information about the early collision process because it belongs to a type of formation in residual seas by basin inversion after the initiation of collision (Yu and Wang 1990), although it is far away from the Yarlung Zangbo suture zone on the northern margin of India continent and had stable sedimentary assemblages, being significantly different from the Cuojiangding Group. The results show that the Jaburashan Formation in Tingri region and the Chitula Formation in the whole southern Tethyan Himalayan belt are probably the products of the initial stage of collision. According to the fossils and stratigraphic relationships of the Jabujishanshan Formation and the Chitulla Formation, the time of the initial collision is also limited to 65–60 Ma.

Wan et al. (2002a, b) conducted detailed studies on the rocks, paleontology and structure near the K/T boundary in Gangba, Zhongba and Jilong, southern Tibet, and found that the K/T boundary was a large unconformity, and continental conglomerate and sandstone were unintegrated above the shallow platform carbonate rocks. Also, the sedimentary facies and micropaleontology groups significantly change crossing the K/T boundary. In the Zhongba and Gamba areas in southern Tibet, the K/T boundary is marked by a major disconformity, separating platform carbonates from overlying terrigenous conglomerates and sandstones. The K/T boundary at the Gamba area is placed between the Orbitoides–Omphalocyclus and Rotalia–Smoutina–Lockhartia faunas. In Tingri, Cretaceous Globotruncana and tertiary Globigerina–Globorotalia microfauna demark the position of the K/T boundary. The occurrence of terrigenous sandstones and boulder-size conglomerates in the early Paleocene is compelling evidence for tectonic uplift and emergence of the southern margin of the Lhasa block and of the northern margin of the Indian plate. Based on the biostratigraphic evidence, Wan et al. (2002a, b) further argued that the uplift is caused by the onset of continental collision during the earliest Paleocene (the Danian). Progressing plate collision resulted in deformation and fragmentation of the Paleocene carbonate platform and deposition of limestone breccias, which were considered as further evidence for tectonic compression as a result of early continental collision during the Thanetian, earlier than indicated by previous studies in the Himalayas. The dramatic change in the sedimentary facies and depositional environment provides the earliest evidence and dating of the initiation of the collision process. All these indicate that the study on Cretaceous-Paleogene boundary strata in both sides of the Yarlung Zangbo suture (i.g., Zhongba, Gamba, Tingri) also prove that Indo-Asia continental collision started at the K/T boundary age (~65 Ma).

Based on the above evidences they believed that the collision of the India and Lhasa continental blocks was initiated at K/T boundary time (ca.65 Ma).

Recently, Hu et al. (2017) summarized and reviewed the studies on the time of the initial India-Asia continental collision based on the sedimentary records on both sides of the Yarlung Zangbo suture zone in Tibet, including Sangdanlin section in Saga region, Gyangze region, Tingri-Gamba region in northern India continent, and Zhongba-Dagin region in southern Asia continent.

They believed that the time of initial collision is the time when the oceanic crust disappeared between continents and the continental crust began to contact, and the following two criteria can be used to constrain the time of collision:

1. The time when the deep water turbidite provenance near the suture zone changed from Indian provenance to Asian provenance. For example, the Zongzhuo Formation is believed to be the earliest geological unit accepting Asian provenance in the Gyantse region, thus can roughly limit the initial Indo-Asian continental collision to the Paleogene.
2. The time when the Himalayan underfilled foreland basin started on both sides of the suture zone and caused the abrupt change of the sedimentary environment or the formation of the unconformity. For example, based on the study of the Sangdanlin profile in Saga area, the time of the initial Indo-Asian continental collision was defined as the middle Paleocene (59 ± 1 Ma) according to the time of provenance change defined by radiolarian fossils (RP6), calcium nannofossils (CNP7) and the youngest age peak of detrital zircon 58.1 ± 0.9 Ma (Hu et al. 2015). In addition, according to the comprehensive stratigraphy and provenance analysis, it is believed that the unconformity between the Qubeiya Formation and the overlying Qixia Formation in the Cuojiangding area of Zhongba County was probably caused by the india-Asia continental initial collision (Hu et al. 2016), and therefore the India-Asia initial collision should be earlier than 58 Ma, but not earlier than 65 Ma. Uplift and rapid exhumation may have occurred in the southern margin of Asia after the initial collision.

They also suggest that the initial collision occurred 20–25 Myrs earlier than the end of normal marine deposition, and that the molasse appeared 30–40 Myrs later than the initial collision. Based on the sedimentary records of the Tethyan Himalayan Paleogene, the initial Indo-Asian continental collision has no obvious diachronicity between the central and western Himalayas.

They also propose that the Himalayan orogeny can be divided into four stages according to the sedimentary records in the Himalayan region: (1) The initial stage of primordial Himalayan period from the middle Paleocene to the early Eocene (59–52 Ma), the initial collision occurred and correspondingly the deep-sea environment occurred in the collision basin, and gentle slope of carbonate developed on the edge of Indian continent; (2) The early stage of primordial Himalayan period from early to middle Eocene (52–41 Ma), characterized by the development of residual shallow marine deposits. Neo-Tethyan Bay gradually disappeared from west to east; (3) The late stage of primordial Himalayan period from late Eocene to Oligocene (41–26 Ma), the entire Himalayan region and southern Tibet were lack of sedimentation; (4) The early stage of the Neo-Himalayan period from late Oligocene to Early Miocene (26–17 Ma), Himalaya uplifted, continental Molasses accumulated rapidly, and Yarlung Zangbo River and India River developed east-west along the suture zone.

6.1.2.4 Evidence from Foreland Basins

Foreland basins around collision zones are direct indicators of continental collision (Ding 2003; Ding et al. 2017). Ding (2003) found that a peripheral foreland basin system (Yarlung Zangbo foreland basin system) was developed to the south of the Yarlung Zangbo suture zone, which was closer to the suture zone than other Cenozoic basins. The Yarlung Zangbo foreland basins formed at about 65–60 Ma (Cai et al. 2011; DeCelles et al. 2014; Ding 2003; Ding et al. 2005; Hu et al. 2015; Wang et al. 2011a, b; Wu et al. 2014). At the same time, the lithologies of the Tethyan Himalayan Belt in the Saga area changed from Indian passive continental margin deposits to foredeep deposits in these peripheral foreland basins (Ding 2003; Ding et al. 2005). A contemporaneous peripheral foreland basin was also developed in the eastern Gyantse Region (Cai et al. 2011).

The Sangdanlin section in the Saga region, located in the northern Tethyan Himalayan, represents the peripheral foreland basin generated by the loading of Eurasia onto the passive margin of the Indian continent, and is a representative section for studying initial time of continental collision. Ding (2003) was one of the first researchers who conducted detailed studies of the structural, paleontological stratigraphy and analysis of sedimentary source in this section, and divided it into the upper Cretaceous Zongzhuo Formation, Paleocene Sangdanlin Formation of Danian-Thanetian Stage and the upper Eocene Zheya Formation. The Zongzhuo Formation is composed of siliceous shale, siliceous limestone and quartz sandstone, of which quartz sandstone is gravity flow deposits of typical passive continental margin in the Indian continent. The Sangdanlin Formation consists of siliceous rocks and siliceous shales, which formed from the Paleocene to the Early Eocene (65–55 Ma) according to radiolarian fossils (Ding 2003). It is a typical foreland basin and belongs to the starvation type peripheral foreland basin due to the lack of detrital material. The Zheya Formation is composed of a set of sandy flysch deposits intercalated with several layers of conglomerate, which are turbidite and channel sandstone conglomerate deposits near the wedge-top slope.

There are different views on the initial collision time in this region. Ding et al. (2005) believed that the collision occurred at about 65 Ma based on the time of transition of the sedimentary environment from passive continental margin of India to starving peripheral foreland basin. Wu et al. (2014) conducted detailed work on detrital zircons in this area, and found that alternation of materials of Indian origin and Eurasian origin in the Sangdanlin Formation, and that the first detrital materials of Eurasian origin appeared earlier than that suggested by Wang et al. (2011a, b). They thus suggested that the collision occurred at 60 Ma in combination with the ages of the youngest detrital zircons (62 ± 1 Ma, 57 ± 2 Ma) and the time of radiolarian fossils (RP4, 63–61 Ma). DeCelles et al. (2014) determined a zircon U-Pb age of 58.5 Ma on the tuff at the top of the Zheya Formation in the Sangdanlin section, which, in combination with the youngest age of 60.6 Ma of detrital zircons from those of Eurasian source appeared for the first time, allow these authors to suggest that the India-Asia collision occurred at 60.6–58.5 Ma. Hu et al. (2015) again conducted the study of the fossil chronology and detrital zircon chronology of the strata near the provenance change, and reached a similar view to DeCelles et al. (2014) and Wu et al. (2014), limiting the India-Asia initial collision time to 59 ± 1 Ma.

Ding et al. (2017) believed that continental collision should have occurred before 60–59 Ma if the first appearance of Eurasia-derived detrital materials in the foredeep of the foreland basin system is used to define the initial collision time. However tectonic deformation is more sensitive than source change to continental collision, so one can use the combination of the tectonic deformation and foreland basin development to constrain the time of the initial collision, which leads to a more reasonable timing of initial collision at 65–63 Ma.

In the Gamba-Tingri area, detrital materials of pure quartz sandstone of the lower Paleocene Jidula Formation show provenance from the Indian continent. The Zongpu Formation of the upper Paleocene is nodular limestone deposits with a southward slump structure (Li et al. 2002; Li and Wan 2003; Wan et al. 2002a, b), indicating that the fore-uplift migrated to this position (i.e., Gamba-Tingri area) during the deposition of the Zongpu Formation (Zhang et al. 2012). The Eocene Jaburai Formation is a set of mudstones intercalated with lithic sandstones (Wang et al. 2002; Zhu et al. 2005), of which detrital materials were derived from the Tethyan Himalayan and the Zangbo suture-Gangdese, involved in the fold-thrust zone from the north (Zhu et al. 2005). It represents the fore-trough deposition after the passage of the fore-uplift. At that time, the Saga-Gyantze foreland basin had been involved in the foreland fold thrust belt, and the fore-trough of the foreland fold thrust belt had migrated to the Gamba-Tingri area in the south of Lhagoi Kangri (Ding et al. 2009).

In general, the earliest peripheral foreland basins developed in the northernmost Tethyan Himalayan belt as accretive wedge and trench material thrust southward into the northern margin of the Passive continent of India at 65–63 Ma. The foredeep sedimentary units corresponding to the Sangdanlin Formation were formed in Saga, and the fore-uplift migrated to the Gamba-Tingri area at 56 Ma and formed nodulous limestones. All these mentioned above indicate that the initial time of the India-Asian collision occurred at ~65 Ma.

6.1.2.5 Paleomagnetic Evidence

Paleomagnetism as an important means to study the continental drift and evolution of plates can quantitatively identify the palaeolatitude and drifting paths of plates, including the change of the relative velocity between two related plates, the kinematics process of plates in geological history, and constrain the timing and locations of continental collision events.

Yin and Harrison (2000) summarized the results of paleomagnetic studies available back then. They pointed out that analysis of Cenozoic magnetic anomalies in the Indian ocean showed that the relative velocity between the Indian and Eurasian plates decreased rapidly from 15–25 to 13–18 cm/year at 50 Ma and this sudden decrease in plate convergent rate had been interpreted as indicating the time for the onset of the India-Asia collision by Patriat and Achache (1984). Paleomagnetic data from sedimentary rocks of the 90°E Ridge in the Indian ocean also show that the northward movement of the Indian plate exhibits slowed down from 18–19 to 4.5 cm/year at about 55 Ma. This was interpreted by Klootwijk et al. (1985, 1992) as indicating the completion of the India-Asia suturing. Thus, the initial contact between India and Asia would have been earlier than 55 Ma. However, Yin and Harrison (2000) argued that it is difficult to determine whether the slowdown of convergence between India and Asia resulted from an increase in tectonic resistance due to contact of the buoyant Indian continental margin with Asia, or whether it simply reflects a sudden decrease in the spreading rate along the Indian mid-oceanic ridge at ~50–55 Ma.

Besse et al. (1984) suggested that the India-Asia collision occurred at ~50 Ma, based on a joint analysis of primary and secondary components of magnetization in late Paleocene sediments from the northernmost margin of India. However, Jaeger et al. (1989) pointed out that the lithologies used in that analysis are in fact older, between 60 and 56 Ma, so that the time of the India-Asia collision should be older than 60 Ma.

Dupont-Nivet et al. (2010) believed that large uncertainties remain in the existing palaeomagnetic data, and particularly those used to constrain the palaeolatitude of the southern margin of Asia (i.e., the Lhasa terrane) before and during the collision. Existing palaeomagnetic data sets from volcanic and sedimentary rocks yield a wide range of palaeolatitude estimates from 5 to 30°N (Westphal and Pozzi 1983; Achache et al. 1984; Lin and Watts 1988; Chen et al. 1993; Liebke et al. 2010; Tan et al. 2010).Therefore, they conducted paleomagnetic studies and provided their paleomagnetic data from the LVS volcanic rocks (54–47 Ma) near Lhasa that the latitude of the southern margin of Asia was 19.5 ± 5.0°N at the time of emplacement. They believed that the onset of collision occurred ca. 48 Ma.

However, it should be noted that their samples were mainly collected from the MNF of LVS volcanic rocks in the Linzhou basin near Lhasa with zircon U-Pb age of 54–47 Ma according He et al. (2007). Only one sample was collected from the UPF without age data. New zircon U-Pb age data of the LVS volcanic rocks show 60.2–58.3 Ma for the LDF, 55.4–52.6 Ma for the MNF (Zhu et al. 2015). It is commonly believed that India-Asia collision was already in the main stage rather than initial stage of collision during the MNF-UPF volcanic eruption.

Almost at the same time, Huang et al. (2010) also reviewed the existing paleomagnetic data on the Tibetan Plateau and pointed out that there are no sufficiently reliable data yet available to answer the key scientific question of when and where the onset time of India-Asia collision took place, and the data interpretations are subject to large uncertainties. For example, the analysis of magnetic anomaly lineation in the Indian Ocean showed that the velocity of the Indian plate with respect to Eurasia has changed significantly in the Cenozoic, leading to large errors in the interpretations. Although the Mesozoic-Cenozoic apparent pole shift curves with a time window of 5 Myrs or even 1 Myr for both the Indian and Eurasian plates had been established, the Apparent Polar Wander Paths (APWP) of India and Eurasia plates could not directly provide constraints on when and where the initial India-Asia collision took place. Therefore, the key to solve the problem is constrained by direct comparison of the fine APWP of the blocks and relatively continuous paleo-latitude variation on both sides of the India-Asia collision zone.

To better understand the onset of the India-Asia collision, Huang et al. (2010) and Chen et al. (2010) conducted a paleomagnetic study on Paleogene volcanic rocks and sediments of the LVS volcanic rocks (64–44 Ma) from the Linzhou and Namling basins of the Lhasa Block at the southern margin of the Asian continent. Their evidence supports that the initial India-Asia collision occurred between 60 and 55 Ma.

Lee and Lawver (1995) believed that the India-Asia collision could happen in two stages: the soft collision (partly collision) at 60–40 Ma and the hard collision (complete collision) after 40 Ma in terms of plate convergence rate data.

6.1.3 Post-collisional Stage

The post-collisional stage is an integrate part of the generalized India-Asia collision process and an important stage for the evolution of the Tibetan Plateau. Abundant post-collisional igneous rocks of varying type are widespread over the entire Plateau, which contain valuable information on the evolution of the Tibetan Plateau (Mo et al. 2006, 2009; Zhao et al. 2006, 2009).

Post-collisional magmatic activities occurred in several distinct geological terranes of the Tibetan Plateau and surrounding regions, including the Qiangtang, Hoh Xil–middle Kunlun, western Kunlun, Karakoram, Gangdese (Lhasa block), western Qinling, and Sanjiang regions, and roughly concentrated in three distinct episodes: ~45–25 Ma, 25–5 Ma, and 5–0 Ma.

The post-collisional volcanic rocks are dominantly potassic-ultrapotassic in composition with SHO and HKCA affinities (in terms of K_2O vs. SiO_2) or SHO or potassic and ultrapotassic affinity (in terms of K_2O vs. Na_2O). In general, those rocks in the Tibetan Plateau can be divided into five types following the classification of Foley et al. (1987): (1) ultrapotassic, including mainly tephrite, phonolitic tephrite, tephritic phonolite, foidite, and some shoshonite; (2) potassic basic-ultrabasic, including kamafugite, carbonatite, absarokite, trachybasalt, and some shoshonites and lamprophyres; (3) potassic intermediate (most common in the Tibetan plateau), including latite, potassic trachyte, and most shoshonites; (4) potassic acidic; and (5) sodic (see Mo et al. 2006 and Chap. 4 of this book). Migration of collisional and post-collisional volcanism with time shows a highly distinctive pattern (Fig. 6.3). Initially, as an initial response to the India-Asia collision, magmatic activity migrated northward between ~65 and 45 Ma, away from the Tsangpo collision suture. Between ~45 and 10 Ma, the volcanic activity migrated outward from the plateau interior, implying wavelike outward propagation of upwelling asthenosphere. Since ~10 Ma to present, the magmatic activity has been migrating to

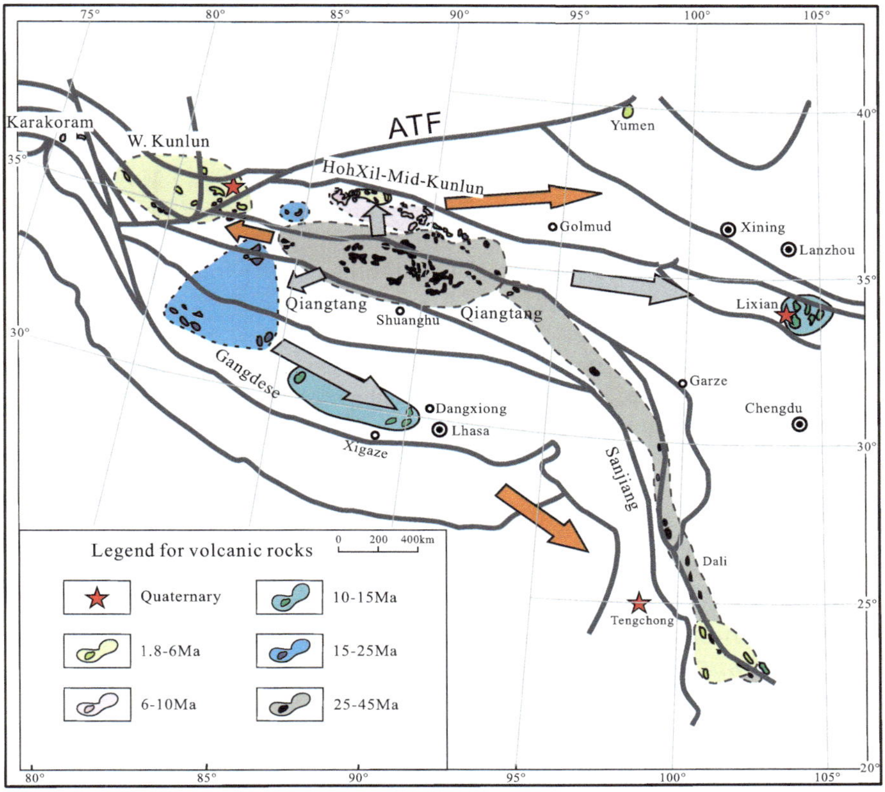

Fig. 6.3 Sketch map showing migration paths of postcollisional volcanism with time in the Tibetan plateau and adjacent regions

northwestern, northeast-eastern, and southeastern peripheral regions of the plateau. Overall, such a highly distinctive pattern of magmatism can be interpreted to reflect lateral asthenospheric mantle flow induced by the approach, and ensuing collision, of relatively thick (India and Eurasia) continental plates.

6.1.4 Conclusive Remarks

In conclusion, the India-Asia collisional process is best described in three stages: the initial stage and the main phase of collision (65 ± 5 Ma to 45 ± 5 Ma) and post-collisional stage (after 45 Ma).

1. **The initiation of India-Asia collision**

As mentioned above, the initiation of the India-Asia collision constrained by the evidence from volcanic rocks and granites took place in the timeframe of 60–68 Ma, which agrees with 60–65 Ma (most 65 Ma, K/T boundary) recorded in the biostratigraphy and sedimentary rocks, and with 60–55 Ma given by the paleomagnetic data. Therefore, we can conclude with confidence for the initial time of India-Asia collision at 65 ± 5 Ma.

2. **The main phase of India-Asia collision**

The volcanic rocks of late MNF and especially UPF are best understood as the products in response to the main phase of the collision (50 ± 5 Ma), characterized by huge volume eruption, extremely high temperature and injection of asthenosphere materials in a rather short period around 50 Ma, with rocks dominated by ~2500 m thick ignimbrites (weld-tuffs) and the volumetrically significant Gangdese batholith.

3. **The post-collisional stage**

The post-collisional stage is important and is closely related to the uplift and growth of the Tibetan Plateau. The post-collisional magmatic activity in the Tibetan Plateau occurred in three distinct episodes: ~45–25 Ma, 25–5 Ma, and 5–0 Ma, dominated by potassic-ultrapotassic types.

6.2 Effects of the India-Asia Continental Collision

6.2.1 Formation of the Extremely Thick Crust of the Tibetan Plateau

6.2.1.1 The Close Relationship Between the Formation of Extremely Thick Crust of the Tibetan Plateau and the India-Asia Collision

The Tibetan Plateau has the thickest crust on Earth, up to twice as thick as normal crust. Seismological data show that the crust thickness of Lhasa Terrane is 65–75 km, up to 80 km at the maximum, with an average thickness of about 70 km, which is characterized by low seismic wave velocity (Vp = 6 km/s, Vs = 3.45 km/s) (Kind et al. 1996; Molnar et al. 1998). The common views on the origin of such thick crust are: (1) double crustal superposition model (Argand 1924; Powell 1986); (2) compressional shortening thickening model (Dewey and Bird 1970; Dewey and Burke 1973; England and McKenzie 1982; England and Houseman 1989); (3) continental injection model, i.e., the crustal thickening is caused by the stronger Indian crust injecting into the weaker lower Tibetan crust (Zhao and Morgen 1987); (4) combined action thickening model of tectonic compression and mantle material injection. All these models suggest that the formation of the huge thickness of the Tibetan crust is closely related to the India-Asia collision (Mo et al. 2007, 2009; Mo 2020).

6.2.1.2 Contribution of Mantle Material to Crustal Thickening

As mentioned above, collisional and post-collisional igneous rocks in the Tibetan Plateau contain significant mantle contributions. In the southern Gangdese belt, (1) the LVS volcanic rocks have $\varepsilon Nd(t)$ value of +9.20 to −3.96, mostly concentrated at +5.43 to −0.52. In terms of Nd crust index (NCI) method (DePaolo 1981, 1985; DePaolo et al. 1992), the LVS volcanic rocks give NCI = 0.16–0.47, with an average of 0.30, which is equivalent to an average of 70% mantle contribution (Mo et al. 2007). (2) the Gangdese Quxu granitoids, with εNd(t) value of +5.21 to +1.64 (Mo et al. 2005a, b), which contain abundant mafic microgranular enclaves (MMEs) and is accompanied by a gabbro-diabase-pyroxenite intrusion belt. (3) During the post-collision period (since 40 Ma), after a ~15 Myr magmatic intermission, potassic- ultrapotassic volcanic rocks (25–8 Ma), Adakite rocks (25–12 Ma, peak at 16 Ma), and peraluminous muscovite granitoids (24–18 Ma) are developed in the Gangdese belt. Chung et al. (2003) and Hou et al. (2004) believed that these potassium-rich adakites were produced by partial melting of the thickened basaltic lower crust being different from the adakites derived by partial melting of the subducting ocean crust (Kay 1978; Defant and Drummond 1990). In brief, mantle-derived materials by means of magmatism contribute to the thickened continental crust of the Tibetan Plateau, especially in the southern Lhasa Terrane.

6.2.1.3 How the Mantle Material Entered into the Crust

In general, magmatism plays an important role in carrying both energy and mass from mantle to the crust. There are several different but mutually compatible models to illustrate this transport process.

Niu et al. (2007, 2013) and Mo et al. (2008) pointed out that the LVS (also SGB) were produced in the broad context of Tethyan ocean closing and India-Asia collision. They believed that volumetrically significant and physically plausible basaltic source material available for generating andesitic magmas parental to the LVS (also SGB) would be ocean crustal materials atop the subducting/underthrusting Tethyan oceanic lithosphere. Partial melting of such basaltic crust can produce "primary" melts of andesitic compositions with inherited asthenospheric isotopic signatures. The LDF may be derived from such "primary" melts of ocean crust melting. Melting of the more felsic materials including terrigenous sediments along with the subducting/underthrusting ocean crust can produce the more felsic melts. The MNF and UPF most likely result from re-melting of rocks equivalent to the older LDF. Thus, mantle material transmit into crustal rocks during the formation of the LVS (also SGB).

Zhu et al. (2015) suggested that an increased mantle input and magmatic flare-up for the LVS volcanic rocks and the coeval Gangdese batholith during India-Asia collision controlled by slab rollback from ~70 Ma and slab breakoff at ~53 Ma. They believed that this inference was consistent with numerical modeling indicating a short duration of slab breakoff followed by intense igneous activity as a result of the enhanced heat input from rising asthenosphere, and supported by other lines of evidence, including: (1) the occurrence of abundant 52–47 Ma mafic enclaves and dykes that suggested significantly increased contributions from the mantle; (2) the presence of ~52.5 Ma bimodal volcanic rocks that point to partial melting of enriched metasomatic layers within lithospheric mantle and crustal melting caused by thermotectonic effects as a result of slab breakoff; (3) high Zr/Y ratio (3.7–6.8) of the basaltic lavas and dykes (~52.9 Ma) in the LVS volcanic rocks that suggest the input of sub-slab asthenospheric mantle; and (4) dramatically increased zircon saturation temperature at ~52.5 Ma recovered by the rhyolitic rocks of the UPF that indicate the anomalously high heat input from the mantle.

Zhao et al. (2009) proposed a two-stage geodynamic model to illustrate the mechanism of mantle metasomatism and the generation of the post-collisional igneous rocks in the Lhasa terrane. With the slab devolatilization processes gradually changed from metamorphic dehydration to melting regime at greater depths, partial melts of carbonate-bearing sediments stemmed from down-going seafloor slab to flux into the overlying mantle wedge, causing clinopyroxene-rich mantle assemblage with isotopically light Mg. The subduction of Indian plate beneath the Lhasa terrane would have absorbed >400 km of shortening and thickening of the crust by tectonic underplating, finally causing elevation of the Tibetan Plateau. Under this tectonic setting, thermal perturbation induced by deep geodynamic processes (e.g., convective thinning and/or delamination of the lithospheric mantle, roll-back, and break-off of the Indian slab) finally gives rise to the partial melting of metasomatized mantle sources and thickening of the lower crust, generating the widespread Cenozoic post-collisional magmatism in the Lhasa terrane.

As for the Miocene adakitic rocks in southern Tibet, Chung et al. (2003) and Hou et al. (2004) believed that these potassium-rich adakites were produced by partial melting of the thickened basaltic lower crust being different from the adakites derived by partial melting of the subducting ocean crust (Kay 1978; Defant and Drummond 1990). Hou et al. (2004) proposed a two-stage model for the generation of the Miocene adakitic rocks. This model emphasizes that the input of mafic magmas originated from a hydrous mantle source with an enrichment of incompatible elements, forming the newly accreted basaltic lower crust during at 65–40 Ma. This lower portion of thickened crust was later partially melted in response to the upwelling and laterally flowing of the asthenospheric mantle and the crustal extension, producing the adakitic magmas during 25–12 Ma.

Besides, some researchers believe that underplating and magma mixing are two important ways of crust-mantle interactions that may also play an important role in the growth and thickening of the Tibetan crust during the India-Asian collision (Jiang et al. 1999; Dong et al. 2005; Mo et al. 2005a, b; Zhu et al. 2011; Hou et al. 2015; Hou and Wang 2018). They found a great deal of evidence of underplating and magma mixing in the Gangdese, and suggested that mafic magma originated from the mantle wedge rose and accumulated at the base of the continental crust for a relatively long time during the India-Asia collision, and their subsequent partial melting produces massive granitic magmas and mixed with the rising mafic magma on one hand, and thickening of the lower crust on the other hand, like many typical tectonic zones elsewhere on Earth (Vernon 1984; Huppert and Sparks 1988; Didier and Baibarin 1991; Altherton and Petford 1993; Petford and Atherton 1996; Dong 1999; Liu et al. 2004).

6.2.1.4 Relative Contribution of the Two Mechanisms to the Crust Thickening

Mo et al. (2007) estimated the crustal thickness during the eruption of the LVS volcanic rocks by using the formula C (km) = 18.2 K_{60} + 0.45 proposed by Condie (1982). The crustal thickness at the time of the LDF (60–58 Ma; Zhu et al. 2015; the same below), the MNF (55–53 Ma) and the UPF (after 53 Ma) emplacement is respectively 35 km, 37 km and 73 km, indicating that the crust had a normal thickness until ~50 Ma, and was significantly thickened after ~50 Ma. Several lines of evidence reveal that potassic adakites (25–12 Ma) and latite/high-P trachyte (25–8 Ma) were both formed in settings with thickened crust (e.g., Wyllie 1977, 1984; Rapp et al. 1999; Chung et al. 2003; Hou et al. 2004; Deng et al. 2004a, b). The crust beneath southern Tibet must have been considerably thickened before 25 Ma. The geophysical data (Kind et al. 1996; Nelson and Zhao 1996; Owens and Zandt 1997) show that the present-day crustal thickness is about 70 km on average in the Lhasa terrane, and confirm the existence of 14–20 km (~15 km on average) thick high-speed layer (Vp = 7.2–7.5 km/s) of the lower crust in southern Tibet. This can be interpreted as a layer of high-pressure garnet-bearing mafic rocks with a density of >3.0 g/cm^3 at the depth of >60 km, or a layer of mafic accumulates of the mantle-derived underplating magma, or eclogite at the base of the thickened crust, representing the contribution of mantle material to crustal thickening. Assuming that the pre-collisional crust was ~35 km thick, thus newly increased thickness of the crust during collision should be another 35 km. It is inferred that the ~15 km of crust thickness in 70 km of the total crustal thickness had been contributed by the mantle material through magmatism, while ~20 km of crustal thickness attributed to structural thickening, and the ratio is about 3:4. Therefore, it is a combination of structural thickening and thickening by the mantle material injection that result in the formation of extremely thick crust of the Lhasa terrane.

6.2.2 *The Formation of Two Types of Crust on the Tibetan Plateau*

What is the nature of the continental crust of the Tibetan Plateau? It is juvenile crust (Jahn et al. 2000; DePaolo 1981; Samson et al. 1989) or reworked mature crust, or coexistence of two types of crust? The detailed work by Mo et al. (2005a, b, 2008), Zhu et al. (2009, 2011), Niu et al. (2013) and Hou et al. (2015) on the igneous rocks in the Lhasa block (Gangdese Belt), which has the thickest crust, proves that two types of continental crust, juvenile crust and reworked mature crust, coexist in the Tibetan Plateau, especially the Lhasa block in southern Tibet.

Mo et al. (2005a, b) used the then available Nd-Sr isotopic data of granites and volcanic rocks from the Gangdese belt, the East Kunlun and the Qiangtang to show the coexistence of two types of crust in the Lhasa block, the juvenile crust and reworked mature crust.

To obtain a comprehensive dataset of bulk-rock geochemical and zircon isotope data on the Mesozoic–early Tertiary magmatic rocks in the Lhasa Terrane, Zhu et al. (2009, 2011) collected a total of 400 samples along the Yanhu–Hor, Coqen–Saga, Xainza–Xigaze, and Nyainrong–Nang traverses with a spatial coverage of ~1100 km × 300 km (from 82 to 93°E longitude, and from 29 to 32°N latitude) across the Lhasa Terrane. All the samples have been analyzed for bulk-rock major and trace element compositions and 120 of the samples have been selected for in situ zircon U–Pb and Hf isotope analysis, and obtained 992 of zircon U-Pb age data and 836 of zircon Hf isotopic data εHf(t). These data show that εHf(t) values of the Central Lhasa subterrane during 210–118 Ma, before the slab break-off and the Qiangtang-Lhasa amalgamation (at ~113 Ma), are successively (−17.3 to −2.5), (−22.0 to +2.0), (−16.7 to +1.4) and (−20.6 to −1.6), (−14.4 to −4.1)at 113 Ma, basically negative; those in the North Lhasa subterrane are (−6.0 to +18.8) during the period of 145–118 Ma and successively (+4.7 to +18.4), (−1.5 to +10.1) and (−10.0 to +0.9) at 113 Ma from west to east, mostly positive; εHf(t) values of the South Lhasa subterrane are successively (−5.0 to +16.2), (−4.7 to +16.7), (+9.6 to +14.9) and (−2.4 to +10.5), exhibiting small negative to large positive.

Hou et al. (2013, 2015), Hou and Wang (2018) conducted Hf isotopic mapping of the Lhasa terrane by using 230 sets of Meso-Cenozoic igneous rock samples (215–12 Ma) and 672 sets of zircon U-Pb ages and Hf isotopic data, including drawing zircon εHf(t) and T_{DM}^{C} isopleth, which directly reflect the spatial distribution and model age of different types of crustal source rocks of igneous rocks. Hf isotopic mapping shows that igneous rocks in the northern, central and southern Lhasa subterranes have different zircon εHf(T) and T_{DM}^{C} values. Most of igneous rocks in the southern and northern Lhasa subterranes have high zircon εHf(T) values (0 to +20) and young T_{DM}^{C} values (0.2–0.8 Ga), while those in the majority of the Central Lhasa subterrane have large εHf(t) negative values (−22 to +5) and ancient T_{DM}^{C} values (1.4–2.4 Ga). Besides, they inferred that the crust of the Central Lhasa is about 70 km thick, that of the South Lhasa is 75 km thick, and that of the North Lhasa, 65 km thick, in combination of Hf isotopic mapping with geophysical observations.

All the work mentioned above reveal that (1) igneous rocks in the northern and southern Lhasa subterranes have mainly positive zircon εHf(t) values and younger Hf model ages (T_{DM}^{C}), that is, the Phanerozoic and Mesoproterozoic ages for the northern, and Phanerozoic age for the southern Lhasa subterranes. Igneous rocks in the central Lhasa subterrane have mainly negative zircon εHf(t), and ancient T_{DM}^{C} (Paleoproterozoic- Archean ages). It indicates that the crust of the northern and southern Lhasa blocks is mainly juvenile crust, while the crust of the central Lhasa block is reworked mature crust. (2) The spatial distribution and nature of the crystalline basement beneath the entire Lhasa terrane has long been speculative and rely on geophysical inference. Zircon Hf isotopes offer a powerful tool for resolving this problem (Flowerdew et al. 2009; Griffin et al. 2002; Kemp et al. 2006). Zhu et al. (2009, 2013) revealed that from the northern and southern subterranes towards the central Lhasa subterrane εHf(t) values change from small negative-large positive to large negative-small positive εHf(t) values. In addition, the inherited zircons of

Archean age (up to ~2877 Ma with a "crustal" model age as old as 3.85 Ga) and the very large negative εHf(t) values of zircons representing the timing of the host rock emplacement (up to −22.0 with Paleoproterozoic to Archean "crustal" model ages) in the central Lhasa subterrane actually require the presence of Paleoproterozoic to Achaean basement. This finding offers convincing evidence for the presence of Paleoproterozoic and Archean basement (up to 2877 Ma) beneath the central Lhasa sub-terrane. It follows that the central Lhasa subterrane once must have been a microcontinent. This continent had acted as a nucleus to which juvenile crust has been accreted in the northern and southern subterranes in the Phanerozoic during its journey of drift across the Tethyan Ocean basins and ultimate continental collision. (3) The syncollisional LVS volcanic rocks and the Gangdese batholith in southern Tibet with isotopically depleted characteristics (εNd(t) > 0) indicate that their primary sources are of mantle origin. The best source candidate in the broad context of Tethyan ocean closing and India–Asia collision is the remaining part of the Tethyan ocean crust. This ocean crust melts when reaching its hydrous solidus during and soon after the collision in the amphibolite facies, producing andesitic melts parental to the LVS volcanic rocks and the coeval batholith with inherited mantle isotopic signatures. Petrologic and geochemical studies of these rocks indicate that syncollisional felsic magmatism may in fact account for much of the net contribution to continental crust growth. These observations and reasoning plus the remarkable compositional similarity between the andesitic LDF and the model bulk continental crust corroborates the proposal that continental collision zones may be the primary sites of net continental crustal growth (juvenile crust) through process of syncollisional felsic magmatism (Mo et al. 2008; Niu et al. 2013).

6.2.3 Continental Collision and Mineralization

Hou and Cook (2009)reviewed the key metallogenic settings and resultant collision-related ore deposits in the Tibetan Orogen in response to the India-Asia collision starting in the early Cenozoic. Their synthesis leads to propose a new conceptual framework for Tibetan metallogenic systems. This framework includes three principal metallogenic epochs in the Tibetan orogen, and metallogenesis: (1) a main-collisional convergent setting (~65–41 Ma); (2) a late-collisional transform structural setting (~40–26 Ma); and (3) a post-collisional crustal extension setting (~25 Ma to the present), each forming more than three distinct types of ore deposits.

Especially, Hou et al. (2015), Hou and Wang (2018) combined the exploration of the lithospheric architecture determined by Hf isotope mapping with quantitative evaluation and prediction of regional metallogenic potential to reveal regional metallogenic regularity. They systematically studied the relationship between lithospheric and crustal structure of the Lhasa Terrance and mineralization, and found that the structure and composition of the upper lithosphere clearly correspond to the spatial distribution of metallogenic belt and major ore deposits. All the porphyry copper deposits are strictly confined to the northern and southern Lhasa subterranes, corresponding to the distribution of juvenile crust, while the Skarn Fe(-Cu) and Pb-Zn deposits and their metallogenic belts are strictly confined to the central Lhasa subterrane its boundary zones, corresponding to ancient mature reworked crust.

Jurassic porphyry Cu-Au deposits (such as Xiongcun) and Miocene porphyry Cu-Mo deposits (such as Qulong, Jiama, Tingong) are all located in the high εHf(t) and low $T_{DM}{}^{C}$ regions of the Southern Lhasa subterrane. Early Cretaceous porphyry Cu deposits (such as Duolong) and related polymetallic deposits are located in the high εHf(T) juvenile crust of the Northern Lhasa subterrane. So far, no porphyry copper deposits have been found in the central Lhasa subterrane. These facts indicate that the juvenile lower crust controls the formation of porphyry Cu deposits in orogenic belts.

The spatial distribution of Pb-Zn and porphyry Mo deposits related to granites is strictly restricted within the ancient central Lhasa subterrane. Large Pb-Zn deposits, such as Mengya'a, are concentrated in crustal blocks with the oldest Hf model ages ($T_{DM}{}^{C}$) or its boundary zones. The known porphyry Mo deposits (e.g., Bangpu) occur near ancient reworked crustal margins. These characteristics indicate that the spatial distribution of Pb-Zn-Mo deposits is related to the compositional heterogeneity and remelting/reworking of the ancient central Lhasa subterrane.

Hou et al. (2015) and Hou and Wang (2018) pointed out that that the nature and composition of the crust controlled the metal association, deposit type, the formation and development of metallogenic systems and the spatial placement of deposits. They further investigated the relationship between crust nature and mineralization from the aspects of mantle/crust affinity of ore-forming elements and redox conditions of mineralization of different types of deposits.

They further point out that Hf-Nd isotope mapping can be developed as a new means of effectively evaluating metallogenic potential and deciding strategic targets, by mean of delineating the spatial distribution of the crust blocks (juvenile, or reworking mature crust) in which different metallogenic systems/metallogenic belts rooted, and reveal the controlling factors of the lithospheric architecture on the spatial distribution of the mineralization. Thus, the control of crustal type and distribution on deposit type and distribution is a first-order control factor, which can quickly determine the prospecting strategic target areas with theoretical understanding built on the lithospheric architecture with Hf-Nd isotope mapping.

6.2.4 Continental Collision and Crustal Shortening

Crustal shortening is one of the effects of the India-Asia continental collision. Based on their research, Yin and Harrison (2000) suggested that at least 1400 km of north-south shortening has been absorbed by the orogen since the onset of the Indo-Asian collision at ~70 Ma. Significant crustal shortening, which leads to eventual construction of the Cenozoic Tibetan plateau, began more or less synchronously in the Eocene (50–40 Ma) in the Tethyan Himalaya in the south, and in the Kunlun Shan and the Qilian Shan some 1000–1400 km to the north; Dupont-Nivet et al. (2010)

believed that 2900 ± 600 km latitudinal convergence between India and Asia could be divided into 1100 ± 500 km within Asia and 1800 ± 700 km within India; Searle et al. (1987) believed that more than 500 km shortening was recorded in the foreland thrust zones of the Indian plate, south of the suture, and there was a 1000 km shortening occurred in Tibet, north of the suture; Jaeger et al. (1989) suggested that about 4000 km of continental shortening in the collision zone since India -Asia collision, probably absorbed jointly by crustal thickening and strike-slip extrusion.

6.3 The Uplift and Growth of the Tibetan Plateau

The surface uplift history of the Tibetan Plateau and Himalaya is among the most important topics in the geosciences because of its effect on regional and global climate in the Cenozoic, its influence on monsoon intensity, its influence on regional vegetation ecological model, biodiversity and watershed model, and its reflection of the dynamics of continental plateaus (Wang et al. 2008; An et al. 2006; Fang et al. 2020).

For a long time, researchers have been studying the uplift and growth of the Tibetan Plateau from different perspectives, and put forward their own understanding of the uplift and growth process of the Plateau, the driving force and environmental and resource effects, such as from sedimentary basin and petrofacies paleongeography (Wang et al. 2008; Zhang et al. 2008, 2010), biostratigraphy (Li et al. 2002; Wan 1990), paleobotany (Xu 1981; Xu et al. 1973), structural geology (Tapponnier et al. 2001; Zhang et al. 2004; Fielding 1996; Li et al. 2006; Wang and Zhang 2010; Wang et al. 2011a, b; Pan et al. 1990, 2004, 2009; Searle and Godin 2003; Zhong and Ding 1996; Pan 1999; Harrison et al. 1992; Yin et al. 1999; Coleman and Hodges 1995; Beaumont et al. 2004; Clark and Royden 2000), igneous petrology (Chung et al. 1998, 2005; Deng 2003; Deng et al. 1996, 2004a, b), geophysics (Zhao and Morgan 1985), quaternary geology—geomorphology (Li 1995; Shi and Liu 1964; Cui et al. 1996; Li et al. 1979), geography (Sun and Zheng 1998), paleoaltimetry (Molnar et al. 2006; Saylor et al. 2009; Garzione et al. 2004) and new technology applied to study the uplift of the Plateau (Polissa et al. 2009; Clark et al. 2010; Rowley et al. 2001; Wang 2007; Fang et al. 2020).

6.3.1 The Uplift History of the Tibetan Plateau

All researchers agree that the uplift and growth of the Tibetan Plateau is a complex and multi-stage process. However, there is considerable divergence in division of stage (Table 6.1).

Li et al. (1979) believed that the India-Asia collision in the Paleogene caused the Gangdese mountain rise and followed by planation. At this time, the greater Qinghai-Tibet region was climatically warm and humid with low altitude. In the

Table 6.1 Representative schemes for the division of uplift stages of the Tibetan Plateau. (Data from Li et al. (1979), Molnar et al. (1993), England and McKenzie (1982), England and Houseman (1989), Zhong and Ding (1996), Chung et al. (1998), Tapponnier et al. (2001), An et al. (2006), Deng et al. (1996, 2004a, b), Wang et al. (2008), Zhang et al. (2010), Wang (2014), Fang (2017))

	Paleogene		Neogene				Quat.
Li et al. (1979)	The Indo-Eurasian collision caused the Gangdese to rise and then flatten, the Tibet was low, hot and humid.		Miocene Himalaya began to uplift and deposited Sivalik group in the south. At the end of Pliocene, the plateau surface was 1000 m high, and another flatting occurred.	The environment in the early Pleistocene was steppe type, and the plateau surface reached to 2000 m high.	During the middle Pleistocene, tectonic movement and climate intensely changed, and MCT was intensely active, reaching to 3000 m high.	In the late Pleistocene, the plateau uplift to 4000 m, blocking the Indian Ocean monsoon. It is the critical period for natural landscape to approach modern pattern.	4300 m high
An et al. (2006)	The important growth events of the Tibetan Plateau are basically consistent with the key stages of Asian monsoon and aridity evolution.	25–20 Ma: The Himalayas and the Gangdese strongly uplifted, the Plateau expanded to the northeast, and the planetary wind system began to transform to a monsoon system.	16~14 Ma: The growth of Plateau accelerated, E-W extension started and the surrounding mountains and basins were deformed.	10–7 Ma: The growth of plateau continuously accelerated, the north and east of the Plateau obviously expanded, inland basins uplifted into mountains, and the monsoon-aridity system was basically established.	4.5–2.6 Ma: The Plateau rapid uplift by isostatic adjustment resulted in the formation of molasses basins around the plateau, and the formation of mountain-river landforms around the plateau and in western China.	1.6–0.7 Ma: The plateau non-uniformly deformed, faulting around the plateau aggravated.	

Zhang et al. (2010), Wang (2014)	The plateau uplift can be divided into five stages	60–35 Ma: continental collision; the Gangdese started uplifting	34–13 Ma: Intracontinental convergence compression, the Gangdese rapid uplift	12–8 Ma: Intracontinental isostatic adjustment, Tethyan Himalaya-Gangdese continued uplift, east-west extension, rapid tectonic uplift at 8 Ma	5 Ma: The plateau as a whole uplifted and the basins within and around the plateau shrink	3.5 Ma: The periphery of the plateau rapidly uplifted, giant conglomerate accumulated
Zhong and Ding (1996)	Four-stage uplift in northeast plateau	45–38 Ma: Compression folding shortening	27–25 Ma: Large block strike-slip rotation	18–13 Ma: Block differential uplift	3 Ma: extension, fast uplift	
Tapponier et al. (2001)	Oblique subduction extrusion, three-stage uplift and expansion towards northeast and east	Eocene plateau: it was bounded by Yarlung Zangbo in the south, Jinsha River in the north	Oligocene–Miocene plateau: bounded by the Jinsha River in the south, Kunlun Mountains in the north	Pliocene–Quaternary plateau: bounded by the Kunlun mountains in the south, Qilian Mountains in the north		Mechanism: Large strike-slip faults played a key role in controlling the development of the Tibetan Plateau from the beginning
Chung et al. (1998)	The Tibetan Plateau uplift diachronically	Rapid uplift began in the east from 40 Ma and then uplift of the west in 20 Ma				Mechanism: convective removal of the lower part of the thickened lithosphere

Miocene the Himalaya began to rise, and the Sivalik group was deposited in the southern foothills. The Tibetan plateau surface was only about 1000 m in elevation until Pliocene. Other phases of planation took place since the late Pliocene. At the end of Pliocene, the plateau rose rapidly evidenced by the formation of conglomerates in the southern foothills of the Himalayas and the plateau interiors. The environment of the Early Pleistocene was of steppe type, and the plateau surface reached to 2000 m in elevation. During the Middle Pleistocene, the tectonic movement and climate change of the Tibetan Plateau became intensified and large-scale glaciation was developed. Strong tectonic movement in late Middle Pleistocene was manifested by both the strong activity of the Main Boundary Thrust (MCT) in the southern foothills of the Himalayas and the faulting in the plateau interiors. The average elevation of the plateau reached 3000 m. Accelerated uplift of the Plateau took place in the late Pleistocene and achieved the present-day elevation of 4500–5000 m, with an average annual rise of >10 mm. Thus, the Plateau became an impassable obstacle to the Indian Ocean monsoon, causing intense dryness in the plateau interiors and glacier shrinking. This stage was the critical period when the natural landscape of the Plateau approached the modern pattern. Except for a few mountainous localities, the plateau is always flat and generally inclines from northwest to southeast. The overall integrity of the plateau is a strong proof that the Tibetan Plateau rises as an integral block.

An et al. (2006) identified six accelerated periods of uplift since the Miocene, i.e., 25–20 Ma, 16–14 Ma, 10–7 Ma, 4.5 Ma, 1.6–1.0 Ma and 0.7 Ma, among which the most important periods of tectonic activity were on 25–20 Ma and 10–7 Ma when the plateau had experienced rapid and large longitudinal and transverse growth. The acceleration around 3.6 Ma may be the result of tectonic movement and climate events.

Zhang et al. (2010, 2013a, b) and Wang et al. (2011a, b) believed that the Tibetan Plateau has experienced five episodes of rapid uplift in the Cenozoic: A (58–53 Ma), B (45–30 Ma), C (25–20 Ma), D (13–7 Ma) and E (since 5 Ma) based on the study of the sedimentary and tectonic deformation of nearly 100 Cenozoic residual basins in the Tibetan Plateau and cryogenic thermochronology.

Zhong and Ding (1996) believed that the plateau uplift was a multi-stage, non-uniform process and eastern Tibetan tectonic junction had undergone four-stage uplift of 45–38 Ma (compressive folding shortening deformation), 25–17 Ma (large-scale block strike-slip and rotation), 13–8 Ma (block differential uplift) and 3 Ma–present (extension and rapid uplift) based on measurements of apatite fission tracks.

Tapponnier et al. (2001) proposed a three-stage model of gradual northward expansion of the Tibetan Plateau (i.e., Eocene, Oligocene-Miocene, Pliocene-Quaternary).

Chung et al. (1998) suggested a diachronous uplift history for the Tibetan plateau, and that the convective removal of the lower lithosphere induced rapid uplift that began in the east at ~40 Ma and in the west ~20 Myrs later.

Deng et al. (1996, 2004a, b) proposed a three-stage (Oligocene, Miocene, and Pleistocene) growth model for the Tibetan Plateau with Gangdese-Qiangtang as the center extending both southwards and northwards based on the spatial and temporal distribution of igneous rocks and pointed out that rapid uplift of the Tibetan plateau was likely associated with the last orogenic episode since the Pleistocene.

Wang et al. (2008) believed that the uplift of the Tibetan Plateau started from the central part (Gangdese-Qiangtang) and formed a proto-plateau at 40 Ma and expanded stepwise throughout the Neogene toward its present southern and northern margins.

Fang (2017) evaluated the existing views and proposed that the uplift could be described in three main stages: 55–30 Ma, 30–25 Ma, and 8 Ma–present, respectively. In the first stage, the Qiangtang and Lhasa terranes may rose close to the present plateau elevation. In the second stage, the Qiangtang and Lhasa terranes may have firstly reached its present elevation accompanied by E-W extension with the Himalaya and Hoh Xil-Kunlun Mountains began to rise. In the last stage since 8 Ma, the southern edge (the Himalayas) and northern edge of the Plateau rapidly uplifted and finally developed into its present appearance.

When and where firstly reached the highest elevation is concerned by many authors. Rowley et al. (2001) established oxygen isotope paleoaltimetry model and believed that the Tethyan Himalaya and southern Tibet already reached their present elevation in the late Miocene (~10 Ma). Currie et al. (2005) estimated the paleoaltitude of the Oiyug Basin in southern Tibet based on the stable isotope composition of carbonate rocks, and believed that its altitude had already reached ~5200 m at 15 Ma. Saylor et al. (2009) recently analyzed the C-O isotopes of aragonite of carbonate rocks and mollusc fossil shells in the Zada Basin, southwest Tibet, and believed that this area was 4000–4500 m high in the time period of 9.2–1 Ma. Spicer et al. (2003) used vegetation altimeter to estimate the ancient elevation of the Namlin Basin and found that it reached 4689 m high at 15 Ma, implying that this area already arrived at its maximum elevation before 15 Ma. Molnar et al. (2010) believed that when the Indian plate and Eurasian plate first collided at 55–45 Ma, the Gangdese belt in southern Tibetan Plateau was just like the Andes mountains today.

6.3.2 The Pattern and Mechanism of Uplift and Growth of the Tibetan Plateau

In general, there are three views for the style and mechanism of uplift and growth of the Tibetan Plateau.

1. ***Crustal and lithospheric shortening, thickening and convective instability caused uplift of the plateau and growth of the plateau from south to center and then to north.***

England and Houseman (1989), Molnar et al. (1993) suggested that India-Asia collision led to successional crustal and lithospheric shortening and thickening of the Tibetan plateau, from southern plateau edge (collision frontal zone), propagating to central and north, and then convective removal of lower lithosphere and replacement by hot asthenosphere beneath the Tibetan Plateau caused and accelerated uplift of the plateau.

England and Houseman (1989) addressed that the tectonics of the plateau were dominated by north-south shortening in the Tertiary. For the last 5 Myrs or so, the plateau has been thinning by the mechanism of extension, resulting in north-south trending normal faults. An explanation for the extension comes from considering the thermal evolution of the thickened continental lithosphere. The lower part of the lithosphere consists of a thermal boundary layer, which, when thickened by horizontal shortening, becomes colder and denser than its surroundings. Convective instability of the thickened thermal boundary layer and replacement by hot asthenosphere would rapidly raise the surface elevation and gravitational potential energy of the overlying part of the lithosphere. The convective instability would happen in a short time compared with the collision time scale but would only occur after there had already been substantial thickening of the lithosphere. Numerical experiments show that for a range of lithospheric parameters, the increase in surface height (as much as 2 km) and of potential energy ($5–10 \times 10^{12}$ N m^{-1}) resulting from convective instability of the lower lithosphere are sufficient for east-west extension to replace north-south compression as the dominant feature of the stress field of the Tibetan plateau.

2. ***The oblique collision between the Indian plate and the Eurasian plate caused successive uplift and growth from south to north.***

Tapponnier et al. (2001) argued that the deformation and uplift of the Plateau were not caused by the continuous shortening and thickening of the crust and lithosphere and the subsequent uniformly rebound and uplift process, but by the oblique collision between the Indian plate and the Eurasian plate, and the deformation occurred mainly along the deep and large shear faults of the plateau margin. They proposed a three-stage model of gradual northward expansion of the Tibetan Plateau (i.e., Eocene Tibet, Oligo-Miocene Tibet and Pliocene-Quaternary Tibet). Firstly, the southern part of the Plateau (south of Tanggula Mountain) was uplifted in the Eocene and extruded southeastward. Then the Qiangtang block, the Hoh Xil and the Kunlun range were uplifted successively in the Oligocene and Miocene. Finally, the northeastern part of the plateau (the area between the Kunlun Mountain and the Qilian Mountain) uplifted in the Pliocene and Quaternary.

3. ***The uplift of Tibetan Plateau started in pre-Eocene time in the Lhasa and Qiangtang terranes and expanded stepwise throughout the Neogene toward its present southern and northern margins by multiple driving forces***

Wang et al. (2008) proposed a model for temporally and spatially differential surface-uplift history of the Tibetan Plateau, i.e., a model of plateau growth varying in time, from pre-India-Asia collision (e.g., 100 Ma ago) to gradual uplift after the India-Asia collision (e.g., 55 Ma ago) and to the more recent abrupt uplift (<7 Ma ago), varying in space, from northward stepwise growth of topography to simultaneous surface uplift across the plateau. The authors presented geologic and geophysical data from north-central Tibet, including magnetostratigraphy, sedimentology, paleocurrent measurements, and $^{40}Ar/^{39}Ar$ and fission-track geochronology, and suggested that the central plateau (the Lhasa and southern Qiangtang terranes) was uplifted to result in the proto-plateau by the Late Paleogene (at 40 Ma), and regions south and north of the central plateau gained elevation significantly later. In the Eocene, the northern boundary of the proto-plateau was in the region of the Tanggula Shan. A high proto-Tibetan Plateau may have contributed to climatic changes farther north in central Asia. Intriguingly, this timing also corresponds to a period of pronounced global cooling and changes in ocean chemistry. With the continuation of India-Asia convergence, the plateau gradually expanded southwards and northwards in two directions: (1) the Himalayas rose in the Neogene and the Hoh Xil Basin rose to the present elevation at ~20 Ma, and (2) the Qilian Mountains rapidly uplifted in the late Cenozoic, forming the present Tibetan Plateau. These mountain ranges constitute the modern southern and northern margins, respectively, of the Tibetan Plateau. That is, the plateau elevation started in pre-Eocene time in the Lhasa and Qiangtang terranes and expanded throughout the Neogene toward its present southern and northern margins in the Himalaya and Qilian Shan (Fig. 6.4).

Deng et al. (1996, 2004a, b) proposed a three-stage growth model of the Tibetan Plateau with Gangdese-Qiangtang as the center extending respectively northward and northward based on the spatial and temporal distribution of igneous rocks (Fig. 6.5). Following the India–Asia 'hard' collision (~45 Ma), the Tibetan Plateau entered an intracontinental stage. An obvious feature is that apparent

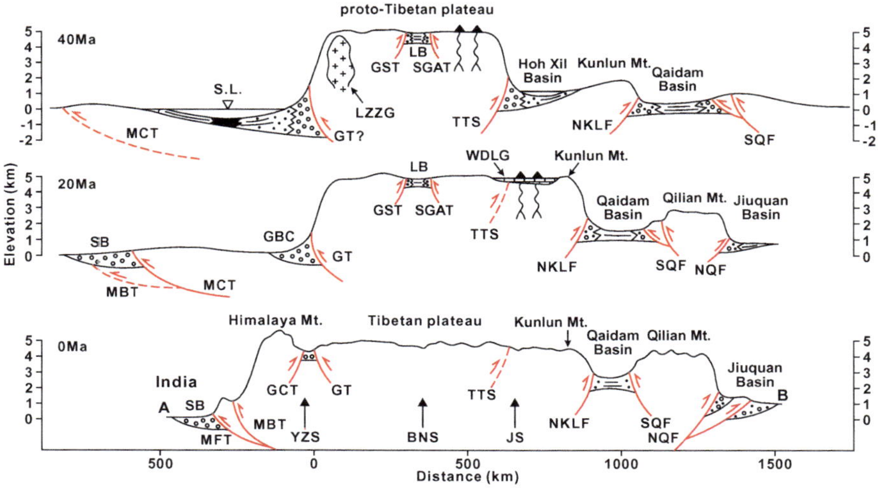

Fig. 6.4 Schematic paleogeographic cross sections (southwest-to-northeast) of the Tibetan Plateau and Himalaya as a function of distance with respect to the present-day Yarlung Tsangpo suture zone (YZS) (from Wang et al. 2009). *MBT* main boundary thrust, *GCT* great counter thrust, *LZZG* Lingzizhong group volcanics, *NKLF* north Kunlun fault, *WDLG* continental Wudaoliang Group, *TTS* Tanggula thrust system, *NQF* north Qilian Mountain fault, *LB* Lunpola basin, *GBC* Gangrinboche conglomerates, *SB* Siwalik foreland basin, *MCT* main central thrust, *GST* Gaize-Siling Tso thrust, *JS* Jinsha River suture zone, *SGAT* Shiquanhe-Gaize-Amdo thrust, *YZS* Yarlung Tsangpo suture zone, *SL* sea level, *MFT* main frontal thrust, *BNS* Bangong-Nujiang suture zone, *SQF* the south Qilian Mountain fault, *GT* Gangdese thrust

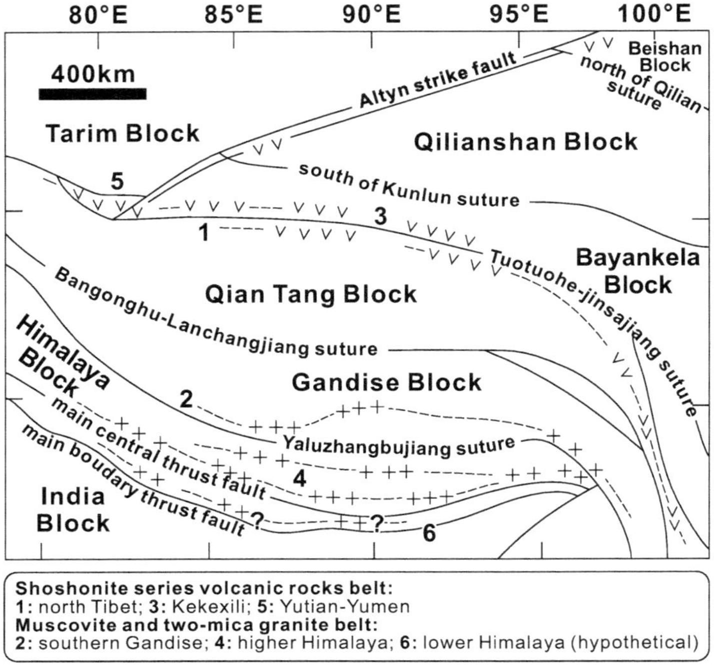

Fig. 6.5 Paired zones of the Qinghai-Tibet-Himalaya intracontinental igneous rocks: shoshonite series belt vs muscovite/ two-mica granite belt (after Deng et al. 1996). Shoshonite series volcanic rocks belt: 1: north Tibet; 3: Kekexili; 5: Yutian-Yumen Muscovite and two-mica granite belt: 2: southern Gandise; 4: higher Himalaya; 6: lower Himalaya (hypothetical). Based on the spatial and temporal distribution of paired zones of intracontinental igneous rocks, Deng et al. (1996, 2004a, b) proposed a three-stage growth model of the Tibetan Plateau with Gangdese-Qiangtang as the center extending respectively northwards and northwards

pairing of igneous belts (shoshonite-muscovite granite pairs) developed on the margin of the Plateau in each growing phase. Thus, these pairing of igneous belts may be used to define the boundaries of the Plateau at that time. The first episode occurred in the Oligocene, with southern and northern boundaries corresponding to the Yarlung Zangbo and Tuotuohe-Jinsha Jiang sutures, respectively. The second episode developed mainly in the Miocene, with southern and northern boundaries corresponding to the MCT (the Main Central thrust) and southern Kunlun suture, respectively, which is important for the formation of the main body of the Plateau. The third episode occurring mainly in the Pleistocene, with southern and northern boundaries corresponding to the MBT (the Main Boundary thrust) and Altyn Tagh fault, respectively, is considered to be the main stage of formation of the Tibetan Plateau. All this shows the migration of plateau boundaries northwards and southwards respectively with time, reflecting north-south lateral growth of the Plateau. However, the south boundaries and the north ones are tectonically different according to the origin and tectonic settings of muscovite/two-mica granites and shoshonitic series. While the south boundaries marked by muscovite/two-mica granites imply the northward intracontinental subduction and a coupling of hot crust with cold mantle as confirmed by recent deep seismic reflection data (Zhao and Nelson 1993), shoshonitic volcanism along the northern margins suggests lateral shortening and distributed thickening as a result of lithosphere convergence and a coupling of cold crust with hot mantle. Rapid uplift of the Tibetan plateau in the last episode was probably associated with the beginning of the collapse in the interior (the Gangdese) and compression on the margins, resulting in the formation of N-S trending extensional basins and lithosphere thinning of 200–120 km. Thus, the mechanism of rapid uplift of the Tibetan Plateau is likely the combination of compressional stress and mountain root buoyance, ultimately controlled by the India-Asia collision and continued convergence.

6.3.3 Environmental Effects of the Tibetan Plateau Uplift

There have been many studies on the environmental effects of the Tibetan Plateau uplift from varying perspectives (Shi et al. 1999; Li 1995; Molnar et al. 1993; Liu et al. 1998; Li and Fang 1998; Li et al. 2001; An et al. 2001, 2006; Harris 2006; Hong et al. 2007).

Li et al. (1979), Li (1995), Li and Fang (1998) pointed out that strong tectonic movements occurred on the Tibetan Plateau at 3.6, 2.4, 1.6 Ma. At ~2.4 Ma, the plateau had reached >2000 m elevation, leading to the formation of East Asian monsoon and the deposition of large amount of loess in the middle reaches of the Yellow River, forming the Loess plateau. The Yellow River is a product of the uplift of the Tibetan Plateau. With the plateau rising intermittently, the Yellow River produced headward erosion and corresponding deep incision. The multi-stage terraces in the upper Yellow River reaches developed at 1.6 Ma, 1.5 Ma, 1.2 Ma, 0.6 Ma and 0.15 Ma are the reflection of the episodic plateau uplift.

Molnar et al. (1993) suggested that convective removal of lower lithosphere beneath the Tibetan Plateau could account for a rapid increase in the mean elevation of the Tibetan Plateau of 1000 m or more within a few million years. The initiation of normal faulting in Tibet at about 8 (±3) Ma suggests that the plateau underwent an appreciable increase in elevation at that time. Meanwhile, the climate of the equatorial Indian Ocean and southern Asia changed at ~6–9 Ma, that is, monsoonal winds apparently strengthened, with northern Pakistan becoming more arid and increased weathering in the eastern Himalaya. Mantle dynamics beneath mountain belts not only profoundly affect tectonic processes near and far from the mountain belts, but also play an important role in modulating regional and global climates.

An et al. (2006) focused on six accelerated periods of uplift since the Miocene, i.e., 25–20 Ma, 16–14 Ma, 10–7 Ma, 4.5–2.6 Ma, 1.6–1.0 Ma and 0.7 Ma, and pointed out that these periods of growth events on the Tibetan Plateau are generally consistent with the key stages of Asian monsoon and aridity evolution. In 24–22 Ma, the Himalayas and the Gangdese uplifted rapidly and the entire plateau grew eastward and northward. This event has caused significant changes: (1) the continental environment of China has changed from the planetary wind system to the monsoon wind system, (2) the inland arid zones and the east Asian monsoon regions began to differentiate, (3) dust accumulation began to appear to form the loess plateau, (4) the summer monsoon developed, (5) the eastern humid region expanded northward and the south China Sea continued to open up. At 16–14 Ma, the arid area in northwest China expanded further and the monsoon intensified, which indicates an important growth event of the Tibetan Plateau. In 10–7 Ma, (1) the major growth of the Tibetan Plateau took place, (2) the east-west extension stated, (3) aeolian red clay accumulation became widely developed on the loess plateau, and (4) the aridity of the inland intensified. The pattern of monsoon-arid environment system in China was established this time and the Indian Ocean summer monsoon strengthened. From 4.5 to 2.6 Ma, the Tibetan Plateau grew northward and northeastward. The plateau growth and the arrival of the Great Ice Age led to the formation of the modern monsoon-arid environment pattern in China, and the Quaternary accumulation of the loess-paleosoil sequences. The pulsating events after 1.3–1.0 Ma and 0.6 Ma showed a significant strengthening of summer monsoon and a trend of inland drying.

In summary, in 24–20 Ma and 16–14 Ma, the Himalayas and the central and southern Tibetan Plateau rose and grew, resulting in the initial formation and development of the Asian monsoon system, and in the appearance and expansion of inland arid areas in northwest China. In the periods of 10–7 Ma and 3.6–2.6 Ma, the Tibetan Plateau, especially its northern and northeastern regions, experienced two significant tectonic uplift and growth events, resulting in the intense development of east Asian winter and summer monsoon, the expansion of inland arid zone and significant intensification of drought. With the advent of the Great Ice Age in the Northern Hemisphere after 3.6 Ma, the modern monsoon-arid environment pattern formed in China (Fig. 6.6).

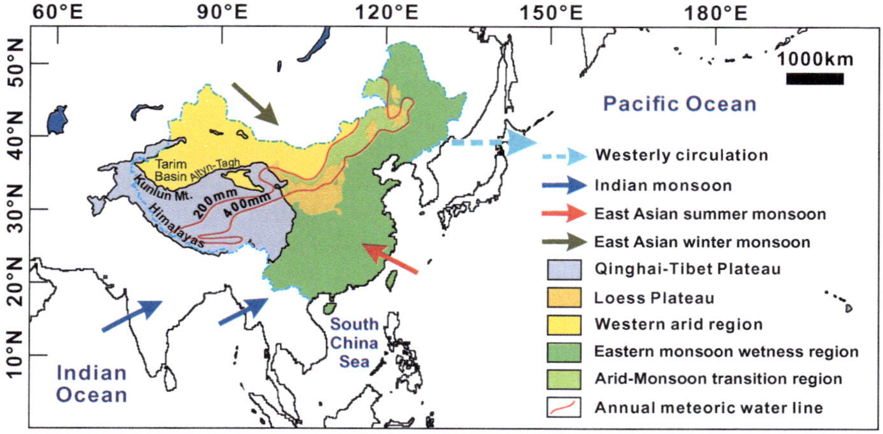

Fig. 6.6 Sketch map of the modern monsoon-arid environment (modified from An et al. 2006). There have been six accelerated periods of uplift of the Tibetan Plateau since the Miocene, i.e., 25–20 Ma, 16–14 Ma, 10–7 Ma, 4.5–2.6 Ma, 1.6–1.0 Ma and 0.7 Ma, which well correspond to the key episodes of Asian monsoon and aridity evolution

6.4 Conclusion

In this chapter we have discussed the processes of the India-Asia continental collision, and the uplift and growth of the Tibetan Plateau.

1. The closure of the Neo-Tethyan ocean and the India-Asia collision are the prerequisite and basis for the formation of the Tibetan Plateau. Based on various lines of evidence, the initial India-Asia collision took place at 65 ± 5 Ma with the main phase of collision at ~45 ± 5 Ma, and then followed by post-collisional processes. The main phase of the collision is characterized by the emplacement of the volumetrically significant Gangdese batholith and widespread of high temperature volcanic eruption with asthenospheric mantle material contribution in terms of mass and energy. The post-collisional stage is closely related to the uplift and growth of the Tibetan Plateau, within which several major crust/mantle deep processes influenced on the uplift and growth of the Plateau.
2. The India-Asia collision and the continued convergence gave long-term effects on the Plateau. The formation of the extremely thickened Tibetan crust is closely related to the India-Asia collision by means of combined contributions of asthenospheric mantle materials and tectonic compression. The collisional and post-collisional igneous rocks in the Tibetan Plateau indeed show significant mantle contributions. It is proved that two types of continental crust coexist in the Tibetan Plateau, the juvenile crust and reworked mature crust. The crust of the northern and southern Lhasa subterranes is dominated by juvenile crust, while the crust of the central Lhasa subterrane is dominated by reworked mature crust. Syncollisional felsic magmatism may in fact account for much of the net contribution to continental crust growth. At least 1400 km of north-south shortening has been absorbed since the onset of the India-Asian collision. Besides, it has been found that the structure and composition of the upper lithosphere show clear controls on the distribution of metallogenic belts and major ore deposits.
3. The surface uplift history of the Tibetan Plateau and Himalaya is among the most important topics in geosciences because of its effects on regional and global climate variation since the Cenozoic, monsoon intensity, regional vegetation ecological models, biodiversity and watershed models. In general, there are three general views on the pattern and mechanism of uplift and growth of the Plateau: (1) crustal and lithospheric shortening, thickening and convective instability caused uplift of the plateau and growth of the plateau from south to center and then to north; (2) the oblique collision between the Indian plate and the Eurasian plate caused successive uplift and growth from south to north; (3) the uplift of the Plateau started in pre-Eocene time from the Lhasa and Qiangtang terranes and expanded stepwise throughout the Neogene toward its present southern and northern margins by multiple driving forces. There have been six accelerated periods of uplift of the Tibetan Plateau since the Miocene, i.e., 25–20 Ma, 16–14 Ma, 10–7 Ma, 4.5–2.6 Ma, 1.6–1.0 Ma and 0.7 Ma, which well correspond to the key episodes of Asian monsoon and aridity evolution.

Acknowledgements Many thanks are due to Drs. Pan Guitang, Xiao Qinghui, Guo Tieying, Zhao Chonghe, Lu Fengxiang, Martin Flower, Don Depaolo, Victor Mocanu, John Mahoney, Paul Robinson, Yildirim Dilek for fruitful discussion on tectonics and magmatism of the Tibetan Plateau and participation of field work, to my students and collaborators Chen Meihua, Badengzhu, Bi Xianmei, Ding Yi, Dong Fnagliu, Dong Xin, Fan li, Fu Xiaofang, Gao Xue, He Changxiang, He Wenyan, Hou Liwei, Hu Xiangsheng, Huang Guicheng, Huang Xingkai, Jiang Wan, Kang Dongyan, Ke Shan, Li Xiaowei, Li Yong, Li Youguo, Liao Zhongli, Lin Peiying, Liu Chengdong, Liu Xiangpin, Liu Yanbin, Liu Yunhua, Luo Mingfei, Luo Zhaohua, Miao Zhuang, Qiu Ruizhao, Shen Shangyue, Tan Jin, Tian Shihong, Wei Qirong, Wang Huichu, Wang Liangliang, Wang Liquan, Wang Min, Wang Yong, Wei Ping, Xiao Run, Xie Guogang, Xue Yingxi, Yang Guilai, Yang Kaihui, Ye Song, Yu Junchuan, Yu Xuehui, Yuan Wanming, Zhai Qingguo, Zhang Aikui, Zhang Baomin, Zhang Bo, Zhang Fengqin, Zhang Shuangquan, Zhang Wenqin, Zhang Xueting, Zhang Zeming, Zhao Haibin, Zhao Xin, Zhou Hongying, Zhou Su, Zhu Qinwen and Zhu Xiangping for their field and laboratory assistance in the Qinghai-Tibet Plateau and Jinshajiang-Nujiang-Lancangjiang region over the past decades. The author also highly appreciate all the researchers' work quoted in the chapter text. This work have been supported by NSFC key project (40830317), NSFC project (J20114), National 973 Projects (2009CB421002), Key project of China Geological Survey (1212010610104,12120112209).

References

Achache J, Courtillot V, Xiu ZY (1984) Paleogeographic and tectonic evolution of southern Tibet since middle cretaceous time: new paleomagnetic data and synthesis. J Geophys Res 89:10311–10339

Aitchison JC, Ali JR, Davis AM (2007) When and where did India and Asia collide. J Geopgys Res 112:1–19

Allégre CJ, Courtillot V, Tapponnier P, Hirn A, Mattauer M, Coulon C, Jaeger JJ, Achache J, Schärer U, Marcoux J, Burg JP, Girardeau J, Armijo R, Gariépy C, Göpel C, Li TD, Xiao XC, Chang CF, Li GQ, Lin BY, Teng JW, Wang NW, Chen GM, Han TL, Wang XB, Deng WM, Sheng HB, Cao YG, Zhou J, Qiu HG, Bao PS, Wang SC, Wang BX, Zhou YX, Xu RH (1984) Structure and evolution of the Himalaya-Tibet orogenic belt. Nature 307:17–22

Altherton MP, Petford N (1993) Generation of sodium- rich magmas from newly underplated basaltic crust. Nature 362(6416):144–146

An ZS, Kutzbach JE, Prell WL, Porter SC (2001) Evolution of Asian monsoons and phased uplift of the Himalaya-Tibetan plateau since late Miocene times. Nature 411:62–66

An ZS, Zhang PZ, Wang EQ et al (2006) Evolution of monsoon-aridity environment in China and growth of the Tibetan plateau since Miocene. Q Res 26(5):678–693

Argand E (1924) La tectonicgue de I'Asie. Proc 13th Int Geol Congr Brussels 7, 171

Barbarin B (1996) Genesis of two main types of peralumineous granitoids. Geology 24:295–298

Beaumont C, Jamieson RA, Nguyen MH et al (2004) Crustal channel flows: 1. Numerical models with applications to the tectonics of the Himalayan-Tibet orogen. J Geophys Res 109(B6)

Beck RA, Burbank DW, Sercomoe WJ et al (1995) Stratigraphic evidence for an early collision between northwest India and Asia. Nature 373:55–58

Ben Othman D, Polve M, Allegre CJ (1984) Nd-Sr isotopic composition of granulites and constraint on the evolution of the lower continental crust. Nature 307:510–515

Besse J, Courtillot V, Pozzi JP, Westphal M, Zhou YX (1984) Palaeomagnetic estimates of crustal shortening in the Himalayan thrusts and Zangbo suture. Nature 311:621–626

Bureau of Geology and Mineral Resources of Xizang Autonomous Region (1993) Regional geology of Xizang (Tibet). Geological Publishing House, Beijing, pp 450–462

Cai F, Ding L, Yue Y (2011) Provenance analysis of upper cretaceous strata in the Tethys Himalaya, southern Tibet: implications for timing of India-Asia collision. Earth Planet Sci Lett 305:195–206

Cao Y, Kang ZQ, Xu JF et al (2020) Geochronology, geochemistry and geological significance of volcanic rocks of the Dianzhong formation, Shiquanhe area, Western Lhasa Block. Earth Sci 45(5):1573–1592

Chappell BW, White AJR (1992) I and S-type granites in the Lachlan fold belt. Trans R Soc Edinburgh: Earth Sci 83:1–26

Chen Y, Cogne JP, Courtillot V, Tapponnier P, Zhou XY (1993) Cretaceous paleomagnetic results from western Tibet and tectonic implications. J Geophys Res 98:17981–18000

Chen JS, Huang BC, Sun LS (2010) New constraints to the onset of the India-Asia collision: paleomagnetic reconnaissance on the Linzizong Group in the Lhasa Block, China. Tectonophysics 489:189–209

Chung SL, Lo CH, Lee TY, Zhang Y, Xie Y, Li X, Wang K, Wang P (1998) Diachronous uplift of the Tibetan plateau starting 40 Myr ago. Nature 394:769–773

Chung SL, Liu DY, Ji JQ, Chu MF, Lee HY, Wen DJ, Lo CH, Lee TY, Yian Q, Zhang Q (2003) Adakites from continental collision zones: melting of thickened lower crust beneath southern Tibet. Geology 31(11):1021

Chung SL, Chu M, Zhang Y, Xie Y, Lo CH, Lee TY, Lan CY, Li X, Zhang YQ, Wang Y (2005) Tibetan tectonic evolution inferred from spatial and temporal variations in post-collisional magmatism. Earth Sci Rev 68:173–196

Clark MK, Royden L (2000) Topographic ooze: building the eastern margin of Tibet by lower crustal flow. Geology 28:703–706

Clark MK, Farley KA, Zheng DW et al (2010) Early Cenozoic faulting of the northern Tibetan plateau margin from apatite (U–Th)/He ages. Earth Planet Sci Lett 296:78–88

Coleman M, Hodges K (1995) Evidence for Tibetan plateau uplift before 14 Myr ago from a new minimum age for east west extension. Nature 374:49–52

Condie KC (1982) Plate tectonics and crustal evolution. Pergamon, New York

Cui ZJ, Gao QZ, Liu GN et al (1996) Planation surface, paleokarst and uplift of the Tibetan plateau. Sci China 26(4):378–385

Currie BS, Rowley DB, Tabor NJ et al (2005) Middle Miocene paleoaltimetry of southern Tibet: implications for the role of mantle thickening and delamination in the Himalayan orogen. Geology 33:181–184

DeCelles PG, Kapp P, Gehrels GE, Ding L (2014) Paleocene-Eocene foreland basin evolution in the Himalaya of southern Tibet and Nepal: implications for the age of initial India-Asia collision. Tectonics 33:824–849

Defant MJ, Drummond MS (1990) Derivation of some modern arc magmas by melting of young subducted lithosphere. Nature 347(6294):662–665

Deng WM (2003) Cenozoic volcanic activity and its tectonic setting in western China—formation mechanism of volcanic rocks in Qinghai-Tibet and adjacent areas. Earth Sci Front 10(2):471–478

Deng JF (2021) Chapter 6. Look at the Tibetan plateau on a larger scale. In: Mo X, Zhu D, Deng J et al (eds) The Mesozoic-Cenozoic tectonomagmatism of the Tibetan plateau. Guangdong Science and Technology Press, Guangzhou, pp 151–166

Deng JF, Zhao HL, Mo X, Luo ZH (1996) Root- column tectonics of the Chinese continent: a key to continental dynamics. Geological Publishing House, Beijing, pp 1–110

Deng JF, Mo X, Zhao HL, Wu ZX, Luo ZH, Su SG (2004a) A new model for the dynamic evolution of Chinese lithosphere: continental roots–plume tectonics. Earth Sci Rev 65:223–275

Deng JF, Luo ZH, Su SG, Mo X, Yu BS, Lai XY, Chen HW (2004b) Petrogenetic, tectonic environment and mineralization. Geological Publishing House, Beijing, 381pp

DePaolo DJ (1981) A neodymium and strontium isotopic study of the Mesozoic calc-alkaline granitic batholiths of the Sierra Nevada and peninsular ranges, California. J Geophys Res Solid Earth 86(B11):10470–10488

DePaolo DJ (1985) Isotopic studies of processes in mafic magma chambers: I. The Kiglapait intrusion, Labrador. J Petrol 26(4):925–951
DePaolo DJ, Perry FV, Baldridge WS (1992) Crustal versus mantle sources of granitic magmas: a two-parameter model based on Nd isotopic studies. Earth Environ Sci Trans R Soc Edinb 83(1/2):439–446
Dewey JF, Bird JM (1970) Mountain belts and the new global tectonics. J Geophys Res 75(14):2625–2647
Dewey JF, Burke K (1973) Tibetan Variscan and Precambrian basement reactivation: products of continental collision. J Geol 81:683–692
Dewey JF, Cande S, Pitman WC (1989) Tectonic evolution of the India-Eurasia collision zone. Ecol Geol Helv 82:717–734
Didier J, Baibarin B (1991) Enclaves and granite petrology. Elsevier, Amsterdam, p 393
Ding L (2003) Discovery of Paleocene deep-water deposits and radiolarian faunas in the Yarlung Zangbo suture in Tibet and constraints on foreland basin evolution. Sci China: Earth Sci 33(1):47–58
Ding L, Kapp P, Wan X (2005) Paleocene-Eocene record of ophiolite obduction and initial India-Asia collision, south central Tibet. Tectonics 24:TC3001
Ding L, Cai FL, Zhang Q, Zhang LY, Xu Q, Yang D, Liu DL, Zhong DL (2009) Foreland basin system and tectonic evolution of Gangdese-Himalayan collision orogenic belt. Chin J Geol 44:1289–1311
Ding L, Maksatbek S, Cai FL, Wang HQ, Song PP, Ji WQ, Xu Q, Zhang LY, Muhammad Q, Upendra B (2017) Processes of initial collision and suturing between India and Asia. Sci China Earth Sci 47:293–309
Ding L, Kapp P, Cai F, Garzione C, Xiong Z, Wang H, Wang C (2022) Timing and mechanisms of Tibetan plateau uplift. Nat Rev Earth Environ 3:652–667. https://doi.org/10.1038/s43017-022-00318-4
Dong SB (1999) Some discussion on the recent development of granite petrogenesis. In: Dong SB (ed) Collected works of Dong Shenbao. Peking University Press, Beijing, pp 246–325
Dong GC (2002) Linzizong volcanic rocks in Linzhou Basin, Tibet and implications for India-Asia continental collision (Ph D dissertation). China University of Geosciences, Beijing
Dong GC, Mo X, Zhao ZD, Guo TY, Wang LL, Chen T (2005) Geochronologic constraints on the magmatic underplating of the Gangdese belt in the India-Eurasia collision: evidence of SHRIMP II zircon U-Pb dating. Acta Geol Sin 79:787–794
Dong GC, Mo X, Zhao ZD, Zhu DC, Song YT, Wang L (2008) Gabbros in southern Gangdese belt of Tibet and the implication for crust-mantle interaction. Acta Petrol Sin 24(2):203–210
Dupont-Nivet G, Lippert P, van Hinsbergen D, Meijers M, Kapp P (2010) Paleolatitude and age of the indo-Asia collision. EGU General Assembly, Geophysical Research Abstracts 12, EGU2010-9648
England P, Houseman G (1989) Extension during continental convergence, with application to the Tibetan plateau. J Geophys Res 94(B12):17561–17579
England P, McKenzie D (1982) A thin viscous sheet model for continental deformation. Geophys J Int 70(2):295–321
Ensele G, Liu B, Durr SB (1994) The Xigaze forearc basin: evolution and facies architecture (Cretacous, Tibet). Sediment Geol 90:1–32
Fang XM (2017) Phase division of uplift of the Tibetan plateau. Sci Technol Rev 35(6):42–50
Fang XM, Dupont-Nivet G, Wang C, Song C, Meng Q, Zhang W, Nie J, Zhang T, Mao Z, Chen Y (2020) Revised chronology of central Tibet uplift (Lunpola Basin). Sci Adv:1–10
Fernandez AN, Barbarin B (1991) Relative reology of coeval mafic and felsic magmas: nature of resulting interaction processes and shape and mineral fabrics of mafic microgranular enclaves. In: Didier J, Barbarin B (eds) Enclaves and granite petrology. Elsevier, Amsterdam, pp 263–276
Fielding EJ (1996) Tibet uplift and erosion. Tectonophysics 260:55–84
Flower M, Tamaki K, Nguyen H (1998) Mantle extrusion: a model for dispersed volcanism and DUPAL-like asthenosphere in East Asia and the Western Pacific. In: Flower MFJ et al

(eds) Mantle dynamics and plate interaction in East Asia. Geodynamics series. American Geophysical Union, pp 67–88
Flowerdew MJ, Chew DM, Daly JS et al (2009) Hidden Archaean and Palaeoproterozoic crust in NW Ireland? Evidence from zircon Hf isotopic data from Granitoid intrusions. Geol Mag 146(6):903–916
Foley S, Venturelli G, Green D, Toscani L (1987) The ultrapotassic rocks: characteristics, classification and constraints for petrogenetic models. Earth Sci Rev 24:81–134
Gaetani M, Garzanti E (1991) Multicyclic history of the northern India continental-margin (northwestern Himalaya). AAPG Bull 75:1427–1446
Garzanti E, Baud A, Mascle G (1987) Sedimentary record of the northward flight of India and its collision with Eurasia (Ladakh Himalaya, India). Geodin Acta 1(4–5):297–312
Garzione CN, Dettman DL, Horton BK et al (2004) Carbonate oxygen isotope paleoaltimetry: evaluating the effect of diagenesis on paleoelevation estimates for the Tibetan plateau. Palaeogeogr Palaeoclimatol Palaeoecol 212:119–140
Gill J (1981) Orogenic andesites and plate tectonics. Springer, Berlin, 390 pp
Griffin W, Wang X, Jackson SE et al (2002) Zircon chemistry and magma mixing, SE China: in-situ analysis of Hf isotopes, Tonglu and Pingtan igneous complexes. Lithos 61(3/4):237–269
Harris N (2006) The elevation history of the Tibetan plateau and its implications for the Asian monsoon. Palaeogeogr Palaeoclimatol Palaeoecol 241:4–15
Harrison TM, Copeland P, Kidd WSF et al (1992) Raising Tibet. Science 255:1663–1670
He S, Kapp P, DeCelles PG, Gehrels GE, Heizler M (2007) Cretaceous-tertiary geology of the Gangdese arc in the Linzhou area, southern Tibet. Tectonophysics 433:15–37
Hong HL, Li ZH, Xue HJ et al (2007) Oligocene clay mineralogy of the Linxia Basin: evidence of paleoclimatic evolution subsequent to the initial-stage uplift of the Tibetan plateau. Clay Clay Miner 55(5):492–505
Hou ZQ, Cook N (2009) Metallogenesis of the Tibetan collisional Orogen: a review and introduction to the special issue. Ore Geol Rev 36:2–24
Hou ZQ, Wang T (2018) Isotopic mapping and deep material probing (II): imaging crustal architecture and its control on mineral systems. Earth Sci Front 25(6):20–41
Hou ZQ, Gao YF, Qu XM, Rui ZY, Mo X, Zhao ZD (2004) Origin of adakitic intrusives generated during mid-Miocene east-west extension in southern Tibet. Earth Planet Sci Lett 220:139–155
Hou ZQ, Zheng Y, Yang ZM, Rui ZY, Zhao ZD, Jiang SH, Qu XM, Sun QZ (2013) Contribution of mantle components within juvenile lower – crust to collisional zone porphyry Cu Systems in Tibet. Mineral Deposita 48(2):173–192
Hou ZQ, Duan LF, Lu YJ, Zheng YC, Zhu DC, Yang ZM, Yang ZS, Wang BD, Pei YR, Zhao ZD, McCuaig TC (2015) Lithospheric architecture of the Lhasa terrane and its control on ore deposits in the Himalayan-Tibetan Orogen. Econ Geol 110(6):1541–1575
Hu XW, Ma RZ, Tao XF et al (2007) Lithogeochemical characteristics and tectonic setting of volcanic rocks of Dianzhong formation in the Coqen area, Tibet, China. J Chengdu Univ Technol 34(1):15–27
Hu X, Garzanti E, Moore T, Raffi I (2015) Direct stratigraphic dating of India-Asia collision onset at the Selandian (middle Paleocene, 59±1Ma). Geology 43:859–862
Hu X, Garzanti E, Wang J, Huang W, An W, Webb A (2016) The timing of India-Asia collision onset-facts, theories, controversies. Earth Sci Rev 160:264–299
Hu X, Wang JG, An W, Garzanti E, Li J (2017) Constraining the timing of the India-Asia continental collision by the sedimentary record. Sci China Earth Sci 47:261–283
Huang BC, Chen JS, Yi ZY (2010) Paleomagnetic discussion of when and where India and Asia initially collided. Chin J Geophys 53(9):2045–2058
Huppert HE, Sparks RSJ (1988) The generation of granitic magmas by intrusion of basalt into continental crust. J Petrol 29(3):599–624
Jaeger JJ, Courtillot V, Tapponnier P et al (1989) Paleontological view of the ages of the Deccan traps, the cretaceous/tertiary boundary and the India-Asia collision. Geology 17:316–319
Jahn BM, Wu FY, Chen B (2000) Massive granitoid generation in Central Asia: Nd isotope evidence and implication for continental growth in the Phanerozoic. Episodes 23(2):82–92

Ji WQ, Wu FY, Chung SL et al (2009) Zircon U-Pb geochronology and Hf isotopic constraints on on petrogenesis of the Gangdese batolith, southern Tibet. Chem Geol 262:229–245

Jiang W, Mo X, Zhao CH, Guo TY, Zhang SQ (1999) Geochemical characteristics of granites and their micromafic enclaves in central Gangdese in the Tibetan plateau. Acta Petrol Sin 15(1):89–97

Kay RW (1978) Aleutian Magnesian andesites: melts from subducted Pacific Ocean crust. J Volcanol Geotherm Res 4(1/2):117–132

Kemp AIS, Hawkesworth CJ, Paterson BA et al (2006) Episodic growth of the Gondwana supercontinent from hafnium and oxygen isotopes in zircon. Nature 439(7076):580–583

Kind R, Ni J, Zhao W et al (1996) Evidence from earthquake data for a partially molten crustal layer in southern Tibet. Science 274(5293):1692–1694

Klootwijk CT, Conaghan PJ, Powell C et al (1985) The Himalayan arc: large scale continental subduction, oroclinal bending and back-arc spreading. Earth Planet Sci Lett 75:167–183

Klootwijk CT, Gee FS, Peirce JW, Smith GM, McFadden PL (1992) An early India contact: paleomagnetic constraints from Ninetyeast ridge, ODP leg 121. Geology 20:395–398

Le Bas MJ, Streckeisen AL (1991) The IUGS systematics of igneous rocks. J Geol Soc 148(5):825–833

Le Fort P (1996) Evolution of the Himalaya. In: Yin A, Harrison TM (eds) The tectonics of Asia. Cambridge University Press, New York, pp 95–106

Le Maitre RW, Streckeisen A, Zanettin B, Le Bas MJ, Bonin B (2020) Igneous rocks: a classification and glossary of terms. Cambridge University Press, Cambridge

Lee TY, Lawver LA (1995) Cenozoic plate reconstruction of Southeast Asia. Tectonophysics 251:85–138

Leech ML, Singh S, Jain AK, Klemperer SL, Manickavasagam RM (2005) The onset of India-Asia continental collision: early, steep subduction required by the timing of UHP metamorphism in the western Himalaya. Earth Planet Sci Lett 234:83–97

Leigeois JP (1998) Some words on the post-collisional magmatism. Lithos 45:15–57

Li J (1995) Three stages of uplift of Tibetan plateau and the age, height, geomorphic environment of planation planes. China Environmental Science Press, Beijing

Liu AL (2020) Spatial and temporal comparison of the Linzizong volcanic composition from typical locations in southern Tibet and geodynamic implications (D). China University of Geosciences, Beijing, pp 85–86

Li J, Fang XM (1998) Uplift of the Tibetan plateau and environmental change. Chin Sci Bull 43(15):1569–1574

Li GB, Wan XQ (2003) Eocene microfossils and the extinction of Tethys in Gamba-Tingri region, Southern Tibet. J Stratigraphy 27(2):99–108

Li J, Wen SX, Zhang QS et al (1979) The age, amplitude and style of Tibetan plateau uplift. Sci China 6:608–616

Li J, Fang XM, Pan BT et al (2001) Late Cenozoic intense uplift of the Tibetan plateau and its influence on the surrounding environment. Q Res 21(5):381–391

Li GB, Wan XQ, Qihe RG, Liang DY, Liu WC (2002) Eocene fossil carbonate microfacies and sedimentary environment in Gamba-Tingri area. Tibet. Geol China 29(4):401–406

Li, H. B., Yang, J. S., Xu, Z. Q. et al. 2006. Constraints of the Altyn- Tagh fault zone on the growth of the north Tibetan plateau. Earth Sci Front 13(4), 59–79

Liao ZL, Mo X, Pan GT et al (2007) Spatial and temporal distribution of peraluminous granites in Tibet and their tectonic significance. J Asian Earth Sci 29(2–3):378–389

Liebke U, Appel E, Ding L, Neumann U, Antolin B, Xu Q (2010) Position of the Lhasa terrane prior to India-Asia collision derived from palaeomagnetic inclinations of 53 Ma old dykes of the Linzhou Basin: constraints on the age of collision and post-collisional shortening within the Tibetan plateau. Geophys J Int 182:1199–1215

Lin J, Watts DR (1988) Palaeomagnetic results from the Tibetan plateau. Philos Trans R Soc A Math Phys Eng Sci 327:239–262

Liu ZS, Wang JM (1994) Geoogy and geochemistry of granites in the south Tibetan plateau. Southwest Jiaotong University Press, Chengdu
Liu, D. S., Zheng, M. P., Guo, Z. T., et al. 1998. The origin and development of the Asian monsoon system and its temporal coupling with polar ice caps and regional tectonic movements. Q Res (3), 194–204
Liu CD, Mo X, Luo ZH, Yu XH, Chen HW, Li SW, Zhao X (2004) Mixing events between the crust- and mantle-derived magmas in eastern Kunlun: evidence from zircon SHRIMP II chronology. Chin Sci Bull 49(8):828–834
Liu AL, Wang Q, Zhu DC, Zhao ZD, Liu SA, Wang R, Dai JG, Zheng YC, Zhang LL (2018) Origin of the ca. 50 Ma Linzizong shoshonitic volcanic rocks in the eastern Gangdese arc, southern Tibet. Lithos 304:374–387
Liu AL, Wang Q, Zhu DC, Cawood PA, Xia Y, Li SM, Liu SA, Huang F, Liu L, Zhao ZD, Mo X (2022) Temporal and spatial variations of enriched source components in Linzizong volcanic succession, Tibet, and implications for the India-Asia collision. J Petrol 63:1–30
Mahoney J, Frei R, Tejada M, Mo X, Leat P, Nagler T (1998) Tracing the Indian Ocean mantle domain through time: isotopic results from old west Indian, east Tethyan, and South Pacific seafloor. J Petrol 39(7):1285–1306
Miller C, Schuster R, Klotzli U, Frank W, Purtscheller F (1999) Post-collisional potassic and ultrapotassic magmatism in SW Tibet: geochemical and Sr-Nd-Pb-O isotopic constraints for mantle source characteristics and petrogenesis. J Petrol 40(9):1399–1424
Mo X (2020) Growth and evolution of the crust of Tibetan plateau from perspective of magmatic rocks. Earth Sci 45(7):2245–2257
Mo X, Zhao Z, Zhou S, Dong G (2002) Evidence for timing of the initiation of India-Asia collision from igneous rocks in Tibet. AGU Fall Meeting Abstracts 83(47), F1003, S62B–1201
Mo X, Zhao ZD, Deng JF, Dong GC, Zhou S, Guo TY, Zhang SQ, Wang LL (2003) Volcanism response to the indo-Asian continental collision. Earth Sci Front 10(3):135–148
Mo X, Dong GC, Zhao ZD, Zhou S, Wang LL, Qiu RZ, Zhang FQ (2005a) Temporal-spatial distribution and characteristics of granites in the Gangdese belt, Tibet and implications for crustal growth and evolution. Geol J China Univ 11(3):281–290
Mo X, Dong G, Zhao Z, Guo TY, Wang LL, Chen T (2005b) Timing of magma mixing in the Gangdisê Magmatic Belt during the India-Asia collision: zircon SHRIMP U-Pb dating. Acta Geol Sin (Eng Ed) 79(1):66–76
Mo X, Zhao Z, Deng J, Flower M, Yu X, Luo Z, Li Y, Zhou S, Dong G, Zhu D, Wang L (2006) Petrology and geochemistry of postcollisional volcanic rocks from the Tibetan plateau: implications for lithosphere heterogeneity and collision-induced asthenospheric mantle flow. In: Dilek Y, Pavlides S (eds) Postcollisional tectonics and magmatism in the Mediterranean region and Asia, Geological Society of America special paper, vol 409, pp 507–530
Mo X, Hou Z, Niu YL, Dong G, Qu X, Zhao Z, Yang Z (2007) Mantle contributions to crustal thickening during continental collision: evidence from Cenozoic igneous rocks in southern Tibet. Lithos 96:225–242
Mo X, Niu Y, Dong G, Zhao Z, Hou Z, Zhou S, Ke S (2008) Contribution of syncollisional felsic magmatism to continental crust growth: a case study of the Paleogene Linzizong volcanic succession in southern Tibet. Chem Geol 250:49–67
Mo X, Zhao ZD, Yu XH, Dong GC, Li YG, Zhou S, Liao ZL, Zhu DC (2009) Cenozoic collision–Postcollisional igneous rocks in the Tibetan plateau. Geological Publishing House, Beijing, 396pp
Molnar P, England P, Martinod J (1993) Mantle dynamics, uplift of the Tibetan plateau and the Indian monsoon. Rev Geophys 31:357–396
Molnar P, Houseman GA, Conrad CP (1998) Rayleigh–Taylor instability and convective thinning of mechanically thickened lithosphere: effects of non-linear viscosity decreasing exponentially with depth and of horizontal shortening of the layer. Geophys J Int 133(3):568–584
Molnar P, Houseman G, England P (2006) Palaeo-altimetry of Tibet: arising from D.B. Rowley and B.S. Currie. Nature 444:E4; discussion E5

Molnar P, Boos WR, Battisti DS et al (2010) Orographic controls on climate and paleoclimate of Asia: thermal and mechanical roles for the Tibetan plateau. Earth Planet Sci Lett 38:77–102

Najman Y, Appel E, Boudagher-Fadel M, Bown P, Carter A, Garzanti E, Godin L, Han J, Liebke U, Oliver G, Parrish R, Vezzoli G (2010) Timing of India-Asia collision: geological, biostratigraphic, and palaeomagnetic constraints. J Geophys Res 115:B12416

Najman Y, Jenks D, Godin L, Boudagher-Fadel M, Millar I, Garzanti E, Horstwood M, Bracciali L (2017) The Tethyan Himalayan detrital record shows that India-Asia terminal collision occurred by 54 Ma in the Western Himalaya. Earth Planet Sci Lett 459:301–310

Nelson KD, Zhao W (1996) Partially molten middle crust beneath southern Tibet: synthesis of project INDEPTH results. Science 274:1684–1688

Niu YL (2005) Generation and evolution of basaltic magmas: some basic concepts and a hypothesis for the origin of the Mesozoic-Cenozoic volcanism in eastern China. Geol J China Univ 11:9–46

Niu YL, Lesher CM (1991) Hydrothermal alteration of mafic metavolcanic rocks and genesis of Fe-Zn-Cu sulfide deposits, Stone Hill district, Alabama. Econ Geol 86:983–1001

Niu YL, Mo X, Dong G, Zhao Z, Hou Z, Zhou S Ke S (2007) Continental collision zones are primary sites of net continental crustal growth: evidence from the Linzizong volcanic succession in southern Tibet. Eos Trans. AGU, 88(52), Fall Meeting, Supplement Abstract V34A-01

Niu YL, Zhao ZD, Zhu DC, Mo X (2013) Continental collision zones are primary sites for net continental crust growth—A testable hypothesis. Earth Sci Rev 127:96–110

Owens TJ, Zandt G (1997) Implications of crustal property variations for models of Tibetan plateau evolution. Nature 387(6628):37–43

Pan Y (1999) Formation and uplift of the Tibetan plateau. Earth Sci Front 6(3):153–163

Pan GT, Wang PS, Xu YR, Jiao SP, Xiang TX (1990) Cenozoic tectonic evolution of the Tibetan plateau. Geological Publishing House, Beijing, 165 pp

Pan GT, Ding J, Yao DS, Wang LQ, Luo JN, Yan YJ, Yong YY, Zheng JK, Liang XZ, Qin DH, Jiang XS, Wang QH, Li RS, Geng QR, Liao ZL, Zhu DC (2004) Geological map of the Tibetan plateau and adjacent areas (1:500 000) and the manual. Chengdu Map Publishing House, Chengdu, 133pp

Pan GT, Xiao QH, Lu SN, Deng JF, Feng YM, Zhang KX, Zhang ZY, Wang FG, Xing GF, Hao GJ, Feng YF (2009) Division of tectonic units in China. China Geol 36(1):1–28

Patriat P, Achache J (1984) India-Eurasia collision chronology has implications for crustal shortening and driving mechanism of plates. Nature 311:615–621

Perry FV, Depaolo DJ, Baldridge WS (1993) Neodymium isotopic evidence for decreasing crustal contributions to Cenozoic ignimbrites of the western United States: implications for the thermal evolution of the Cordilleran crust. Geol Soc Am Bull 105:872–882

Petford N, Atherton M (1996) Na-rich partial melts from newly underplated basaltic crust: the Cordillera Blanca Batholith, Peru. J Petrol 37(6):1491–1521

Polissa PJ, Freeman KH, Rowley DB et al (2009) Paleoaltimetry of the Tibetan plateau from D/H ratios of lipid biomarkers. Earth Planet Sci Lett 287:64–76

Powell CM (1986) Continental underplating model for the rise of the Tibetan plateau. Earth Planet Sci Lett 81(1):79–94

Rapp RP, Shimizu N, Norman MD, Applegate GS (1999) Reaction between slab-derived melts and peridotite in the mantle wedge: experimental constraints at 3.8 GPa. Chem Geol 160(4):335–356

Rowley DB (1996) Age of initiation collision between India and Asia: a review of the stratigraphic data. Earth Planet Sci Lett 145:1–13

Rowley DB (1998) Minimum age of initiation of collision between India and Asia north of Everest based on the subsidence history of the Zhepure Mountain section. J Geol 106:229–235

Rowley DB, Pierrehumbert RT, Currie BS et al (2001) A new approach to stable isotope-based paleoaltimetry: implications for paleoaltimetry and paleohypsometry of the high Himalaya since the late Miocene. Earth Planet Sci Lett 188:253–268

Rudnick RL, Fountain DM (1995) Nature and composition of the continental crust: a lower crustal perspective. Rev Geophys 33(3):267–309

Rudnick RL, Gao S (2003) Composition of continental crust. Treatise Geochem 3:1–64
Samson SD, McClelland WC, Patchett PJ et al (1989) Evidence from neodymium isotopes for mantle contributions to Phanerozoic crustal genesis in the Canadian Cordillera. Nature 337(6209):705–709
Saylor JE, Quade J, Dettman DL et al (2009) The late Miocene through present paleoelevation history of southwestern Tibet. Am J Sci 309:1–42
Searle MP, Godin L (2003) The south Tibetan detachment and the Manaslu leucogranite: a structural reinterpretation and restoration of the Annapurna-Manaslu Himalaya, Nepal. J Geol 111:505–523
Searle MP, Windley BF, Coward MP, Cooper DJW, Rex AJ, Rex D, Tingdong L, Xuchang X, Jan MQ, Thakur VC, Kumar S (1987) The closing of Tethys and the tectonics of the Himalaya. Geol Soc Am Bull 98:678–701
Shi YF, Liu DS (1964) Preliminary report of scientific investigation in Shishapangma area. Chin Sci Bull 10:928–938
Shi YF, Li J, Li B et al (1999) Uplift of the Tibetan plateau in the late Cenozoic and environmental changes in East Asia. Acta Geograph Sin 54(1):10–20
Sparks RSJ, Marshall LA (1986) Thermal and mechanical constraints on mixing between mafic and silicic magmas. J Volcan Geotherm Res 29:99–124
Spicer RA, Harris NBW, Widdowson M et al (2003) Constant elevation of southern Tibet over the past 15 million years. Nature 421:622–624
Sun SS, McDonough WF (1989) Chemical and isotopic systematics of oceanic basalts: implications for mantle composition and processes. Geol Soc Lond Spec Publ 42(1):313–345
Sun HL, Zheng D (1998) Formation and evolution of the Tibetan plateau. Guangdong Science and Technology Press, Guangzhou, 357pp
Sylvester PJ (1998) Post-collision strongly peralumineous granits. Lithos 45:29–44
Tan X, Gilder S, Kodama KP, Jiang W, Han Y, Zhang H, Xu H, Zhou D (2010) New paleomagnetic results from the Lhasa block: revised estimation of latitudinal shortening across Tibet and implications for dating the India-Asia collision. Earth Planet Sci Lett 293:396–404
Tapponnier P, Xu ZQ, Roger F, Meyer B, Arnaud N, Wittlinger G, Yang JS (2001) Oblique stepwise rise and growth of the Tibet plateau. Science 294(5547):1671–1677
Vernon RH (1984) Microgranitoid enclaves in granites—globules of hybrid magma quenched in a plutonic environment. Nature 309(5967):438–439
Wan XQ (1990) Cretaceous-Paleogene foraminifera of Tibet and the evolution of the Tethyan Himalayan Sea. Acta Micropalaeontol Sin 7(2):169–186
Wan XQ, Ding L, Li JG, Cai HW (2001) Marine strata from late cretaceous to early Eocene in Zhongba area, Tibet. J Stratigraphy 25(4):268–272
Wan XQ, Jansa LF, Sarti M (2002a) Cretaceous and Paleogene boundary strata in southern Tibet and their implication for the India-Eurasia collision. Lethaia 35(2):131–146
Wan X, Liang D, Li G (2002b) Palaeocene strata in Gamba, Tibet and influence of tectonism. Acta Geol Sin 76(2):155–162
Wang Q (2007) Geochemistry and zircon U-Pb chronology of Linzizong group volcanic rocks in Western Gangdese, Tibet, Master thesis. China University of Geosciences, Beijing
Wang GC (2014) On the geodynamic mechanism of episodic uplift of the Tibetan plateau during the Cenozoic era. Acta Geol Sin (Eng Ed) 88(2):699–716
Wang EC (2017) A discussion on the timing of the initial collision between the Indian and Asian continents. Sci China Earth Sci 47:284–292
Wang GC, Zhang KX (2010) The expansion and formation process of the Tibetan plateau discussed from the temporal and spatial differences of Cenozoic tectonic uplift. Geoscience 35(5):713–727
Wang CS, Liu ZF, Li XH, Wan XQ (1999) Xigaze Fore-Arc Basin and Yarlung Zangbo suture zone, Tibet, Geological Publishing House, Beijing, 237pp
Wang C, Li X, Hu X, Jansa LF (2002) Latest marine horizon north of Qomolangma (Mt Everest): implications for closure of Tethys seaway and collision tectonics. Terra Nova 14:114–120

Wang CS, Li XH, Hu XM (2003) Age of initial collision of India with Asia: review and constraints from sediments in southern Tibet. Acta Geol Sin 77(1):16–24

Wang GC, Xiang SY, Wang A, Garver JI, Wintsch RP, Zhang KX (2007) Thermochronology of early Mesozoic-Cenozoic tectonic processes in east Kunlun and its adjacent areas. Geoscience 32(05):605–614

Wang CS, Zhao X, Liu Z, Lippert PC, Graham SA, Coe RS, Yi HS, Zhu LD, Liu S, Li YL (2008) Constraints on the early uplift history of the Tibetan plateau. PNAS 105(13):4987–4992

Wang CS, Dai JG, Liu ZF, Zhu LD, Li YL, Jia GD (2009) The uplift history of the Tibetan plateau and Himalaya and its study approaches and techniques: a review. Earth Sci Front 16(3):1–30

Wang GC, Cao K, Zhang KX, Wang A, Liu C, Meng YN, Xu YD (2011a) Spatio-temporal framework of tectonic uplift stages of the Tibetan plateau in Cenozoic. Sci China Earth Sci 54(1):29–44

Wang J, Hu X, Jansa L, Huang Z (2011b) Provenance of the upper cretaceous-Eocene deep-water sandstones in Sangdanlin, southern Tibet: constraints on the timing of initial India-Asia collision. J Geol 119:293–309

Wesott WA (1980) Fan-delta sedimentary and tectonie setting. AAPG Bull 64(3):1236–1275

Westphal M, Pozzi JP (1983) Paleomagnetic and plate tectonic constraints on the movement of Tibet. Tectonophysics 98:1–10

Willems H, Zhou Z, Zhang B, Gräfe KU (1996) Stratigraphy of the upper cretaceous and lower tertiary strata in the Tethyan Himalayas of Tibet (Tingri area, China). Geol Rundsch 85:723–754

Wu ZH (2003) 1:250000 Dangxiong regional geological survey report. Institute of Geomechanics, Chinese Academy of Geological Sciences, Beijing

Wu FY, Ji WQ, Wang JG, Liu CZ, Chung SL, Clift PD (2014) Zircon U-Pb and Hf isotopic constraints on the onset time of India-Asia collision. Am J Sci 314:548–579

Wyllie PJ (1977) Crustal anatexis: an experimental review. Tectonophysics 43(1/2):41–71

Wyllie PJ (1984) Constraints imposed by experimental petrology on possible and impossible magma sources and products. Philos Trans R Soc Lond Ser A Math Phys Sci 310(1514):439–456

Xing LY, Zhao ZD, Qi NY, Tang Y, Liu D, Tong X, Wang Q, Zhu DC (2020) Geochemistry and petrogenesis of sandstones and their basaltic interlayers of Shexing formation from Linzhou basin, South Tibet. Acta Petrol Sin 36(9):2729–2750

Xiong SQ, Zhou FH, Yao ZX et al (2001) Aeromagnetic survey of central and Western Tibetan plateau. Geological Publishing House, Beijing, pp 75–87

Xu R (1981) Palaeobotanical evidence for continental drift and Himalayan ascent: the age range and form of Tibetan plateau uplift. Science Press, Beijing, pp 8–18

Xu R, Tao JR, Sun XJ et al (1973) The discovery of fossil beds of Quercus altissima from Mount Shishapangma and its botanical and geological significance. Acta Bot Sin 15(1):103–119

Yin A, Kapp PA, Murphy MA et al (1999) Significant late Neogene east-west extension in northern Tibet. Geology 27(9):787

Yin A, Harrison TM (2000) Geologic evolution of the Himalayan-Tibetan Orogen. Annu Rev Earth Planet Sci 28(1):211–280

Yu GM, Wang CS (1990) Tibetan sedimentary geology, Tibet. Geological Publishing House

Zhai QG, Li C, Li HM, Wang TW (2005) Zircon U-Pb ages of leucogranites in central Gangdese, Tibet. Geol Bull China 24(4):349–353

Zhang YQ, Dai TM, Hong AS (1981) Isotope chronology of granitoids in the south Tibetan plateau. Geochemistry 1:8–17

Zhang PZ, Shen Z, Wang M et al (2004) Continuous deformation of the Tibetan plateau from global positioning system data. Geology 32(9):809–812

Zhang KX, Wang GC, Cao K, Liu C, Xiang SY, Hong HL, Kou XH, Xu YD, Chen FN, Meng YN, Chen RM (2008) Cenozoic sedimentary records and geochronological constraints of differential uplift of the Qinghai-Tibet plateau. Sci China 51(11):1658–1672

Zhang KX, Wang GC, Ji JL, Luo MS, Kou XH, Wang YM, Xu YD, Chen FN, Chen RM, Song BW, Zhang JY, Liang YP (2010) Paleogene-Neogene stratigraphic realm and sedimentary sequence

of the Qinghai -Tibet plateau and their response to uplift of the plateau. Sci China Earth Sci 53(9):1271–1294
Zhang Q, Willems H, Ding L, Gräfe KU, Appel E (2012) Initial India-Asia continental collision and foreland basin evolution in the Tethyan Himalaya of Tibet: evidence from stratigraphy and paleontology. J Geol 120:175–189
Zhang KX, Wang GC, Xu YD, Luo MS, Ji JL, Xiao GQ, Wang A, Song BW, Liang YP, Jiang SS, Cao K, Chen FN, Chen RM, Yang YF (2013a) Sedimentary evolution oi the Tibetan plateau in Cenozoic and its response to the uplift of the plateau. Acta Geol Sin (Eng Ed) 87(2):555–575
Zhang KX, Wang GC, Hong HL, Xu YD, Wang A, Cao K, Luo MS, Ji JL, Xiao GQ, Lin X (2013b) The study of the Cenozoic uplift in the Tibetan plateau: a review. Geol Bull China 32(1):1–18
Zhao W, Morgan WJ (1985) Uplift of Tibetan plateau. Tectonics 4:359–369
Zhao RL, Morgan WJ (1987) Injection of Indian crust into Tibetan lower: a two dimensional finite element model study. Tectonics 6:489–504
Zhao W, Nelson KD (1993) Deep seismic reflection evidence for continental underthrusting beneath southern Tibet. Nature 366(9):557–559
Zhao ZD, Mo X, Nomade P, Renne PR, Zhou S, Dong GC, Wang LL, Zhu DC, Liao ZL (2006) Post-collisional ultrapotassic rocks in the Lhasa block, the Tibetan plateau: spatial and temporal distribution and its implications. Acta Petrol Sin 22(4):787–794
Zhao Z, Mo X, Dilek Y, Niu Y, DePaolo DJ, Robinson P, Zhu D, Sun C, Dong G, Zhou S, Luo Z, Hou Z (2009) Geochemical and Sr–Nd–Pb–O isotopic compositions of the post-collisional ultrapotassic magmatism in SW Tibet: petrogenesis and implications for India intra-continental subduction beneath southern Tibet. Lithos 113:190–212
Zhong DL, Ding L (1996) Discussion on the uplifting process and mechanism of the Tibetan plateau. Sci China 26(4):289–295
Zhou S (2002) Study on geochronology of magmatic rocks and ophiolites in the key regions of Gangdese and Yarlung Zangpo belts, Tibet (PhD dissertation). China University of Geosciences, Beijing
Zhou S, Mo XX, Dong GC et al (2004) 40Ar–39Ar geochronology of Cenozoic Linzizong volcanic rocks from Linzhou Basin, Tibet, China, and their geological implications. Chin Sci Bull 49(18):1970–1979
Zhu DC, Pan GT, Mo X, Duan LP, Liao ZL (2004) Timing of collision between India and Eurasia. Adv Earth Sci 19(4):564–571
Zhu B, Kidd WSF, Rowley DB et al (2005) Age of initiation of the India-Asia collision in the east-central Himalaya. Geology 113:265–285
Zhu DC, Mo XX, Niu YL, Zhao ZD, Wang LQ, Liu YS, Wu FY (2009) Geochemical investigation of early cretaceous igneous rocks along an east-west traverse throughout the Central Lhasa terrane, Tibet. Chem Geol 268(3/4):298–312
Zhu DC, Zhao ZD, Niu YL, Mo X, Chung SL, Hou ZQ, Wang LQ, Wu FY (2011) The Lhasa terrane: record of a microcontinent and its histories of drift and growth. Earth Planet Sci Lett 301(1/2):241–255
Zhu DC, Zhao ZD, Niu YL, Dilek Y, Hou ZQ, Mo X (2013) The origin and pre-Cenozoic evolution of the Tibetan plateau. Gondwana Res 23:1429–1454
Zhu DC, Wang Q, Zhao ZD, Chung SL, Cawood PA, Niu YL, Liu SA, Wu FY, Mo X (2015) Magmatic record of India-Asia collision. Sci Rep 5:14289
Zhu DC, Wang Q, Chung SL, Cawood PA, Zhao ZD (2019) Gangdese magmatism in southern Tibet and India–Asia convergence since 120 Ma. From: Treloar PJ, Searle MP (eds) Himalayan tectonics: a modern synthesis. Geol Soc Lond, Spec Publ 483:583–604

Chapter 7
Accretional and Collisional Processes of the Lhasa Terrane towards the Formation of the Tibetan Plateau

Abstract This chapter evaluates the different scenarios of subduction polarity, closure timing, and geodynamic processes related to the accretion of the Lhasa Terrane based on abundant geological and geochemical observations. The accretion of the Lhasa Terrane to the southern Tibetan Plateau involves the closure of the Shiquanhe-Nam Tso Ocean between Central and Northern Lhasa subterranes and the Bangong-Nujiang Tethys between the Lhasa and Qiangtang terranes. The closure of the Bangong-Nujiang Tethys is clear, whereas the evolution of the Shiquanhe-Nam Tso Ocean remains debated and requires further investigation. The tectonic affinity of the Shiquanhe-Nam Tso mélange belt has been proposed to represent an ophiolite nappe or an imbricate of the Bangong-Nujiang suture, the south branch or back-arc basin of the Bangong-Nujiang Tethys, and an embryonic oceanic basin within the Lhasa Terrane. A Jurassic-Early Cretaceous arc-trench system, termed Beila-Nagqu in the eastern segment and Rutog-Gerze in the western segment, has been documented in the Southern Qiangtang subterrane, resulting from the northward subduction of the Bangong-Nujiang Tethys. Several distinct observations, e.g., ophiolitic formation, angular unconformities, and rock association of adakites and ocean island basalt-type rocks, as well as magmatic gap, likely reflect a series of geodynamic events including subparallel mid-ocean ridge-trench collision (ca. 170–160 Ma), slab window formation (ca. 160–148 Ma), and subsequent subduction termination (ca. 148–125 Ma). The Early Cretaceous Baingoin-Yanhu magmatic arc identified in the Northern Lhasa subterrane cannot be attributed to the low-angle northward subduction of the Neo-Tethys but related to the southward subduction of the Bangong-Nujiang Tethys. It is most likely that the Bangong-Nujiang Tethys experienced a divergent double-sided subduction before ca. 140 Ma followed by a soft collision between opposing arcs that may have occurred at ca. 140–130 Ma. A magmatic flare-up (ca. 120–110 Ma) following a short magmatic lull (ca. 130–120 Ma) in both the Southern Qiangtang and Northern Lhasa subterranes likely recorded the detachment and break-off of the Bangong-Nujiang Tethyan oceanic slab.

Prof. Di-Cheng Zhu has fully contributed to this chapter.

X. Mo et al., *Magmatism and Crustal Evolution of the Tibetan Plateau*, Modern Approaches in Solid Earth Sciences 25,
https://doi.org/10.1007/978-981-97-2109-2_7

Keywords Shiquanhe-Nam Tso mélange zone · Bangong-Nujiang suture zone · Ridge-trench collision · Divergent double-sided subduction · Soft Lhasa–Qiangtang collision

7.1 Closure of the Shiquanhe-Nam Tso Ocean

The Shiquanhe-Nam Tso mélange zone (SNMZ) extends for ~2000 km from Lhasa in the eastern segment, western Nam Tso, Yunzhug, Asa, Lagkor Tso in the central segment, and Shiquanhe River in the western segment (Fig. 7.1; BGMRXAR 1993; Girardeau et al. 1985; Pan 2006; Zhu et al. 2013). The ophiolite sequences in the SNMZ are strongly dismembered, but are generally composed of assemblages of peridotite, gabbro, diabase, and pillow basalt (Girardeau et al. 1985; Wang et al. 2008; Xu et al. 2014). The gabbros and plagiogranites yield Jurassic and Cretaceous zircon U-Pb ages (Fig. 7.1; Xu et al. 2014; Tang et al. 2020), consistent with radiolarite ages (Baxter et al. 2009). The tectonic affinity of the SNMZ remains controversial, and the evolution of the Shiquanhe-Nam Tso Ocean is debated.

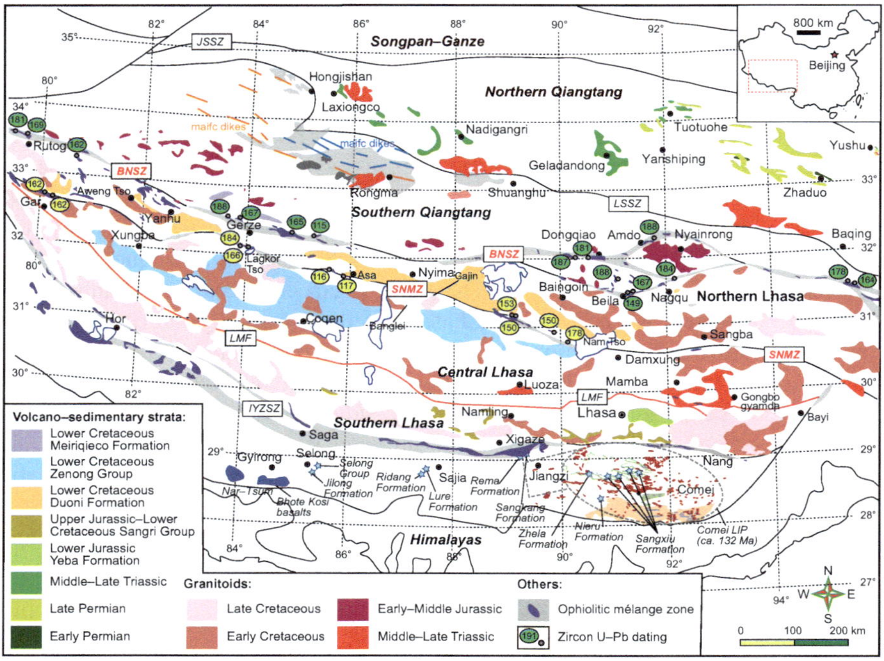

Fig. 7.1 Distributions of the Shiquanhe-Nam Co mélange zone (SNMZ) and Bangong-Nujiang suture zone (BNSZ), showing zircon U-Pb ages of mafic rocks within the BNSZ (modified from Zhu et al. 2013). See text for details

7.1.1 *Tectonic Affinity of the Shiquanhe-Nam Tso Ocean*

Four contrasting hypotheses have been proposed for the origin of the SNMZ, including (1) an ophiolite nappe or imbricate of the BNSZ, (2) the south branch of the Bangong-Nujiang Tethys, (3) a back-arc basin related to southward subduction of the Bangong-Nujiang Tethyan lithosphere, and (4) an embryonic oceanic basin within the Lhasa Terrane.

7.1.1.1 Ophiolite Nappe or Imbricate of the BNSZ

Based on the petrological and geochemical similarities between ophiolites from Yunzhug in the SNMZ and Dongqiao in the BNSZ, previous studies suggested that the SMNZ represents a nappe of the BNSZ (Girardeau et al. 1985; Zhang et al. 2007; Baxter et al. 2009). However, recent zircon U-Pb ages and geochemical data reveal that the Yunzhug ophiolites were formed at ca. 150 Ma (Fig. 7.1) and show enriched mid-ocean ridge basalt-type (E-MORB) compositions (Zeng et al. 2018a), whereas the Dongqiao ophiolites were formed at ca. 188 Ma (Fig. 7.1) and belong to typical SSZ-type ophiolites (Liu et al. 2016). Although there are some Middle Jurassic (ca. 170–160 Ma; Fig. 7.1; Tang et al. 2020) SSZ-type ophiolites in both the BNSZ and SNMZ, SSZ-type geochemical signatures are common in many Phanerozoic ophiolites in the Tethyan system (Dilek and Furnes 2009). Furthermore, Mugagangri Complex is widespread in the BNSZ, but are absent in the SNMZ, arguing against the nappe proposal.

Based on regional mapping and geochronologic data, Kapp et al. (2003) suggested that the SNMZ in Shiquanhe represent imbricates of the BNSZ in Rutog. This model differs from the nappe model in two aspects: (1) the SNMZ represents tectonic blocks within a mélange rather than a single sheet of oceanic lithosphere; (2) the SNMZ ophiolites were structurally beneath Jurassic flysch (Kapp et al. 2003). However, this hypothesis is too simplistic. It cannot explain the presence of eastern segment of the Northern Lhasa subterrane between the SNMZ and BNSZ (Zhu et al. 2013), which has a width of >200 km without presence of any mélange and ophiolites in it. Besides, southward-directed thrusts are absence in the eastern SNMZ segment.

7.1.1.2 South Branch of the Bangong-Nujiang Tethys

Recent studies on Ren Co ophiolites in the middle segment of the SNMZ documented a typical normal-MORB type ophiolitic sequence (Tang et al. 2020). The rock associations include peridotite, gabbro, sheeted dike, basalt, and plagiogranite, with U-Pb zircon ages of ca. 169 to 147 Ma (Fig. 7.1). The identification of N-MORB ophiolite indicates the presence of a large oceanic basin, which is linked to the Bangong-Nujiang Tethys and likely represents the main suture zone of the Meso-Tethys Ocean. This model suggests that the Northern Lhasa subterrane was formed by accreted terranes, oceanic arcs, subduction complexes, and fragments of oceanic plateaus (Zhang et al. 2014b; Tang et al. 2020; Yan and Zhang 2020; Chen et al. 2021).

Zeng et al. (2018a) suggested that the SNMZ was not the south branch of the Bangong-Nujiang Tethys for following reasons. (1) Paleozoic strata are present on both sides of eastern segment of the SNMZ, indicating that the eastern segment of the Northern Lhasa subterrane likely has basement similar to the Central Lhasa subterrane. (2) The latitudinal convergence within the Lhasa Terrane is insignificant (~3.2°) as revealed by paleomagnetic data (Li et al. 2017b). Besides, the south branch proposal requires multi-subduction zones within the Bangong-Nujiang Ocean, which has not been identified yet. These observations support that the SNMZ represent a boundary between the Central and Northern Lhasa subterranes (Zhu et al. 2011, 2013; Kapp and DeCelles 2019).

7.1.1.3 Embryonic Oceanic Basin within the Lhasa Terrane

Zeng et al. (2018a) suggested that the SNMZ is remnant of an embryonic oceanic basin within the Lhasa Terrane based on the identification of E-MORB-type ophiolites (ca. 150 Ma; Fig. 7.1) around Yunzhug in the SNMZ. This conclusion is supported by the presence of abundant argillaceous components embedded in the basalts of ophiolites, which indicates a continental source and thus a narrow oceanic basin between continental margins (Xu et al. 2014). Furthermore, the Rila Formation along the SNMZ likely represents Late Jurassic-Early Cretaceous bathyal–abyssal deposits as supported by the presence of peridotite-bearing coarse-grained sandstone and conglomerate, siliceous siltstone, pelsparite, calcirudite, and marble (Zhang et al. 2011; Volkmer et al. 2014). These sedimentary records suggest that the oceanic realm was narrow and is consistent with embryonic oceanic basin.

However, this embryonic oceanic basin model is problematic. Firstly, a typical embryonic ocean is characterized by absence of sheeted dikes and mafic plutonic rocks, but there are large-scale sheeted dikes, gabbros, pillow basalts with N-MORB features in Yunzhug ophiolites (Xu et al. 2014; Tang et al. 2020). Besides, tectono-sedimentary breccias and top-to-basement detachment faults have been identified in the SNMZ, which is a typical signature of embryonic ocean (Manatschal and Müntener 2009; Piccardo et al. 2014). Secondly, the ophiolites within the SNMZ yield mantle-like zircon $\delta^{18}O$ values and Nd-Hf isotopes, arguing against an embryonic oceanic basin within continental lithosphere (Tang et al. 2020). Thirdly, Jurassic SSZ-type (ca. 170–160 Ma) ophiolites have been reported within the SNMZ (Xu et al. 2014), which is older than the E-MORB-type ophiolites (ca. 150 Ma) and generally indicate subduction initiation.

7.1.1.4 Back-Arc Basin

The basalt within the SNMZ shows back-arc basalt characteristics, such as high Th and low Ti, likely indicating that the SNMZ is a back-arc basin developed in the Early Cretaceous by southward subduction of the Bangong-Nujiang Tethyan lithosphere (Zhu et al. 2009, 2011; Pan et al. 2012; Zhong et al. 2015). This interpretation is supported by zircon Hf isotopic mapping, which reveals that the SNMZ is a tectonic boundary separating juvenile crust (i.e., Northern Lhasa subterrane) to the north from ancient basement (i.e., Central Lhasa subterrane) to the south (Zhu et al. 2011). This proposal is also challenged because of the similarity of rock assemblages and formations ages between ophiolites within the SNMZ and BNSZ (Shi et al. 2008; Huang et al. 2017), and the presence of N-MORB and E-MORB-type basalt within the SNMZ ophiolites (Zeng et al. 2018a; Tang et al. 2020). However, basalts in mature back-arc basin can share similar geochemical signatures to mid-ocean ridge basalts (e.g., Pearce 2014).

7.1.2 Closure Timing of the Shiquanhe-Nam Tso Ocean

The evolution of the Shiquanhe-Nam Tso Ocean remains controversial. Various models have been proposed for the cause of the closure of the Shiquanhe-Nam Tso Ocean because the tectonic affinity of the SNMZ is debated. In the framework of the south branch of the Bangong-Nujiang Tethys, the closure of the Shiquanhe-Nam Tso Ocean may have resulted from double-polarity subduction of the Tethyan lithosphere beneath the Southern Qiangtang subterrane to the north and the Central Lhasa subterrane to the south (Tang et al. 2020). The subduction terminated at ca. 120 Ma by Lhasa-Qiangtang collision. The embryonic oceanic basin model suggests that the Shiquanhe-Nam Tso Ocean was formed by continental break-up of the Lhasa Terrane driven by pull-apart effects of northward subduction of the Neo-Tethyan and Bangong–Nujiang Tethyan lithosphere and the closure mechanism requires further investigation (Zeng et al. 2018a). In the framework of back-arc basin, the Lhasa-Qiangtang collision can result in back-arc spreading cessation (Zhu et al. 2011).

The radiolarian cherts of the Lagkor Tso ophiolitic mélange within the SNMZ have Middle Jurassic to Early Cretaceous ages (Baxter et al. 2009), in phase with emplacement of gabbro and plagiogranite, which yield zircon U-Pb ages of ca. 170–160 Ma, ca. 150 Ma, and ca. 120–115 Ma (Xu et al. 2014; Zeng et al. 2018a, b; Tang et al. 2020). All these ages indicate that the Shiquanhe-Nam Tso Ocean remain opened until Early Cretaceous (Zhu et al. 2013).

In summary, the closure timing and mechanism of the Shiquanhe-Nam Tso Ocean remains controversial and requires further investigation. A possible model is that the Shiquanhe-Nam Tso Ocean is a back-arc basin related to southward subduction of the Bangong-Nujiang ocean and is closed by back-arc spreading cessation as the result of the Lhasa-Qiangtang collision in the Early Cretaceous.

7.2 Closure of the Bangong-Nujiang Tethys

The Bangong-Nujiang suture zone (BNSZ; Fig. 7.1) extends for ~2000 km from Dengqen and Amdo in the eastern segment, via Dongqiao to Gerze in the central segment, to Rutog in the western segment, separating the Lhasa Terrane to the south and Southern Qiangtang subterrane (SQ) to the north (Zhu et al. 2013, 2016; Wang

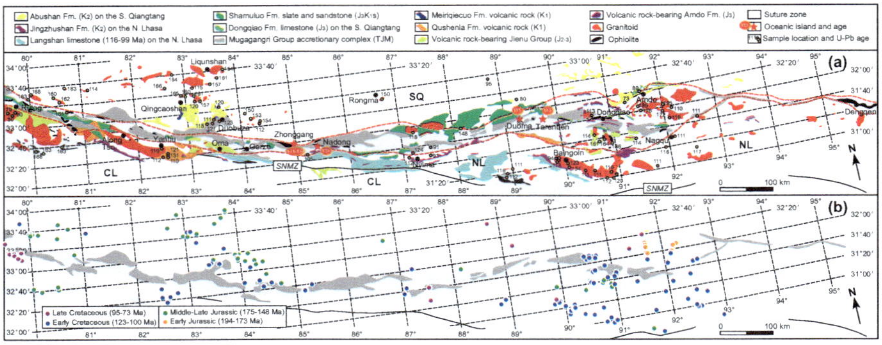

Fig. 7.2 (**a**) Simplified geological map showing the distinct Jurassic–Cretaceous geological records within Lhasa–Qiangtang collision zone (adapted from Wang et al. 2013; Zhu et al. 2016). (**b**) Simplified map of (**a**) showing time and space distribution of magmatic rocks within the Lhasa–Qiangtang collision zone

et al. 2016). The BNSZ are composed of ophiolitic fragments, Mugagangri complex, and Jurassic and Early Cretaceous igneous rocks. The ophiolites are well exposed and include ultramafic rocks, gabbros, basalts, and cherts (Wang et al. 2016) with formation ages of 190–180 and 170–160 Ma dated using samples from the mafic rocks (Fig. 7.1; Liu et al. 2014; Tang et al. 2018). The Mugagangri complex is characterized by heterogeneous blocks of limestone, sandstone, and volcanic rock in a highly deformed slate and fine sandstone matrix (Kapp et al. 2003, 2005, 2007; Wang et al. 2013; Zeng et al. 2016a). Igneous rocks within the BNSZ are formed at ca. 166–160 Ma and 125–110 Ma (Fig. 7.2; Fan et al. 2018; Wu et al. 2018; Li et al. 2020).

7.2.1 *Subduction Polarity of the Bangong-Nujiang Tethyan Lithosphere*

The BNSZ represents a remnant of the Bangong-Nujiang Tethyan lithosphere, which may subduct northward beneath the SQ (Li et al. 2014a, b) or southward beneath the Lhasa Terrane (Zhu et al. 2009, 2011; Pan et al. 2012; Li et al. 2018), or both (Zhu et al. 2016). A detailed evaluation is discussed below.

7.2.1.1 Northward Subduction beneath the Qiangtang Terrane

In the western segment of the Southern Qiangtang subterrane (SQ), a Jurassic Rutog-Gerze arc-trench system has been documented by numerous studies (Figs. 7.2 and 7.3). From north to south, a magmatic arc, a forearc basin, and an accretionary prism have been identified. A broad Rutog-Gerze magmatic belt develops in the SQ,

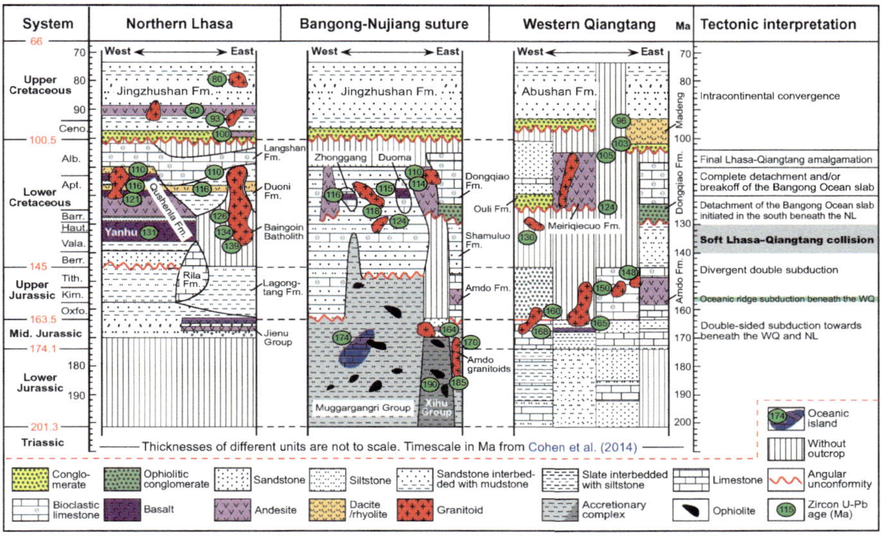

Fig. 7.3 Generalized tectonostratigraphic columns for the Lhasa–Qiangtang collision zone showing the lithostratigraphical units and their relations (from Zhu et al. 2016)

extending approximately 400 km in E-W length and approximately 130 km in N-S width (Fig. 7.2). This magmatic belt was active from ca. 170 to 148 Ma (Li et al. 2014a, b) and succeeded by a prolonged magmatic gap (ca. 148–125 Ma) (Fig. 7.4; Li et al. 2020). It consists of four rock associations of low-Mg series, high-Mg series, medium-K adakite and OIB-type rock, and high-K adakite (See Chap. 2). All of these are common above subduction zones, indicating that the magmatic belt was a continental arc. The forearc basin lies south of the magmatic arc, including the Sewa, Shaqiaomu and Jiebuqu formations (Huang et al. 2016). Detrital zircons from sedimentary rocks of these formations yield abundant Jurassic ages of ca. 170–146 Ma, which is identical to ages of the Rutog-Gerze magmatic arc (Fig. 7.4; Li et al. 2020). Further to the south, an accretionary prism along the southern margin of the Qiangtang Terrane is identified by the Wuga and Gamulong formations with rapid deposition of conglomerates in a trench environment (Li et al. 2017a; Sun et al. 2019). A Jurassic age group of ca. 184–148 Ma is also reported for detrital zircons in rocks from the Wuga and Gamulong formations. Thus, a Jurassic magmatic arc-trench system is identified (Fig. 7.4).

In the eastern segment of the BNSZ, a Jurassic Beila-Nagqu arc-trench system was identified and linked to the Rutog-Gerze arc-trench system (Fig. 7.4; Li et al. 2020). From south to north, a Beila-Nagqu accretionary prism represented by the Beila-Nagqu mélange belt within the BNSZ, a Daru Tso magmatic arc within the BNSZ, and retro-arc sedimentary units represented by the Daoban (Fig. 7.4; Li et al. 2020) and Xiali formations (Li et al. 2019) in the SQ were reported. The Daru Tso magmatic arc was active at ca. 166–160 Ma and followed by a magmatic gap (ca. 160–120 Ma). This arc extended E-W for about 40 km in length with about 20 km

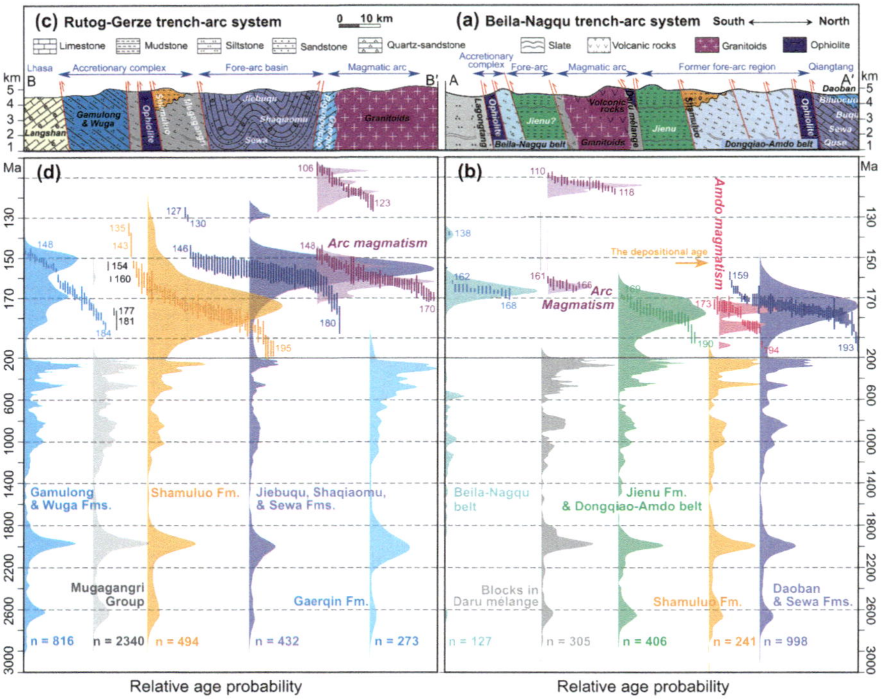

Fig. 7.4 Simplified cross sections around (**a**) Beila and (**c**) Gerze. (**b** and **d**) Probability density plots of U-Pb ages of detrital zircon from mélange and Jurassic strata along these two sections, as well as ages of Jurassic-Early Cretaceous magmatic rocks in the SQ and BNSZ (from Li et al. 2020)

in N-S width (Hu et al. 2017; Wu et al. 2018). Detrital zircon ages reveal that the accretionary prism and retro-arc sedimentary units are time-equivalent to the Daru Tso magmatic arc (Li et al. 2020).

These observations document that a single arc-trench system developed along the SQ and BNSZ. This arc-trench system can be divided into two distinct segments along the strike, which was active at ca. 166–160 Ma in the eastern segments and ca. 170–148 Ma in the western segment, and is succeeded by prolonged magmatic gaps of ca. 160–120 Ma and ca. 148–125 Ma, respectively (Fig. 7.4). This arc-trench system was likely generated by northward subduction of the Bangong-Nujiang Tethyan lithosphere during the Middle-Late Jurassic.

7.2.1.2 Southward Subduction Beneath the Lhasa Terrane

Recently, an Early Cretaceous magmatic arc along the Northern Lhasa subterrane (NL) has been identified (Figs. 7.2 and 7.3; Li et al. 2018). This arc includes granites of the Baingoin Batholith in the west segment and the volcanic rocks of Qushenla Formation. These igneous rocks result from active magmatism at ca. 139–100 Ma.

The Baingoin granites are high-K calc-alkaline rocks and belong to I-type granites (ca. 134–127 Ma) in the eastern NL (Fig. 7.3; Zhu et al. 2016). The volcanic rocks around Yanhu in the western NL are basalt and andesite (Li et al. 2018) and are divided into two stages. The early stage (ca. 131–116 Ma) basalts belong to medium-K calc-alkaline series and show geochemical signatures of typical arc basalts. The late stage (ca. 117–103 Ma) magmatism is dominated by high-K calc-alkaline rocks and also show arc signatures (e.g., Sui et al. 2013; Wu et al. 2015). These observations argue for the existence of the Baingoin-Yanhu arc in the NL.

Previous studies have documented an Early Cretaceous arc-trench system in the Southern Lhasa subterrane (Li et al. 2018), including from south to north of the ophiolites at ca. 130–120 Ma (Dai et al. 2013), Xigaze forearc basin (ca. 123–80 Ma; Wang et al. 2017), and Gangdese magmatic rocks (ca. 137–108 Ma; Zhu et al. 2019). This arc-trench is widely attributed to northward subduction of the Neo-Tethys along the southern margin of the Lhasa Terrane (Zhu et al. 2019). The Baingoin–Yanhu identified in the northern Lhasa subterrane is either linked to the Gangdese arc-trench system, indicating low-angle northward subduction of the Neo-Tethys (Kapp et al. 2005, 2007; Zeng et al. 2020), or is attributed to south-dipping subduction of the Bangong Tethys along the northern margin of the Lhasa Terrane (Pan 2006; Zhu et al. 2009, 2011; Sui et al. 2013). We favor that two opposed subduction zones developed in the Lhasa Terrane during the Early Cretaceous because of the following three observations:

1. The Early Cretaceous magmatism in the Lhasa Terrane is not readily explained by one arc-trench system unless it was unusually broad. Figure 7.5 displays spatial and temporal distributions of the Mesozoic magmatism in the Lhasa Terrane (Li et al. 2018). This figure reveals that the Early Cretaceous igneous rocks cover a broad region with a maximum N-S width of ~310 km in the Lhasa Terrane. Furthermore, the Lhasa Terrane has undergone strong north-south shortening since Early Cretaceous as evidenced by numerous folding and faulting, such as the Gangdese thrust, Great Counter thrust, and Gaize-Seling Tso backthrust (Murphy et al. 1997; Kapp et al. 2007; Volkmer et al. 2014). About 50% of north-south shortening is estimated by structural studies (Volkmer et al. 2014). Therefore, the Early Cretaceous igneous rocks covered likely over 600 km from north to south, broader than the arcs above low-angle subduction zones.
2. A high-angle subduction of the Neo-Tethys lithosphere beneath the Southern Lhasa subterrane rather than low-angle subduction beneath the entire Lhasa Terrane was suggested based on extension of the Southern Lhasa subterrane. The evidence for extension includes (a) formation of ophiolites and intrusion of mafic dykes (ca. 130–120 Ma), which indicates subduction initiation and extension (Dai et al. 2013; Xiong et al. 2016), (b) formation of garnet amphibolite blocks in the mélange (Guilmette et al. 2012), (c) paleomagnetic and field evidence for forearc hyperextension (Maffione et al. 2015), and (e) onset of sedimentation within the Xigaze forearc basin (ca. 123 Ma; Wang et al. 2017).
3. The model of two opposed arc systems (Fig. 7.6) can well explain the rock associations and spatial and temporal distributions of Early Cretaceous magmatism

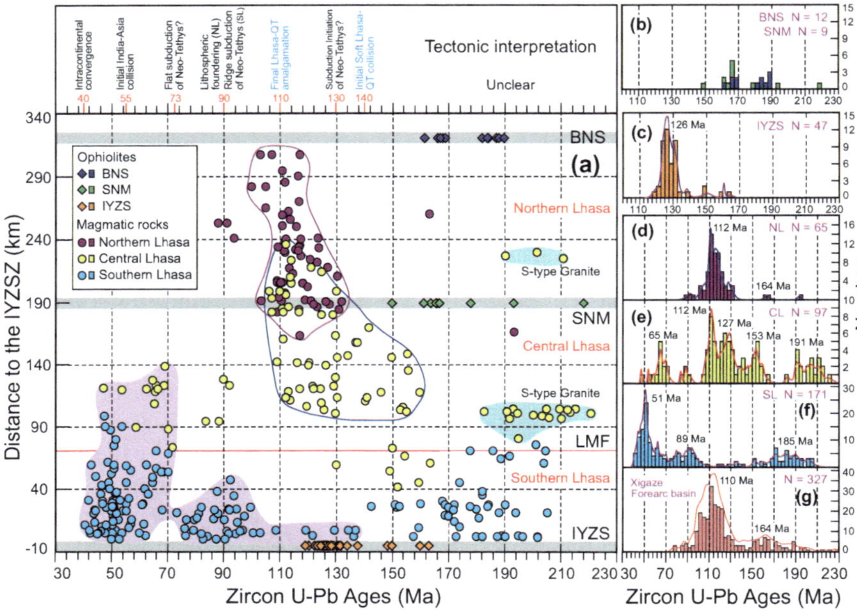

Fig. 7.5 (**a**) Distances (km) from sample locations to the Indus-Yarlung Zangbo suture (IYZS) vs. zircon U-Pb ages (Ma) for Late Triassic to the Late Eocene magmatism of the Lhasa Terrane (from Li et al. 2018). Histograms of ages (Ma) of ophiolites in the Bangong-Nujiang suture (BNS) and the Shiquanhe River-Nam Tso mélange (SNM) (**b**), in the IYZS (**c**), of magmatic rocks in the northern (**d**), central (**e**), and southern (**f**) Lhasa subterranes, and of detrital zircons for sediments from the Xigaze forearc basin (**g**)

in the Lhasa Terrane. The magmatic rock associations during ca. 139–116 Ma are medium-K calc-alkaline basalt and I-type granite in the southern and northern Lhasa subterranes, but high-K basalt and A-type granite in the central Lhasa subterrane, suggesting the existence of two opposed arc systems and an intervening extensional region (see Chap. 2). Besides, the N-S width of the Lhasa Terrane during the Early Cretaceous is likely >800 km based on estimates of shorting and original arc-trench width, which can provide enough space for two subduction zones (Li et al. 2018).

In summary, the Baingoin-Yanhu magmatic arc in the northern Lhasa subterrane was likely generated by southward subduction of the Bangong-Nujiang Tethyan lithosphere in the Early Cretaceous (Fig. 7.6).

7.2.1.3 A Divergent Double-Sided Subduction Model

The observations and interpretations presented above indicate that the Bangong-Nujiang Tethys likely experienced pre-Cretaceous double-sided subduction (Zhu et al. 2016), resulting in arc-trench system of Rutog-Gerze-Beila-Nagqu in the SQ and BNSZ (Fig. 7.7; Li et al. 2020), and Yanhu-Baingoin magmatic arc in the

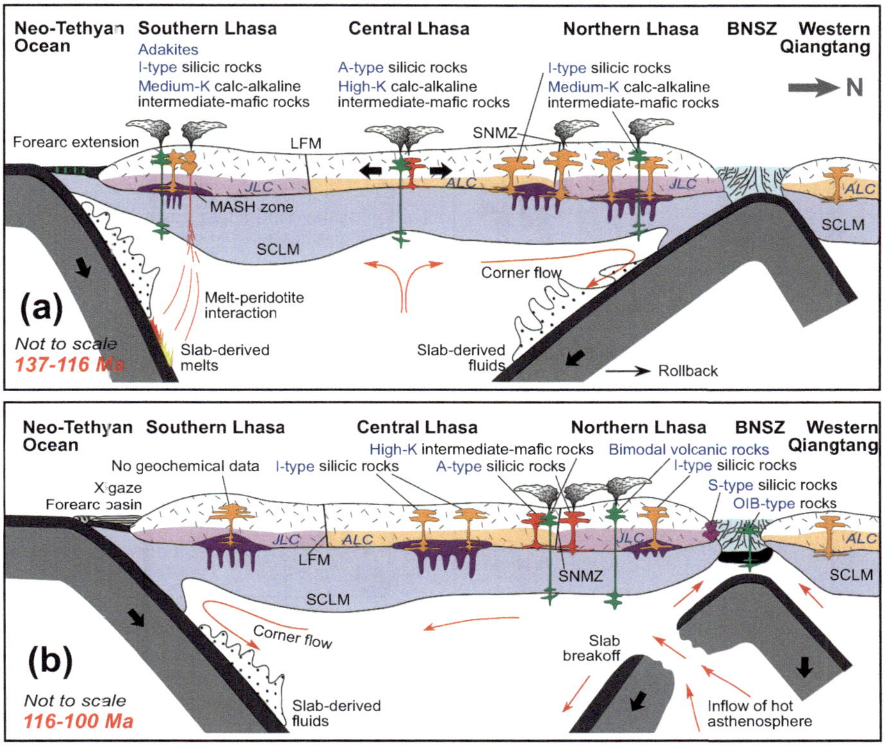

Fig. 7.6 Cartoon showing the development of two opposed subduction zones beneath the Lhasa Terrane (from Li et al. 2018)

Northern Lhasa subterrane (Fig. 7.7; Li et al. 2018). This model is analogous to the modern Pacific Ocean, where oceanic lithosphere is subducting beneath the western South American plate to the east and eastern Australian and Eurasian plate to the west (Schellart et al. 2006; Cawood et al. 2009).

7.2.2 *Closure Timing of the Bangong-Nujiang Tethys*

The closure timing of Bangong-Nujiang Tethys remain controversial and four hypotheses have been proposed, varying from Middle Jurassic to Late Cretaceous.

7.2.2.1 Closure Termination in the Middle Jurassic

The angular unconformity separating the Biluocuo Formation from underlying Buqu Formation in the eastern SQ segment indicates a major tectonic event at ca. 166 ± 1 Ma (Ma et al. 2017). A similar tectonic event of rapid deposition of

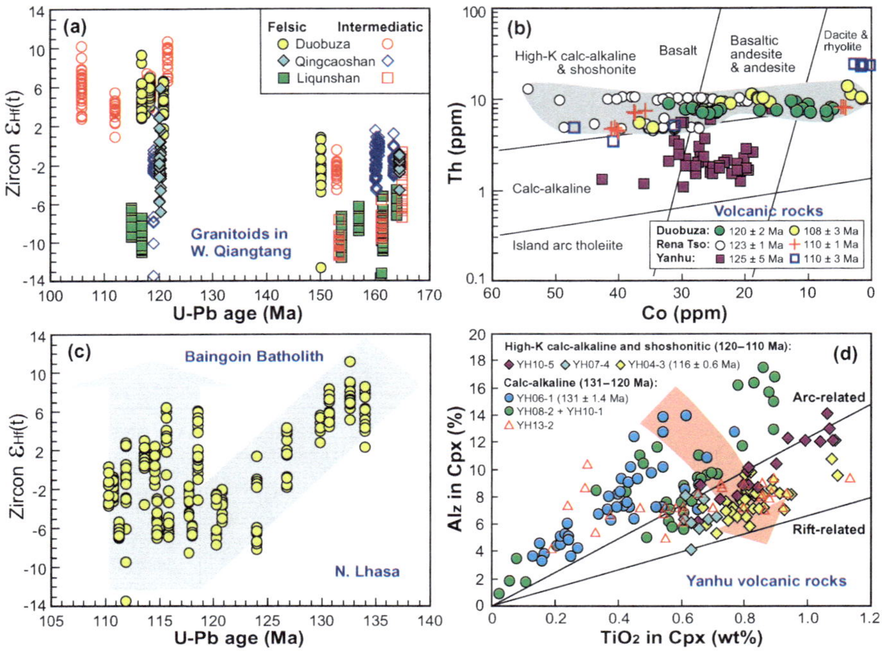

Fig. 7.7 Diagrams for arc magmatic rocks in the SQ and northern Lhasa Terrane (from Zhu et al. 2016). (**a**) Plot of $\varepsilon_{Hf}(t)$ vs. U–Pb ages for granites in the SQ; (**b**) Plot of Th vs. Co for volcanic rocks from the SQ and northern Lhasa subterrane; (**c**) Plot of $\varepsilon_{Hf}(t)$ vs. U-Pb ages of Baingoin granitoids in the northern Lhasa subterrane; (**d**) AlZ vs. TiO_2 of clinopyroxene from the Yanhu basalts in the northern Lhasa subterrane

Gamulong conglomerates in a trench environment at ca. 166 Ma is also emphasized in the western SQ segment (Sun et al. 2019). A slightly younger angular unconformity between the Shamuluo Formation and the underlying Mugagangri complex (mélange) is identified in the western SQ segment before ca. 153 Ma based on studies of detrital zircon ages (Li et al. 2017a). All these observations indicate a compressional setting and are interpreted as the result of the Lhasa-Qiangtang collision (Ma et al. 2017; Li et al. 2017a; Sun et al. 2019; Ma et al. 2018a). This Middle Jurassic collisional event can also explain the magmatic gap from ca. 148 to 125 Ma in a post-collisional environment.

However, there are lines of evidence arguing against Middle Jurassic closure of the Bangong-Nujiang Tethys. Firstly, numerous Middle Jurassic (ca. 170–160 Ma; e.g., Tang et al. 2018) and Late Jurassic (ca. 148 Ma, Zhong et al. 2017) SSZ-type ophiolites have been reported along the BNSZ. These ophiolites can be generated by subduction related settings, such as subduction initiation, or ridge subduction, rather than collisional settings (Dilek and Furnes 2014). Secondly, paleomagnetic data are also inconsistent with closure of the Bangong-Nujiang Tethys. During ca. 180 to 170 Ma, the Lhasa Terrane was separated from the Southern Qiangtang subterrane with paleolatitude difference of 20°–23° (Ma et al. 2018b; Cao et al. 2019). Assuming closure of the Bangong-Nujiang Tethys occurred at ca. 166 Ma, it requires an unrealistically high northward velocity of ~157 to ~625 mm/y for the Lhasa Terrane.

7.2.2.2 Closure Termination in the Late Cretaceous

Some researchers suggest that the Bangong-Nujiang Tethys has not been closed until Late Cretaceous. This conclusion is based on the existence of ca. 120–108 Ma OIB-type basaltic rocks and limestones within the BNSZ, which is interpreted to represents the presence of ocean islands (thus oceanic crust; Zhu et al. 2006; Fan et al. 2014; Liu et al. 2014; Zhang et al. 2014b). However, these OIB-type rocks can also be generated by decompression melting of asthenospheric materials (Zhu et al. 2016).

Several geological observations indicate that the Lhasa-Qiangtang collision has likely terminated before the Late Cretaceous. Firstly, the Duba Formation from in the Lhasa Terrane was likely deposited from ca. 122 to 110 Ma and sourced from the Jurassic magmatic arc in the SQ (Lai et al. 2019), suggesting that the Bangong-Nujiang Tethys has closed. Secondly, paleomagnetic data imply that the paleolatitude of the southern margin of the Lhasa Terrane is virtually indistinguishable from that of the southern margin of the Qiangtang Terrane. This inference is based on paleomagnetic data from the Lhasa Terrane (~20 ± 4°N during ca. 110–50 Ma; Lippert et al. 2014) and SQ (29.3 ± 5.7°N at ca. 110–104 Ma; Chen et al. 2017) and the understanding of the wide (>600 km) Lhasa Terrane (Zhu et al. 2016). Thirdly, A-type granites (ca. 118–110 Ma; Wu et al. 2018; Fan et al. 2018), which is coeval with the OIB-type rocks, are also reported within the BNSZ, suggesting a within-plate setting.

7.2.2.3 A Diachronous Closure Commencing in the Middle Jurassic and Termination in the Late Cretaceous

A diachronous collisional model was proposed based on changing sedimentary environments along the BNSZ, involving Middle Jurassic collision in the east and Late Cretaceous collision in the west (Pan et al. 1983). The presence of Amdo gneiss with metamorphism at ca. 171 Ma was further regarded as evidence for the Lhasa-Qiangtang collision (Xu et al. 1985). As discussed above, the presence of Middle-Late Jurassic ophiolites and paleomagnetic data argue against the conclusion that the Bangong-Nujiang Tethys has closed before the Late Jurassic. In fact, recent studies have revealed that the Amdo micro-terrane (or named Amdo micro-continent) is a micro block within the Bangong-Nujiang Tethys and does not represent the Lhasa Terrane (e.g., Zhu et al. 2013). Furthermore, subsequent studies on metamorphic rocks from the Amdo micro-terrane indicated that the Amdo was accreted to the SQ during ca. 190–170 Ma (Guynn et al. 2006, 2013; Zhang et al.

2014a). Besides, the Late Cretaceous collision model in the west is also contradicted by subsequent observations mentioned above (Sect. 7.2.2.2), such as paleomagnetic, sedimentary, and magmatic records.

7.2.2.4 A Possible Closure Timing of Late Jurassic to Early Cretaceous

Stratigraphic and structural works along the BNSZ suggest a closure timing of Late Jurassic to Early Cretaceous (Zhu et al. 2016). From east to west, the key observations include (Fig. 7.3): (1) the angular unconformity between conglomerate of the Dongqiao Formation and the Early Cretaceous fossiliferous strata around Amdo (Fig. 7.1; ca. 143–131 Ma; Sun 2005), (2) the angular unconformity between the underlying ophiolites and overlying limestone in the Dongqiao Formation around Amdo (Aptian-Albian; ca. 125–100 Ma; Wang and Dong 1984; Yin et al. 1988); (3) a changing from marine to non-marine environments at ca. 125–118 Ma in the Nyima basin as indicated by geological mapping and geochronological data (Kapp et al. 2007); (4) a Late Jurassic–Early Cretaceous shortening event as revealed by structural mapping and detrital zircon ages around Domar (Raterman et al. 2014). Besides, the Upper Cretaceous terrestrial molasse (Fig. 7.3) are widespread in both the Lhasa Terrane (Jingzhushan Formation) and Southern Qiangtang subterrane (Abushan Formation), indicating that the Lhasa-Qiangtang collision had occurred by this time.

In summary, a combination of previous studies on various geological records, such as sedimentary environments, angular unconformities, crustal shorting, magmatic gap (ca. 140–130 Ma; Fig. 7.4) in the Southern Qiangtang subterrane, and magmatic rock associations in the Early Cretaceous (Fig. 7.4; e.g., A-type granites and OIB-type rocks in the Northern Lhasa subterrane and BNSZ), support that the Lhasa–Qiangtang collision most likely occurred at ca. 140–130 Ma (Zhu et al. 2016).

7.2.3 *Implications for Geodynamic Scenarios*

The evolution of the Bangong-Nujiang Tethys remains controversial, including the timing of subduction initiation to the north and south, the timing and direction of mid-ocean ridge subduction, identification of oceanic plateau subduction, slab shallowing and steeping processes. Among them, two geodynamic scenarios have been well documented: ridge subduction and Lhasa-Qiangtang collision.

7.2.3.1 Northward Subduction of a Subparallel Mid-Ocean Ridge in the Middle-Late Jurassic

The geodynamic scenarios for northward subduction of the Bangong-Nujiang Tethys during the Jurassic to Early Cretaceous remains elusive. Several key observations over 1200 km along strike in the SQ and BNSZ, which must be accounted for in any geodynamic model, are emphasized here (Fig. 7.8; Li et al. 2020). (1) The SSZ-type ophiolite within the BNSZ were formed at two stages of ca. 190–180 Ma and ca. 170–160 Ma (Shi 2007; Shi et al. 2008; Li et al. 2013; Liu et al. 2014, 2016; Wang et al. 2016; Huang et al. 2017; Zhong et al. 2017; Tang et al. 2018). (2) It is a compressional setting during ca. 166–160 Ma in the SQ and BNSZ as inferred from the uplift of the Daru Tso arc, angular unconformities, and rapid deposition of trench deposits, indicating a collisional event (Li et al. 2020). (3) The magmatism in the SQ and BNSZ can be divided into four stages (See Chap. 2). The first stage (ca. 170–148 Ma) is represented by high-K calc-alkaline I-type granites (Li et al. 2014a, b), high-Mg andesites, dacites, and rhyolites (Zeng et al. 2016b; Li et al. 2020), and high-K calc-alkaline adakites (Hao et al. 2016). The second stage is characterized by a special rock association of medium-K calc-alkaline adakites (slab-derived) and OIB-type rocks during ca. 160–148 Ma (Li et al. 2016a, b; Fan et al. 2016). A magmatic gap (the third stage), extending from ca. 148 to 125 Ma (Li et al. 2014a; Fan et al. 2016; Zhu et al. 2016; Wu et al. 2018), has been documented over 1200 km along strike. Then Early Cretaceous magmatic rocks (ca. 125–100 Ma; i.e., the four stage) with various rock types are widespread in both the SQ and BNSZ (Zhu et al. 2016).

Here we propose a subparallel ridge subduction model (Fig. 7.9) to explain the geological observations above for the following reasons (Li et al. 2020). Paleogeography studies indicate that the ridge-transform system of the

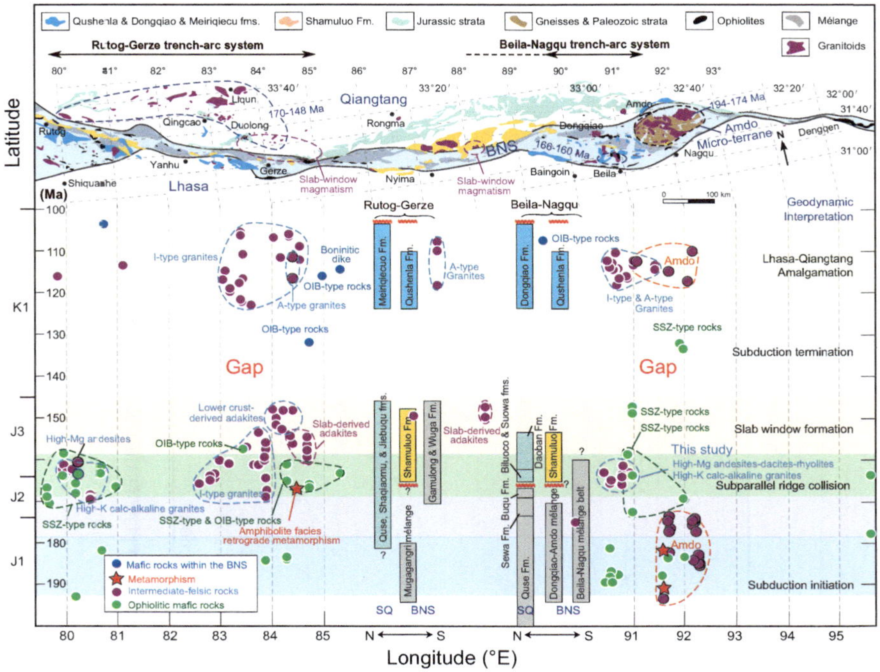

Fig. 7.8 The time-space distributions of Jurassic-Cretaceous geological records, showing the Jurassic-Early Cretaceous magmatism, generation of ophiolites, metamorphism, sedimentation, and tectonism in the SQ, Amdo microterrane, and BNS (from Li et al. 2020)

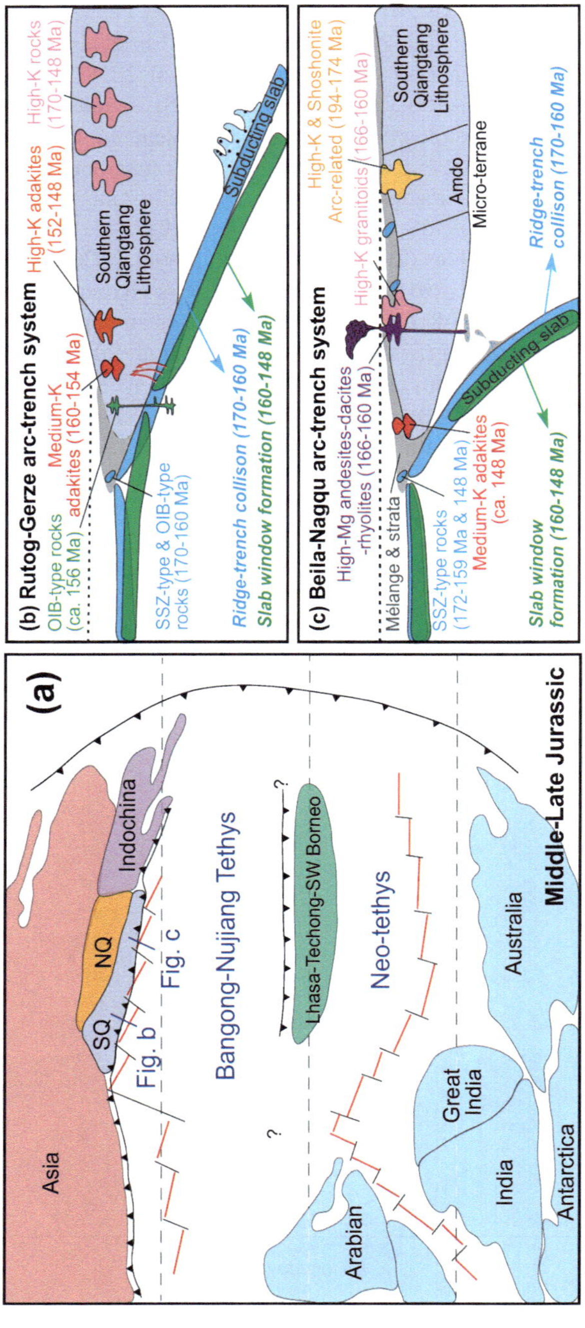

Fig. 7.9 Cartoons of plate reconstructions in the Middle-Late Jurassic timeframe (from Li et al. 2020)

Bangong-Nujiang Tethys was E-W trending and subparallel to the convergent margin (Metcalfe 2013; Gibbons et al. 2015). A collisional event between subparallel mid-ocean ridge and trench can well explain the formation of SSZ-type ophiolites and development of a coeval compressional setting (ca. 170–160 Ma). Subsequently, the mid-ocean ridge would subduct beneath the overlying plate, resulting in the formation of slab windows beneath the forearc and a normal subducting slab beneath the magmatic arc (Li et al. 2020). This geometry of subparallel ridge subduction is consistent with the distribution of slab window magmatic rocks (medium-K adakites and OIB-type rocks) in the southern margin of the SQ (Li et al. 2016a, b) and normal magmatic arc rocks (high-K series) to the north (Li et al. 2014a, b). The magmatic gap (ca. 148–125 Ma) likely reflects subduction termination, because subduction would likely be interrupted by the detachment of the subducted slab and coupling of the divergent ocean-floor slab to the former convergent margin through transform faults (Aragón et al. 2013).

In summary, we propose sequential tectonic events to explain the diverse geological records of the BNS and SQ (Fig. 7.9), including subparallel ridge-trench collision (ca. 170–160 Ma), slab window formation (ca. 160–148 Ma), and subsequent subduction termination (ca. 148–125 Ma).

7.2.3.2 An Arc–Arc "Soft" Collision during Late Jurassic to Early Cretaceous

Two arc-trench systems have been identified on the Southern Qiangtang subterrane and Northern Lhasa subterrane, resulting in a divergent double-sided subduction model for the Bangong-Nujiang Tethys (Zhu et al. 2016). Therefore, the Bangong-Nujiang Tethys were likely closed by soft arc-arc collision (e.g., Soesoo et al. 1997). This inference is further supported by the absence of high-grade metamorphic rocks along the BNSZ during the Early Cretaceous when the ocean closed at ca. 140–130 Ma. This absence is less likely caused by presence of a sedimentary cover or low degrees of exhumation, because (1) the sedimentary cover for the southern Qiangtang subterrane is only about 3.5 km thick (Haines et al. 2003), and (2) the Lhasa-Qiangtang collision zone experienced rapid exhumation from mid-crustal levels at ca. 108–90 Ma (Kapp et al. 2007; Volkmer et al. 2014). Under the framework of arc-arc collision, Zhu et al. (2016) proposed a four-stage collisional model to account for the geological records along the Lhasa-Qiangtang collision zone (Fig. 7.10).

Stage A (>140 Ma): Northward subduction of the Bangong-Nujiang Tethys likely initiated in the Mid-Late Triassic, which is triggered by collision between the Southern-Northern Qiangtang subterranes (Zhu et al. 2013; Zeng et al. 2016a). A well-developed Jurassic arc-trench system and various geological observations likely recorded sequential ridge-trench interactions, including subparallel ridge-trench collision (ca. 170–160 Ma), slab window formation (ca. 160–148 Ma), and subsequent subduction termination (<148 Ma) (see above; Li et al. 2020). Southward subduction of the Bangong-Nujiang Tethys likely initiated in the Middle Permian triggered by collision of the Lhasa Terrane with northern Australia (Zhu et al. 2011,

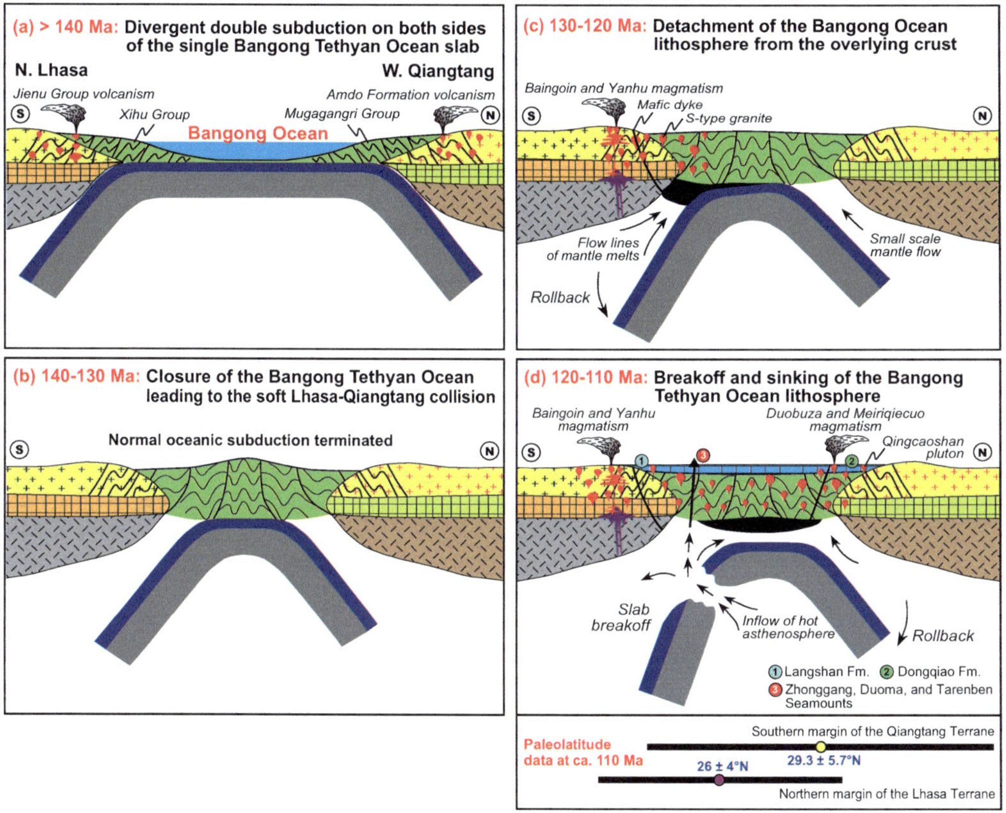

Fig. 7.10 Schematic illustrations showing the closure of the Bangong-Nujiang Ocean driven by a divergent double-sided subduction system in the central Tibet during the Jurassic–Early Cretaceous (not to scale; from Zhu et al. 2016)

2013). The Triassic and Jurassic igneous rocks in the Central Lhasa subterrane (Cao et al. 2016) likely represented the magmatic arc generated by southward subduction of the Bangong-Nujiang Tethys, although more work remain required to document an arc-trench system (arc, fore-arc, and trench).

Stage B (ca. 140–130 Ma): The closure of the Bangong-Nujiang Tethys likely occurred at ca. 140–130 Ma, resulting in soft collision between opposing arcs in the Lhasa and Qiangtang terranes (Fig. 7.10). This collision probably accounted for the angular unconformities between the Shamuluo and Qushenla formations with their underlying strata (Yin et al. 1988; Kapp et al. 2005, 2007; Wang et al. 2013). The scarcity of magmatism during this stage in both northern Lhasa and southern Qiangtang subterranes can be attributed to the termination of normal subduction (Soesoo et al. 1997).

Stage C (ca. 130–120 Ma): In the Southern Qiangtang subterrane, the magmatism was re-activated with weak intensity. This may reflect small-scale mantle flow and limited convective heat (Magni et al. 2012), indicating oceanic lithosphere remains attached to the overlying sedimentary section. In the Northern Lhasa subterrane, various rock types have been identified (Fig. 7.7), including (1) ca. 134–130 Ma metaluminous compositions of the Baingoin Batholith, which yield positive zircon $\varepsilon_{Hf}(t)$ values and indicate a juvenile lower crust source (Zhu et al. 2016); (2) ca. 131–120 Ma medium-K basalts and andesites around Yanhu, implying a mantle wedge source (Li et al. 2018); and (3) ca. 125–120 Ma peraluminous granitoids in the Baingoin Batholith, which have negative zircon $\varepsilon_{Hf}(t)$ values and require sources of sedimentary rocks (Zhu et al. 2016). These rocks suggest decompression and dehydration melting of mantle wedge materials and overlying crust south of the collision zone, reflecting upwelling of materials from the mantle wedge. This inference is consistent with ruptures and detaches of the Bangong-Nujiang Tethyan lithosphere due to gravitational instability (Soesoo et al. 1997) below the "soft" arc–arc collision zone.

Stage D (ca. 120–110 Ma): Magmatism in this stage show a compositional diversity (Fig. 7.7), including from north to south of the ca. 120–104 Ma granitoids around Duobuza in the Southern Qiangtang subterrane, ca. 120–110 Ma OIB-type rocks within the BNSZ, ca. 118–110 Ma granitoids from the Baingoin Batholith, ca. 116–110 Ma high-K basalt, ca. 110 Ma a bimodal suit of basalts and rhyolites, and ca. 112 Ma A-type granites and rhyolites (Sui et al. 2013; Chen et al. 2014; Zhu et al. 2016; Li et al. 2018). These rocks require partial melting of the asthenosphere, metasomatized lithosphere, and overlying crustal rocks (Van de Zedde and Wortel 2001) and generally reflect a slab-window setting. Continued arc-arc collision would lead to sinking, followed by detachment and break-off, of the Bangong-Nujiang Tethyan lithosphere. Such processes will create a gap that is filled with upwelling hot asthenosphere, and can explain the formation of the magmatic rocks through the collision zone (Fig. 7.10).

Acknowledgements Discussions with Peter A. Cawood and Roberto Weinberg are gratefully appreciated. This work was financially supported by the Second Tibetan Plateau Scientific Expedition and Research Program (2019QZKK0702), the Chinese National Natural Science Foundation (grants 42121002 and 91755207), and the 111 project (B18048).

References

Aragón E, Pinotti L, Fernando D, Castro A, Rabbia O, Coniglio J, Demartis M, Hernando I, Cavarozzi CE, Aguilera YE (2013) The Farallon-Aluk ridge collision with South America: implications for the geochemical changes of slab window magmas from fore-to back-arc. Geosci Front 4(4):377–388

Baxter AT, Aitchison JC, Zyabrev SV (2009) Radiolarian age constraints on Mesotethyan Ocean evolution, and their implications for development of the Bangong-Nujiang suture, Tibet. J Geol Soc 166(4):689–694

BGMRXAR (Bureau of Geology Mineral Resources of Xizang Autonomous Region) (1993) Regional geology of Xizang (Tibet) autonomous region. Geological Publishing House, Beijing, pp 1–450. (in Chinese with English abstract)

Cao M, Qin K, Li G, Li J, Zhao J, Evans NJ, Hollings P (2016) Tectono-magmatic evolution of late Jurassic to early cretaceous granitoids in the west Central Lhasa subterrane, Tibet. Gondwana Res 39:386–400

Cao Y, Sun Z, Li H, Pei J, Liu D, Zhang L, Ye X, Zheng Y, He X, Ge C (2019) New paleomagnetic results from middle Jurassic limestones of the Qiangtang terrane, Tibet: constraints on the evolution of the Bangong-Nujiang Ocean. Tectonics 38(1):215–232

Cawood PA, Kröner A, Collins WJ, Kusky TM, Mooney WD, Windley BF (2009) Accretionary orogens through earth history. Geol Soc Lond Spec Publ 318(1):1–36

Chen S-S, Fan W-M, Shi R-D, Xu J-F, Yang K, Pan Z (2021) Early cretaceous volcanic rocks in Yunzhug area, Central Tibet, China, associated with arc–continent collision in the Tibetan plateau? Lithos 380:105827

Chen W, Zhang S, Ding J, Zhang J, Zhao X, Zhu L, Yang W, Yang T, Li H, Wu H (2017) Combined paleomagnetic and geochronological study on cretaceous strata of the Qiangtang terrane, Central Tibet. Gondwana Res 41:373–389

Chen Y, Zhu D-C, Zhao Z-D, Meng F-Y, Wang Q, Santosh M, Wang L-Q, Dong G-C, Mo X-X (2014) Slab breakoff triggered ca. 113 Ma magmatism around Xainza area of the Lhasa terrane, Tibet. Gondwana Res 26(2):449–463

Dai J, Wang C, Polat A, Santosh M, Li Y, Ge Y (2013) Rapid forearc spreading between 130 and 120 Ma: evidence from geochronology and geochemistry of the Xigaze ophiolite, southern Tibet. Lithos 172:1–16

Dilek Y, Furnes H (2009) Structure and geochemistry of Tethyan ophiolites and their petrogenesis in subduction rollback systems. Lithos 113(1–2):1–20

Dilek Y, Furnes H (2014) Ophiolites and their origins. Elements 10(2):93–100

Fan J-J, Li C, Sun Z-M, Xu W, Wang M, Xie C-M (2018) Early Cretaceous MORB-type basalt and A-type rhyolite in northern Tibet: evidence for ridge subduction in the Bangong–Nujiang Tethyan Ocean. J Asian Earth Sci 154:187–201

Fan J-J, Li C, Wu H, Zhang T-Y, Wang M, Chen J-W, Xu J-X (2016) Late Jurassic adakitic granodiorite in the Dong Co area, northern Tibet: Implications for subduction of the Bangong–Nujiang oceanic lithosphere and related accretion of the Southern Qiangtang terrane. Tectonophysics 691:345–361

Fan J-J, Li C, Xie C-M, Wang M (2014) Petrology, geochemistry, and geochronology of the Zhonggang ocean island, northern Tibet: implications for the evolution of the Banggongco–Nujiang oceanic arm of the Neo-Tethys. Int Geol Rev 56(12):1504–1520

Gibbons A, Zahirovic S, Müller R, Whittaker J, Yatheesh V (2015) A tectonic model reconciling evidence for the collisions between India, Eurasia and intra-oceanic arcs of the central-eastern Tethys. Gondwana Res 28(2):451–492

Girardeau J, Marcoux J, Fourcade E, Bassoullet J, Youking T (1985) Xainxa ultramafic rocks, Central Tibet, China: tectonic environment and geodynamic significance. Geology 13(5):330–333

Guilmette C, Hébert R, Dostal J, Indares A, Ullrich T, Bédard É, Wang C (2012) Discovery of a dismembered metamorphic sole in the Saga ophiolitic mélange, South Tibet: assessing an early

cretaceous disruption of the neo-Tethyan supra-subduction zone and consequences on basin closing. Gondwana Res 22(2):398–414
Guynn J, Tropper P, Kapp P, Gehrels GE (2013) Metamorphism of the Amdo metamorphic complex, Tibet: implications for the Jurassic tectonic evolution of the Bangong suture zone. J Metamorph Geol 31(7):705–727
Guynn JH, Kapp P, Pullen A, Heizler M, Gehrels G, Ding L (2006) Tibetan basement rocks near Amdo reveal "missing" Mesozoic tectonism along the Bangong suture, Central Tibet. Geology 34(6):505–508
Haines SS, Klemperer SL, Brown L, Jingru G, Mechie J, Meissner R, Ross A, Wenjin Z (2003) INDEPTH III seismic data: from surface observations to deep crustal processes in Tibet. Tectonics 22(1):1305
Hao L-L, Wang Q, Wyman DA, Ou Q, Dan W, Jiang Z-Q, Wu F-Y, Yang J-H, Long X-P, Li J (2016) Underplating of basaltic magmas and crustal growth in a continental arc: evidence from late Mesozoic intermediate–felsic intrusive rocks in southern Qiangtang, Central Tibet. Lithos 245:223–242
Hu P-Y, Zhai Q-G, Jahn B-M, Wang J, Li C, Chung S-L, Lee H-Y, Tang S-H (2017) Late Early Cretaceous magmatic rocks (118–113 Ma) in the middle segment of the Bangong–Nujiang suture zone, Tibetan Plateau: evidence of lithospheric delamination. Gondwana Res 44:116–138
Huang Q-T, Liu W-L, Xia B, Cai Z-R, Chen W-Y, Li J-F, Yin Z-X (2017) Petrogenesis of the Majiari ophiolite (western Tibet, China): implications for intra-oceanic subduction in the Bangong–Nujiang Tethys. J Asian Earth Sci 146(Supplement C):337–351
Huang T-T, Xu J-F, Chen J-L, Wu J-B, Zeng Y-C (2016) Sedimentary record of Jurassic northward subduction of the Bangong–Nujiang Ocean: insights from detrital zircons. Int Geol Rev 59:1–19
Kapp P, DeCelles PG (2019) Mesozoic–Cenozoic geological evolution of the Himalayan-Tibetan orogen and working tectonic hypotheses. Am J Sci 319(3):159–254
Kapp P, DeCelles PG, Gehrels GE, Heizler M, Ding L (2007) Geological records of the Lhasa-Qiangtang and indo-Asian collisions in the Nima area of Central Tibet. Geol Soc Am Bull 119(7–8):917–933
Kapp P, Murphy MA, Yin A, Harrison TM, Ding L, Guo JH (2003) Mesozoic and Cenozoic tectonic evolution of the Shiquanhe area of western Tibet. Tectonics 22(4):1029–1054
Kapp P, Yin A, Harrison TM, Ding L (2005) Cretaceous-tertiary shortening, basin development, and volcanism in Central Tibet. Geol Soc Am Bull 117(7–8):865–878
Lai W, Hu X, Garzanti E, Sun G, Garzione CN, BouDagher Fadel M, Ma A (2019) Initial growth of the northern Lhasaplano, Tibetan plateau in the early late cretaceous (ca. 92 Ma). Bulletin 131(11–12):1823–1836
Li J, Xia B, Xia L, Xu L, Liu W, Cai Z, Yang Z (2013) Geochronology of the Dong Tso ophiolite and the tectonic environment. Acta Geol Sin 87(6):1604–1616
Li J-X, Qin K-Z, Li G-M, Richards JP, Zhao J-X, Cao M-J (2014a) Geochronology, geochemistry, and zircon Hf isotopic compositions of Mesozoic intermediate–felsic intrusions in Central Tibet: petrogenetic and tectonic implications. Lithos 198:77–91
Li L, Garzione CN, Fan M, Li X, Li X (2019) Jurassic sedimentation in the south-central Qiangtang terrane reveals successive terrane collisions in Central Tibet. Geosphere 15(2):433–449
Li S, Guilmette C, Ding L, Xu Q, Fu J-J, Yue Y-H (2017a) Provenance of Mesozoic clastic rocks within the Bangong-Nujiang suture zone, Central Tibet: implications for the age of the initial Lhasa-Qiangtang collision. J Asian Earth Sci 147(Supplement C):469–484
Li S-M, Wang Q, Zhu D-C, Cawood PA, Stern RJ, Weinberg R, Zhao Z, Mo X-X (2020) Reconciling orogenic drivers for the evolution of the Bangong-Nujiang Tethys during middle-late Jurassic. Tectonics 39(2):e2019TC005951
Li S-M, Wang Q, Zhu D-C, Stern RJ, Cawood PA, Sui Q-L, Zhao Z (2018) One or two early cretaceous arc systems in the Lhasa Terrane, southern Tibet. J Geophys Res Solid Earth 123(5):3391–3413

Li SM, Zhu DC, Wang Q, Zhao Z, Zhang LL, Liu SA, Chang QS, Lu YH, Dai JG, Zheng YC (2016a) Slab-derived adakites and subslab asthenosphere-derived OIB-type rocks at 156 ± 2 Ma from the north of Gerze, central Tibet: records of the Bangong–Nujiang oceanic ridge subduction during the Late Jurassic. Lithos 262:456–469

Li SM, Zhu DC, Wang Q, Zhao ZD, Sui QL, Liu SA, Liu D, Mo XX (2014b) Northward subduction of Bangong–Nujiang Tethys: Insight from Late Jurassic intrusive rocks from Bangong Tso in western Tibet. Lithos 205:284–297

Li Y, He J, Han Z, Wang C, Ma P, Zhou A, Liu S-A, Xu M (2016b) Late Jurassic sodium-rich adakitic intrusive rocks in the southern Qiangtang terrane, central Tibet, and their implications for the Bangong–Nujiang Ocean subduction. Lithos 245:34–46

Li Z, Ding L, Song P, Fu J, Yue Y (2017b) Paleomagnetic constraints on the paleolatitude of the Lhasa block during the Early Cretaceous: implications for the onset of India–Asia collision and latitudinal shortening estimates across Tibet and stable Asia. Gondwana Res 41:352–372

Lippert PC, van Hinsbergen DJ, Dupont-Nivet G (2014) Early cretaceous to present latitude of the central proto-Tibetan plateau: a paleomagnetic synthesis with implications for Cenozoic tectonics, paleogeography, and climate of Asia. Geol Soc Am Spec Pap 507:1–21

Liu T, Zhai Q-g, Wang J, Bao P-s, Qiangba Z, Tang S-h, Tang Y (2016) Tectonic significance of the Dongqiao ophiolite in the north-central Tibetan plateau: evidence from zircon dating, petrological, geochemical and Sr–Nd–Hf isotopic characterization. J Asian Earth Sci 116:139–154

Liu W-L, Xia B, Zhong Y, Cai J-X, Li J-F, Liu H-F, Cai Z-R, Sun Z-L (2014) Age and composition of the Rebang co and Julu ophiolites, Central Tibet: implications for the evolution of the Bangong Meso-Tethys. Int Geol Rev 56(4):430–447

Ma A, Hu X, Garzanti E, Han Z, Lai W (2017) Sedimentary and tectonic evolution of the southern Qiangtang basin: implications for the Lhasa-Qiangtang collision timing. J Geophys Res Solid Earth 122(7):4790–4813

Ma A, Hu X, Kapp P, Han Z, Lai W, BouDagher-Fadel M (2018a) The disappearance of a late Jurassic remnant sea in the southern Qiangtang block (Shamuluo formation, Najiangco area): implications for the tectonic uplift of Central Tibet. Palaeogeogr Palaeoclimatol Palaeoecol 506:30–47

Ma Y, Yang T, Bian W, Jin J, Wang Q, Zhang S, Wu H, Li H, Cao L (2018b) A stable southern margin of Asia during the cretaceous: paleomagnetic constraints on the Lhasa-Qiangtang collision and the maximum width of the neo-Tethys. Tectonics 37(10):3853–3876

Maffione M, Van Hinsbergen DJ, Koornneef LM, Guilmette C, Hodges K, Borneman N, Huang W, Ding L, Kapp P (2015) Forearc hyperextension dismembered the south Tibetan ophiolites. Geology 43(6):475–478

Magni V, Van Hunen J, Funiciello F, Faccenna C (2012) Numerical models of slab migration in continental collision zones. Solid Earth 3(2):293–306

Manatschal G, Müntener O (2009) A type sequence across an ancient magma-poor ocean–continent transition: the example of the western Alpine Tethys ophiolites. Tectonophysics 473(1–2):4–19

Metcalfe I (2013) Gondwana dispersion and Asian accretion: tectonic and palaeogeographic evolution of eastern Tethys. J Asian Earth Sci 66:1–33

Murphy M, Yin A, Harrison T, Durr S, Ryerson F, Kidd W (1997) Did the Indo-Asian collision alone create the Tibetan plateau? Geology 25(8):719–722

Pan GT (2006) Spatial temporal framework of the Gangdese Orogenic Belt and its evolution. Acta Petrol Sin 22:521–533

Pan GT, Wang LQ, Li RS, Yuan SH, Ji WH, Yin FG, Zhang WP, Wang BD (2012) Tectonic evolution of the Qinghai-Tibet plateau. J Asian Earth Sci 53:3–14

Pan GT, Zheng HX, Xu YR, Wang PS, Jiao SP (1983) A preliminary study on Bangong Co-Nujiang Suture. Geological Memoirs of Qinghai–Xizang Plateau (12)—Geological Tectonics of "Sanjiang". Geological Publishing House, Beijing, pp 229–242. (in Chinese with English abstract)

Pearce JA (2014) Immobile element fingerprinting of ophiolites. Elements 10(2):101–108

Piccardo GB, Padovano M, Guarnieri L (2014) The Ligurian Tethys: mantle processes and geodynamics. Earth Sci Rev 138:409–434

Raterman NS, Robinson AC, Cowgill ES (2014) Structure and detrital zircon geochronology of the Domar fold-thrust belt: evidence of pre-Cenozoic crustal thickening of the western Tibetan plateau. Geol Soc Am Spec Pap 507:89–104

Schellart W, Lister G, Toy V (2006) A late cretaceous and Cenozoic reconstruction of the Southwest Pacific region: tectonics controlled by subduction and slab rollback processes. Earth Sci Rev 76(3–4):191–233

Shi R, Yang J, Xu Z, Qi X (2008) The Bangong Lake ophiolite (NW Tibet) and its bearing on the tectonic evolution of the Bangong–Nujiang suture zone. J Asian Earth Sci 32(5–6):438–457

Shi RD (2007) SHRIMP dating of the Bangong Lake SSZ-type ophiolite: constraints on the closure time of ocean in the Bangong Lake-Nujiang River, northwestern Tibet. Chin Sci Bull 52(7):936–941

Soesoo A, Bons PD, Gray DR, Foster DA (1997) Divergent double subduction: tectonic and petrologic consequences. Geology 25(8):755–758

Sui QL, Wang Q, Zhu DC, Zhao ZD, Chen Y, Santosh M, Hu ZC, Yuan HL, Mo XX (2013) Compositional diversity of ca. 110 Ma magmatism in the northern Lhasa terrane, Tibet: implications for the magmatic origin and crustal growth in a continent–continent collision zone. Lithos 168:144–159

Sun G, Hu X, Xu Y, BouDagher-Fadel MK (2019) Discovery of middle Jurassic trench deposits in the Bangong-Nujiang suture zone: implications for the timing of Lhasa-Qiangtang initial collision. Tectonophysics 750:344–358

Sun LX (2005) Late Jurassic–Cretaceous sedimentary response to collision process in middle Bangonghu–Nujiang suture. China University of Geosciences, Beijing, pp 1–121. (in Chinese with English abstract)

Tang Y, Zhai Q-G, Chung S-L, Hu P-Y, Wang J, Xiao X-C, Song B, Wang H-T, Lee H-Y (2020) First mid-ocean ridge-type ophiolite from the Meso-Tethys suture zone in the north-central Tibetan plateau. GSA Bull 132:2202–2220. https://doi.org/10.1130/b35500.1

Tang Y, Zhai Q-G, Hu P-Y, Wang J, Xiao X-C, Wang H-T, Tang S-H, Lei M (2018) Rodingite from the Beila ophiolite in the Bangong–Nujiang suture zone, northern Tibet: new insights into the formation ofophiolite-related rodingite. Lithos 316:33–47

Van de Zedde D, Wortel M (2001) Shallow slab detachment as a transient source of heat at midlithospheric depths. Tectonics 20(6):868–882

Volkmer JE, Kapp P, Horton BK, Gehrels GE, Minervini JM, Ding L (2014) Northern Lhasa thrust belt of Central Tibet: evidence of cretaceous–early Cenozoic shortening within a passive roof thrust system. Geol Soc Am Spec Pap 507:59–70

Wang B-D, Wang L-Q, Chung S-L, Chen J-L, Yin F-G, Liu H, Li X-B, Chen L-K (2016) Evolution of the Bangong–Nujiang Tethyan ocean: Insights from the geochronology and geochemistry of mafic rocks within ophiolites. Lithos 245:18–33

Wang J-G, Hu X, Garzanti E, Ji W-Q, Liu Z-C, Liu X-C, Wu F-Y (2017) Early cretaceous topographic growth of the Lhasaplano, Tibetan plateau: constraints from the Damxung conglomerate. J Geophys Res Solid Earth 122(7):5748–5765

Wang LQ, Pan GT, Ding J, Yao DS (2013) Guidebook of 1:1,500,000 geological map of the Qinghai–Xizang (Tibet) Plateau and adjacent areas (in Chinese with English abstract). Geological Publishing House, Beijing, pp 1–288

Wang MZ, Dong DY (1984) Stromatoporoids from the Dongqiao Formation (Upper Jurassic–Lower Cretaceous) in northern Xizang (Tibet). Acta Palaeontol Sin 23:343–352. (in Chinese with English abstract)

Wang W-L, Aitchison JC, Lo C-H, Zeng Q-G (2008) Geochemistry and geochronology of the amphibolite blocks in ophiolitic mélanges along Bangong-Nujiang suture, Central Tibet. J Asian Earth Sci 33(1–2):122–138

Wu H, Li C, Xu M, Li X (2015) Early Cretaceous adakitic magmatism in the Dachagou area, northern Lhasa terrane, Tibet: Implications for slab roll-back and subsequent slab break-off of the lithosphere of the Bangong–Nujiang Ocean. J Asian Earth Sci 97:51–66

Wu H, Qiangba Z, Li C, Wang Q, Gesang W, Ciren O, Basang D (2018) Geochronology and geochemistry of early cretaceous granitic rocks in the Dongqiao area, Central Tibet: implications for magmatic origin and geological evolution. J Geol 126(2):249–260
Xiong Q, Griffin WL, Zheng J-P, O'Reilly SY, Pearson NJ, Xu B, Belousova EA (2016) Southward trench migration at ~130–120 Ma caused accretion of the neo-Tethyan forearc lithosphere in Tibetan ophiolites. Earth Planet Sci Lett 438:57–65
Xu M, Li C, Zhang X, Wu Y (2014) Nature and evolution of the neo-Tethys in Central Tibet: synthesis of ophiolitic petrology, geochemistry, and geochronology. Int Geol Rev 56(9):1072–1096
Xu R-H, Schärer U, Allègre CJ (1985) Magmatism and metamorphism in the Lhasa block (Tibet): a geochronological study. J Geol 93(1):41–57
Yan L-L, Zhang K-J (2020) Infant intra-oceanic arc magmatism due to initial subduction induced by oceanic plateau accretion: a case study of the Bangong Meso-Tethys, Central Tibet, western China. Gondwana Res 79:110–124
Yin J, Xu J, Liu C, Li H (1988) The Tibetan plateau: regional stratigraphic context and previous work. Philos Trans R Soc Lond Ser A Math Phys Sci 327(1594):5–52
Zeng M, Zhang X, Cao H, Ettensohn FR, Cheng W, Lang X (2016a) Late Triassic initial subduction of the Bangong-Nujiang Ocean beneath Qiangtang revealed: stratigraphic and geochronological evidence from Gaize, Tibet. Basin Res 28(1):147–157
Zeng X-W, Wang M, Fan J-J, Li C, Xie C-M, Liu Y-M, Zhang T-Y (2018b) Geochemistry and geochronology of gabbros from the Asa ophiolite, Tibet: implications for the early cretaceous evolution of the Meso-Tethys Ocean. Lithos 320:192–206
Zeng Y-C, Chen J-L, Xu J-F, Wang B-D, Huang F (2016b) Sediment melting during subduction initiation: geochronological and geochemical evidence from the Darutso high-Mg andesites within ophiolite melange, Central Tibet. Geochem Geophys Geosyst 17(12):4859–4877
Zeng Y-C, Xu J-F, Chen J-L, Wang B-D, Kang Z-Q, Huang F (2018a) Geochronological and geochemical constraints on the origin of the Yunzhug ophiolite in the Shiquanhe–Yunzhug–Namu Tso ophiolite belt, Lhasa terrane, Tibetan plateau. Lithos 300:250–260
Zeng Y-C, Xu J-F, Huang F, Li M-J, Chen Q (2020) Generation of the 105–100 Ma Dagze volcanic rocks in the North Lhasa terrane by lower crustal melting at different temperature and depth: implications for tectonic transition. GSA Bull 132(5–6):1257–1272
Zhang K-J, Xia B, Zhang Y-X, Liu W-L, Zeng L, Li J-F, Xu L-F (2014b) Central Tibetan Meso-Tethyan oceanic plateau. Lithos 210:278–288
Zhang Q-H, Ding L, Cai F-L, Xu X-X, Zhang L-Y, Xu Q, Willems H (2011) Early cretaceous Gangdese retroarc foreland basin evolution in the Selin Co basin, Central Tibet: evidence from sedimentology and detrital zircon geochronology. Geol Soc Lond Spec Publ 353(1):27–44
Zhang X, Shi R, Huang Q, Liu D, Gong X, Chen S, Wu K, Yi G, Sun Y, Ding L (2014a) Early Jurassic high-pressure metamorphism of the Amdo Terrane, Tibet: constraints from zircon U–Pb geochronology of mafic granulites. Gondwana Res 26(3–4):975–985
Zhang Y, Zhang K, Li B, Wang Y, Wei Q, Tang X (2007) Zircon SHRIMP U–Pb geochronology and petrogenesis of the plagiogranites from the Lagkor Lake ophiolite, Gerze, Tibet, China. Chin Sci Bull 52(5):651–659
Zhong Y, Liu W-L, Xia B, Liu J-N, Guan Y, Yin Z-X, Huang Q-T (2017) Geochemistry and geochronology of the Mesozoic Lanong ophiolitic mélange, northern Tibet: implications for petrogenesis and tectonic evolution. Lithos 292:111–131
Zhong Y, Xia B, Liu W-L, Yin Z-X, Hu X-C, Huang W (2015) Geochronology, petrogenesis and tectonic implications of the Jurassic Namco–Renco ophiolites, Tibet. Int Geol Rev 57(4):508–528
Zhu D, Pan G, Mo X, Wang L, Zhao Z, Liao Z, Geng Q, Dong G (2006) Identification for the Mesozoic OIB-type basalts in Central Qinghai-Tibetan Plateau: geochronology, geochemistry and their tectonic setting. Acta Petrol Sin 09:1312–1328
Zhu D-C, Li S-M, Cawood PA, Wang Q, Zhao Z-D, Liu S-A, Wang L-Q (2016) Assembly of the Lhasa and Qiangtang Terranes in Central Tibet by divergent double subduction. Lithos 245(Supplement C):7–17

Zhu DC, Mo XX, Niu YL, Zhao ZD, Wang LQ, Liu YS, Wu FY (2009) Geochemical investigation of early cretaceous igneous rocks along an east-west traverse throughout the Central Lhasa Terrane. Tibet Chem Geol 268(3–4):298–312
Zhu D-C, Wang Q, Chung S-L, Cawood PA, Zhao Z-D (2019) Gangdese magmatism in southern Tibet and India–Asia convergence since 120 Ma. Geol Soc Lond Spec Publ 483(1):583–604
Zhu DC, Zhao ZD, Niu YL, Dilek Y, Hou ZQ, Mo XX (2013) The origin and pre-Cenozoic evolution of the Tibetan Plateau. Gondwana Res 23(4):1429–1454
Zhu DC, Zhao ZD, Niu YL, Mo XX, Chung SL, Hou ZQ, Wang LQ, Wu FY (2011) The Lhasa Terrane: record of a microcontinent and its histories of drift and growth. Earth Planet Sci Lett 301(1–2):241–255

Chapter 8
Crustal Growth in Continental Collision Zones and Crustal "Destruction" in Continental Interiors by Granitoid Magmatism

Abstract The standard island arc model has served the community well for understanding continental crust accretion because island arc andesites satisfy the andesitic composition of the bulk continental crust and because both continental crust and island arc rocks share a common geochemical "arc signature" characterized by relative enrichment in fluid-soluble elements (e.g., U, K, Pb) and depletion in fluid-insoluble elements (e.g., Nb, Ta, Ti). However, rigorous analysis of geological, petrological and geochemical observations can readily reveal fundamental shortcomings of the island arc model, which, among others, include: (1) primary arc magmas are *not* andesitic, but basaltic; (2) magmatic crust production at arcs is mass-balanced by subduction erosion and sediment recycling with no net crust accretion; (3) arc magmatic rocks are overall more depleted in incompatible elements than needed for continental crust; (4) island arc magmatism cannot explain the episodic continental crust growth over earth's history. Our collaborative studies on the ~55 ± 5 Ma syncollisional granitoids associated with the India-Asia collision in southern Tibet (Mo et al. 2007, 2008, 2009) led to the hypothesis that continental collision zones are primary sites of net continental crustal growth (Niu et al. 2013). Our test on syncollisional granitoids in older orogens in the vast Tibetan Plateau and the adjacent regions (i.e., the ~220 Ma West Kunlun, ~240 Ma East Kunlun, ~440 Ma Qilian and ~220 Ma West Qinling orogens) fully support this hypothesis. We also test the hypothesis that the Cretaceous granitoids along the southeast coastline of continental China mark the end of paleo-Pacific subduction because of the trench jam caused by the arrival and collision of the Chinese continental shelf (the basement of East and South China Seas) of exotic origin at ~100 Ma (Niu et al. 2015). The latter may be an oceanic plateau, but is more likely a continental mass probably affiliated with western Australia in its history that is to be further tested. Our study of the Jurassic-Cretaceous granitoids in the interiors of continental China supports the hypothesis that these granitoids represent crustal response to the lithosphere thinning as the result of basal hydration weakening with the water coming from the paleo-Pacific slab stagnant in the mantel transition zone (Niu et al. 2015).

Prof. Yaoling Niu has fully contributed to this chapter.

X. Mo et al., *Magmatism and Crustal Evolution of the Tibetan Plateau*, Modern Approaches in Solid Earth Sciences 25,
https://doi.org/10.1007/978-981-97-2109-2_8

Accompanying the widespread lithosphere thinning in the vast region of eastern continental China is the generation of basaltic magmas whose intrusion and underplating caused the deep crustal anatexis and widespread granitoids randomly distributed throughout eastern China in time (~190–88 Ma with a peak at ~140 Ma) and space (in a wide E-W zone in excess of 1000 km). These granitoids of deep crustal origin contribute no new crustal accretion, but "destruct" the existing lower crust with the mafic granulite residues becoming the sub-Moho constituents, which resolves the puzzle (Gao et al. 1998a, b) why eastern China has thinner and less mafic lower crust that is also detected seismically (Li et al. 2006). Granitoid magmatism can thus both contribute to juvenile crustal accretion and destruct the existing crust in response to varying aspects of plate tectonics.

Keywords Shortcomings of island arc model · Syncollisional granitoids and net continental crust accretion · Intra-continental granitoids and crustal destruction as a response to lithosphere thinning

8.1 The History and Context

Compared with other planets in the solar system, e.g., our immediate neighbors Mars and Venus, the Earth is unique in many ways. It has (1) continents and ocean basins (bimodal topography), (2) granitic continental crust and basaltic oceanic crust, (3) plate tectonics, (4) massive hydrosphere, (5) free oxygen-bearing atmosphere, (6) numerous life forms of varying sophistication and (7) advanced forms of life like the mankind sustained by the hydrous sphere and oxygen-rich atmosphere. The *Goldilocks principle* has been conveniently used to explain all these unique features. That is, the Earth is at right distance from the sun with Earth's surface temperature being just about right—warm enough to maintain water bodies (oceans) without being evaporated if it were too hot and without being frozen if it were too cold (i.e., the "habitable zone"). Importantly, the earth's magnetic field would be required to protect surface water and atmosphere from being blown away by the solar wind. If the latter is proven to be true, then the *Goldilocks principle* would be at least insufficient because the formation and presence of the metallic core as the geodynamo for the magnetic field must be considered. Much research in all natural science subject areas is needed to sort out the causality among these unique Earth features, but "No water, no granites—no oceans, no continents" (Campbell and Taylor 1983) is a logical argument for the geologist in terms of the understood experimental petrology, geological observations and the earth's plate tectonics that is absent elsewhere in the solar system. Therefore, a genuine understanding of the origin and evolution of continental crust, which is the most accessible part of the solid earth on which we live, is imperative to address the above whole-earth problems as well as to develop novel strategies for mineral exploration.

Indeed, it has long been the quest of the geologist for the origin of the continental crust (CC) and this quest continues as we witness the mounting literature with new data, new ideas and debate. Despite the expanding literature, answers to fundamental questions remain unsatisfactory. These include when and where on earth the first continental crust was generated, the way and rate of its growth, and how it has acquired the "andesitic" composition from mantle derived magmas. The first-order complementarity in incompatible element abundances between continental crust and ocean crust led to the speculation that the incompatible-element enriched bulk continental crust (BCC) represents the extract of a small amount of very low-degree (~1.5% melting) melt from the primitive mantle in Earth's early history (e.g., Hofmann 1988), which left an incompatible-element depleted residue in the upper mantle that became the source of the present-day mid-ocean ridge basalts (MORB), widely known as depleted MORB mantle (DMM) (e.g., Gast 1968; O'Nions et al. 1979; Allègre et al. 1983; Zindler and Hart 1986). However, in what sort of geological setting and how such very low-degree melting took place in Earth's early history is illusive and is at best a hypothesis that cannot yet be tested (Niu and O'Hara 2009).

When the first CC formed/appeared on Earth has attracted increasing interest in recent years (e.g., Hawkesworth et al. 2019; Cawood and Hawkesworth 2019), but uncertainties remain because of the likely preservation problem (Clift and Vannucchi 2004; Scholl and von Huene 2007; Niu and O'Hara 2009; Hawkesworth et al. 2009; Condie et al. 2017; Cawood and Hawkesworth 2019). The oldest preserved crustal rocks we know of at present are the Hadean age ~4.0–4.2 Ga tonalitic Acasta Gneiss (Bowring and Housh 1995; Iizuka et al. 2007) in northern Canada, and granitoid rocks may have existed even earlier at ~4.35 Ga as inferred from "granitic" zircons from Jack Hills in western Australia (Cavosie et al. 2004; Harrison et al. 2005). Hadean rocks are rare and the more abundant early granitoids are Archean age TTG-granite-greenstone belt complexes (see Condie and Aster 2009; Ernst 2009). The TTG (tonalite-trondhjemite-granodiorite) assemblages are considered by many as the primary architecture of the CC (e.g., Rollinson 2008; Condie and Aster 2009). The origin of the TTGs has been extensively studied and discussed over the last decades with ideas including shallow mantle melting and fractional crystallization, partial melting of eclogite/amphibolite, and incipient melting of hydrous, fertile upper mantle peridotite etc. (see review by Ernst 2009) However, the compositional similarity (Fig. 8.1) between TTGs and the Phanerozoic subduction-related adakites (e.g., Defant and Drummond 1990; Castillo 2006), sanukitoids (e.g., Tatsumi 2008) or high-Mg andesites (e.g., Kelemen 1995) has led to the popular acceptance that the TTGs must have formed in subduction settings (e.g., Condie 2005; Martin et al. 2005), suggesting the existence of plate tectonics back then despite the continued debate. While the latter proposal is interesting, it is misleading because if the Archean TTGs were indeed the primary architecture of CC, then the BCC composition (see Rudnick and Gao 2003) would closely resemble that of the TTGs, but this is not the case. Figure 8.1 clearly shows that the TTGs and modern adakites have a conspicuous "garnet signature" characterized by depletion of heavy rare earth elements (REEs) and of progressively heavier REEs ($[Dy/Yb]_N > 1.5$) that the BCC and most crustal rocks do not have ($(Dy/Yb)_N \approx 1.0$), a simple fact recognized 45 years ago (Taylor 1977). This observational discrepancy does not deny the possible significance of TTG production in the Archean but is informative that our primary effort to understand CC accretion must be focused on magmatism that

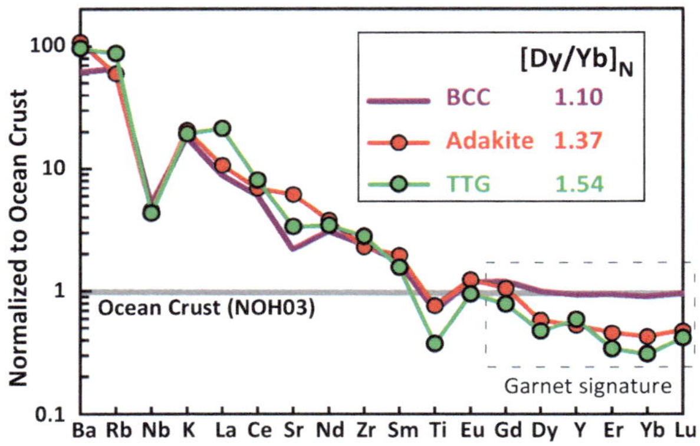

Fig. 8.1 Multi-element diagram normalized to the model composition of the ocean crust (Niu and O'Hara 2003) to show the compositional similarity of the Archean TTGs and the Phanerozoic adakites (using averaged data of Martin et al. 2005) with remarkable depletion of Y and other heavy rare earth elements (HREEs), which is known as the "garnet signature", indicating the presence of garnet as a residual phase during melting. Model bulk continental crust (BCC; Rudnick and Gao 2003) is also plotted for comparison, showing that the BCC does not have the "garnet signature". This comparison demonstrates in simple clarity that Archean TTG rocks, albeit geologically important, do not contribute significantly to the mass of the present-day bulk continental crust. Modified from Niu et al. (2013)

produces the mean composition of the present-day BCC with no significant "garnet signature" (i.e., $Dy/Yb]_N \approx 1.0 \pm 0.1$; Fig. 8.1).

Taylor (1967) proposed with wisdom the "andesite" model for the origin and growth of the CC based on andesitic composition of the BCC and the abundant phanerozoic andesites along the Pacific margin of North America and island arcs such as the Aleutians, Marianas and Solomons in the northern and western Pacific. The fact that these island-arc andesites are geographically removed far from continental masses led to the idea of juvenile andesitic crust accretion of mantle origin. The discovery of the plate tectonics and the improved understanding of magmatism above subduction zones has led to the much more rigorous "island-arc model" for the origin and growth of the CC (Taylor 1977), which has since become the standard CC growth model widely accepted to this day. This model is particularly convincing because compared with mantle-derived melts from all other tectonic settings on earth, island-arc magmas show remarkable similarity to the BCC in incompatible element patterns characteristically depleted in fluid-insoluble elements (e.g., Nb, Ta and Ti) and enriched in fluid-soluble elements (e.g., U, K and Pb) (Fig. 8.2), pointing to possible genetic connection between seafloor subduction and CC accretion. Most studies agreed that the CC continues to grow since the Archean (Fig. 8.3; See Campbell 2003; Condie 2005; Dhuime et al. 2012; Hawkesworth et al. 2019), while also recognizing that such continuum represents a cumulative growth dominated by a number of magmatic episodes (Fig. 8.3; McCulloch and Bennett 1994; Condie 1998, 2000; Condie et al. 2017), interpreted to have resulted from pulses of mantle derived melts, perhaps in the form of mantle plumes (Stein and Hofmann 1994; Abbott and Mooney 1995; Abbott et al. 1997; Polat et al. 1998; Albarède 1998;

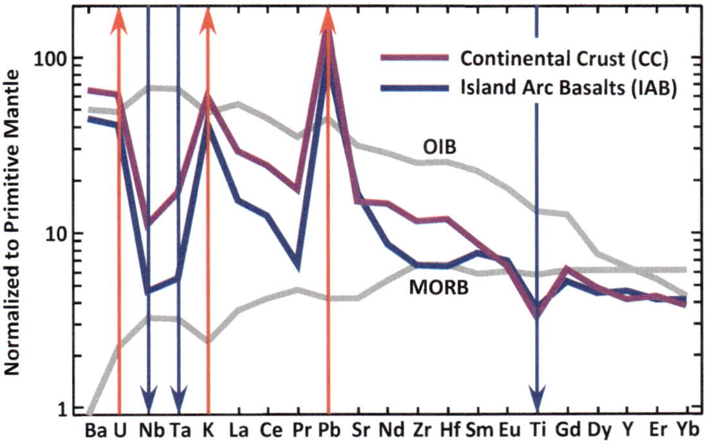

Fig. 8.2 Multi-element diagram normalized against the primitive mantle to show that compared to mantle derived melts from mid-ocean ridges (MORB) and intra-plate ocean islands (OIB), island arc basalts (IAB; data from Elliott 2003) and the model composition of bulk continental crust (CC; Rudnick and Gao 2003) display similar patterns—both are relatively depleted in "fluid-insoluble" elements (e.g., Nb, Ta and Ti) and enriched in "fluid-soluble" elements (e.g., Ba, U, K, Pb), which is consistent with the popular model for arc basalt petrogenesis resulting from subducting slab dehydration induced mantle wedge peridotite melting (e.g., Tatsumi and Eggins 1995). The compositional similarity between the bulk CC and IAB was the very reason for the standard "island-arc model" for the CC (Taylor 1977). Data used for primitive mantle, MORB and OIB are from Sun and McDonough (1989). Modified from Niu et al. (2013)

Condie 2000; Niu et al. 2003; Kemp et al. 2006; Hawkesworth and Kemp 2006; Parman 2007; Pearson et al. 2007; Condie et al. 2017; Puetz 2018) although these episodic peaks may not be so striking since 3.0 Ga (Dhuime et al. 2012; Hawkesworth et al. 2019).

As the title of the book states, the magmatism and crustal evolution of the Tibetan Plateau, a unique Earth's surface structure as the best testing ground, preserves material evidence to address fundamental aspects of CC growth. Indeed, our collaborative studies on syncollisional granitoids, especially those in the southern Lhasa Terrane (Fig. 8.4) associated the India-Asia collision at ~55 ± 5 Ma (e.g., Mo et al. 2007, 2008, 2009; Niu and O'Hara 2009), offered insights into the CC accretion and led to the testable hypothesis that continental collision zones are primary sites of net continental crust (Niu et al. 2013), which overcomes obvious shortcomings of the "island-arc" model without the need of mantle plumes that would create more difficulties (see below). With the support of Chinese NSF, we, together with our students, have been able to test this hypothesis with new insights by sampling and studying syncollisional granitoids in older orogens of the Greater Tibetan Plateau and the adjacent regions (i.e., the ~225 Ma West Kunlun, ~240 Ma East Kunlun, ~440 Ma Qilian and ~220 West Qinling orogens: Fig. 8.4). To further verify the uniqueness of syncollisional granitoids as net material contribution to crustal growth, we have also sampled and compared with "intra-plate" granitoids in eastern China (e.g., the Cretaceous "Laoshan" and "Xiao Qingling" granitoids) and

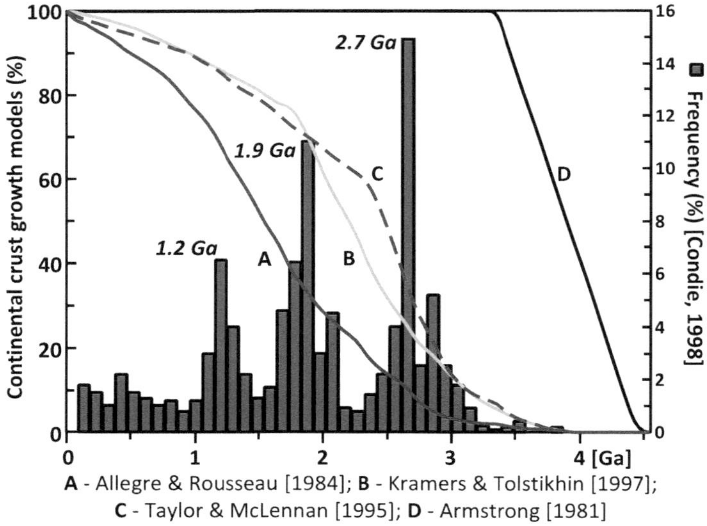

Fig. 8.3 Histogram of "juvenile" continental crust with low-background frequencies interpreted as resulting from plate tectonic processes whereas the peaks as caused by volluminous mantle melting anomalies associated with mantle plumes that are episodic (Condie 2000; Hawkesworth and Kemp 2006; McCulloch and Bennett 1994; Stein and Hofmann 1994). There are numerous published models for continental crust growth, but the selected curves are endmembers that well capture the spectrum: A (Allègre and Rousseau 1984), B (Kramers and Tolstikhin 1997) and C (Taylor and McLennan 1995) and D (Armstrong 1981). Most models favor gradual crustal growth over much of Earth's history, including the Phanerozoic era (Cawood and Hawkesworth 2019; Dhuime et al. 2012; Hawkesworth et al. 2019). Modified from Niu et al. (2013)

granitoids associated with active continental margins (e.g., those along the southeast coastline; Fig. 8.4) as well as granitoids in the northeast and in Hainan Island to the south. While comprehensive studies of all these large sample suites are expected in coming years, the results obtained so far (1) verify the hypothesis of net crustal growth in continental collision zones, (2) offer new perspectives that intraplate granitoid magmatism may in fact modify and even "destruct" the existing deep CC, and (3) further confirm the hypothesis that the Chinese continental shelf (i.e., the shelf of East and South China Seas) is of exotic origin representing an oceanic plateau or more likely a land mass that drifted with the paleo-Pacific plate and collided with continental China at ~100 Ma with the trench/suture lying off the arc-curved southeast coastline (Fig. 8.4; Niu et al. 2015).

In this chapter, we review the merits and shortcomings of the current models on CC growth (Niu et al. 2013). This review is important because much of the community remains unaware of the inadequacies of the standard models. This will allow the community to understand the need of new hypothesis and to better appreciate our understandings on the significance of syncollisional granitoid magmatism that contributes to net crustal growth and the effect of intraplate granitoid magmatism that may potentially destruct the existing crust, both of which are consequences of plate tectonics. Models for CC growth need to consider both syncollisional and "intra-plate" granitoid magmatism.

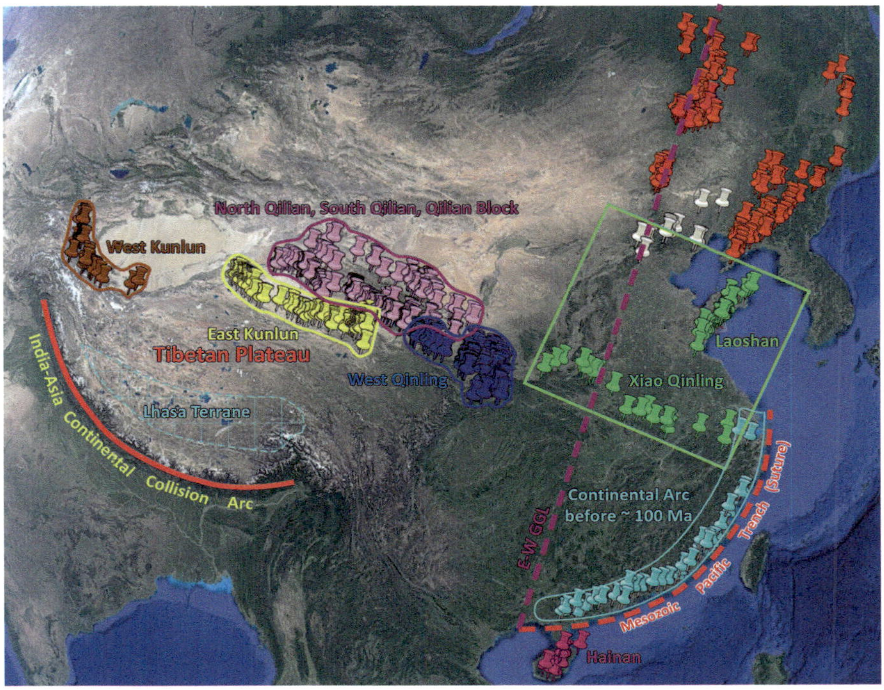

Fig. 8.4 Portion of the global google map highlighting continental China and adjacent regions, showing the Lhasa Terrane in southern Tibet where our syncollisional granitoid studies in a global context has led to the hypothesis "continental collisional zones are primary sites of net continental crust growth (Mo et al. 2007, 2008, 2009; Niu and O'Hara 2009; Niu et al. 2013). To further test this hypothesis, we have sampled and studied syncollisional granitoids in older orogens of West Kunlun (~225 Ma), East Kunlun (~240 Ma), Qilian (~440 Ma North Qilian + South Qilian + Qilian Block) and West Qinling (~220 Ma). For comparison, we also sampled "intra-plate" "Xiao Qinling" (~128–138 Ma) and "Laoshan" (~115–140 Ma) granitoids and magmatic arc granitoids along the southeast coastline (~102 Ma) in eastern continental China plus other granitoids in the Hainan Island and North and Northeast China for future studies. Note that the arc-curved southeast coastline resembles the shape and scale of the India-Asia collision arc, pointing to its nature of active continental magmatic arc setting and marking the fossil trench of paleo-Pacific subduction and suture of the Chinese continental shelf of exotic origin (microcontinent or oceanic plateau origin) colliding with continental China at ~100 Ma (Niu et al. 2015; Niu and Tang 2016; also see Fig. 8.10). Also note the east-west great gradient line (GGL) that marks the contrast between the plateaus in the west and the hilly lowland in the east in terms of crustal thickness, lithosphere thickness, gravity, mantle heat flow as well as surface elevation (see Niu 2005, 2014)

8.2 The Need of a New Hypothesis on Continental Crust Accretion

A new working hypothesis is needed if the existing models become inadequate in accounting for new observations. As mentioned above, the standard island-arc model for continental crust accretion (Taylor 1977) is a reinterpretation of the "andesite model" (Taylor 1967) in terms of the plate tectonics theory in which seafloor subduction must be the very cause of island arc and continental margin andesite magmatism (Ringwood and Green 1966; Green and Ringwood 1966, 1968). The "island arc model" has gained general acceptance because it agreed with the observations and geological understanding 50 years ago. These include (1) andesites erupted on island arcs are consistent with the andesitic BCC composition (Taylor 1967, 1977) and (2) island arc andesites could be generated through partial melting of subducting ocean crust (Ringwood and Green 1966; Green and Ringwood 1966, 1968). Importantly, this model has been further substantiated subsequently by the geochemical "arc signature" of BCC and CC rock compositions characteristically enriched in fluid-soluble elements (e.g., Rb, U, K, Pb) and depleted in fluid-insoluble elements (e.g., Ti, Nb and Ta) (Fig. 8.2). These fundamental observations that led to the island-arc model are the very geochemical footing despite the shortcomings (Niu et al. 2013). In addition, mantle plume hypothesis has also been advocated to explain the episodic growth of the CC as recorded in magmatic rocks (Fig. 8.3) because there was no understandable reason how plate tectonics/seafloor subduction would have been episodic since its inception although it remains debated on when the modern-day style plate tectonics began on Earth (e.g., Stern 2007; van Hunen and Moyen 2012).

Because accretion of juvenile CC is accomplished through magmatism with materials ultimately coming from the mantle consisting of peridotites, any successful model for CC growth must be able to answer three basic questions: (1) Can partial melting of mantle peridotites produce volumetrically significant melts of andesitic composition for the CC? (2) If not, what kind of processes must have converted mantle derived basaltic melts into andesitic composition? (3) What sort of geological processes must have maintained net CC growth, especially the episodic growth (Fig. 8.3)? One can readily note from the recent literature that much of the community remains unfamiliar with the shortcomings of the standard models.

8.2.1 *Bulk Arc Crust Is Too Mafic for the Andesitic Bulk Continental Crust*

As subduction-zone magmatism is interpreted to have resulted from slab-dehydration induced mantle wedge melting (e.g., Gill 1981; Tatsumi and Eggins 1995; Pearce and Peate 1995), the primary arc magmas must be compositionally basaltic. So far, the most primitive arc melts show high MgO (~10–12 wt%) and low SiO_2 (~47–49 wt%) (Nye and Reid 1986; Eggins 1993; Bacon et al. 1995). Therefore, the bulk island arc crust (AC) must be basaltic (e.g., Gill 1981; Arculus 1981; Pearcy et al. 1990). This is in marked contrast from the andesitic BCC composition with low MgO (~4–5 wt%) and high SiO_2 (~57–64 wt%) (Taylor 1967, 1977; Taylor and McLennan 1985, 1995; Wedepohl 1995; Rudnick and Gao 2003). Nevertheless, the Island-arc model remains popular for CC growth largely because of the characteristic "arc signature" (see Fig. 8.2). One of the major efforts over the last 30 years has been to improve the island-arc model by searching for mechanisms to convert the basaltic bulk AC composition to the andesitic bulk CC composition. Removing the more mafic component from the AC such as the mafic/ultramafic lower AC cumulate beneath an active arc or during orogenesis, called delamination, has been a welcome idea (e.g., Kay and Kay 1993; Rudnick 1995; Kelemen et al. 2003; Plank 2005; Lee and Anderson 2015) to explain why only felsic arc lithologies contribute to CC growth, although the perceived physical mechanisms for such removal are physically not straightforward (Jull and Kelemen 2001; Lee and Anderson 2015) because this assumes the absence of lithospheric mantle and free mass exchange between the forming magmatic arc crust and the melt-producing asthenosphere. This is unlikely to be true because mantle peridotite xenoliths have been observed and studied in some island arc volcanic suites (e.g., Parkinson and Arculus 1999; Ionov et al. 2012; Tollan et al. 2015). The presence of mantle lithosphere beneath active volcanic arcs is consistent with compositional systematics of global volcanic arcs (see Niu 2021) and with the development of back-arc spreading centers as evidenced by chemical and isotopic systematics of the Northern Mariana Trough basalts (Chen et al. 2021a, b). That is, the delamination model has physical difficulties in explaining how the mafic deep arc crust founders through the compositionally (major elements) depleted and physically buoyant mantle lithosphere into the denser and melting asthenosphere.

On the other hand, it is not necessary to physically separate the mafic/ultramafic cumulates from the more evolved andesitic rocks because the apparent problem lies in the popular view of the felsic crustal rocks above the Moho and the mafic/ultramafic cumulates (e.g., chromite-bearing dunite, wehrlite, amphibole-rich pyroxenites) below the Moho because the Moho is a seismic discontinuity, not a lithological division (Davidson and Arculus 2006). If this argument is true, then the mantle lithosphere right beneath the continental crust should be dominated by such mafic/ultramafic cumulates. However, xenolith studies show no evidence for this claim.

Importantly, recent seismic studies along the Izu-Bonin-Mariana Island arc system suggest that the mid-upper AC is characterized by rocks of andesitic composition (e.g., diorite), which is interpreted as convincing evidence for continental crust formation at island arcs (Takahashi et al. 2007; Tatsumi 2008; Tatsumi et al. 2008). However, andesitic (even rhyolitic) rocks are common in all island arc systems because hydrous basaltic melts have lowered liquidus temperatures and can undergo advanced degrees of cooling and evolution towards andesitic, dacitic and even rhyolitic melts. This has in fact been used as evidence for the "andesite" (Taylor 1967) and "island arc" (Taylor 1977) models for CC. Furthermore, melting or assimilation of country (existing arc) rocks of magma chambers can also lead to more felsic melts (Worthington et al. 1999; Tamura et al. 2011). Nevertheless, bulk AC is well understood to be of basaltic composition because primary arc magmas are derived from high extents of hydrous mantle wedge melting (Tatsumi and Eggins 1995; Niu 2005). The latter is thus expectedly consistent with the 3-D seismic investigation at the Mariana arc that the bulk AC is indeed basaltic (Calvert et al. 2008). Subducting ocean crust may melt in warm subduction zones (e.g., Kelemen 1995; Tatsumi

2008), but interaction of such melts with the mantle wedge peridotites during ascent can still make primary arc magmas to be basaltic.

8.2.2 Lower Arc Crust Is Too Depleted in Incompatible Elements for Lower Continental Crust

Recently, Kelemen and Behn (2016) demonstrate that the deep AC of magmatic cumulate origin is too depleted in incompatible element abundances to match the required abundance levels for the lower CC, emphasizing the difficulty of the island arc model for CC growth. To reconcile this problem, these authors suggest that it is re-lamination, not delamination, in the arc setting that determines the composition of the deep CC. In this model, they speculate that buoyant magmatic rocks generated at arcs are subducted and upon heating at depth will ascend and re-laminate at the base of the overlying crust. Such re-laminated lower AC will have elevated abundances of incompatible elements to match the lower CC composition. How such re-lamination physically works as a globally general process is unclear, but the importance of this work lies in its recognition of the deficiency of the island arc model for CC growth, i.e., the lower AC is too depleted in incompatible elements to satisfy the lower CC.

8.2.3 Felsic Arc Magmas Are Too Depleted in Incompatible Elements for Upper Continental Crust

In fact, we show here that silicic arc magmas with $SiO_2 = 67.42 \pm 4.8$ wt% (N = 233), equivalent to model upper CC (e.g., 66.62 wt% of Rudnick and Gao 2003), are also too depleted in incompatible elements for the upper CC. Figure 8.5 shows that relative to the BCC, felsic arc magmas (dacite, rhyodacite and rhyolite) are depleted in the progressively more incompatible elements whereas the upper CC is enriched in the progressively more incompatible elements. This observation demonstrates in simple clarity that arc magma differentiation will not produce incompatible element abundances and patterns required by the upper CC.

If basal "re-lamination" is required to enhance the incompatible element abundances of the lower AC to account for the lower CC, then an additional process or processes would be required to raise incompatible element abundances in felsic arc magmas to account for the upper CC. Figure 8.5b–d shows clearly the more evolved high SiO_2 syncollisional granitoids readily explain the abundances and patterns of incompatible elements of the upper CC without invoking any more complex process. We can reiterate with demonstrations that island arc magmatic rocks are not suitable materials for making BCC despite the similar incompatible element patterns with relative enrichment in fluid-soluble elements and depletion in fluid-insoluble elements (see Fig. 8.2).

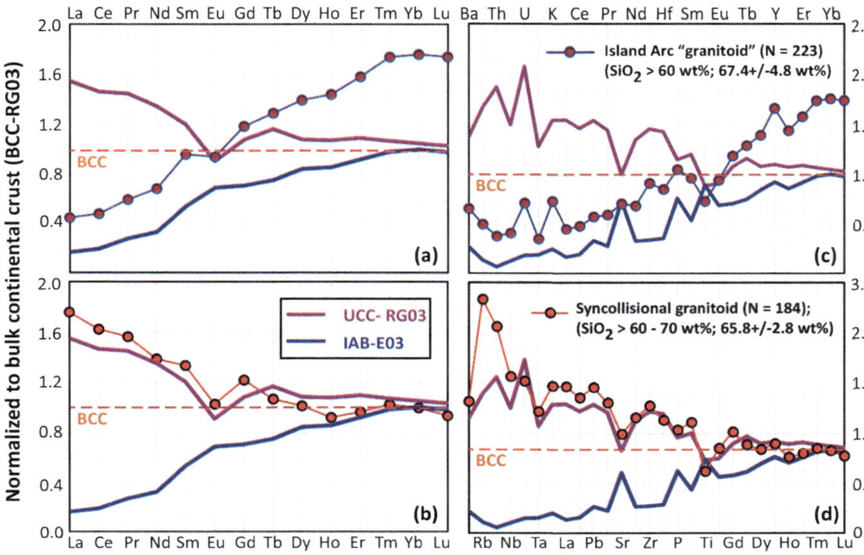

Fig. 8.5 Rare earth element (REE; **a**, **b**) and multiple element (**c**, **d**) patterns to show that normalized to the model bulk continental crust (BCC; Rudnick and Gao 2003), the "granitoids" (felsic volcanic rocks) of global island arcs (**a**, **c**) are depleted in the progressively more incompatible elements than model upper continental crust (UCC-RG03; Rudnick and Gao 2003) whereas the felsic syncollisional granitoids (**b**, **d**) match the upper continental crust well, demonstrating in simple clarity that island arc rocks cannot make continental crust as we propose and continue to test (Niu et al. 2013). For comparison, average composition of island arc basalts (IAB; Elliott 2003) is also plotted. The global arc "granitoid" data are from the literature (Miller et al. 1992; Elburg and Foden 1998; Woodhead et al. 1998, 2010; Ishizuka et al. 2002; Haraguchi et al. 2003; Leat et al. 2003, 2007; Jicha et al. 2004; Lindsay et al. 2005; Finney et al. 2008; Reagan et al. 2008; Schuth et al. 2009; Tamura et al. 2009; Petterson et al. 2011; Cunningham et al. 2012; Hanyu et al. 2012; Turner et al. 2012; Barker et al. 2013). The syncollisional granitoid data are our own published and unpublished (Chen et al. 2015, 2016, 2018; Duan et al. 2016; Hu et al. 2016; Huang et al. 2014, 2015, 2016, 2017; Kong et al. 2017, 2019, 2020; Mo et al. 2007, 2008; Shao et al. 2017)

8.2.4 Arc Settings Have No Net Crustal Mass Addition

If we accept that the CC has been growing over Earth's history (Fig. 8.3), then the rate of new crust production must exceed the rate of crustal loss to maintain the net growth on a global scale. However, studies have suggested that on modern Earth, at least in the Phanerozoic, AC production through subduction-zone magmatism (at both island arcs and Andean-type continental arcs; ~2.3 km^3/yr) is mass-balanced by subduction erosion (~1.5 km^3/yr) and sediment recycling (~1.0 km^3/yr) (von Huene and Scholl 1991; Clift and Vannucchi 2004; Scholl and von Huene 2007; Clift et al. 2009; Stern and Scholl 2010), suggesting that subduction-zone related magmatism is not responsible for net continental crustal growth. Furthermore, isotopic studies suggest that crustal recycling at subduction zones has been important since ~2.0 Ga (Armstrong 1968; Kramers and Tolstikhin 1997). This does not mean that no new crust is forming at arcs, but does suggest that no net crustal buildup is kept at arcs. This observation, together with the fact that the bulk AC is too mafic and too depleted in incompatible elements in both lower and upper crust as elaborated above, demonstrates that active subduction zones are not the sites for net CC growth, thus contributing no new mass to the growing CC (Niu and O'Hara 2009; Niu et al. 2013), at least in the phanerozoic (Scholl and von Huene 2007).

8.2.5 Adakites and High-Mg Andesites Associated with Subduction Zones Are Volumetrically Unimportant

It is common to refer to the Archean TTGs (see above) and the hotly studied adakites, high-Mg andesites and sanukitoids in discussing CC accretion in recent years (see Rollinson and Martin 2005; Kelemen 1995; Kelemen et al. 2003; Castillo 2006, 2012; Tatsumi 2008; Moyen 2009, 2011). Slab melting and melt-mantle interactions are interpreted to be important in producing these rocks that give rise to the BCC composition (see Fig. 8.1). However, formation of these magmas requires both warm slabs (Defant and Drummond 1990; Castillo 2006, 2012) and unusually hot mantle wedges that are rare in modern Earth (Tatsumi 2008). In addition, all these magmas/rocks have the heavy rare earth element (HREE) depleted "garnet signature" as do the most Archean TTGs, but the model BCC composition does not have this "garnet signature" (Fig. 8.1). That is, TTGs (plus the Phanerozoic equivalents all together) are volumetrically insignificant in building the present-day bulk CC mass. This is a simple and clear demonstration that places an undeniable constraint on the validity of any model on continental crust accretion. That is, models for CC growth must be constrained by juvenile granitoid magmatism with no "garnet signature" (i.e., $[Dy/Yb]_N \approx 1.0$; Fig. 8.1).

8.2.6 Arc-Continent Collision Is Relevant, But Again Arc Crust Is Too Mafic

Following the emphasis on juvenile granitoid preservation in the continental collision hypothesis (Niu et al. 2007; Mo et al. 2008; Niu and O'Hara 2009), others also emphasized the significance of "arc-continent collision" and "preservation" for CC growth (e.g., Condie and Aster 2010; Hawkesworth et al. 2009, 2010) as originally proposed for continental amalgamation in response to plate tectonic cycles (or "Wilson Cycle"; Dewey and Bird 1970; Burke and Dewey 1975). However, the arc-continent collision or terrane collision models face the same difficulties: (1) while this facilitates lateral growth of the continent, such as the development of the giant *Central Asian Orogenic Belt* (CAOB) (e.g., Kröner et al. 2007; Liu et al. 2012; Xiao et al. 2013) and Mesozoic/Cenozoic growth in eastern Eurasia (Niu et al. 2015,

2017), AC or oceanic plateaus rocks are still too mafic for the andesitic BCC; (2) AC melting during arc-continent collision could be invoked to explain juvenile crust addition, but arcs have positive topography, and are too shallow and too cold to melt (see Niu et al. 2003) unless the arc rocks were subducted and heated up subsequently; (3) even if the AC can melt, its unusually high Sr makes such melt inappropriate for BCC (Niu et al. 2013).

In brief, arc-continent collision is probably common in the geological record and can add BCC growth in both volume and lateral extent, but cannot contribute to the andesitic BCC composition (Niu and O'Hara 2009). Recently, Condie and Kroner (2013) suggested a major change in the tectonic setting of continental growth at the end of the Archean from "oceanic plateau settings" in the Archean to continental arc settings thereafter. Their key argument is that the oceanic arc settings are not primary sites for crustal growth as we have been demonstrating (Mo et al. 2008; Niu and O'Hara 2009; Niu et al. 2013), and that the continental arc settings are important. But as we argue, the arc crust is mass balanced at both oceanic and continental arcs (Clift and Vannucchi 2004; Scholl and von Huene 2007) and preservation that is key to net crustal growth requires continental collision (Niu and O'Hara 2009; Niu et al. 2013).

In this context, it is worth to note that modern example for arc-continent collision is rare (Cunningham et al. 2012), and our relevant knowledge is based on interpretations of lithological assemblages associated with, and preserved within, orogenic suture zones. These interpretations are mostly reasonable but need substantiation by considering likely geological scenarios as essential elements of tectonic reconstruction. For example, if island arcs are developed at and evolved from prior continental margins because of trench retreat and overriding plate rifting and separation (Niu et al. 2003; Niu 2014), then such arcs will not collide with the prior continental masses without subduction zones developed in back-arc regions. That is, it is unlikely that the present-day island arcs in the western Pacific will collide with Eurasian and Australian continents but would collide with American continents when the Pacific Ocean basin with subduction zones diminishes in the future. In the latter scenario, one would expect juxtaposed lithological assemblages of island arc rocks, continental arc rocks, intraplate ocean islands and possible fragments of ocean crust in the orogenic suture zones. This would contribute to the crustal masses, but not compositionally andesitic materials required for the crustal growth except for possible ocean crust melting during the collision as we argue (Niu et al. 2003, 2013; see below).

8.2.7 *Mantle Plume Contributions Are Also Too Mafic If Any*

Mantle plumes have been invoked to explain the episodic crustal growth (see Fig. 8.3 and discussion above). Mantle plume melts, either as flood basalts on land or underplated beneath, are too mafic for the andesitic BCC composition. Volumetrically large oceanic plateaus can be amassed to existing continents (e.g.,

Niu et al. 2003, 2015, 2017; Song et al. 2017), but they are again too mafic, and difficulties remain on how they can be converted to the andesitic BCC composition. We thus cannot avoid the conclusion that compositionally magmatism associated with mantle plumes is not a major contributor to CC growth although mantle plumes could have played an important role in raising the proto-continental crust from submarine to standing above the sea level in the early Archean (Campbell and Davies 2017) and in facilitating the stabilization of cratonic mantle lithosphere (Niu 2020a, b; Niu 2021).

8.3 The Continental Collision Hypothesis

The testable hypothesis that "continental collision zones are primary sites of net continental crust growth" (Niu et al. 2013) was developed following our collaborative research on syncollisional granitoid magmatism in southern Tibet associated with the India-Asia collision at 55 ± 5 Ma (Mo et al. 2007, 2008, 2009). This hypothesis has been tested favorably in recent years through sampling and studying syncollisional granitoids of older orogenic belts in the greater Tibetan Plateau and the adjacent regions (Fig. 8.4; Huang et al. 2014, 2015, 2019; Chen et al. 2015, 2016, 2018; Duan et al. 2016; Hu et al. 2016; Li et al. 2016, 2017; Zhang et al. 2016; Kong et al. 2017, 2019, 2020; Shao et al. 2017, 2021). This hypothesis can be concisely summarized as follows:

> Upon ocean basin closing and continental collision, partial melting of the unfully subducted upper ocean crust under amphibolite facies conditions produces primitive andesite melt compositionally closely resembling the bulk continental crust (BCC). As the ocean crust is derived from the mantle in no distant past, the andesite melt (parental to the syncollisional granitoids) with mantle isotopic signatures would contribute as juvenile material to continental crustal growth (Fig. 8.5). It is the collision that preserves and thus protects such juvenile crust from being lost through subduction erosion.

Fig. 8.6). It is the collision that preserves and thus protects such juvenile crust from being lost through subduction erosion.

Clift et al. (2009) argue for no new crust formation in continental collision zones because of surface erosion and sedimental recycling. Such material loss is expected, but we maintain that much of the juvenile crust is well preserved. For example, the 65–45 Ma syncollisional granitoids, especially the subaerial volcanic rocks in southern Tibet are well preserved (Mo et al. 2007, 2008). The preservation of the volcanic sequence manifests rather limited erosion. It should be noted that the present-day high seawater Sr isotope values ($^{87}Sr/^{86}Sr$ = ~0.709) has been interpreted to be largely associated with erosion of the Himalaya-Tibetan Plateau region (Richter et al. 1992). This interpretation is reasonable, but we must emphasize that the primary source of the erosion is the peaky Himalayan mountains of ancient Indian crustal rocks with high $^{87}Sr/^{86}Sr$ (>0.72), rather than the syncollisional granitoids ($^{87}Sr/^{86}Sr \leq 0.706$) in the flat southern Tibetan Plateau, whose surface elevation remains unchanged with fossil leaves well preserved over the past ~15 Myrs at least

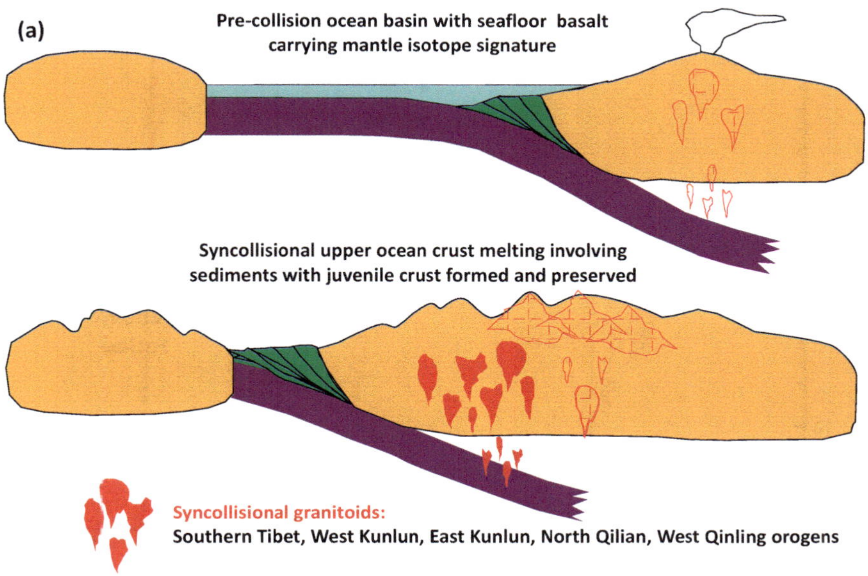

Fig. 8.6 (**a**) Cartoon showing the paleo-ocean basin with passive continental margin (e.g., Tethys oceans) and Andean-type active volcanic margin at the edge of overriding continental plates prior to and during the collision to show the concept of subducting ocean crust melting to produce the syncollisional andesitic melts parental to the syncollisional granitoids (modified from Mo et al. 2008; Niu et al. 2013; Shao et al. 2017). (**b**) Simplified phase diagram showing hydrous (H_2O-saturated) solidi of basalts and the more felsic lithologies (granite and tonalite) modified from Niu (2005) using the experimental data (e.g., Stern et al. 1975; Wyllie 1977; Wyllie and Wolfe 1993). The three major metamorphic facies fields (eclogite, amphibolite and granulite) are drawn after Peacock (2003). The two geothermal gradients approximating two extreme subduction zone thermal scenarios are inferred from models of Peacock and Wang (1999). The star and the red solid curve with arrow illustrate the concept of the thermal evolution of the underthrusting ocean crust along a high T/P path because of retarded subducting and enhanced heating upon and during continental collision. It is expected that the highly hydrated (altered/weathered) ocean crust begins to melt when intersecting the hydrous basaltic solidus under amphibolite facies conditions. Sediments or the more felsic rocks (i.e., granitic or tonalitic compositions) will melt under similar conditions. The yellow oval (labeled "A") indicates P–T conditions, where melting can take place, beneath active arcs constrained by petrologic and seismic observations (see Kelemen et al. 2003). The stippled rectangle (labeled "B") indicates the likely melting conditions for the syncollisional magmatism. Note that the melting reactions are such that the melting residues (pl + opx + cpx ± gnt) would become the well-recognized mafic granulite assemblage (modified after Mo et al. 2008; Niu et al. 2013; Shao et al. 2017)

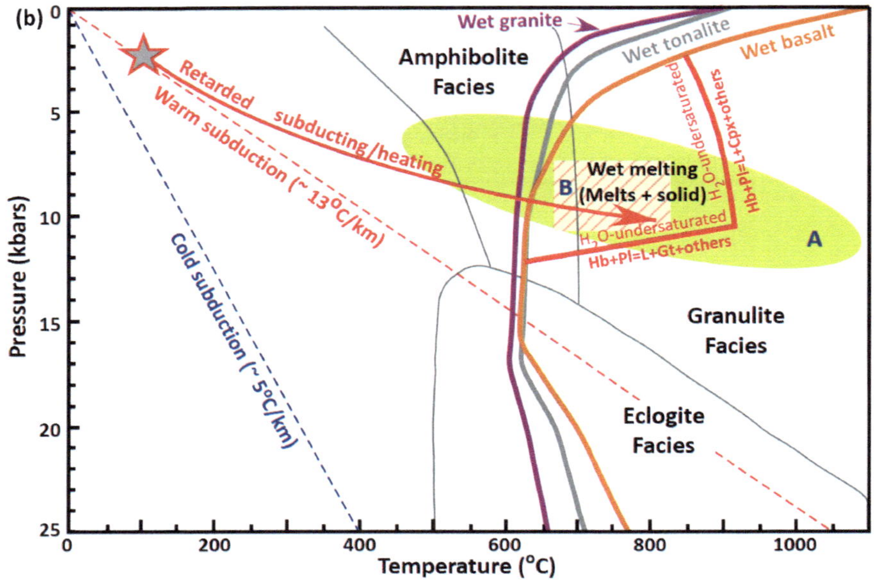

Fig. 8.6 continued

in southern Tibet (Spicer et al. 2003). Furthermore, volumetrically significant syncollisional granitoids are well preserved in much older orogens (e.g., The ~225 Ma West Kunlun, ~240 Ma East Kunlun, ~440 Ma Qilian and ~220 West Qinling), indicating effective preservations of these juvenile crustal materials for CC growth.

We emphasize that the continental collision hypothesis readily explains why CC growth is episodic. There is no understood reason why magmatism associated with active seafloor subduction (island arc are continental arc) is episodic, but continental collision is understood to be episodic. Hence, syncollisional juvenile granitoids and their preservation contribute to the episodic CC growth without invoking mantle plumes that are inappropriate materials in the first place. Supercontinent assemblage is thus expected to represent pulses of CC growth (Fig. 8.3). We should note, however, the peaks in Fig. 8.3 have been revised to varying extent and will continue to be revised, depending on the datasets used and interpretations of tectonic settings, but peaks remain (e.g., Campbell and Allen 2008; Voice et al. 2011; Cawood et al. 2012; Puetz 2018). Further efforts are useful to investigate any possible match of the CC growth peaks with zircon data, but exact match is unexpected because, for example, syncollisional juvenile granitoids are known to have been produced and preserved in all the phanerozoic orogens we studied, especially the giant Central Asian Orogenic Belt (CAOB) with abundant juvenile granitoids (Xiao et al. 2013), but all such "local" or "regional" collision events are predicted to happen more frequently in time and space "scattered" globally, yet they are not parts of supercontinent formation.

In the following, we compare the collision-zone hypothesis with the island-arc model in terms of observables and understood concepts before elaborating some key elements.

8.3.1 *The Advantages of the Continental Collision-Zone Hypothesis over the Island Arc Model*

This comparison is best done in the following score-card table (modified from Niu et al. 2013):

	Standard Island-arc model	Collision-zone hypothesis
Magma source	Mantle wedge peridotite with slab components	Amphibolite of upper ocean crust (MORB) protolith with terrigenous sediments
Primary melt composition relative to BCC	Basaltic ≠ BCC (too mafic)	Andesitic ≈ BCC
Incompatible element abundances relative to BCC	Depleted deep crustal cumulate for lower CC; depleted highly evolved felsic rocks for upper CC	Match reasonably well
Conversion to BCC composition	Required to (1) remove mafic or ultra-mafic cumulate into the mantle (unknown processes); and to (2) enrich progressively more incompatible elements for upper crust	No need: Andesitic melt for BCC and residue of mafic granulite already in the mantle
Net contribution to BCC growth	No, mass balanced by subduction erosion and sediment recycling	Yes, preserved in continental collision zones as syncollisional andesites and massive granitoids
Juvenile crustal material	Yes, but overall mass-balanced	Yes, well preserved
Mantle isotopic signature	Yes, from mantle wedges	Yes, inherited from ocean crust formed at ocean ridges in no distant past
Explanation for Nb-ta-Ti depletion ("arc signature")	Yes, but the mechanism not well understood	Yes, melting of amphibolite with sediment
Explanation for weak Sr depletion relative to Pr and Nd in BCC	No, variably high Sr excess, but yes in the more evolved felsic rocks	Yes
Explanation for weak Eu depletion relative to Sm and Gd In BCC	No or weak	Yes
Explanation for sub-chondritic Nb/ta ratio in BCC	No	Yes

	Standard Island-arc model	Collision-zone hypothesis
Excess Sr and Eu in melting residue	No	Yes, leading to excess Sr and Eu in the residue entering the mantle
Excess Nb and ta in melting residue	Unknown (yes in the subducting ocean crust)	Yes, leading to excess Nb and ta in the residue entering the mantle
Episodic crustal growth	No	Yes, non-constant continental collision and juvenile crustal preservation

8.3.2 Continent-Continent Collisions Are Essential for Continental Crust Growth

Kröner et al. (2007) concluded that the juvenile crust along the Phanerozoic *Central Asian Orogenic Belt* (CAOB) formed through collision of complex "arc-and-backarc" systems with the North China Craton to the south and with the Siberian Craton to the north. Obviously, continental collision is indeed the very process that can achieve continental growth by collaging seafloor geological features with positive topography such as island arcs, scattered ocean islands and fragments of oceanic plateaus. All this explains why blocks of ocean island basalts can be found in orogenic sutures, why most ophiolites have geochemical features of island arc basalts, and why the CC possesses geochemical signatures of island arc signatures (Fig. 8.2) (Niu et al. 2003). Nevertheless, all these collaged land masses are still too mafic for the andesitic BCC composition, and additional or different processes are needed for the "conversion" (Niu et al. 2013).

8.3.3 Syncollisional Partial Melting of Upper Ocean Crust Produces Andesites of BCC Composition

Continental collision results from the closure of an ocean basin and thereafter subduction ceases also, but convergence of the "subducting" seafloor may continue for some time, perhaps up to 10 Myrs if not more (Mo et al. 2008). One can readily estimate that for a convergence rate of ~50 mm/yr. over a ~5 Myr period with ~20% upper ocean crust melting can produce ~30–41 km^3 granitoid melts per km strike length per Myr, which is similar to and slightly greater than the model average of ~30 km^3 km^{-1} arc length/Myr for Phanerozoic arc-crust production (e.g., Reymer and Schubert 1984). There are large uncertainties with all these estimates, but the key point is that it is not physically difficult to produce volumetrically significant juvenile crustal mass associated with syncollisional granitoid magmatism. While the present-day Pacific-style subduction beneath American plates to the east and beneath marginal seas in the western Pacific is possible, the scenario in which subduction of a seafloor with a passive margin beneath Andean-type active continental margin is predicted to be more common in explaining many observations in the geological record. This concept is illustrated in Fig. 8.6, which can effectively explain the origin and petrogenesis of the syncollisional granitoids such as the 55 ± 5 Ma batholiths associated with the India-Asia collision in southern Tibet (Mo et al. 2007, 2008, 2009; Zhu et al. 2015), the ~440 Ma North Qilian (e.g., Chen et al. 2015, 2016, 2018; Huang et al. 2015, 2016), ~240 Ma East Kunlun (e.g., Huang et al. 2014; Hu et al. 2016; Shao et al. 2017; Kong et al. 2020), ~225 Ma West Kunlun (Zhang et al. 2016), and ~220 Ma West Qinling (e.g., Duan et al. 2016; Kong et al. 2017) orogens (Fig. 8.4).

In this context, we should emphasize that convergence will continue upon continental collision and shortly afterwards for some time, depending on the geometry and configuration of the colliding continental margins and other factors. But the continued India-Asia convergence at present at a rate of ~50 mm/year since the collision at 55 ± 5 Ma is "anomalous" because this continued convergence has been driven by subduction of the giant India-Australia Plate at the Sumatra-Java Trench and the convergence will continue until this giant plate disintegrates into smaller plates (see Niu 2020a, b). That is, the convergence will last for no more than a few Myrs (<10 Myrs) if the collision is complete without assistance of residual or along-strike trailing subduction. The latter is manifested by the collision of Chinese continental shelf of exotic origin with continental China at ~100 Ma marked by the continental arc granitoids (Xue et al. 2018) with arc-shaped suture off the southeast coastline (Fig. 8.4; Niu et al. 2015).

8.3.3.1 Syncollisional Andesites and Granitoids Show Remarkable Compositional Similarity to BCC

The ~55 ± 5 Ma syncollisional andesites from southern Tibet (STSA) show remarkable compositional similarity to the BCC (Rudnick and Gao 2003) on one-to-one plots for both major and trace element abundances as well as characteristic trace element ratios (Fig. 8.7a, b; Mo et al. 2008; Niu et al. 2013). This compositional observation points to a genetic link between BCC formation and syncollisional andesites as a result of partial melting of the ocean crust (see Fig. 8.6 and discussion below). Note the compatible elements (oxide) like MgO, Ni and Cr are lower in the average STSA than in the BCC. This is expected because the STSA are melt compositions with certain extent of mafic mineral (i.e., the amphibole is on the liquidus) fractionation that depletes these compatible elements. We should also note that the BCC is the best model estimate that must consider all the lithologies in the crust, including mafic lithologies like ophiolite in orogenic belts and basaltic rocks erupted on land etc., which are generally low in incompatible elements, but strongly enriched in the "mafic" elements like Mg, Cr and Ni. Such compositional similarity shown by the STSA is further substantiated by the ~440 Ma North Qilian syncollisional granitoids with conspicuous mantle isotopic signature (Fig. 8.7c, d; see below).

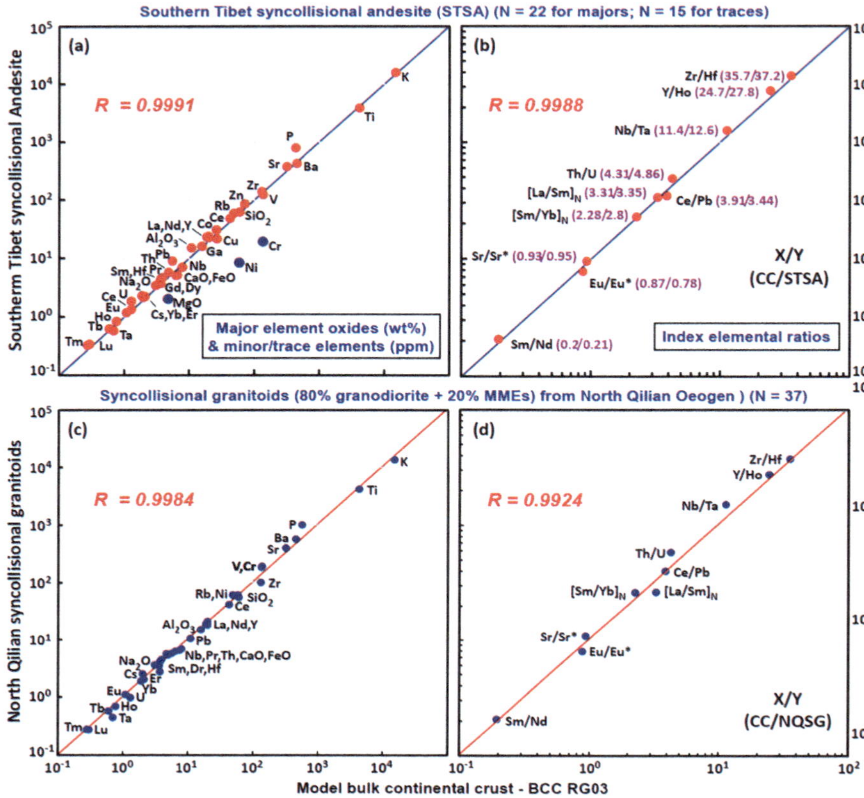

Fig. 8.7 (**a–b**) (after Mo et al. 2008; Niu et al. 2013) demonstrate explicitly that the ~55 ± 5 Ma syncollisional andesites in southern Tibet (STSA) show remarkable resemblance to the model bulk continental crust (BCC; Rudnick and Gao 2003), which together with mantle isotopic characteristics (Mo et al. 2007, 2008, 2009), offer strong evidence for melting of the ocean crust for juvenile continental crust growth. Note that the lower "mafic" elements Mg, Ni and Cr in the STSA reflect mafic mineral (e.g., amphibole) fractionation. (**c–d**) (after Chen et al. 2015, 2016, 2018) demonstrate further that the ~440 Ma North Qilian (Fig. 8.4) syncollisional granitoids of mantle isotopic compositions match well the compositions of the model BCC, an independent verification of our hypothesis

8.3.3.2 Syncollisional Granitoids Represent Juvenile Crust Derived from the Mantle

Figure 8.8 shows syncollisional granitoids from the four older orogenic belts (Fig. 8.4) in εHf(t) vs. εNd(t) space. As is the case for the ~55 ± 5 Ma syncollisional granitoids and volcanic successions associated with the India-Asia collision in southern Tibet (see Fig. 8.7a, b; Mo et al. 2007, 2008, 2009), all these syncollisional granitoids are isotopically far more depleted (high εHf(t) and εNd(t) values) than the mature continental rocks represented by the Mesozoic intra-plate granitoids derived from deep crust in eastern China (see below). Simple mass balance

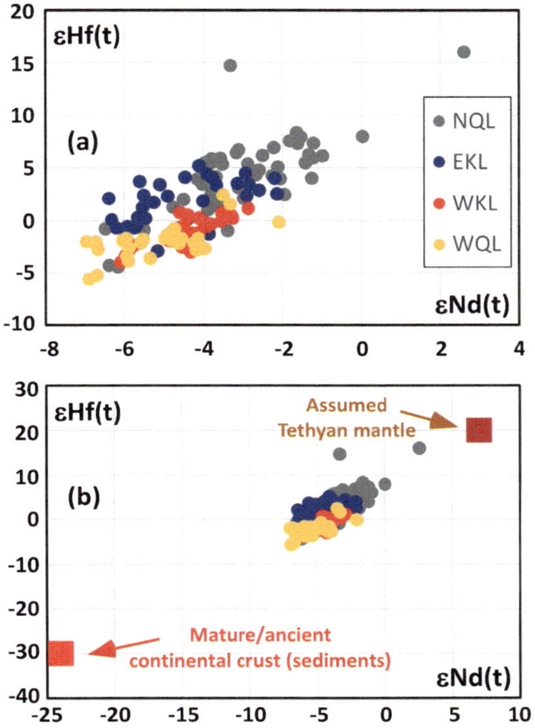

Fig. 8.8 (**a**) Hf and Nd isotope compositions of syncollisional granitoids from North Qilian (NQL; N = 51), East Kunlun (EKL, N = 31), West Kunlun (WKL, N = 27) and West Qinling (WQL, N = 27) orogens, showing significant mantle contributions. (**b**) By assuming a mantle source for Tethyan seafloor basalts with εNd(t) = 7 and εHf(t) = 20 and mature continental crust with εNd(t) = −25 and εHf(t) = −30 as inferred for eastern China deep crust (see Fig. 8.14 below), we can calculate that isotopically these syncollisional granitoids have significant mantle contributions: on average ~96% for NQL, ~80% for WKL, ~80% for EKL and ~50–60% for WQL (Chen et al. 2015, 2016, 2018; Huang et al. 2014; Kong et al. 2017; Shao et al. 2017; Zhang et al. 2016)

calculations show that isotopically these rocks represent 65–95% mantle contributions with only 5–35% crustal contributions perhaps due to magma assimilation in crustal magma chambers or melt contributions from terrigenous sediments as part of the underthrusting ocean crust. Hence, these syncollisional granitoids represent juvenile crust with primary materials coming from the mantle.

8.3.3.3 "Mantle Origin" of the Syncollisional Granitoids—Partial Melting of the Ocean Crust

The dominance of mantle isotopic signature of the syncollisional granitoids suggest that these rocks must be derived from the mantle. The common interpretation in the literature would be partial melting of the mantle peridotite to explain the mantle isotopic composition. It is conceptually important to note, however, that melting of wet mantle peridotite could produce high-Mg andesitic melts (O'Hara 1965; Kushiro et al. 1968; Hirose 1997), but the amount of melt so produced is volumetrically insignificant (<1% mass fraction) and unlikely to produce the syncollisional granitoids, which in the case of the southern Lhasa Terrane, can amount to ~350,000–500,000 km^3 (Fig. 8.4; Niu et al. 2007; Mo et al. 2008) and is likely similar in these older orogenic belts.

To produce volumetrically significant granitoids of andesitic composition, the source rocks must be basaltic in composition (see Niu 2005; Niu et al. 2013). It is convenient to speculate that to partially melt basaltic AC could produce andesitic melts with inherited "arc signature" (Fig. 8.2). However, there are three obvious problems with this view: (1) the AC with positive topography is shallow and is too cold to melt during collision; (2) arc rocks have variably too much Sr to explain the BCC (e.g., Niu and O'Hara 2009; Niu et al. 2013); (3) Arc deep crust cumulate (Kelemen and Behn 2016), arc basalts (Fig. 8.5) and highly evolved felsic arc rocks (Fig. 8.5) are all too depleted in incompatible elements for the CC. These are strong lines of evidence against the island arc model for the BCC. Melting of the AC will not produce the BCC composition.

The best source candidate in the broad context of ocean basin closing and continental collision is the remaining part of the ocean crust (Fig. 8.6; Mo et al. 2008; Niu et al. 2013). Partial melting of this ocean crust can produce volumetrically significant melts of andesitic composition. Because the ocean crust was derived from the depleted MORB mantle from ocean ridges in no distant past, the syncollisional granitoids will have mantle isotopic signatures inherited from the ocean crust (Fig. 8.8), thus forming the juvenile crust.

The question is whether this remaining underthrusting ocean crust is hot enough to melt. The answer is "yes" with caveat because it is both "wet" enough and "hot" enough to melt. Figure 8.6 (after Mo et al. 2008; Niu et al. 2013) is a simplified phase diagram, showing that the remaining underthrusting ocean crust takes a high T/P evolution path upon collision and then undergoes hydrous melting under amphibolite facies conditions (see photomicrographs of amphibolite of seafloor basalt crust in Fig. 8.9a; Niu and Lesher 1991). This melting produces primary andesitic melts with mantle isotopic signatures, either erupted with limited fractionation as syncollisional andesites (e.g., Mo et al. 2007, 2008) or intruded, slowly cooled, and solidified as diorites/granodiorites, i.e., the volumetrically significant syncollisional granitoids in the four older orogens (see Fig. 8.4 and above). Several questions need addressing for this concept:

Why is it wet and hydrous melting? This is straightforward as the ocean crust is hydrothermally altered during its accretion at ocean ridges and pervasively weathered/hydrated subsequently on the seafloor. This is in fact the premise on which the widely accepted subduction-zone flux-melting for arc magmatism is based (e.g., Gill 1981; Tatsumi et al. 1986; McCulloch and Gamble 1991).

Is it possible that the underthrusting crust would take such a high T/P path (Fig. 8.6b)? This is possible and likely because the slab "subduction" becomes retarded (underthrusting) upon collision, and tends to reach thermal equilibrium with the ambience, especially the upper ocean crust. Also, the ambience is hot because it was a hot active continental margin magmatic regime (e.g., the Andean-type) prior to the collision. Importantly, the continued underthrusting upon and during collision means that its induced mantle wedge flow must still have continued (until, perhaps, the slab break-off), which should supply heat and maintain the hot regime for some time (see above). In fact, petrologic and seismic data as well as temperature-dependent viscosity mantle wedge flow models all consistently indicate P-T conditions beneath many active arcs are much hotter (Kelemen et al. 2003) as indicated by the yellow oval area with dT/dP = ~20–50 °C (Fig. 8.6b). We note that slab breakoff has been widely advocated as the major mechanism that may provide heat for the syncollisional magmatism (e.g., Davies and von Blanckenburg 1995; Zhao et al. 2009; van Hunen and Allen 2011; Zhu et al. 2015). This is possible as the slab breakoff can potentially create a window for the ascent of subjacent hot asthenosphere, but the efficacy of this model needs further evaluation because the amount of asthenosphere upwelling is restricted by the finite "space" left by the broken slab (Niu 2017).

Why is the melting taking place under amphibolite facies conditions? The above analysis indicates that with the high dT/dP path of ~20 °C, the underthrusting ocean crust will enter the amphibolite facies conditions and cross the wet solidus of basaltic rocks in P-T space (Fig. 8.6b). Importantly, this analysis is corroborated by the geochemical data of all the syncollisional granitoids, which resembles the composition of the BCC with no HREE-highly depleted "garnet signature" (i.e., $[Dy/Yb]_N = 1.0 \pm 0.1$; Figs.8.1, 8.7). If the underthrusting ocean crust were melted under eclogitic conditions with garnet presence as a residual phase (Fig. 8.6b), the composition of the melt would differ from that of the BCC, but resembles those of adakites (also sanukitoids or high-Mg andesites) and TTGs with highly depleted HREEs (i.e., $[Dy/Yb]_N > 1.3$; see Fig. 8.1).

8.3.3.4 Composition of Ocean Crustal Melts Produced under Amphibolite Facies Conditions

High degree melting of amphibolite of basaltic protolith can produce andesitic melts as shown experimentally (Stern et al. 1975; Wyllie and Wolfe 1993; Wolf and Wyllie 1994; Rapp 1995). The question is whether ocean crust (MORB) melting under amphibolite facies conditions can produce andesites with trace element systematics that match that of syncollisional granitoids of andesitic composition or BCC (Fig. 8.7). Well-designed melting experimental studies are in urgent need, but simple calculations show that this match is highly probable (Niu and O'Hara 2009). The calculations can be done by using (1) amphibolite of MORB protolith with modes of ~66.4 wt% hornblende, ~29.2 wt% plagioclase (plus minor quartz) and ~4.4 wt% ilmenite/pseudobrookite (Niu and Lesher 1991; see photomicrographs in Fig. 8.9a), (2) composition of upper ocean crust (lavas + sheeted dikes) with 75% N-type MORB and 25% E-type MORB (see Niu and O'Hara 2003), (3) relevant partition coefficients compiled in the literature (see Table 1 of Niu and O'Hara

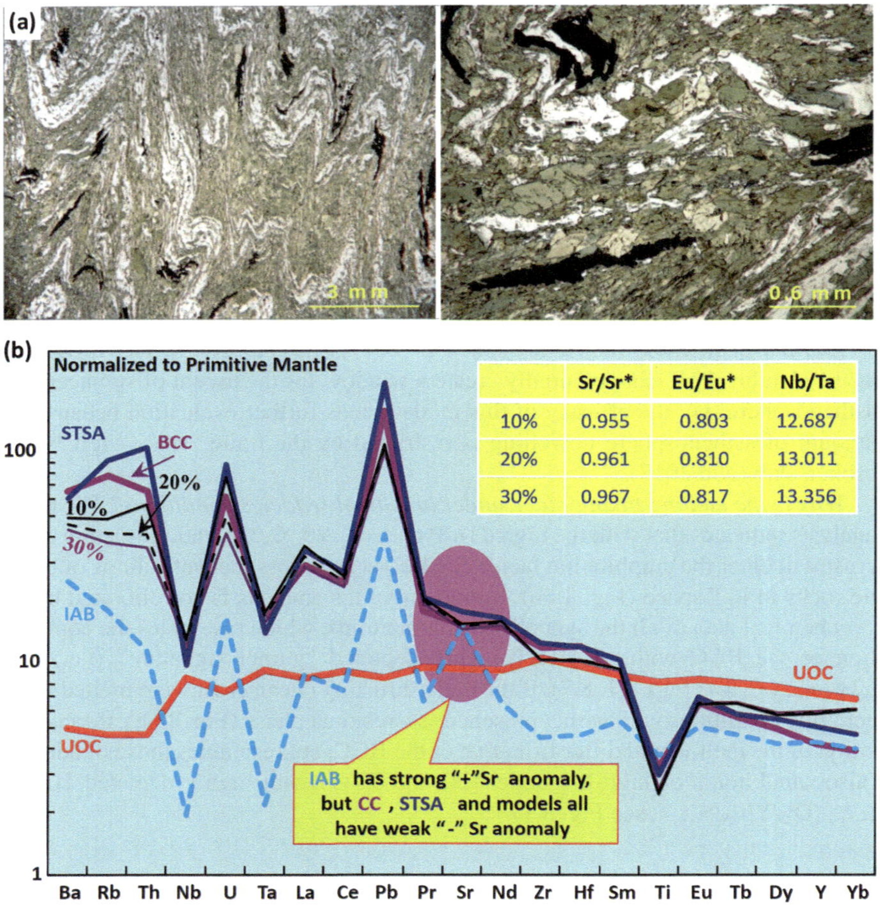

Fig. 8.9 (**a**) Photomicrographs of amphibolite of ocean crust (MORB) protolith with ~66.4 wt% hornblende, ~29.2 wt% plagioclase (plus minor quartz) and ~4.4 wt% ilmenite/pseudobrookite (Niu and Lesher 1991). (**b**) Primitive mantle (PM; Sun and McDonough 1989) normalized multi-element diagram to compare compositions of the bulk CC (Rudnick and Gao 2003), the 55 ± 5 Ma southern Tibet syncollisional andesite (STSA; Mo et al. 2008), upper ocean crust (UOC) consisting of 75% N-type MORB and 25% E-type MORB (Niu and O'Hara 2003, 2008, 2009), and average island arc basalts (IAB; Elliott 2003). We also show model calculations of aggregated melt compositions of 10%, 20% and 30% accumulated fractional melting of the UOC under amphibolite facies conditions using the mineral modes of amphibolite given above and the relevant partition coefficients in the literature (see Table 1 of Niu and O'Hara 2009) plus 30% terrigenous of GLASS composition (Plank and Langmuir 1998) required by the Nd and Hf isotope data (Fig. 8.8). The results well capture the overall characteristics of the bull CC and STSA (and the syncollisional granitoids shown in Fig. 8.7) except for Ba, Rb and Th (modified after Niu and O'Hara 2009; Niu et al. 2013)

2009), and (4) 10% to 30% accumulated fractional melting (similar to equilibrium batch melting in this case). The abundances of the progressively more incompatible elements of the model melts do not well match those of the syncollisional andesites and BCC, but they do capture the key features such as the relative depletion of Nb, Ta and Ti and relative enrichment of Th, U, La and Pb as well as the more sub-chondritic Nb/Ta ratio of <13 (vs. 16.1 of the upper ocean crust and chondritic value of 17.4) characteristic of continental crust (Rudnick and Gao 2003). The major discrepancy lies in the smaller amplitude of the negative Ti-Nb-Ta anomalies.

This imperfect match is not ideal, but isotopically, the syncollisional granitoids of andesitic composition dominated by mantle signatures with elevated $\varepsilon_{Nd(t)}$ and $\varepsilon_{Hf(t)}$ values (Fig. 8.8) requires 5 to 35% input of mature crustal material (Mo et al. 2007, 2008; Niu and O'Hara 2009; Niu et al. 2013; Fig. 8.8b), either assimilated through crustal magma chamber processes or melt contribution of terrigenous sediments as part of the subducting/underthrusting ocean crust upon continental collision. So far, some sediment melting experiments have been carried out for phase stability (Nichols et al. 1994; Johnson and Plank 1999), no melting reactions have been established to quantify sediment melting under amphibolite facies conditions (Hermann and Spandler 2008). However, for conceptual illustration on the effect of sediment involvement, we can simply add ~30% of the global subducting sediment (GLOSS) (Plank and Langmuir 1998) to the model amphibolite melts with implicit assumption of total sediment melting (Fig. 8.9b), which captures both the abundances and patterns of BCC reasonably well for all elements with the exception of Ba, Rb Th and U that can be matched by varying sediment (crustal) compositions (see Niu and O'Hara 2009; Niu et al. 2013).

Overall, the above exercises (see Table 1 of Niu and O'Hara 2009), although not yet ideal without adequate experimental data on incongruent melting reactions of amphibolite and sediments, can illustrate the concept that partial melting of amphibolite of MORB protolith with sediment addition can reasonably well produce the BCC composition and syncollisional andesite such as STSA and syncollisional granodiorite (Figs. 8.7, 8.9b). That is, this syncollisional ocean crust melting under amphibolite facies conditions (Figs. 8.6, 8.9) can produce the required "arc signature" of the BCC (Fig. 8.2) without invoking active subduction. Note also in Fig. 8.9 that arc rocks have a strong positive Sr anomaly while this anomaly does not exist, or if anything, a weak negative one for both syncollisional granitoids and the BCC, emphasizing that arc rocks themselves or as magma source are inappropriate for contributing to the BCC composition. One may argue that arc rock melting may leave significant plagioclase as a residual phase, which may give rise to none or weak Sr anomaly in the melt for the BCC, but if so, there would be strong negative Eu anomaly in the BCC, but there is none or a rather weak one in the BCC (Niu and O'Hara 2009).

8.3.3.5 The Mass of the Juvenile Crust Produced by Ocean Crust during Continental Collision

The syncollisional granitoids in the case of southern Tibet amount to ~350,000–500,000 km^3. The question is such large volumes of juvenile melts can be generated during continental collision. We have done the calculations following Mo et al. (2008) and Niu et al. (2013). We consider (1) the syncollisional granitoids are melting product of the underthrusting ocean crust; (2) a convergence rate of ~50 mm/yr. (or 50 km/Myr) over a ~5 Myr period for the syncollisional granitoids; (3) only the upper ~3 km basaltic portion of the ~7 km ocean crust is involved in melting; (4) the extent of melting is up to 30%; and (5) the syncollisional granitoids are about 3% less dense than basalt, then we obtain 5 Myr * 50 km/Myr * 3.0 km * 0.3 * 1.03 = 232 km^3 juvenile crust per km strike length produced in this 5 Myr period. That is, the syncollisional juvenile crust production rate is ~46 km^3 km^{-1} strike length per Myr. By assuming 30% melting of the upper ~4 km ocean crust, we would get ~62 km^3 km^{-1} strike length per Myr. Therefore, to produce the observed mass of the juvenile crust associated with continental collision zones is highly plausible.

8.4 Similarities and Differences of Syncollisional vs. Intraplate and Continental Arc Granitoids in Eastern China

Continental crust grows if the juvenile material accretion exceeds the potential loss of existing crustal mass (vs. preservation). The latter is often attributed to surface erosion and sediment recycling as well as possible subduction erosion (e.g., Clift et al. 2009). However, granitoid magmatism is not restricted to zones of seafloor subduction and continental collision but can occur in the continental interiors. The best examples are Jurassic-Cretaceous granitoids widespread in eastern continental China (Fig. 8.10a). Two important questions must be addressed: (1) What is the origin of such "intra-plate" granitoids and (2) if they contribute to continental crust growth. With these questions in mind in our studying the petrogenesis of syncollisional granitoids and continental crust growth, we carried out a geological and literature survey on eastern China granitoids (Fig. 8.10a). Figure 8.11 presents one important aspect of this survey. Compared to the syncollisional granitoids with significant mantle contribution in terms of Nd isotope compositions with a mean $\varepsilon Nd(t) = -2.62$ (Fig. 8.11b), the eastern China granitoids are dominated by ancient crustal source rocks with a mean $\varepsilon Nd(t) = -9.60$ (Fig. 8.11a), suggesting that "intraplate" granitoids largely result from anataxis of ancient deep crust and represent crustal re-working with no new "juvenile" mass addition. It is thus reasonable to state that the non-orogenic intraplate granitoid magmatism in general contributes little to crustal accretion.

However, we should emphasize that some of the eastern China granitoids do have mantle input with high $\varepsilon Nd(t)$, about ~30% of the surveyed data have

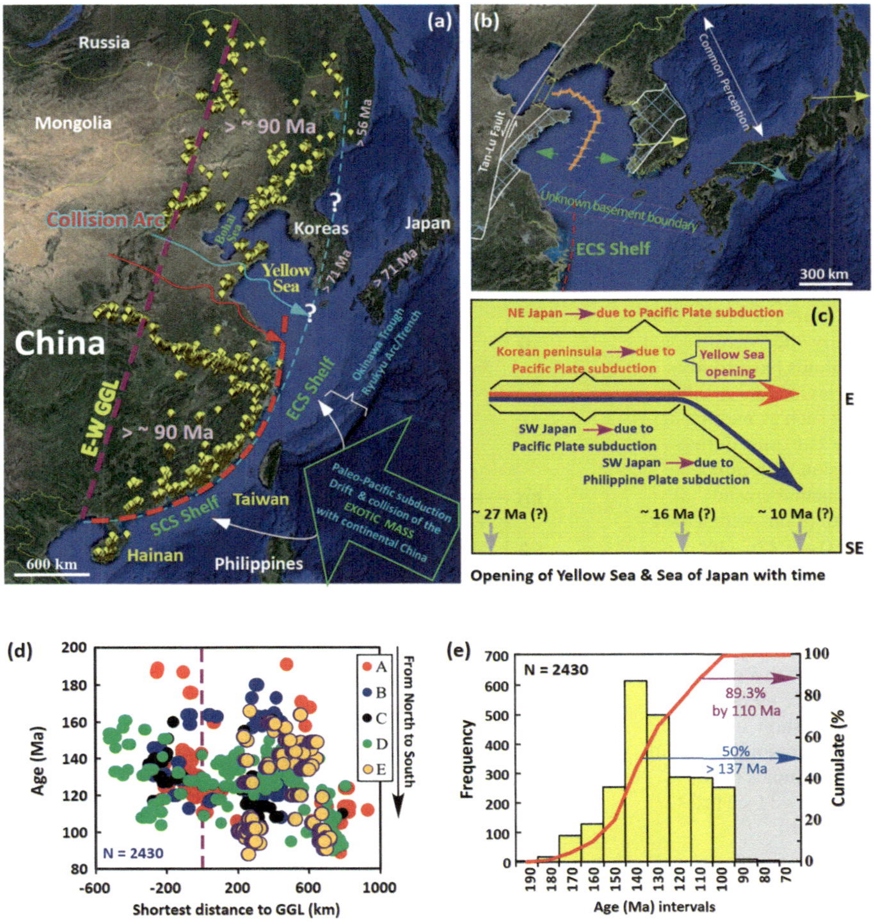

Fig. 8.10 (**a**) after Niu et al. (2015) showing the widespread Jurassic-Cretaceous granitoids in eastern continental China and the Chinese continental shelf of exotic origin (microcontinent or oceanic plateau of mantle plume origin detailed in Fig. 8.15). Collision of the "Chinese continental shelf" (comprising the basement of South China Sea (SCS) and East China Sea (ECS) at ~100 Ma jammed the trench indicated by the arc-shaped dashed red line, produced the last episode of the granitoid magmatism along the southeast coast (Figs. 8.4, 8.12, 8.13, 8.14). The latter is in nature the same as syncollisional granitoids. This also represents the termination of the intraplate granitoid magmatism in the interiors of eastern continental China. (**b**, **c**) show that the Yellow Sea is of continental rift origin with the Korean Peninsula separated from the Jiaodong Peninsula to form the present-day configuration probably no earlier than ~57 Ma (Niu and Tang 2016; Tang et al. 2016). That is, as are the "Xiao Qinglin" granitoids, the "Laoshan granitoids" along the Jiaodong coast (Fig. 8.4) are of intraplate origin at the time of emplacement. (**d**) plots the age of the Jurassic-Cretaceous granitoids as in (**a**) as a function of shortest distance with respect to the GGL: A—samples north of 46°N; B—samples between 41°N and 46°N; C—samples between 36°N and 41°N, D—samples between 29°N and 36°N; E—samples south of 29°N. Such random areal granitoid distribution reflects areal mantle-derived basaltic magmas as heat source intruding and underplating the deep crust for the anatexis and the observed granitoids in the vast region. (**e**) shows that the granitoid magmatism in eastern continental China peaks at ~140 Ma, indicating the time of the peak mantle lithosphere thinning and the magmatism is essentially stopped at ~100 Ma, which represents the time of the collision of the arriving Chinese continental shelf of exotic origin (see Fig. 8.15; Niu et al. 2015)

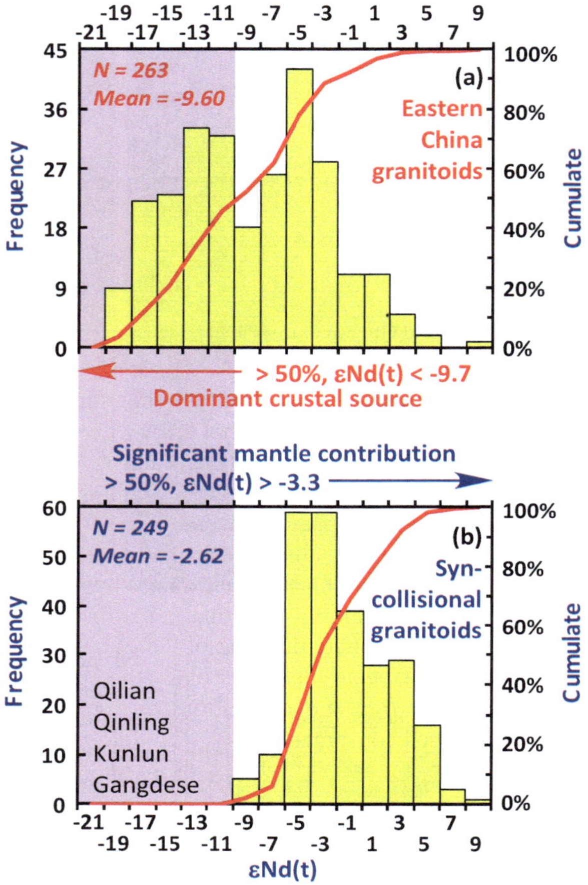

Fig. 8.11 Nd isotope (εNd(t)) histograms to compare the Jurassic–Cretaceous granitoids in eastern continental China (**a**) (literature data compiled in Niu et al. 2015; see Figs. 8.10, 8.15 below) with syncollisional granitoids from Qilian, Qinling, Kunlun and Gangdese granitoids (**b**; data source as in Fig. 8.8). Clearly, the syncollisional granitoids have sources with significant mantle input such as melting of the ocean crust during collision with minor continental crustal contributions with a mean εNd(t) = −2.62 (see Figs. 8.6, 8.7, 8.8, 8.9), whereas the eastern China granitoids are dominated by ancient deep crustal sources with a mean (εNd(t)) = −9.60; some samples show important mantle isotopic input with ~30% samples have εNd(t) > −2 (see text for discussion)

εNd(t) > −3 and as high as +7 (Fig. 8.11a). All this important observation needs understanding in the broad context of granitoid magmatism and continental crust accretion. Because the literature data are spotty with no coherent documentation, we thus sampled granitoids in eastern China from Northeast all the way down south to the Hainan Island (Fig. 8.4) with detailed petrological and geochemical studies completed on samples from the intraplate "Xiao Qinling" (Hong et al. 2018) and "Laoshan" (Gao et al. 2019) granitoids and the continental margin arc granitoids along the southeast coastline (Xue et al. 2018). While detailed work needs completing on granitoid samples from Northeast China and the Hainan Island, the data available have led to the testable hypothesis that the Jurassic-Cretaceous intraplate granitoid magmatism may in fact be a process of existing crustal destruction whereas the continental arc granitoids, similar to syncollisional granitoids, contribute juvenile materials for crustal growth.

8.4.1 Major and Trace Element Compositions

Figure 8.12 compares in SiO_2 variation diagrams the syncollisional granitoids (*Syn*; see above) with Jurassic-Cretaceous granitoids in the continental interiors (*Int*) of eastern China and granitoids along its southeast coastline associated with continental arc magmatism (*Arc*) (Fig. 8.4; Niu et al. 2015; Hong et al. 2018; Gao et al. 2019; Xue et al. 2018). The systematic trends in these diagrams, though similar to liquid lines of descents, are defined by the bulk-rock composition-determined modal mineralogy. The important message is that the *Syn* granitoids have, on average, low SiO_2 and high MgO relative to *Arc* granitoids and especially *Int* granitoids. However, for all these samples at a given SiO_2 they have similar TiO_2, Al_2O_3, MgO and CaO with higher Cr and Ni for *Syn* than *Arc* and *Int* samples.

All these granitoids have similar rare earth element (REE) abundances levels (not shown) and show (Fig. 8.13a–c) varying similar REE patterns enriched in the lighter REEs relative to heavier ones, i.e., $[La/Sm]_N \gg 1.0$ and $[Sm/Yb]_N \gg 1.0$

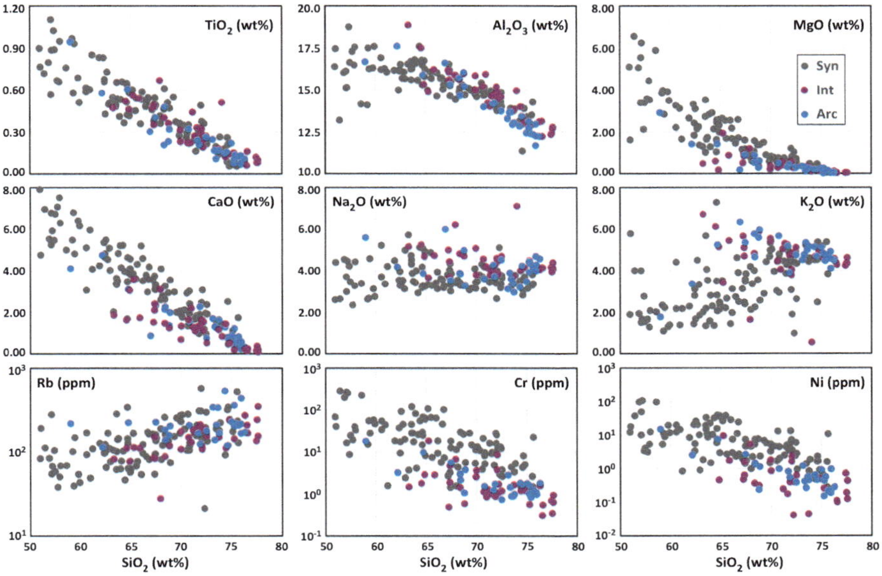

Fig. 8.12 SiO_2 variation diagrams to compare eastern China intraplate granitoids and subduction-related arc granitoids along the southeast coastline with the syncollisional granitoids (see Fig. 8.4 and above). This comparison offers three insights: (1) The syncollisional granitoids are on average more mafic with lower SiO_2 and higher MgO; (2) The compositional trends, although resembling the liquid lines of descent as expected, are controlled by bulk-rock compositions and the mineralogy; (3) at a given SiO_2 level, all these granitoids show varying abundances of other elements, but there are overall no obvious between-setting differences. These suggest that the magma source rocks are compositionally heterogeneous, but lithologically similar and are most consistent with mafic deep crustal melting under amphibolite facies conditions (see Fig. 8.6)

while having essentially flat heavy REE patterns with $[Dy/Yb]_N$ close to unity. Some more felsic samples with ~70–73 wt% SiO_2 may give an adakitic signature mostly because of combined fractionation of amphibole, titanite and zircons instead of the garnet effect (Chen et al. 2018; Xiao et al. 2020). Figure 8.13d–f show that all these samples have similar Th and U variation levels or trends with respect to SiO_2 and similarly varying negative Nb (vs. Th) and Ta (vs. U) anomalies (i.e., $[Nb/Th]_N < 1$ and $[Ta/U]_N < 1$) characteristic of all granitoids and continental crustal rocks.

All these data (Figs. 8.12, 8.13) suggest that although their tectonic settings of magma generation may differ, they have similar source rocks of heterogeneous compositions and similar melting conditions. The source rocks are most consistent with being amphibolite of basaltic protolith of varying compositions. That is, all these granitoids are broadly consistent with variably evolved "I-type granite" typically consisting of varying modes of plagioclase, alkali-feldspar, quartz, amphibole and biotite plus accessary minerals, which differ from "S-type granite" because they define significant inverse Al_2O_3-SiO_2 trend (Fig. 8.12) with little or no muscovite and without characteristic Al-rich mineral phases (e.g., cordierite, sillimanite, garnet). In terms of REE abundances and patterns (Fig. 8.13a–c), the melting conditions are largely under amphibolite facies without garnet signature except for a few *Syn* and *Int* samples with relatively high $[Sm/Yb]_N$ that could suggest garnet present as a residual phase but can be readily explained by combined fractionation of amphibole, titanite and zircons (Chen et al. 2018; Xiao et al. 2020). There is certainly no convincing evidence for eclogite melting.

8.4.2 Nd-Hf Isotope Compositions

Despite the similar major and trace element compositions between these granitoid suites and the inferred source rock types and amphibolite facies melting conditions, they have marked difference in Nd-Hf isotope compositions. The syncollisional granitoids (*Syn*) are dominated by mantle isotopes of ocean crustal melting (see above). The continental arc granitoids (*Arc*) also have important mantle isotope input. However, the intraplate granitoids (*Int*) are dominated by ancient crustal Nb-Hf isotopes (Fig. 8.14).

We should note the Nd isotope contrast between syncollisional granitoids dominated by our own data and the literature data on eastern China granitoids. The latter does include samples with high εNd(t) that reflect important mantle contribution (Fig. 8.11; Niu et al. 2015), but we show here that these granitoids with important mantle input are from the southeast coastline genetically belonging to continental arc (*Arc*) granitoids. The comparison in Fig. 8.14 offers insights into the granitoid magmatism in eastern China with simple clarity (see below).

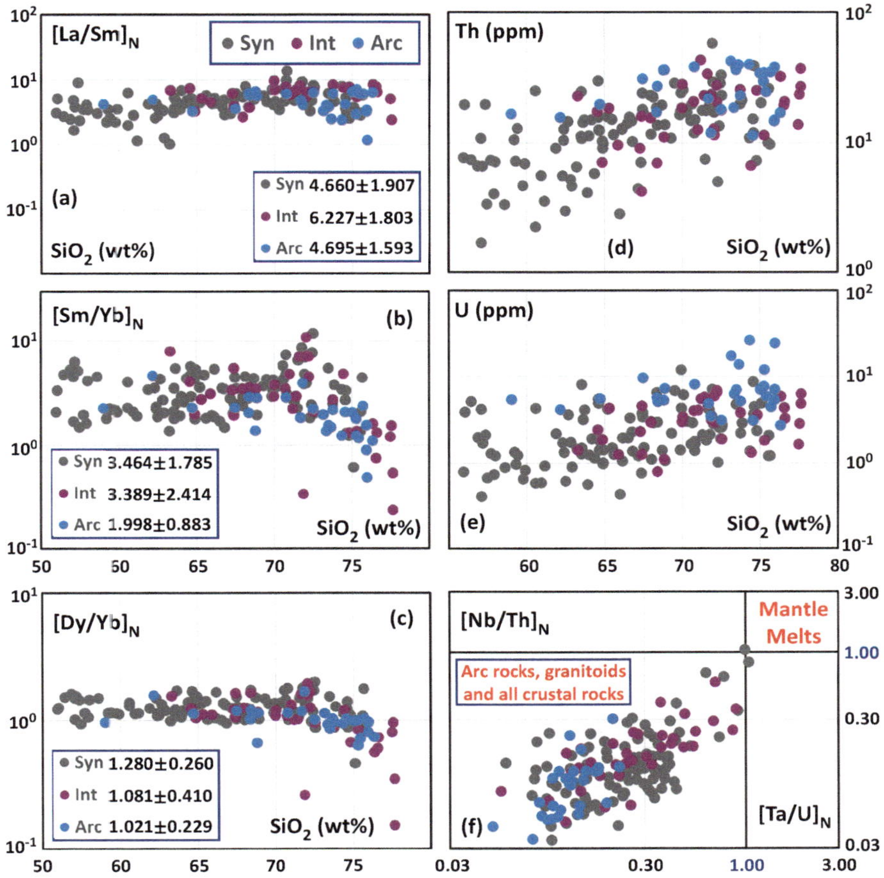

Fig. 8.13 SiO_2 variation diagrams (as in Fig. 8.12) showing index REE ratios (normalized against primitive mantle of Sun and McDonough 1989) $[La/Sm]_N$ (**a**), $[Sm/Yb]_N$ (**b**) and $[Dy/Yb]_N$ (**c**), index element Th (**d**) and U (**e**) plus $[Nb/Th]_N$) vs. $[Ta/U]_N$ covariation (f). (**a**) $[La/Sm]_N \gg 1.0$, indicating light REE enriched patterns for all samples and sample suites; (**b**) $[Sm/Yb]_N \gg 1.0$ for all *Syn* samples and most of the *Int* and *Arc* samples, indicating depleted heavy REEs (e.g., Yb) relative to middle REEs (e.g., Sm); (**c**) $[Dy/Yb]_N \approx 1.0$ for most samples, indicating the flat heavy REE patterns without the garnet signature (see Fig. 8.1), i.e., the absence of garnet as a residual phase in the magma source regions. A few *Syn* samples with $[Dy/Yb]_N > 1.0$ could be interpreted as adakites, but these can be caused by amphibole fractionation with the aid of titanite crystallization determined by the parental magmas and magma source composition (Chen et al. 2018; Xiao et al. 2020). A few *Int* and *Arc* samples with low $[Sm/Yb]_N$ and $[Dy/Yb]_N$ at high SiO_2 result from extensive feldspar crystallization that also gives rise to significant Rb-Sr isochron ages (Gao et al. 2019; Xue et al. 2018). (**d**–**e**) Th and U show similar SiO_2 variation trends with no distinction between sample settings, reflecting similar source rocks (e.g., amphibolite) of inhomogeneous compositions. (**f**) shows that all the granitoids have negative anomalies of Nb and Ta (i.e., $[Nb/Th]_N < 1$ and $[Ta/U]_N < 1$) characteristic of arc rocks, granitoids and all the continental crustal rocks vs. asthenospheric mantle derived basaltic melts in all settings (see Niu et al. 1999). Again, there is no between-setting compositional distinction with the broad positive correlation reflection similar geochemical behavior of Nb with Ta and of Th with U as expected (**f**)

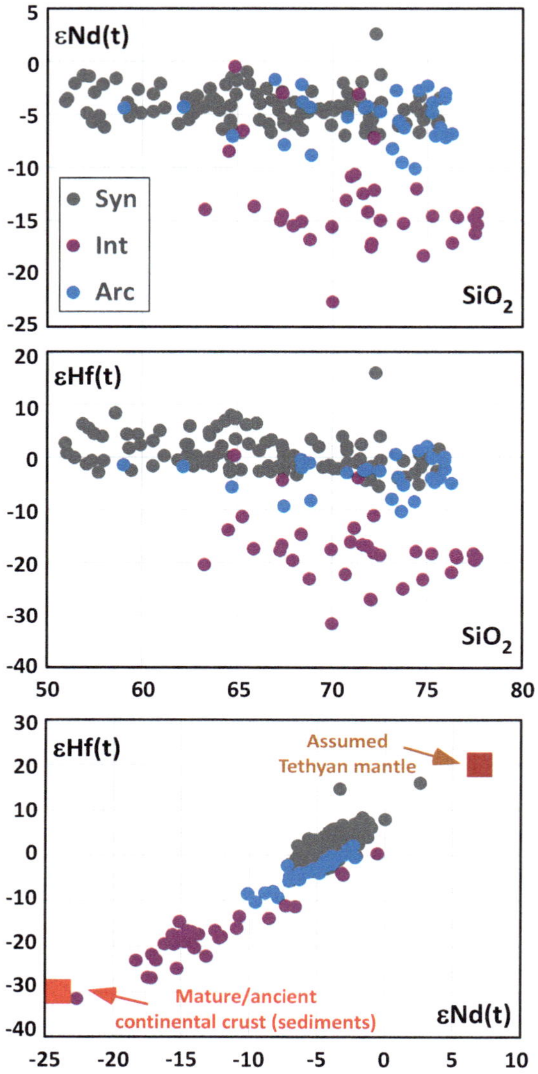

Fig. 8.14 SiO_2 variation diagrams of εNd(t) and εHf(t) to compare syncollisional granitoids (see Figs. 8.12, 8.13) with eastern continental China intra-plate granitoids (see Fig. 8.4 of "Xiao Qinling" and "Laoshan"; Hong et al. 2018; Gao et al. 2019) and the continental arc granitoids along the southeast coastline (Xue et al. 2018) (see Figs. 8.4, 8.10). Clearly, the granitoid magma source for the intraplate is best understood as resulting from melting of ancient deep crust (εNd(t) < 1 and εHf(t) < 0) in contrast with the syncollisional granitoids of ocean crust melting (Figs. 8.6, 8.9). The continental arc granitoids show similarity to syncollisional granitoids, reflecting isotopically significant mantle input. These "Arc" granitoids along the coastline are the youngest and are very likely of the same origin as the syncollisional granitoids discussed above with more mature crustal assimilation (Fig. 8.6; Xue et al. 2018) at the time of ocean closing and subduction termination due to trench jam at ~100 Ma (see Figs. 8.10a, 8.15; Niu et al. 2015)

8.4.3 *Granitoid Magma Generation in Varying Tectonic Settings*

The century-old debate on the origin of granites and granitoids was settled by experimental demonstrations (Tuttle and Bowen 1958) that granitoids are of magmatic origin through partial melting of existing crustal rocks. Wyllie (Wyllie 1977; Wyllie et al. 1976) further synthesized experimental data on possible sources and concluded that granitoid magmas are produced by deep crustal melting. The continued studies accumulated over the past 40 years, especially with the aid of trace element and isotope geochemistry, have consolidated our understudying that except for minor A-type granitoids that could be produced by protracted fractional crystallization of alkali basaltic melts, all other granitoids must be produced by crustal melting with the melt undergoing varying extent of fractional crystallization (see Niu 2005). Volumetrically significant granitoid magmas are understood to result from partial melting of basalts or basaltic rocks at the presence water under suitable pressure and temperature conditions (e.g., Fig. 8.6). The process is often termed "anataxis" or deep crustal melting with the familiar felsic veins/veinlets or layers (leucosomes) showing the process of melt segregation from the mafic residues (melanosomes) on outcrops of exposed deep crustal terranes. This is likely to be true in all tectonic settings for *Syn*, *Int* and *Arc* granitoids we discuss here, but the source of heat and water may vary and need understanding.

8.4.3.1 Syncollisional Granitoid Magma (Syn) Generation

This is discussed in Sect. 8.3 (also see Fig. 8.6; Mo et al. 2008; Niu et al. 2013). Partial melting of basaltic ocean crust formed at ocean ridges in no distant past produces I-type granitoids of dominantly dioritic-granodioritic (volcanically andesitic) compositions with inherited mantle isotopic signatures (e.g., high εNd(t) and εHf(t) values; Figs. 8.6, 8.8). While this interpretation is logical and rigorous, experimental melting experiments on amphibolite (Fig. 8.9) are needed to quantify the extent of melting, to characterize trace element characteristics and to assess whether the heat source is adequate for the melting and for producing the volumetrically significant syncollisional granitoid batholiths.

8.4.3.2 Intraplate Granitoid Magma (Int) Generation

The Jurassic-Cretaceous granitoid magmatism is widespread in eastern continental China, especially east of the E-W GGL (see Niu et al. 2015; Figs. 8.4, 8.10a). While such magmatism can be interpreted as resembling that of the present-day Andean-type active continental margin genetically associated with paleo-Pacific seafloor subduction (e.g., Deng et al. 1998, 2004, 2007; Zhou et al. 2006; Zhang 2010), there is fundamental difference. Except for the granitoids along the southeast coastline (Fig. 8.4; Xue et al. 2018), much of the granitoid magmatism occurred in the continental interiors without any well-defined narrow continental magmatic arc as does along the Andes at present. Indeed, the Jurassic–Cretaceous granitoids in eastern China span emplacement time for ~100 Myrs from ~190 to ~88 Ma and spread diffusively in a wide E-W zone in excess of more than 1000 km (see Fig. 8.10a). Importantly, these granitoids show no systematic NW-to-SE age variation of any sort as demonstrated by plotting the emplacement age as a function of longitude or as a function of the shortest distance (perpendicular) to the GGL (Fig. 8.10d). That is, the granitoid magmatism could take place in coastal regions as well as anywhere in the continental interiors as far as ~1400 km to the west at any latitude and at any given time (Fig. 8.10a, d). The exception is, if any, that the granitoids along the southeast coastline are among the youngest (Fig. 8.10d).

Thus, the Jurassic-Cretaceous granitoid magmatism in the continental interiors of eastern China cannot be explained by the Andean-type active continental margin model, but requires the presence of an areally vast mantle heat source in space and time echoed by the very granitoids at the crustal level (Niu 2014; Niu et al. 2015). The Nd-Hf isotopically enriched composition of these granitoids points to, in simple clarity, the deep continental crust rocks (Fig. 8.14). Depp crustal melting (anatexis or "re-working") cannot happen without perturbed geotherm, or excess heat and such excess heat must be available areally beneath entire eastern continental China in space and time. This is in fact a straightforward crustal response to the mantle lithosphere thinning as a result of basal hydration weakening with the water coming from the dehydration of the paleo-Pacific slab lying stagnant in the mantle transition zone, whose dehydration provides the water that percolated upwards in the form of hydrous melt and converted the basal lithosphere into the asthenosphere, thus thinned the lithosphere accompanied by hydrous basaltic magmatism (Niu 2005, 2014). Intrusion and underplating of such hydrous basaltic magmas caused the deep crustal melting under amphibolite facies conditions, resulting the Jurassic-Cretaceous granitoids (Fig. 8.15; Niu et al. 2015).

It is important to note that the Jurassic-Cretaceous granitoids show a skewed histogram distribution with the abundances peaking at 140 Ma, with 50% granitoids emplaced before ~137 Ma, 90% emplaced by 110 Ma, and the youngest granitoids are ~100–88 Ma mostly found at the southeast coastline (Fig. 8.10d–e). Such sudden granitoid magmatism termination at ~90 Ma along the coastline indicate the termination of the paleo-Pacific subduction beneath continental China, which can only occur if the trench is jammed by volumetrically large and physically buoyant block carried by the paleo-Pacific plate—either a micro-continent landmass or an oceanic plateau (Niu et al. 2003). The latter is most consistent with being the mass of continental shelf of East and South China Seas. The arc-shaped southeast coastline of continental China marked the volcanic arc with the jammed trench off the coastline (Niu et al. 2015).

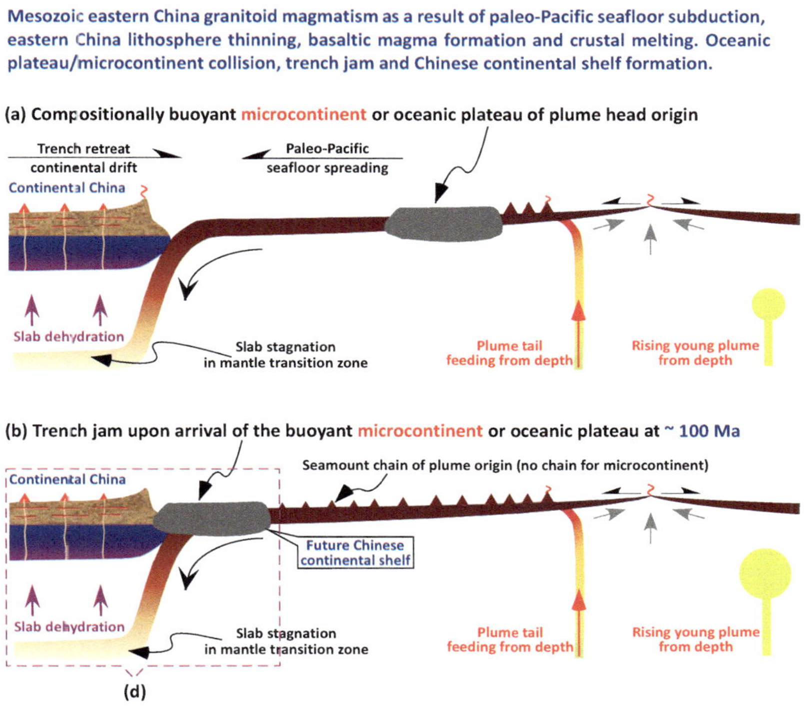

Fig. 8.15 This set of illustrations explains that the widespread Jurassic-Cretaceous granitoids in eastern continental China (Fig. 8.4) coincide in space and time with the coeval lithosphere thinning in the region. Paleo-Pacific seafloor subduction beneath continental China is expected to produce the Andean-type continental arc granitoid magmatism—the youngest suite along the southeast coastline (**b, d**; also see Fig. 8.4). The widespread Jurassic-Cretaceous granitoids in the continental interiors in space and time are best understood as resulting from the stagnation of the paleo-Pacific slab lying in the mantel transition zone where its dehydration released water to have formed hydrous melt (incipient mantle melt) that rose/weakened the base of the lithosphere and converted it into the asthenosphere to have caused the lithosphere thinning accompanied by widespread basaltic magmas. The intrusion and underplating of the basaltic magmas in the deep crust led to the anatexis and the intraplate granitoid magmatism with ancient crustal isotopic signature (Fig. 8.13) as described in (**a–b, d**). Arrival of a massive and compositionally buoyant microcontinent or oceanic plateau of mantle plume origin drifted with the paleo-Pacific plate collided with continental China, jammed the trench indicated by the arc-curved coastal line (Figs. 8.4,8.10). This is equivalent to continental collision without continued convergence because of absence of the slab pull of the same plate elsewhere (vs. India-Asia continental collision and continued convergence since the collision at ~55 Ma (Niu 2020a, b). The youngest (~102 Ma) "arc" granitoids along the southeast coastline are equivalent to syncollisional granitoids with significant mantle contributions (Figs. 8.11, 8.12, 8.13). (Modified from Niu et al. 2015)

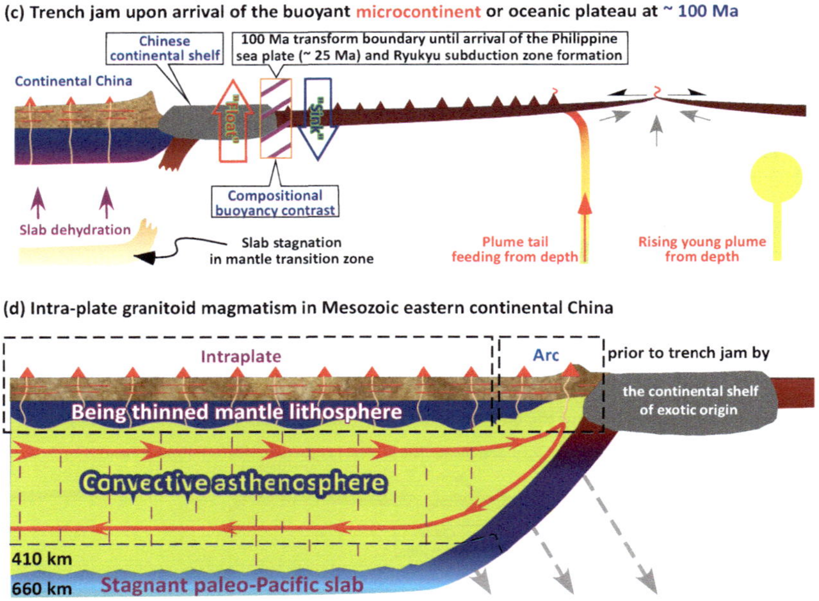

Fig. 8.15 continued

Crustal Destruction Associated with the Int Granitoid Magma Generation

The data and discussion above offer a novel perspective that explains why eastern continental China lacks expected mafic lower crust revealed by geochemical studies of eastern China crustal xenoliths and outcrops (see Fig. 8.16a; Gao et al. 1998a, b) as echoed by the seismic velocity structure (Li et al. 2006). Because magma generation is always partial melting (Niu 2005), in which the fusible components entering the granitoid melts leaving and segregated from the more refectory mafic residues left behind. Partial melting of amphibolite (both vapor-saturated and dehydration melting) produces granitoid magmas and mafic residues with granulite mineral assemblage (Fig. 8.6b; Wolf and Wyllie 1994; Wyllie and Wolfe 1993). This crustal melting has two consequences: (1) destruction of lower crust with the mafic refractory residues of granulite mineral assemblage becoming seismically sub-Moho constituents, i.e., the crust becomes thinned because of the elevation of the Moho depth (Fig. 8.16); (2) the granitoid intrusions and volcanic rocks are subject to erosion and sedimental recycling, contributing to crustal loss.

No Evidence for Thick or Thickened Crust in the Mesozoic Eastern China

There has been the view that eastern China in the Mesozoic may be a plateau such as portions of the present-day Andes because of paleo-Pacific subduction (Zhang et al. 2001) by interpreting some selected Cretaceous granitoids in eastern China as adakites resulting from eclogite melting in the deep crust. It is possible that some granitoids may indeed have relative depletion in heavy REEs such as the Qumushan

(a)	Lower Crust				Bulk Crust			
	Global	East China	EC/Global		Global	East China	EC/Global	
SiO_2 (wt%)	53.4	58.18	1.090	9.0%	60.6	61.57	1.016	1.60%
TiO_2	0.82	0.77	0.939	-6.1%	0.72	0.69	0.958	-4.17%
Al_2O_3	16.9	14.85	0.879	-12.1%	15.9	14.06	0.884	-11.6%
FeO^T	8.57	7.33	0.855	-14.6%	6.71	5.96	0.888	-14.6%
MgO	7.24	5.05	0.698	-30.2%	4.66	3.77	0.809	-19.1%
CaO	9.59	6.51	0.679	-32.1%	6.41	5.03	0.785	-21.5%
Na_2O	2.65	2.86	1.079	7.9%	3.07	2.94	0.958	-4.23%
K_2O	0.61	1.51	2.475	148%	1.81	2.12	1.171	17.1%
P_2O_5	0.10	0.14	1.400	40.0%	0.13	0.16	1.231	23.1%
Thickness (km)	14.38	11.96	0.832	-16.8%	37.1	35.59	0.959	-4.1%
V_P (km/s)	7.14	6.78	0.950	-5.0%	6.67	6.39	0.958	-4.2%

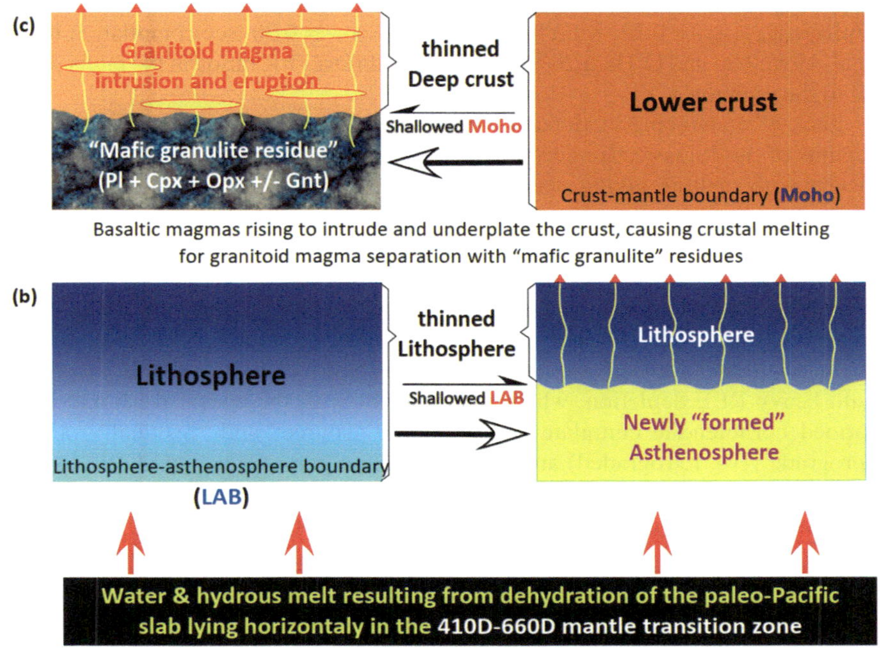

Fig. 8.16 (**a**) is summarized from Gao et al. (1998a, b) to emphasize that compared with the global continental crust data, the lower crust of eastern China (EC) is about 16.8% thinner with ~5% lower seismic low Vp velocity because of the more evolved and less mafic compositions. This was interpreted as resulting from deep crust under eclogite facies conditions being too dense and thus delaminated into the asthenospheric mantle. With the presence of the lithospheric mantle, it is unlikely that the deep crust would have passed through the mantle lithosphere into the asthenosphere. (**b, c**) illustrate that lithosphere thinning related basaltic magmas that acted as the heat source (intruding and underplating the deep crust) to have caused the deep crustal melting and production of the observed granitoid magmatism in the vast region in the Jurassic-Cretaceous (Figs. 8.10, 8.11, 8.12, 8.13, 8.14). The mafic granulite melting residue would have elevated seismic velocity becoming the "mantle" materials, giving rise to the elevation of the seismic Moho. Consequently, the eastern China continental crust would be thinner and have more felsic deep crust (felsic granulite) composition. As a result, accompanied by the lithosphere thinning, the lithosphere-asthenosphere boundary (LAB) shallows and the Moho also shallows. This process essentially destructs the existing crust by melting at depth with the granitoid erosion that may also lost the crustal masses. (Modified from Niu 2005)

pluton from the North Qilian Orogen (Chen et al. 2016; Xiao et al. 2020), which have relatively higher $[Dy/Yb]_N$ (>1.280 ± 0.260; Fig. 8.13c). The latter could be interpreted as adakitic signature, but they are better understood as resulting from combined fractionation of amphibole + plagioclase + biotite + apatite + zircon + titanite determined by parental magmas derived from melting seafloor basalts of Ti-rich composition without requiring garnet present as a residual phase in magma source regions (Chen et al. 2016; Xiao et al. 2020). Figure 8.13c shows that while $[Dy/Yb]_N$ varies for both *Int* (1.081 ± 0.410) and *Arc* (1.028 ± 0.229) granitoids, the average ratios are close to unity, which is significantly lower than ratios claimed for adakites (> ~1.37) or Archean TTGs (> ~1.51) (see Fig. 8.1), thus providing no convincing evidence for eclogite melting or melting with garnet present as a residual phase (i.e., partial melting of eclogite or garnet-bearing amphibolite) because among all the possible minerals involved garnet is the only phase that can cause significant Yb-Dy fractionation (e.g., $Kd[Yb/Dy]_{GARNET\text{-}MELT} = 3.5$; $Kd[Yb/Dy]_{AMPHIBOLE\text{-}MELT} = 0.73$; $Kd[Yb/Dy]_{PLAGIOCLASE\text{-}MELT} = 0.30$; see compilation in Niu et al. 1996; Niu and O'Hara 2009) with elevated $[Dy/Yb]_N$ ratios in the melts parental to the granitoids.

Lack of convincing evidence of garnet as a residual phase in magma source regions of the Jurassic-Cretaceous granitoids in eastern continental China is consistent with melting taking place in the amphibolite facies conditions, which is probably not even at conditions of garnet amphibolite, let alone of eclogite facies conditions (see Fig. 8.6b). These observations and understanding suggest that the continental crust at the time of the magmatism is no thicker than 40 km, which thus does not support the speculation for the presence of continental plateau in eastern China in the Mesozoic. In this context, it is worth to note that the ~2.8–2.5 Ga old Taishan granitoids in eastern North China Craton are not TTGs with heavy REE depletion, which is consistent with the coeval fresh and undeformed hornblendite cumulate and preserved "source" and "residual" rocks of "prograde" (vs. retrograded) amphibolite with no garnet (Shan et al. 2021). This suggests that the eastern China deep crust must not have ever been under greater depths of thickened crust in its history since the Archean and there is no evidence for the presence of a continental plateau in eastern China in the Mesozoic. In terms of straightforward physics, plateau formation is not a simple consequence of seafloor subduction, but a consequence of compositionally buoyant subducting plate such as the underthrusting continental lithospheric root beneath the Tibetan Plateau and the subduction of seamount chains of the Nazca plate beneath the Andes (Niu 2019),

8.4.3.3 Active Continental Arc Granitoid Magma (Arc) Generation

The data of eastern China granitoids used in the Nd isotope comparison (Fig. 8.11; Niu et al. 2015) are the literature data from throughout out the region (Fig. 8.10a), including spotty data on granitoids along the coastline. Our new data on the coastal granitoids are the youngest (Xue et al. 2018) and have significant mantle isotope signature (Fig. 8.14) compared to the *Int* granitoids, pointing to remelting of juvenile basaltic crust of mantle wedge melting. It is possible that such continental *Arc* granitoids can contribute to continental crust growth locally here because of the preservation as a result of collision, but is likely mass balanced globally because of sediment recycling and subduction erosion (see above).

The important message is that these youngest granitoids with strong mantle isotope signature at the arc-curved southeast coastline represent the last episode of paleo-Pacific seafloor subduction-zone related granitoid magmatism at the time or before subduction cessation. The latter resulted from trench jam caused by the arrival and collision of the physically buoyant Chinese continental shelf of exotic origin (Niu et al. 2015). The basement of the Chinese continental shelf (the single coherent mass of East and South China Seas) is physically consistent with an oceanic plateau or more likely a continental landmass with the basement, as exposed in Hainan Island, consisting of lithologies resembling those of western Australia, rather than South China (Xu et al. 2014; Niu et al. 2015; personal communication with Prof. Shu Sun in 2016 cited in Niu 2020a, b). We can further hypothesize that the Island of Taiwan resulted from the basement uplift associated with the arrival and subduction of the Philippines Sea plate. Confirmation of this hypothesis is of far-reaching significance in reconstructing the Mesozoic plate tectonics concerning the vast hemisphere involving Eurasian-Pacific-Indo-Australian plates and their evolution to the present.

8.5 Summary

The key messages concerning granitoid magmatism and continental crust accretion in a global context presented in this chapter are summarized with clarifications as follows:

(1) The standard island arc model (Taylor 1967, 1977) has served the community well for understanding the origin and growth of continental crust over the past half century. It has received general acceptance because of its petrological and geochemical footing. The bulk continental crust has an andesitic composition resembling those of andesites erupted from volcanoes along island arcs above subduction zones. All such andesites as well as island arc basalts are relatively enriched in fluid-soluble elements (e.g., U, K, Pb) but depleted in fluid-insoluble elements (e.g., Nb, Ta, Ti) (Fig. 8.2; e.g., Tatsumi and Eggins 1995), which is a compositional characteristic called "arc-signature" well shared by both upper and bulk continental crust (e.g., Rudnick and Gao 2003). Hence, subduction-zones are sites of continental crust growth as per the island-arc model.

(2) Studies have shown over the years, however, that the island arc model has several difficulties that much of the geoscience community needs to be well informed: (1) the primary arc magmas are not andesites, but the more mafic basalts resulting from subducting slab-dehydration induced mantle wedge

peridotite melting (e.g., Gill 1981); (2) there is no net crust accretion at subduction-zone settings because magmatic arc crust is mass balanced by subduction erosion and sediment recycling (Clift and Vannucchi 2004; Scholl and von Huene 2007); (3) the lower arc crust of cumulate origin is too depleted in incompatible elements to meet the required level for the lower continental crust (Kelemen and Behn 2016); (4) in fact, highly evolved felsic arc magmas of andesitic-rhyolitic compositions are also depleted in incompatible elements to explain the upper continental crust as we show (Fig. 8.5); (5) island-arc magmatism cannot explain the episodic crustal growth (Fig. 8.3) and the mantle plume hypothesis proposed to overcome this shortcoming is also difficult because plume magmas are too mafic (vs. andesitic); (6) the Archean TTG assemblages are considered by many as primary architecture of the continental crust (e.g., Rollinson 2008; Condie and Aster 2009) with highly depleted heavy REEs, resembling the modern subduction-related adakites, which seems to strengthen the island arc (or subduction) model for continental crust, but the bulk continental crust shows no such TTG- or adakite-signature (Fig. 8.1).

(3) Mafic deep arc crust delamination (removal) has been widely accepted as an effective process to produce "andesitic" upper arc crust suitable for continental crust growth (e.g., Lee and Anderson 2015), but this requires the absence of lithospheric mantle so that the forming mafic arc crust can founder into the melting mantle wedge asthenosphere, which is implausible. Re-lamination of deep arc crust by melts derived from melting of subducted arc rocks is proposed to enrich incompatible elements of the lower arc crust to the level required for the lower continental crust (Kelemen and Behn 2016), but it is questionable if such process, if true, is of globally general significance. Mantle plume hypothesis has been proposed to explain the episodic crustal growth (e.g., Stein and Hofmann 1994; Parman 2007; Pearson et al. 2007), but mantle plume magmas are too mafic (vs. andesitic). The heavy REE mismatch between the bulk continental crust and TTGs/adakites indicates that the latter rocks are volumetrically insignificant in building the continental crust.

(4) Because of the observations and understanding given in (1–3) above, and following our studies on syncollisional granitoids in southern Tibet associated with the India-Asia collision at ~55 ± 5 (Mo et al. 2007, 2008, 2009), we hypothesized that *continental collision zones are primary sites for net continental crust growth*, or simply *collision hypothesis* (Niu et al. 2013). This hypothesis overcomes all the difficulties of the standard island arc model in terms of straightforward geological and petrological processes concisely summarized below.

(5) Upon ocean basin closing and continental collision, partial melting of the unfully subducted upper ocean crust, along with seafloor sediments, under amphibolite conditions produces primitive andesite melt compositionally closely resembling the bulk continental crust. As the ocean crust is derived from the mantle in no distant past, the andesite melt (parental to the syncolli-

sional granitoids) with inherited mantle isotopic signatures contribute as juvenile material to continental crustal growth (Fig. 8.6). It is the collision that preserves and thus protects such juvenile crust from being lost through subduction erosion. Since the onset of plate tectonics, seafloor subduction continues, but continental collision is episodic globally, thus explaining the episodic crustal growth. All are geologically, petrologically and geochemically straightforward processes with no need to invoke physically implausible scenarios.

(6) We have successfully tested the *collision hypothesis* by sampling and studying syncollisional granitoids in older orogens in the vast Tibetan Plateau and the adjacent regions (Fig. 8.4), including the West Kunlun (~225 Ma), East Kunlun (~240 Ma), Qilian (~440 Ma North Qilian + South Qilian + Qilian Block) and West Qinling (~220 Ma) orogens (Figs. 8.6, 8.7, 8.8, 8.9, 8.10, 8.12, 8.13, 8.14).

(7) To further test the *collision hypothesis* and for comparison, we have also sampled and studied granitoids in intra-plate continental interiors (*Int*) of eastern China (the "Xiao Qinling" and "Laoshan" granitoid belts) and our hypothesized continental arc granitoids (*Arc*) along the coastline of southeast China (Fig. 8.4; Niu et al. 2015). The *Arc* granitoids are among the youngest (~102 Ma; Xue et al. 2018) with significant mantle input (high εNd(t)-εHf(t); Fig. 8.14). They are thus best understood as representing the last episode of Jurassic-Cretaceous granitoids marking the ending of the paleo-Pacific seafloor subduction because of the arrival and collision of the Chinese continental shelf of exotic origin with the arc-curved coastline manifesting the continental magmatic arc and the suture off the coast (Figs. 8.4, 8.10). The basement of the Chinese continental shelf may be an oceanic plateau or more likely a continental landmass (Figs. 8.4, 8.10a), probably related to landmass of western Australia. Careful study of basement rocks and rock histories of the Hainan Island and the island of Taiwan is expected to be revealing in plate tectonics reconstruction in the vast hemisphere involving Eurasian-Pacific-Indo-Australian plates since the Mesozoic (Niu et al. 2015). It is worth to mention the personal communication with Professor Shu Sun (2016) "*Sanya, Hainan Island, has Middle Cambrian strata, which contain phosphorites and fossils similar to those in Australia and completely different from those in mainland China, which we paid attention to in the past but have never resolved it.*" quoted in Niu (2020b).

(8) The mechanism for "Xiao Qinling" and "Laoshan" granitoid magmatism in the continental interiors (*Int*) is entirely different. We should note that although the "Laoshan" granitoids are at the coastline at present, they were in Chinese continental interiors at the time of Jurassic-Cretaceous emplacement long before the Yellow Sea opened at ~57 Ma (Fig. 8.10; Niu and Tang 2016). We should also note that while the "Xiao Qinling" granitoids of Jurassic-Cretaceous age show distribution parallel to the Triassic Qinling-Dabie orogenic belt (Fig. 8.4), they are over 100 Myrs younger and have no genetic relationship with the latter. Their trend is simply a matter of advanced uplift, erosion and outcropping in response to continued convergence since the

South-North China craton collision in the Late Triassic as manifested by the ~400 km strike-slip offset of the Tan-Lu fault zone. The highly enriched Nd-Hf isotope compositions of the *Int* granitoids are consistent with melting of ancient deep crust (Fig. 8.14).

(9) The Jurassic-Cretaceous granitoids are widespread in eastern continental China (Figs. 8.4, 8.10; Niu et al. 2015), spanning emplacement time for ~100 Myrs (from ~190 to ~88 Ma) randomly in space and time in a wide E-W zone in excess of 1000 km (Fig. 8.10a). That is, the granitoid magmatism could take place in the coastal regions as well as anywhere in the continental interiors as far as ~1400 km to the west at any latitude and at any given time (Fig. 8.10a, d). The Nd-Hf isotopically enriched composition of these granitoids points to, in simple clarity, the deep continental crust to be the magma source (Fig. 8.14). Depp crustal melting (anatexis or "re-working") requires excess heat that can only be provided by mantle derived basaltic magmas through intrusion and underplating. This further means that the basaltic magmas supplying the heat must be available areally beneath entire eastern continental China in space and time. This is in fact a straightforward crustal response to the mantle lithosphere thinning as a result of basal hydration weakening with the water coming from dehydration of the paleo-Pacific slab lying stagnant in the mantle transition zone, whose dehydration provides the water that percolated upwards in the form of hydrous melt and converted the basal lithosphere into the asthenosphere, thus thinned the lithosphere accompanied by hydrous basaltic magmatism (Niu 2005, 2014). Intrusion and underplating of such hydrous basaltic magmas caused the deep crustal melting under amphibolite facies conditions, resulting in the Jurassic-Cretaceous granitoids (Fig. 8.15; Niu et al. 2015). This understanding results from our research in the framework of North China Craton destruction (see Zhu et al. 2012), but we emphasize the lithosphere thinning and granitoid magmatism throughout entire eastern continental China (Niu 2014; Sun et al. 2021).

(10) The deep crustal melting produced the granitoids in question in eastern China, which destroyed the preexisting deep crust with the more refractory mafic granulite facies residues left behind (see Fig. 8.6b). Theoretically, the more mafic residues with enhanced seismic velocity can be detected as becoming the sub-Moho mantle constituent. The xenoliths representing the resultant crust would become the less mafic lithologies. All these resulted in the thinned deep crust and elevated/shallowed Moho that is coeval with the lithosphere thinning and the elevated lithosphere-asthenosphere boundary (LAB) in eastern China (Niu 2005, 2014). This can effectively address the puzzle attempted to resolve by Gao et al. (1998a, b) why the eastern China continental crust is thin and is missing the mafic deep crust (Li et al. 2006) as depicted in Fig. 8.16.

(11) While the origin of granites and granitoids that has been debated for many decades was elegantly resolved experimentally as the result of partial melting of crustal rocks (Tuttle and Bowen 1958), some clarifications remain in demand, especially because of the applications of radiogenic isotopes in the petrogenesis granitoids in recent decades. Mantle-like Nd-Hf isotope compo-

sitions of some granitoids (e.g., the syncollisional granitoids discussed here) have been correctly interpreted as representing juvenile crust, but mistaken by many as such granitoid magmas resulting from mantle melting. The latter is fundamentally incorrect. Partial melting of mantle peridotites cannot produce granitoid magmas. Protracted fractional crystallization of alkali basaltic magmas can produce peralkaline rhyolites and some A-type granites with no or weak negative Nb and Ta anomalies, but with significant negative Sr and Eu anomalies (e.g., Shao et al. 2015, 2021). Such granitoids are volumetrically rare. Most granitoids belong to the I-Type granites resulting from partial melting of basaltic rocks in varying tectonic settings. In many orogens highly deformed metasedimentary rocks can melt to produce the S-Type granites. These terminologies are familiar to everyone, but the key point is that all types of granitoids are, material-wise, ultimately coming from the mantle, but not directly and undergo at least two stages of melting and re-processing (re-working). For example, the syncollisional granitoids (*Syn*; Figs. 8.4, 8.12, 8.13, 8.14) we study result from melting of the ocean crust (with sediments) derived in no distant past from the mantle at ocean ridges, carrying mantle isotope signatures, and thus qualifying them as juvenile crustal contributions. This includes granitoids form southeast coastline China (*Arc*; Figs. 8.4, 8.12, 8.13, 8.14). Re-melting of underplated basaltic rocks in no distant past can also produce juvenile granitoids for crustal growth if the conditions permit melting. The granitoids in the interiors of eastern continental China (*Int*; e.g., "Xiao Qinling" and "Laoshan" suites) are also largely I-type granitoids derived from partial melting of mafic lower crust under amphibolite facies conditions, representing re-processed or re-worked existing crustal rocks, making no new crust materials. The magma source of the *Int* granitoids is most consistent with basaltic compositions of complex tectonic and metamorphic histories. We should point out that using detrital zircon ages of all sorts without clear knowledge of their granitoid petrogenesis may lead to unreliable models of continental crustal growth unless the age data are filtered with zircon Hf-O isotope compositions.

(12) One of the most extensively discussed issues on granitoid petrogenesis concerns the origin of mafic magmatic enclaves (MMEs), which has also been used by some as the acronym for mafic microgranular enclaves. The MMEs have been widely used as evidence for magma mixing between mantle derived basaltic magmas represented by the MMEs and the crust-derived felsic magmas represented by the host granitoids. This magma mixing interpretation has been used in all textbooks and has thus become fact in almost all the literature over the last decades. Based on field observations and detailed petrography, we hypothesized (Niu et al. 2013) that the MMEs dominated by amphibole, biotite and plagioclase (plus accessory phases) are early cumulate of the same magmatic systems. We have tested and thoroughly proved this hypothesis on syncollisional granitoids with no exception (Huang et al. 2014, 2015, 2016, 2019; Chen et al. 2015, 2016, 2018; 2021a, b; Xiao et al. 2020). We should also note the fact that the mafic MMEs is dominated by amphiboles with vary-

ing smaller amount of plagioclase and biotite, which gives the bulk composition resembling "gabbros" in the TAS ($Na_2O + K_2O$) vs. SiO_2) diagram. The latter has been widely interpreted as gabbros representing basaltic magmas. This is incorrect and has led to incorrect interpretations on the petrogenesis and crustal and mantle processes. Amphibole + plagioclase + biotite are typical liquidus phases of andesite melts and crystallization and cumulation of these phases give the MMEs. If the parental melts are basalts, then the liquidus phases would be clinopyroxene + plagioclase ± olivine ± titanomagnetite, but these are not the mineralogy of MMEs. It is possible that some lesser abundant MMEs in intraplate granitoids may be of mixing origin although evidence is sparce and further investigation is warranted.

Acknowledgements Yaoling Niu thanks his research students/collaborators Chen Shuo, Cui Huixia, Duan Meng, Gao Yajie, Gong Hongmei, Guo Pengyuan, Hong Di, Hu Yan, Hu Zhenxing, Huang Hui, Kong Juanjuan, Li Jiyong, Liu Jinjun, Liu Yi, Ma Yuxin, Shao Fengli, Sun Pu, Sun Wenli, Wang Xiaohong, Ye Lei, Xue Qiqi, Zhang Guorui and Zhang Yu for their field company and laboratory assistance over the past decade. Some of the data used in Figs. 8.12–8.14 are yet to be published by Yaoling Niu and some of these collaborators. This work and the on-going research have been supported by National Natural Science Foundation of China (NSFC grants: 91014003, 41130314, 41630968, 91958215, U1606401).

References

Abbott D, Mooney W (1995) The structural and geochemical evolution of the continental crust support for the oceanic plateau model of continental growth. Rev Geophys 33:231–242

Abbott DH, Drury R, Mooney WD (1997) Continents as lithological icebergs: the importance of buoyant lithospheric roots. Earth Planet Sci Lett 149:15–27

Abbott D, Mooney W (1995) The structural and geochemical evolution of the continental crust support for the oceanic plateau model of continental growth. Rev Geophys 33:231–242

Albarède F (1998) The growth of continental crust. Tectonophysics 296:1–14

Allègre CJ, Rousseau D (1984) The growth of the continent through geological time studied by Nd isotope analysis of shales. Earth Planet Sci Lett 67:19–34

Allègre CJ, Hart SR, Mister JF (1983) Chemical-structure and evolution of the mantle and continents determined by inversion of Nd and Sr isotopic data, I. Theoretical methods. Earth Planet Sci Lett 66:177–190

Arculus RJ (1981) Island arc magmatism in relation to the evolution of the crust and mantle. Tectonophysics 75:113–133

Armstrong RL (1968) A model for the evolution of strontium and lead isotopes in a dynamic earth. Rev Geophys Space Phys 6:175–200

Armstrong RL (1981) Radiogenic isotopes: the case for crustal recycling on a near-steady state no-continental-growth. Philos Trans R Soc Lond A301:443–472

Bacon CR, Bruggman PE, Christiansen RL, Clynne MA, Donnelly-Nolan JM, Hildreth W (1995) Primitive magmas at five cascade volcanic fields: melts from hot, heterogeneous sub-arc mantle. Can Mineral 35:397–423

Barker SJ, Wilson CJN, Baker JA, Millet M-A, Rotella MD, Wright IC, Wysoczanski RJ (2013) Geochemistry and petrogenesis of silicic magmas in the intra-oceanic Kermadec arc. J Petrol 54:351–391

Bowring SA, Housh T (1995) The Earth's early evolution. Science 269:1535–1540

Burke K, Dewey JF (1975) The Wilson cycle. In: Geological Society of America, Northeastern Section, 10th Annual Meeting, Syracuse, NY, Abstracts with Pro-grams, Boulder, CO, 48
Calvert AJ, Klemperer SL, Takahashi N, Kerr BC (2008) Three-dimensional crustal structure of the Mariana Island arc from seismic tomography. J Geophys Res 113:B01406. https://doi.org/10.1029/2007JB004939
Campbell IH (2003) Constraints on continental growth models from Nb/U ratios in the 3.5 Ga Barberton and other Archaean basalt-komatiite suites. Am J Sci 303:319–351
Campbell IH, Allen CM (2008) Formation of supercontinents linked to increases in atmospheric oxygen. Nat Geosci 1:554–558
Campbell IH, Davies DR (2017) Raising the continental crust. Earth Planet Sci Lett 460:112–122
Campbell IH, Taylor SR (1983) No water, no granites-no oceans, no continents. Geophys Res Lett 10:1061–1064
Castillo PR (2006) An overview of adakite petrogenesis. Chin Sci Bull 51:257–268
Castillo PR (2012) Adakites petrogenesis. Lithos 134(135):304–316
Cavosie AJ, Valley JW, Wilde SA (2004) The oldest terrestrial mineral record: a review of 4400–4000 Ma detrital zircons from Jack Hills, Western Australia. In: Van Kranendonk MJ, Smithies RH, Bennett VC (eds) (Condie, K.C., Series Editor) Developments in Precambrian geology, vol 15, pp 91–112
Cawood PA, Hawkesworth CJ (2019) Continental crustal volume, thickness and area, and their geodynamic implications. Gondwana Res 66:116–125
Cawood PA, Hawkesworth CJ, Dhuime B (2012) Detrital zircon record and tectonic setting. Geology 40:875–878
Chen S, Niu YL, Sun WL, Zhang Y, Li JY, Guo PY, Sun P (2015) On the origin of mafic magmatic enclaves (MMEs) in syn-collisional granitoids: evidence from the Baojishan pluton in the north Qilian Orogen, China. Mineral Petrol 109:577–596
Chen S, Niu YL, Li JY, Sun WL, Zhang Y, Hu Y, Shao FL (2016) Syncollisional adakitic granodiorites formed by fractional crystallization: insights from their enclosed mafic magmatic enclaves (MMEs) in the Qumushan pluton, north Qilian Orogen at the northern margin of the Tibetan plateau. Lithos 248(251):455–468
Chen S, Niu YL, Xue QQ (2018) Syn-collisional felsic magmatism and continental crust growth: a case study from the north Qilian Orogenic Belt at the northern margin of the Tibetan plateau. Lithos 308(309):53–64
Chen S, Niu YL, Wang XH, Xue QQ, Sun WL (2021a) Identifying crystal accumulation in granitoids through amphibole composition and in situ zircon O isotopes in north Qilian Orogen. J Petrol 62:egab096. (in press)
Chen YH, Niu YL, Xue QQ, Gao YJ, Castillo PR (2021b) An iron isotope perspective on back-arc basin development: messages from Mariana trough basalts. Earth Planet Sci Lett 572:117133
Clift P, Vannucchi P (2004) Controls on tectonic accretion versus erosion in subduction zones: implications for the origin and recycling of the continental crust. Rev Geophys 42:31. https://doi.org/10.1029/2003RG000127
Clift P, Vannucchi P, Morgan JP (2009) Crustal redistribution, crust-mantle recycling and phanerozoic evolution of the continental crust. Earth Sci Rev 97:80–144
Condie KC (1998) Episodic continental growth and supercontinents: a mantle avalanche connection? Earth Planet Sci Lett 163:97–108
Condie KC (2000) Episodic continental growth models: afterthoughts and extensions. Tectonophysics 322:153–162
Condie KC (2005) TTGs and adakites: are they both slab melts? Lithos 80:33–44
Condie KC, Aster RC (2009) Zircon age Episodicity and growth of continental crust. EOS 90(41):364
Condie KC, Aster RC (2010) Episodic zircon age spectra of orogenic granitoids: the supercontinent connection and continental growth. Precambrian Res 180:227–236

Condie KC, Kroner A (2013) The building blocks of continental crust: evidence for a major change in the tectonic setting of continental growth at the end of the Archean. Gondwana Res 23:394–402
Condie KC, Arndt N, Davaille A, Puetz SJ (2017) Zircon age peaks: production or preservation of continental crust? Geosphere 13:227–234
Cunningham H, Gill J, Turner S, Caulfield J, Edwards L, Day S (2012) Rapid magmatic processes accompany arc–continent collision: the western Bismarck arc, Papua New Guinea. Contrib Mineral Petrol 164:789–804
Davidson JP, Arculus RJ (2006) The significance of Phanerozoic arc magmatism in generating continental crust. In: Brown M, Rushmer T (eds) Evolution and differentiation of the continental crust. Cambridge Press, pp 135–172
Davies JH, von Blanckenburg F (1995) Slab breakoff: a model of lithosphere detachment and its test in the magmatism and deformation of collisional orogens. Earth Planet Sci Lett 129:85–102
Defant MJ, Drummond MS (1990) Derivation of some modern arc magmas by melting of young subducted lithosphere. Nature 347:662–665
Deng JF, Zhao HL, Luo ZH, Guo ZF, Mo XX (1998) Mantle plumes and lithosphere motion in East Asia. Am Geophys Union Geodyn Ser 27:59–66
Deng JF, Mo XX, Zhao HL, Wu ZX, Luo ZH, Guo SG (2004) A new model for the dynamic evolution of Chinese lithosphere: 'continental roots-plume tectonics'. Earth Sci Rev 65:223–275
Deng JF, Su SG, Niu YL, Liu C, Zhao GC, Zhao XG, Zhou S, Wu ZX (2007) A possible model for the lithospheric thinning of North China craton: evidence from the Yanshanian (Jura-cretaceous) magmatism and tectonic deformation. Lithos 96:22–35
Dewey JF, Bird JM (1970) Mountain belts and the new global tectonics. J Geophys Res 75:2625–2647
Dhuime B, Hawkesworth CJ, Cawood PA, Storey CD (2012) A change in the geodynamics of continental growth 3 billion years ago. Science 335:1334–1336
Duan M, Niu YL, Kong JJ, Sun P, Hu Y, Zhang Y, Chen S, Li JY (2016) Zircon U–Pb geochronology, Sr–Nd–Hf isotopic composition and geological significance of the late-Triassic Baijiazhuang and Lvjing granitic plutons in western Qinling Orogen. Lithos 260:443–456
Eggins SM (1993) Origin and differentiation of picritic arc magmas, Ambae (Aoba), Vanuatu. Contrib Mineral Petrol 114:79–100
Elburg M, Foden J (1998) Temporal changes in arc magma geochemistry, northern Sulawesi, Indonesia. Earth Planet Sci Lett 163:381–398
Elliott T (2003) Tracers of the slab. Geophys Monogr 238:23–45
Ernst WG (2009) Archean plate tectonics, rise of proterozoic supercontinentality and onset of regional, episodic stagnant-lid behaviour. Gondwana Res 15:243–253
Finney B, Turner S, Hawkesworth C, Larsen J, Nye C, George R, Bindeman I, Eichelberger J (2008) Magmatic differentiation at an Island-arc caldera: Okmok volcano, Aleutian Islands, Alaska. J Petrol 49:857–884
Gao S, Luo TC, Zhang BR, Zhang HF, Han YW, Zhao ZD, Hu YK (1998a) Chemical composition of the continental crust as revealed by studies in East China. Geochim Cosmochim Acta 62:1959–1975
Gao S, Zhang BR, Jin ZM, Hern H, Luo TC, Zhao ZD (1998b) How mafic is the lower continental crust? Earth Planet Sci Lett 161:101–117
Gao YJ, Niu YL, Duan M, Xue QQ, Sun P, Chen S, Xiao YY, Guo PY, Wang XH, Chen YH (2019) The petrogenesis and tectonic significance of the early cretaceous intraplate granites in eastern China: the Laoshan granite as an example. Lithos 328(329):200–211
Gast PW (1968) Trace element fractionation and the origin of tholeiitic and alkaline magma types. Geochim Cosmochim Acta 32:1055–1086
Gill JB (1981) Orogenic Andesites and plate tectonics. Springer, New York, p 390
Green TH, Ringwood AE (1966) Origin of the calc-alkaline igneous rock suite. Earth Planet Sci Lett 1:307–316

Green TH, Ringwood AE (1968) Genesis of the calc-alkaline igneous rock suite. Contrib Mineral Petrol 18:105–162

Hanyu T, Gill J, Tatsumi Y, Kimura J-I, Sato K, Chang Q, Senda R, Miyazaki T, Hirahara Y, Takahashi T, Zulkarnain I (2012) Across- and along-arc geochemical variations of lava chemistry in the Sangihe arc: various fluid and melt slab fluxes in response to slab temperature. Geochem Geophys Geosyst 13:Q10021. https://doi.org/10.1029/2012GC004346

Haraguchi S, Ishii T, Kimura J-I, Ohara Y (2003) Formation of tonalite from basaltic magma at the Komahashi-Daini seamount, northern Kyushu-Palau ridge in the Philippine Sea, and growth of Izu-Ogasawara (Bonin)-Mariana arc crust. Contrib Mineral Petrol 143:151–168

Harrison TM, Blichert-Toft J, Müller W, Albarède F, Holden P, Mojzsis SJ (2005) Heterogeneous hadean hafnium: evidence of continental crust at 4.4 to 4.5 Ga. Science 310:1947–1950

Hawkesworth CJ, Kemp AIS (2006) The differentiation and rates of generation of the continental crust. Chem Geol 226:134–143

Hawkesworth C, Cawood P, Kemp A, Storey C, Dhuime B (2009) A matter of preservation. Science 323:49–50

Hawkesworth CJ, Dhuime B, Pietranik AB, Cawood PA, Kemp AIS, Storey CD (2010) The generation and evolution of the continental crust. J Geol Soc 167:229–248

Hawkesworth CJ, Cawood PA, Dhuime B (2019) Rates of generation and growth of the continental crust. Geosci Front 10:165–173

Hermann J, Spandler CJ (2008) Sediment melts at sub-arc depths: an experimental study. J Petrol 49:717–740

Hirose K (1997) Melting experiments on lherzolite KLB-1 under hydrous conditions and generation of high-magnesian andesitic melts. Geology 25:42–44

Hofmann AW (1988) Chemical differentiation of the earth: the relationship between mantle, continental crust, and oceanic crust. Earth Planet Sci Lett 90:297–314

Hong D, Niu YL, Xiao YY, Sun P, Kong JJ, Guo PY, Shao FL, Wang XH, Duan M, Xue QQ, Gong HM, Chen S (2018) Origin of the Jurassic-cretaceous intraplate granitoids in eastern China as a consequence of paleo-Pacific plate subduction. Lithos 322:405–419

Hu Y, Niu YL, Li JY, Ye L, Kong JJ, Chen S, Zhang Y, Zhang GR (2016) Petrogenesis and tectonic significance of the late Triassic mafic dikes and felsic volcanic rocks in the east Kunlun Orogenic Belt, northern Tibet plateau. Lithos 245:205–222

Huang H, Niu YL, Nowell GN, Zhao ZD, Yu XH, Zhu DC, Mo XX, Ding S (2014) Geochemical constraints on the petrogenesis of granitoids in the east Kunlun orogenic belt, northern Tibetan plateau: implications for the continental crust growth through syncollisional felsic magmatism. Chem Geol 370:1–18

Huang H, Niu YL, Nowell G, Zhao ZD, Yu XH, Mo XX, Ding S (2015) The nature and history of the Qilian block in the context of the development of the greater Tibetan plateau. Gondwana Res 28:209–224

Huang H, Niu YL, Mo XX (2016) Syn-collisional granitoids in the Qilian block on the northern Tibetan plateau: a long-lasting magmatism since continental collision through slab steepening. Lithos 246(247):99–109

Huang H, Niu YL, Mo XX (2017) Garnet effect on Nd-Hf isotope decoupling: Evidence from the Jinfosi batholith, Northern Tibetan Plateau. Lithos 274/275:31–38

Huang H, Niu YL, Teng FZ, Wang SJ (2019) Discrepancy between bulk-rock and zircon Hf isotopes accompanying Nd–Hf isotope decoupling. Geochim Cosmochim Acta 259:17–36

Niu YL, Collerson KD, Batiza R, Wendt I, Regelous M (1999) The origin of E-type MORB at ridges far from mantle plumes: the East Pacific Rise at 11°20'N. J Geophys Res 104:7067–7087

von Huene R, Scholl DW (1991) Observations at convergent margins concerning sediment subduction, subduction erosion, and the growth of continental crust. Rev Geophys 29:279–316

van Hunen J, Allen MB (2011) Continental collision and slab break-off: a comparison of 3-D numerical models with observations. Earth Planet Sci Lett 302:27–37

van Hunen J, Moyen J-F (2012) Archean subduction: fact or fiction. Annu Rev Earth Planet Sci 40:195–219

Iizuka T, Komiya T, Ueno Y, Katayama I, Uehara Y, Maruyama S, Hirata T, Johnson SP, Dunkley DJ (2007) Geology and zircon geochronology of the Acasta gneiss complex, northwestern Canada: new constraints on its tectonothermal history. Precambrian Res 153:179–208
Ionov DA, Benard A, Plechov PY, Sjcherbakov VD (2012) Along-arc variations in lithospheric mantle compositions in Kamchatka, Russia: first trace element data on mantle xenoliths from the Klyuchevskoy group volcanoes. Chem Geol 263:122–131
Ishizuka O, Yuasa M, Uto K (2002) Evidence of porphyry copper-type hydrothermal activity from a submerged remnant back-arc volcano of the Izu-Bonin arc implications for the volcanotectonic history of back-arc seamounts. Earth Planet Sci Lett 198:381–399
Jicha BR, Singer BS, Brophy JG, Fournelle JH, Johnson CM, Beard BL, Lapen TJ, Mahlen NJ (2004) Variable impact of the subducted slab on Aleutian Island arc magma sources: evidence from Sr, Nd, Pb, and Hf isotopes and trace element abundances. J Petrol 45:1845–1875
Johnson MC, Plank T (1999) Dehydration and melting experiments constrain the fate of subducted sediments. Geochem Geophys Geosyst 1:29
Jull M, Kelemen PB (2001) On the conditions for lower crustal convective stability. J Geophys Res 106:6423–6446
Kay RW, Kay SM (1993) Delamination and delamination magmatism. Tectonophysics 219:177–189
Kelemen PB (1995) Genesis of high Mg andesites and the continental crust. Contrib Mineral Petrol 120:1–19
Kelemen PB, Behn MD (2016) Formation of lower continental crust by relamination of buoyant arc lavas and plutons. Nat Geosci 9:197–205
Kelemen PB, Yogodzinski GM, Scholl DW (2003) Along-strike variation in lavas of Aleutian Island arcs: genesis of high mg# andesite and implications for continental crust. Geophys Monogr 238:223–276
Kemp AIS, Hawkesworth CJ, Paterson BA, Kinny PD (2006) Episodic growth of the Gondwana supercontinent from hafnium and oxygen isotopes in zircon. Nature 439:580–583
Kong JJ, Niu YL, Duan M, Zhang Y, Hu Y, Li JY, Chen S (2017) Petrogenesis of Luchuba and Wuchaba granitoids in western Qinling: geochronology and geochemical evidence. Mineral Petrol 111:887–908
Kong JJ, Niu YL, Duan M, Xiao YY, Zhang Y, Guo PY, Sun P, Gong HM (2019) The syncollisional granitoid magmatism and crust growth during the west Qinling orogeny, China: insights from the Jiaochangba pluton. Geol J 54:4014–4033
Kong JJ, Niu YL, Hu Y, Zhang Y, Shao FL (2020) Petrogenesis of the Triassic granitoids from the east Kunlun Orogenic Belt, NW China: implications for continental crust growth from syncollisional to post-collisional setting. Lithos 364(365):105513
Kramers JD, Tolstikhin IN (1997) Two terrestrial lead isotope paradoxes, forward transport modelling, core formation and the history of the continental crust. Chem Geol 139:75–110
Kröner A, Windley BF, Badarch G, Tomurtogoo O, Ehgner E, Jahn BM, Gruschka S, Khain EV, Demoux A, Wingate MTD (2007) Accretionary growth an crust formation in the central Asian Orogenic Belt and comparison with the Arabian-Nubian shield. Geol Soc Am Memoir 200:181–209
Kushiro I, Syono Y, Akimoto S (1968) Melting of a peridotite nodule at high pressures and high water pressures. J Geophys Res 73:6023–6029
Leat PT, Smellie JL, Millar IL, Larter RD (2003) Magmatism in the South Sandwich arc. Geol Soc Lond Spec Publ 219:285–313
Leat PT, Larter RD, Millar IL (2007) Silicic magmas of protector shoal, south sandwich arc: indicators of generation of primitive continental crust in an Island arc. Geol Mag 144:179–190
Lee C-T, Anderson DL (2015) Continental crust formation at arcs, the arclogite "delamination" cycle, and one origin for fertile melting anomalies in the mantle. Sci Bull 60:1141–1156
Li SL, Mooney WD, Fan JC (2006) Crustal structure of mainland China from deep seismic sounding data. Tectonophysics 420:239–252

Li JY, Niu YL, Hu Y, Chen S, Zhang Y, Duan M, Sun P (2016) Origin of the late early cretaceous granodiorite and associated dioritic dikes in the Hongqilafu pluton, northwestern Tibetan plateau: a case for crust-mantle interaction. Lithos 260:300–314

Li JY, Niu YL, Chen S, Sun WL, Zhang Y, Liu Y, Ma YX, Hu ZX, Zhang GR (2017) Petrogenesis of granitoids in eastern section of the central Qilian block: evidence from geochemistry and zircon U–Pb geochronology. Mineral Petrol 111:23–41

Lindsay JM, Trumbull RB, Siebel W (2005) Geochemistry and petrogenesis of late Pleistocene to recent volcanism in southern Dominica, lesser Antilles. J Volcanol Geotherm Res 148:254–294

Liu W, Liu XJ, Xiao WJ (2012) Massive granitoid production without massive continental-crust growth in the Chinese Altay: Insight into the source rock of granitoids using integrated zircon U–Pb age, Hf–Nd–Sr isotopes and geochemistry. Am J Sci 312:629–684

Martin H, Smithies RH, Rapp R, Moyen JF, Champion D (2005) An overview of adakite, tonalite-trondhjemite-granodiorite (TTG) and sanukitoid: relationships and some implications for crustal evolution. Lithos 79:1–24

McCulloch MT, Bennett VC (1994) Progressive growth of the Earth's continental crust and depleted mantle: geochemical constraints. Geochim Cosmochim Acta 58:4717–4738

McCulloch MT, Gamble JA (1991) Geochemical and geodynamical constraints on subduction zone magmatism. Earth Planet Sci Lett 102:358–374

Miller DM, Langmuir CH, Goldstein SL, Franks AL (1992) The importance of parental magma composition to calc-alkaline and tholeiitic evolution: evidence from Umnak Island in the Aleutians. J Geophys Res 97:321–343

Mo X, Hou Z, Niu YL, Dong G, Qu X, Zhao Z, Yang Z (2007) Mantle contributions to crustal thickening in South Tibet in response to the India-Asia collision. Lithos 96:225–242

Mo X, Niu Y, Dong G, Zhao Z, Hou Z, Zhou S, Ke S (2008) Contribution of syncollisional felsic magmatism to continental crust growth: a case study of the Paleogene Linzizong volcanic succession in southern Tibet. Chem Geol 250:49–67

Mo XX, Dong GC, Zhao ZD, Zhu DC, Zhou S, Niu YL (2009) Mantle input to the crust in southern Gangdese, Tibet, during the Cenozoic: zircon Hf isotopic evidence. J Earth Sci 20:241–249

Moyen JF (2009) High Sr/Y and La/Yb ratios: the meaning of the "adakitic signature". Lithos 112:556–574

Moyen JF (2011) The composite Archean grey gneisses: petrological significance, and evidence for a non-unique tectonic setting for Archean crustal growth. Lithos 123:21–36

Nichols GT, Wyllie PJ, Stern CR (1994) Subduction zone melting of pelagic sediments constrained by melting experiments. Nature 371:785–788

Niu YL (2005) Generation and evolution of basaltic magmas: some basic concepts and a hypothesis for the origin of the Mesozoic-Cenozoic volcanism in eastern China. Geol J China Univ 11:9–46. http://www.dur.ac.uk/yaoling.niu/MyReprints-pdf/2005NiuGXDZH.pdf

Niu YL (2014) Geological understanding of plate tectonics: basic concepts, illustrations, examples and new perspectives. Glob Tectonics Metallogeny 10:23–46

Niu YL (2017) Slab breakoff: a causal mechanism or pure convenience? Sci Bull 62:456–461

Niu YL (2019) The rise of the Tibetan plateau as a consequence of progressive underthrusting of the Indian lithosphere driven by subduction of the giant India-Australia plate at the Sumatra-Java trench. In: AGU Fall Meeting 2019

Niu YL (2020a) What drives the continued India-Asia convergence since the collision at 55 Ma? Sci Bull 65:169–172

Niu YL (2020b) On the cause of continental breakup: a simple analysis in terms of driving mechanisms of plate tectonics and mantle plumes. J Asian Earth Sci 194:104367

Niu YL (2021) Lithosphere thickness controls the extent of mantle melting, depth of melt extraction and basalt compositions in all tectonic settings on earth—a review and new perspectives. Earth Sci Rev 217:103614

Niu YL, Lesher CM (1991) Hydrothermal alteration of mafic metavolcanic rocks and genesis of Fe–Zn–Cu sulfide deposits, Stone Hill district, Alabama. Econ Geol 86:983–1001

Niu YL, O'Hara MJ (2003) The origin of ocean Island basalts (OIB): a new perspective from petrology, geochemistry and mineral physics considerations. J Geophys Res 108:2209. https://doi.org/10.1029/2002JB002048
Niu YL, O'Hara MJ (2009) MORB mantle hosts the missing Eu (Sr, Nb, Ta and Ti) in the continental crust: new perspectives on crustal growth, crust-mantle differentiation and chemical structure of oceanic upper mantle. Lithos 112:1–17
Niu YL, O'Hara MJ (2008) Global correlations of ocean ridge basalt chemistry with axial depth: a new perspective. J Petrol 49:633–664
Niu YL, Tang J (2016) Origin of the Yellow Sea: an insight. Sci Bull 61:1076–1080
Niu YL, Waggoner G, Sinton JM, Mahoney JJ (1996) Mantle source heterogeneity and melting processes beneath seafloor spreading centers: the East Pacific rise 18°–19°S. J Geophys Res 101:27711–27733
Niu YL, O'Hara MJ, Pearce JA (2003) Initiation of subduction zones as a consequence of lateral compositional buoyancy contrast within the lithosphere: a petrologic perspective. J Petrol 44:851–866
Niu, Y.L., Mo, X., Dong, G., Zhao, Z., Hou, Z., Zhou, S., Ke, S., 2007. Continental collision zones are primary sites of net continental crustal growth: evidence from the Linzizong volcanic succession in southern Tibet. In: Eos trans. AGU, 88(52), fall meeting, supplement abstract V34A-01
Niu YL, Zhao ZD, Zhu DC, Mo XX (2013) Continental collision zones are primary sites for net continental crust growth—a testable hypothesis. Earth Sci Rev 127:96–110
Niu YL, Liu Y, Xue QQ, Shao FL, Chen S, Duan M, Guo PY, Gong HM, Hu Y, Hu ZX, Kong JJ, Li JY, Liu JJ, Sun P, Sun WL, Ye L, Xiao YY, Zhang Y (2015) Exotic origin of the Chinese continental shelf: new insights into the tectonic evolution of the western Pacific and eastern China since the Mesozoic. Sci Bull 60:1598–1616
Niu YL, Shi XF, Li TG, Wu SG, Sun WD, Zhu RX (2017) Testing the mantle plume hypothesis: an IODP effort to drill into the Kamchatka-Okhotsk Sea basement. Sci Bull 62:1464–1472
Nye CJ, Reid MR (1986) Geochemistry of primary and least fractionated lavas from Okmok volcano, central Aleutians—implications for arc magma genesis. J Geophys Res 91:271–287
O'Nions RK, Evensen NM, Hamilton PJ (1979) Geochemical modeling of mantle differentiation and crustal growth. J Geophys Res 84:6091–6101
O'Hara MJ (1965) Primary magmas and the origin of basalts. Scott J Geol 1:19–40
Parkinson IJ, Arculus RJ (1999) The redox state of subduction zones: insights from arc-peridotites. Chem Geol 160:409–423
Parman SW (2007) Helium isotopic evidence for episodic mantle melting and crustal growth. Nature 446:900–903
Peacock SM (2003) Thermal structure and metamorphic evolution of subducting slabs, in inside the subduction factory. Geophys Monogr 238:7–22
Peacock SM, Wang K (1999) Seismic consequences of warm versus cool subduction zone metamorphism: examples from northeast and Southeast Japan. Science 286:937–939
Pearce JA, Peate DW (1995) Tectonic implications of the composition of volcanic arc magmas. Annu Rev Earth Planet Sci 23:251–285
Pearcy JG, DeBari SM, Sleep NH (1990) Mass balance calculations for two sections of Island arc crust and implications for the formation of continents. Earth Planet Sci Lett 96:427–442
Pearson DG, Parman SW, Nowell GM (2007) A link between large mantle melting events and continent growth seen in osmium isotopes. Nature 449:202–205
Petterson MG, Haldane MI, Smith DJ, Billy D, Jordan NJ (2011) Geochemistry and petrogenesis of the Gallego volcanic field, Solomon Islands, SW Pacific and geotectonic implications. Lithos 125:915–927
Plank T (2005) Constraints from thorium/lanthanum on sediment recycling at subduction zones and the evolution of continents. J Petrol 46:921–944
Plank T, Langmuir CH (1998) The chemical compositions of subducting sediments and its consequences for the crust and mantle. Chem Geol 145:325–394

Polat A, Kerrich R, Wyman DA (1998) The late Archean Schreiber-Hemlo and White River-Dayohessarah greenstone belts, Superior Province: collages of oceanic plateaus, ocean arcs, and subduction-accretion complexes. Tectonophysics 289:295–326

Puetz S (2018) A relational database of global U–Pb ages. Geosci Front 9:877–891

Rapp RP (1995) Amphibolite-out phase boundary in partially melted metabasalt, its control over liquid fraction and composition, and source permeability. J Geophys Res 100:15601–15610

Reagan MK, Hanan BB, Heizler MT, Hartman BS, Hickey-Vargas R (2008) Petrogenesis of volcanic rocks from Saipan and Rota, Mariana Islands, and implications for the evolution of nascent Island arcs. J Petrol 49:441–464

Reymer A, Schubert G (1984) Phanerozoic addition rates to the continental crust and crustal growth. Tectonics 3:63–77

Richter FM, Rowley DB, DePalo DJ (1992) Sr isotope evolution of seawater: the role of tectonics. Earth Planet Sci Lett 109:11–23

Ringwood AE, Green DH (1966) An experimental investigation of the gabbro-eclogite transformation and some geophysical implications. Tectonophysics 3:383–427

Rollinson H (2008) Secular evolution of the continental crust: implications for crust evolution models. Geochem Geophys Geosyst 9:Q12010. https://doi.org/10.1029/2008GC002262

Rollinson H, Martin H (eds) (2005) Geodynamic controls on adakite, TTG and sanukitoids genesis: implications for models of crust formation (special issue). Lithos 79:1–269

Rudnick RL (1995) Making continental crust. Nature 378:571–578

Rudnick RL, Gao S (2003) Composition of the continental crust. Treatise Geochem 3:1–64

Scholl DW, von Huene R (2007) Crustal recycling at modern subduction zones applied to the past: issues of growth and preservation of continental basement crust, mantle geochemistry, and supercontinent reconstruction. Geol Soc Am Memoir 200:9–32

Schuth S, Münker C, König S, Qopoto C, Basi S, Garbe-Schönberg D, Ballhaus C (2009) Petrogenesis of lavas along the Solomon Island arc, SW Pacific: coupling of compositional variations and subduction zone geometry. J Petrol 50:781–811

Shan L, Niu YL, Ma SL (2021) Petrogenesis of the Neoarchean diorite and hornblendite in the Taishan area, western Shangdong: constraints on crustal evolution. Geol Bull China 40:1149–1177. (in Chinese with English abstract)

Shao FL, Niu YL, Regelous R, Zhu DC (2015) Petrogenesis of peralkaline rhyolites in an intra-plate setting: Glass House Mountains, Southeast Queensland, Australia. Lithos 216(217):196–210

Shao FL, Niu YL, Liu Y, Chen S, Kong JJ, Duan M (2017) Petrogenesis of Triassic granitoids in the east Kunlun Orogenic Belt, northern Tibetan plateau and their tectonic implications. Lithos 282(283):33–44

Shao FL, Niu YL, Kong JJ, Liu Y, Wang GD, Zhang Y (2021) Petrogenesis and tectonic implications of the Triassic rhyolites in the east Kunlun orogenic belt. Geosci Front 12:101234

Song SG, Yang LM, Zhang YQ, Niu YL, Wang C, Su L, Gao YL (2017) Qi–Qin accretionary belt in Central China Orogen: accretion by trench jam of oceanic plateau and formation of intra-plate arc in the early Paleozoic Qin-qi-Kun Ocean. Sci Bull 62:1035–1038

Spicer RA, Harris NW, Widdowson M, Herman AB, Guo SX, Valdes PJ, Wolfe JA, Kelley SP (2003) Constant elevation of southern Tibet over the past 15 million years. Nature 421:622–624

Stein M, Hofmann AW (1994) Mantle plumes and episodic crustal growth. Nature 372:63–68

Stern RJ (2007) When and how did plate tectonics begin? Theoretical and empirical considerations. Chin Sci Bull 52:578–591

Stern RJ, Scholl D (2010) Yin and Yang of continental crust creation and destruction by plate tectonics. Int Geol Rev 52:1–31

Stern CR, Huang WL, Wyllie PJ (1975) Basalt–andesite–rhyolite–H_2O: crystallization intervals with excess H_2O and H_2O—undersaturated liquidus surfaces to 35 kb, with implications for magma genesis. Earth Planet Sci Lett 28:189–196

Sun S-S, McDonough WF (1989) Chemical and isotopic systematics of ocean basalt: implications for mantle composition and processes. Geol Soc Spec Pub 42:323–345

Sun P, Guo PY, Niu YL (2021) Eastern China continental lithosphere thinning is a consequence of paleo-Pacific subduction: a review and new perspectives. Earth Sci Rev 218:103680

Takahashi N, Kodaira S, Klemperer LS, Tatsumi Y, Kaneda Y, Suyehiro K (2007) Crustal structure and evolution of the Mariana intra-oceanic Island arc. Geology 35:203–206
Tamura Y, Gill JB, Tollstrup D, Kawabata H, Shukuno H, Chang Q, Miyazaki T, Takahashi T, Hirahara Y, Kodaira S, Ishizuka O, Suzuki T, Kido Y, Fiske RS, Tatsumi T (2009) Silicic magmas in the Izu-Bonin oceanic arc and implications for crustal evolution. J Petrol 50:685–723
Tamura Y, Ishizuka O, Stern RJ, Shukuno H, Kawabata H, Embley RW, Hirahara Y, Chang Q, Kimura JI, Tatsumi Y, Nunokawa A, Bloomer SH (2011) Two primary basalt magma types from Northwest Rota-1 volcano, Mariana arc and its mantle diapir or mantle wedge plume. J Petrol 52:1143–1183
Tang, J., Xu, W.L., Niu, Y.L., Wang, F., Ge, W.C., Sorokin, A.A., Chekryzhov, I.Y., 2016. Geochronology and geochemistry of Late Cretaceous–Paleocene granitoids in the Sikhote-Alin Orogenic Belt: Petrogenesis and implications for the oblique. Lithos 266/267, 202–212
Tatsumi Y (2008) Making continental crust: the sanukitoid connection. Chin Sci Bull 53:1620–1633
Tatsumi Y, Eggins SM (1995) Subduction zone magmatism. Blackwell Science, Cambridge
Tatsumi Y, Hamilton DL, Nesbitt RW (1986) Chemical characteristics of fluid phase released from a subducted lithosphere and origin of arc magmas—evidence from high-pressure experiments and natural rocks. J Volcanol Geotherm Res 29:293–309
Tatsumi Y, Shukuno H, Tani K, Takahashi N, Kodaira S, Kogiso T (2008) Structure and growth of the Izu-Bonin-Mariana arc crust: 2. Role of crust-mantle transformation and the transparent Moho in arc crust evolution. J Geophys Res 113:B02203. https://doi.org/10.1029/2007JB005121
Taylor SR (1967) The origin and growth of continents. Tectonophysics 4:17–34
Taylor SR (1977) Island arc models and the composition of the continental crust. In: Talwani M, Pitman WC III (eds) Island arcs, deep sea trenches, and back-arc basins, Maurice Ewing Series, vol 1. AGU, Washington, D.C., pp 325–335
Taylor SR, McLennan SM (1985) The continental crust: its composition and evolution. Blackwell, Oxford, p 312
Taylor SR, McLennan SM (1995) The geochemical evolution of the continental crust. Rev Geophys 33:241–265
Tollan P, O'Neill H, Hermann J, Benedictus A, Arculus RJ (2015) Frozen melt–rock reaction in a peridotite xenolith from sub-arc mantle recorded by diffusion of trace elements and water in olivine. Earth Planet Sci Lett 422:169–181
Turner S, Caulfield J, Rushmer T, Turner M, Cronin S, Smith I, Handley H (2012) Magma evolution in the primitive, intra-oceanic Tonga arc: rapid petrogenesis of dacites at Fonualei volcano. J Petrol 53:1231–1253
Tuttle OF, Bowen NL (1958) Origin of granite in the light of experimental studies in the system $NaAlSi_3O_8$–$KAlSi_3O_8$–SiO_2–H_2O. Geol Soc Am Memoir 74:153
Voice PJ, Kowalewski M, Eriksson KA (2011) Quantifying the timing and rate of crustal evolution: global compilation of radiometrically dated detrital zircon grains. J Geol 119:109–126
Wedepohl H (1995) The composition of the continental crust. Geochim Cosmochim Acta 59:1217–1239
Wolfe MB, Wyllie PJ (1994) Dehydration melting of amphibolite at 10 kbar: effects of temperature and time. Contrib Mineral Petrol 115:369–383
Woodhead JD, Eggins SM, Johnson RW (1998) Magma genesis in the New Britain Island arc: further insights into melting and mass transfer processes. J Petrol 39:1641–1668
Woodhead JD, Hergt J, Sandiford M, Johnson RW (2010) The big crunch: physical and chemical expressions of arc/continent collision in the western Bismarck arc. J Volcanol Geotherm Res 190:11–24
Worthington TJ, Gregory MR, Bondarenko V (1999) The Denham caldera on Raoul volcano: dacitic volcanism in the Tonga-Kermadec arc. J Volcanol Geotherm Res 90:29–48
Wyllie PJ (1977) Crustal anatexis: an experimental review. Tectonophysics 13:41–71

Wyllie PJ, Wolfe MB (1993) Dehydration-melting of amphibolite: sorting out the solidus. Geol Soc Spec Pub 76:405–416

Wyllie PJ, Huang WL, Stern CR, Maaloe S (1976) Granitic magmas: possible and impossible sources, water contents, and crystallization sequences. Can J Earth Sci 13:1007–1019

Xiao WJ, Windley BF, Allen MB, Han C (2013) Paleozoic multiple accretionary and collisional tectonics of the Chinese Tianshan orogenic collage. Gondwana Res 23:1316–1341

Xiao YY, Chen S, Niu YL, Wang XH, Xue QQ, Wang GD, Gao YJ, Gong HM, Kong JJ, Shao FL, Sun P, Duan M, Hong D, Wang D (2020) Mineral compositions of syn-collisional granitoids and their implications for the formation of juvenile continental crust and adakitic magmatism. J Petrol 61:1–19

Xu YJ, Cawood PA, Du YS, Zhong Z, Hughes NC (2014) Terminal suturing of Gondwana along the southern margin of South China Craton: evidence from detrital zircon U–Pb ages and Hf isotopes in Cambrian and Ordovician strata, Hainan Island. Tectonics 33:2490–2504

Xue QQ, Niu YL, Chen S, Sun P, Duan M, Gao YJ, Hong D, Xiao YY, Wang XH, Guo PY (2018) Tectonic significance of cretaceous granitoids along the southeast coast of continental China. Geol J 55:173–196

Zhang Q (2010) Is eastern China an integral part of the circum Pacific tectonic belt? Granitoid Controversy (3). http://www.ysxb.Ac.Cn/ysxb/ch/reader/view_news.Aspx?Id=20100301829260001. (in Chinese)

Zhang Q, Qian Q, Wang WQ, Wang Y, Zhao TP, Hao J, Guo GJ (2001) An East China plateau in mid-late Yanshanian period: implication from adakites. Chin J Geol 36:248–255

Zhang Y, Niu YL, Hu Y, Liu JJ, Ye L, Kong JJ, Duan M (2016) The syncollisional granitoid magmatism and continental crustal growth in the west Kunlun Orogen, China—evidence from geochronology and geochemistry of the Arkarz pluton. Lithos 245:191–204

Zhao ZD, Mo XX, Dilek Y, Niu YL, DePaolo DJ, Robinson P, Zhu DC, Sun CG, Dong GC, Zhou S, Luo ZH, Hou ZQ (2009) Geochemical and Sr–Nd–Pb–O isotopic compositions of the post-collisional ultrapotassic magmatism in SW Tibet: petrogenesis and implications for India intra-continental subduction beneath southern Tibet. Lithos 113:190–212

Zhou XM, Sun T, Shen WZ, Shu LS, Niu YL (2006) Petrogenesis of Mesozoic granitoids and volcanic rocks in South China: a response to tectonic evolution. Episodes 29:26–33

Zhu RX, Xu YG, Zhu G, Zhang HF, Xia QK, Zheng TY (2012) Destruction of the North China craton. Sci China Earth Sci 55:1565–1587

Zhu DC, Wang Q, Zhao ZD, Chung S-L, Cawood PA, Niu YL, Liu SA, Wu FY, Mo XX (2015) Magmatic record of India-Asia collision. Sci Rep 5:14289

Zindler A, Hart SR (1986) Chemical geodynamics. Annu Rev Earth Planet Sci 14:493–571

Chapter 9
Diversity and Evolution of the Tibetan Plateau Lithosphere

Abstract In this chapter, we discuss the nature of the upper mantle beneath the southern Tibetan Plateau on the basis of mantle xenolith hosted in the Sailipu ultrapotassic volcanic rocks (trachyandesite, ~17Ma). Sailipu mantle xenolith differs from cratonic lithosphere in having high temperature (990~1140°C) and highly metasomatized composition as illustrated by the magmatism associated with well-developed N-S normal fault systems on the plateau surface. Three geochemical components (reservoirs) are readily identified in terms of Sr-Nd-Pb-O isotopes, i.e., (A) the North Tibet geochemical province, (B) the Neo-Tethyan mantle reservoir, and (C) the Himalayan continental crust component. These three reservoirs control the composition of the magmatism in the Lhasa terrane as shown by the geochemical mixing relationships recorded in the magmatic rocks.

Keywords Mantle xenolith · Upper mantle · Crust · Geochemical end-members · Tibetan plateau

9.1 The Nature of the Lithospheric Mantle in Southern Tibet

The nature of the subcontinental lithospheric mantle beneath the Tibetan Plateau is not well-known although post-collisional volcanic rocks scattered throughout the plateau have been inferred to result from low degree melting of deep portions of the lithospheric mantle. The primary difficulty is the rarity of mantle-derived xenoliths and megacrystals in potassic and ultrapotassic volcanic rocks in the hinterland of the plateau.

Zhao et al. (2008a, b) reported for the first time the mantle xenoliths hosted in the Sailipu volcanic rocks (trachyandesite, with an age of ~17 Ma), which are typical ultrapotassic rocks similar to many of those throughout southern Tibet. The authors provide detailed petrography and mineral compositions of the xenoliths.

Prof. Zhidan Zhao has fully contributed to this chapter.

X. Mo et al., *Magmatism and Crustal Evolution of the Tibetan Plateau*, Modern Approaches in Solid Earth Sciences 25,
https://doi.org/10.1007/978-981-97-2109-2_9

Fig. 9.1 (**a**, **b**) Sailipu lherzolite hosted in the ultrapotassic lava

The Sailipu xenoliths have varying size (~ 0.5–1.5 cm in diameter; Fig. 9.1a, b) and have two lithological groups. The first group is pyroxenites (Opx + Cpx), and the second is lherzolite (Ol + Opx + Cpx ± Phl ± Sp). The compositions of olivine ($Mg^{\#}$ = 89–90, CaO = 0.0–0.12%, TiO_2 < 0.03%, NiO = 0.29–0.80%), Cpx ($Mg^{\#}$ = 88–91, Al_2O_3 = 5.5–7%), Opx (TiO_2 = 0.05–0.15%, Al_2O_3 = 2–5%) and spinel ($Mg^{\#}$ = 58–76, $Cr^{\#}$ = 10–44, Cr_2O_3 = 9–35%, MnO = 0.09–0.24%, FeO = 11–18%, Al_2O_3 = 29–57%, MgO = 16–21%) are similar to those of mantle xenoliths hosted in Cenozoic alkali basalts in eastern China.

Calculations give the xenolith equilibration P-T conditions of 990–1140 °C and 16–20 kb (Fig. 9.2; Zhao et al. 2011). This geotherm is similar to that in eastern China, Kenya and other rift-related lithospheric mantle thermal structure, implying regional scale extension beneath southern Tibet in the Miocene while the India-Asia convergence continues since the collision at 55 ± 5 Ma. The xenolith samples show evidence of multi-stage metasomatic processes, including phlogopite and quartz in pyroxenites, and the rim-core compositional variation in spinels. The secondary spinel along with neighboring phlogopite suggest that a potassic, host-rock like metasomatic agent enriched in K, Si and H_2O must have played an important role in modifying the mantle lithosphere beneath southern Tibet. Further study on the Sailipu mantle xenoliths will help reveal the composition, thermal regime, and processes of the upper mantle beneath southern Tibet, and in discussing the origin of the orogenic ultrapotassic rocks.

Trace elements of clinopyroxene (Cpx) from the Sailipu mantle xenoliths provide important information (Liu et al. 2011). Cpx displays upward-convex REE

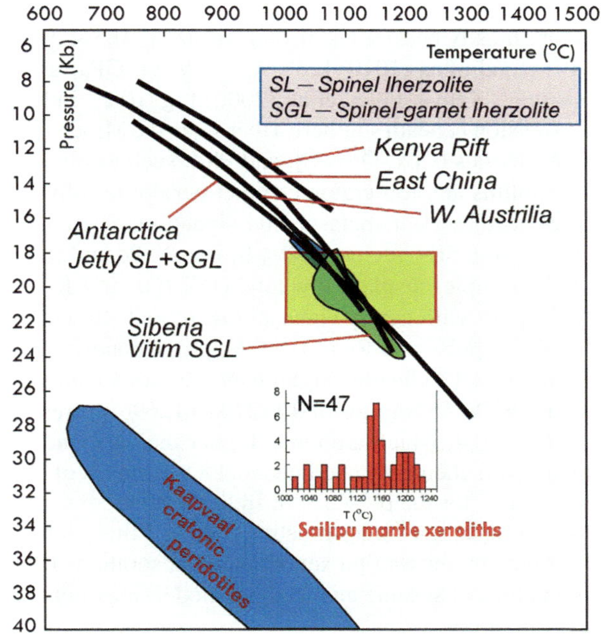

Fig. 9.2 Thermobarometry of Sailipu lherzolite compared to other regions in all typical rifting settings. Sailipu samples in yellow box from Zhao et al. (2011); Eastern China from Xu et al. (1998, 1999) and Huang et al. (2004); others regions from Foley et al. (2006)

patterns with a small negative Eu anomaly and varying ratios of $(La/Yb)_N$ (3.0–7.8), $(La/Sm)_N$ (0.3–1.6) and $(Sm/Yb)_N$ (2.6–15.5). Clinopyroxene also is strongly enriched in LILEs (e.g., Rb, Ba, Th, and U) and depleted in HFSEs (e.g., Zr, Hf, Nb, and Ta). The negative Sr, Nb, Ta, Zr, Hf and Ti anomalies in Cpx echo the compositions of the ultrapotassic lavas from southern Tibet. Such an enriched Cpx, along with reaction textures and compositional features preserved in olivine, phlogopite and spinel, suggests modal metasomatism by K-rich hydrous melts. These observations support the model of delamination described by Zhao et al. (2009). A hot, highly metasomatized lithospheric mantle beneath southern Tibet in the Miocene is best understood as lithospheric removal, which triggers the generation of the ultrapotassic magmatism. Liu et al. (2014) reported in-situ oxygen isotope data on olivine crystals in these Sailipu mantle xenoliths, and identified a metasomatic event with anomalously heavy oxygen isotope compositions ($\delta^{18}O = +8.03 \pm 0.28‰_0$). They interpretated such oxygen isotope anomaly as resulting from sub-arc lithospheric mantle metasomatism, which does not contribute significantly to typical island arc basalts of mantle wedge origin, but may contribute to the high-$\delta^{18}O$ signature of arc magmas through assimilation.

Further studies of Os isotopes and highly siderophile elements (HSEs) of the Sailipu mantle xenoliths were carried out to examine the possible presence of

Indian cratonic mantle lithosphere beneath southern Tibet (Zhao et al. 2011). The studied xenoliths are mostly peridotites of varying size (~ 0.5–1.5 cm in diameter). The calculated P-T conditions (1010–1230 °C, ~1.6–2.0 GPa) are similar to rift-related geotherms (e.g., Kenya, Foley et al. (2006) Fig. 10.2), indicating the influence of regional extension beneath southern Tibet since the Miocene. Compositional discriminations of minerals (Cpx, Opx, Ol, and Phl), such as olivine with Fo < 90, suggest that the xenoliths are non-cratonic spinel-peridotite (olivine Fo > ~91 for cratonic peridotites) with a clear metasomatic signature. The analysis of Os isotopes and abundances of HSEs (including Os, Ir, Ru, Pt, Pd and Re) on six olivine-dominated peridotite samples displays low total HSE (Os+Ir + Ru + Pt + Pd + Re) abundances (8.7–25 ppb) with nearly constant Os, Ir, and Ru, but rather varied Pt (2–13 ppb), Pd (0.4–5.2 ppb), and Re (0.01–0.5 ppb). Chondrite-normalized Pd/Ir ratios range from 0.2 to 2.4, reflecting significant metasomatism in some samples. The xenoliths exhibit $^{187}Os/^{188}Os$ ratios of 0.12213–0.12696, corresponding to varying γOs from −4.2 to −0.4, which are much higher than ancient cratonic mantle. Thus, the data on mineral chemistry and bulk-rock Os isotopes of the Sailipu xenoliths provide no evidence for the presence of Indian cratonic mantle lithosphere. It is probable, however, that the underthrusting Indian lithosphere may be much deeper than represented by the Sailipu xenoliths under southern Tibet. In this case, testing the interpretations of seismic and tectonic models may not be possible without garnet-facies peridotites.

9.2 Pb Isotopic Geochemistry of Tibetan Plateau and Its Implications

Pb isotopic compositions of basement, granitoids and volcanic rocks with a total of 486 data sets were compiled and analyzed by Zhao et al. (2007). These authors identified three major crust/mantle geochemical end-members using Pb isotopes (Table 9.1; Fig. 9.3). The first one with low Pb isotopes is the Neo-Tethyan mantle reservoir shown by basaltic rocks in the Yarlung Zangpo ophiolite; the second one with the most enriched Pb compositions is the Himalayan continental crust represented by basement and granitoids from Tethyan and Higher Himalayas; the third one is the North Tibetan Plateau enriched mantle source region, showing moderately enriched Pb isotope compositions.

In the frame of above-mentioned three end-members, rocks from the Lhasa terrane can be divided into two groups (Table 9.2; Fig. 9.4). The first group, including the I-type Gangdese granitoid plutons, Linzizong volcanic rocks and ore-bearing adakitic porphyries, shows strong Tethyan oceanic crust affinity with significant source contribution of recycled Tethyan ocean crust or depleted mantle. The second group, represented by the ultrapotassic volcanic rocks in the western Lhasa terrane, exhibits obvious input of the Himalayan continental crust to their mantle source regions. Therefore, the processes of Tethyan seafloor subduction and

Table 9.1 Average Pb isotopic compositions of the volcanic rocks from Tibetan Plateau

Tectonic unit/location	Rock type	Age/Ma	Number	$^{206}Pb/^{204}Pb$	$^{207}Pb/^{204}Pb$	$^{208}Pb/^{204}Pb$	Reference
Postcollisional rocks in central S. Lhasa terrane							
Majiang, Nyemo	Shoshonite	10 ~ 16	5	18.577	15.628	38.848	Turner et al. (1996)
Yangying, Damxung	Shoshonite	9 ~ 11	3	18.631	15.729	39.339	Zhao et al. (2009); Nomade et al. (2004)
Wuyu, Namling	Granite porphyry	11	5	18.499	15.603	39.021	Zhao et al. (2001)
Wuyu, Namling	Dacite	12 ~ 14	9	18.520	15.614	38.983	Zhao et al. (2009)
Jiama, Dagze	Basalt-andesite	38 ~ 42	9	18.458	15.598	38.650	Gao et al. (2006)
Linzizong, Linzhou	Felsic rocks	43 ~ 65	13	18.653	15.599	38.751	Mo et al. (2008)
Potassic-Ultrapotasic volcanic rocks in W. Lhasa terrane							
Zabuye salt Lake	Trachyandensite	16.0 ~ 16.2	6	18.813	15.759	39.548	Zhao et al. (2009)
Shiquanhe	PVR, UPVR		2	18.707	15.775	39.654	Turner et al. 1996
Bangba, Gegya	PVR	18 ~ 23	6	18.588	15.712	39.489	Miller et al. (1999)
Wenbu, Nima	Trachyte, phonolite	23	5	18.531	15.726	39.363	Ding et al. (2003)
Chazi, Ngamring	Trachyandensite	13	5	18.864	15.741	39.531	Ding et al. (2003)
Western Lhasa terrane	UPV	11 ~ 17	14	18.748	15.767	39.736	Zhao et al. (2009)
Na-volcanic rocks in N. Tibet							
Bangdacuo, W Qiangtang	Sodic basalt	44	2	18.416	15.522	38.692	Deng (1998)
Bangdacuo, W Qiangtang	Sodic Hawaiite	45 ~ 60	3	18.338	15.501	38.534	Ding et al. (2003)
Jingyu Lake, Hohxil area	Sodic latite		6	18.613	15.612	37.510	Yang et al. (2002)
Lagala, Qiangtang	Olivine basalt	60	2	18.851	15.581	38.927	Ding et al. (1999, 2003)
Potassic-Ultrapotasic volcanic rocks in N. Tibet							
Yanghu, Qiangbaqian, W Kunlun	Latite	15 ~ 16.5	4	18.701	15.688	39.075	Miller et al. (1999), Deng (1998)
Heishibeishan, W Kunlun	Latite	0.07	2	18.713	15.671	38.992	Turner et al. (1996); Deng (1998)
Pulu, W Kunlun	PVR	1 ~ 1.4	9	18.662	15.644	38.758	Xie (1992, 1989); Guo et al. (2006)
Ashkule, W Kunlun	Trachyandensite	0.1 ~ 2.8	7	18.779	15.674	39.065	Cooper et al. (2002)
Quanshuigou, W Kunlun	Latite, tephrite	3.6 ~ 5.6	9	18.745	15.690	39.106	Arnund et al. (1992)

(continued)

Table 9.1 (continued)

Tectonic unit/location	Rock type	Age/Ma	Number	$^{206}Pb/^{204}Pb$	$^{207}Pb/^{204}Pb$	$^{208}Pb/^{204}Pb$	Reference
Hohxil	Latite		9	18.735	15.695	39.173	Turner et al. (1996)
Songpan-Ganzi	PVR, UPVR	0.3 ~ 17	14	18.735	15.678	38.926	Guo et al. (2006)
Yulinshan, Qiangtang	UPVR	30	19	18.986	15.716	39.374	Ding et al. (2003); Ding et al. (1999)
Ulugh Muztagh, Qiangtang	Felsic rocks	4 ~ 16	13	18.996	15.647	35.779	McKenna and Walker (1990)
N Qiangtang	PVR, UPVR	27 ~ 45	16	18.896	15.710	39.205	Turner et al. (1996); Guo et al. (2006)
N Qiangtang	High-K CA rocks	39 ~ 42	17	18.749	15.667	38.999	Yi et al. (2004)
Tethyan Himalaya volcanic rocks							
Tethyan Himalaya	Basalt	T-J	13	18.416	15.560	38.558	Zhu (2003)
Tethyan Himalaya	Dacite	133	3	19.239	15.730	40.295	Liao (2003)

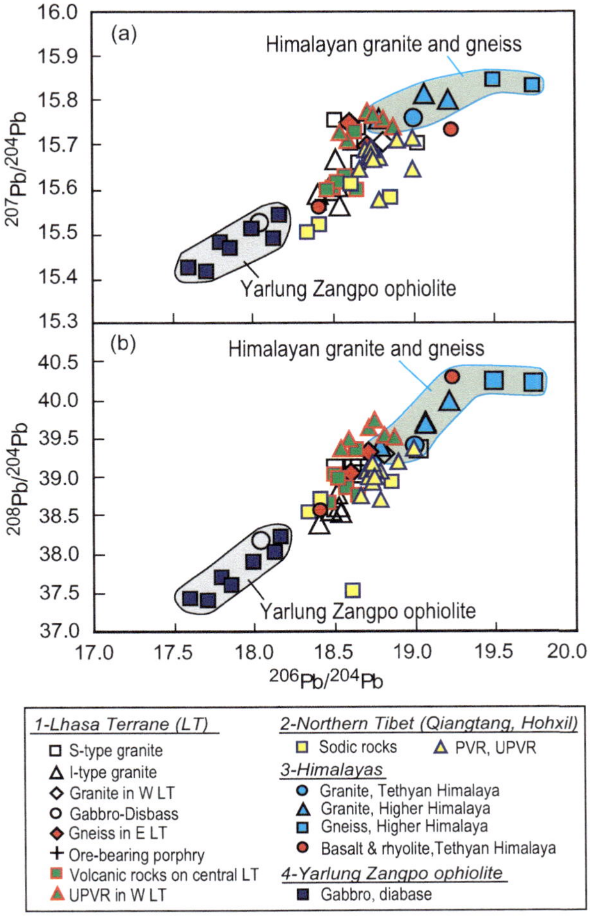

Fig. 9.3 Plots of $^{207}Pb/^{204}Pb$ versus $^{206}Pb/^{204}Pb$ (**a**) and $^{208}Pb/^{204}Pb$ versus $^{206}Pb/^{204}Pb$ (**b**) of rocks from the Tibetan plateau. See Table 9.1 and Table 9.2 for data sources

subsequent India-Asia collision/underthrust are well recorded by Pb isotopic geochemistry in the Lhasa terrane magmatism. The crust-mantle interaction among the above end-members have played an important role in the magmatism associated with the evolution and uplift of the Tibetan plateau. These three end-member identification agrees with the former results using Sr-Nd isotopic studies as discussed below.

Table 9.2 Pb isotopic compositions of the basement and granitoid from Lhasa block and Himalayan belts, basic rokcs from Yalung Zangpo ophiolite, Tibetan Plateau

Tectonic unit/location/rock type (Age/Ma)	Rock/minerals	Number	$^{206}Pb/^{204}Pb$	$^{207}Pb/^{204}Pb$	$^{208}Pb/^{204}Pb$	Reference
W Lhasa terrane basement						
Yangbajain, gneiss	Kf	3	18.713	15.698	39.325	Gariépy et al. (1985)
Amdo, gneiss	Kf	4	18.603	15.750	39.043	Gariépy et al. (1985)
Central Lhasa terrane, S-type granite						
Yangbajain, granite	Kf	7	18.607	15.701	39.087	Gariépy et al. (1985)
Yangbajain, 2-mica granite (53 ma, K-Ar)	WR	2	18.658	15.660	39.033	Liao (2003)
Baingoin-Gulu, granite	Kf	12	18.664	15.738	39.095	Gariépy et al. (1985)
Amdo, granite	Kf	10	18.513	15.754	39.101	Gariépy et al. (1985)
Lhozhag, muscovite granite (116 ~ 127 ma, U-Pb)	WR	4	19.031	15.703	39.337	Liao (2003)
S Lhasa terrane, I-type granite						
Quxu, granodiorite (41.1 ma, U-Pb)	Kf	7	18.518	15.669	38.784	Gariépy et al. (1985); Schärer et al. (1984)
Quxu, granodiorite and granite	Kf, Pl	8	18.471	15.608	38.544	Li et al. (1987)
Lhasa, granitoid	Kf	5	18.524	15.605	38.603	Gariépy et al. (1985); Schärer et al. (1984)
Dazhuka, diorite (67 ~ 94.2 ma, U-Pb)	Kf se	3	18.403	15.588	38.391	Gariépy et al. (1985); Schärer et al. (1984)
Renbu-Quxu, granitoid	WR	11	18.536	15.564	38.533	Zhao et al. (2009)
S Lhasa, ore-bearing porphyry	WR	22	18.455	15.577	38.597	Qu et al. (2002)
W Lhasa, Jiangba-Bangba-Xungba, granitoid	WR	WR	18.803	15.708	39.294	Zhao et al. (2007)
Xigaze-Quxu, gabbro-diorite	WR	13	18.040	15.525	38.168	Zhao et al. (2007)
Tethyan Himalaya, leucogranite	Kf/WR	20	19.009	15.757	39.403	Gariépy et al. (1985); Liao (2003); Schärer et al. (1986)

Tethyan Himalaya rocks						
Nyalamm, migmatitic granite (17 ma, U-Pb)	Kf	8	19.062	15.814	39.715	Gariépy et al. (1985); Schärer et al. (1986)
Mount Everest, granitoid (14.3 ma, U-Pb)	Kf	8	19.068	15.813	39.692	Gariépy et al. (1985); Schärer et al. (1986)
Manaslu, leucogranite	Kf	Kf	19.213	15.801	39.994	Gariépy et al. (1985);
Manaslu, leucogranite (17 ~ 24 ma)	WR	11	18.780	15.758	39.406	Vidal et al. (1982); Harrison et al. (1999)
High Himalaya, gneiss	Kf	4	19.736	15.833	40.212	Gariépy et al. (1985)
High Himalaya, gneiss	Kf	4	19.736	15.833	40.212	Gariépy et al. (1985)
High Himalaya, gneiss	Kf	4	19.736	15.833	40.212	Gariépy et al. (1985)
High Himalaya, gneiss	WR	7	19.500	15.843	40.241	Harrison et al. (1999)
Yalung Zangpo ophiolite						
Xigaze, gabbro (120 ma, U-Pb)	WR	11	17.607	15.426	37.413	Göpel et al. (1984)
Xigaze, Xialu, gabbro	WR	6	17.716	15.415	37.396	Mahoney et al. (1998)
Luobusha, gabbro (177 ma, Sm-Nd)	WR/Pl/Px	5	17.795	15.482	37.681	Zhou (2002)
Xigaze, Decun, basalt and diabase	WR	6	17.862	15.469	37.581	Niu et al. (2006)
Xigaze, Jiding, basalt and gabbro	WR	WR	17.994	15.513	37.885	Niu et al. (2006)
Xigaze, Ngamring, gabbro	WR	4	18.164	15.543	38.222	Niu et al. (2006)
Dangqiong, Dajiweng, Dazhuka, gabbro	WR	19	18.134	15.491	38.019	Zhang et al. (2005)

Abb: *WR* whole rocks, *Kf* orthoclase, *Pl* plagioclase, *Px* Pyroxenite

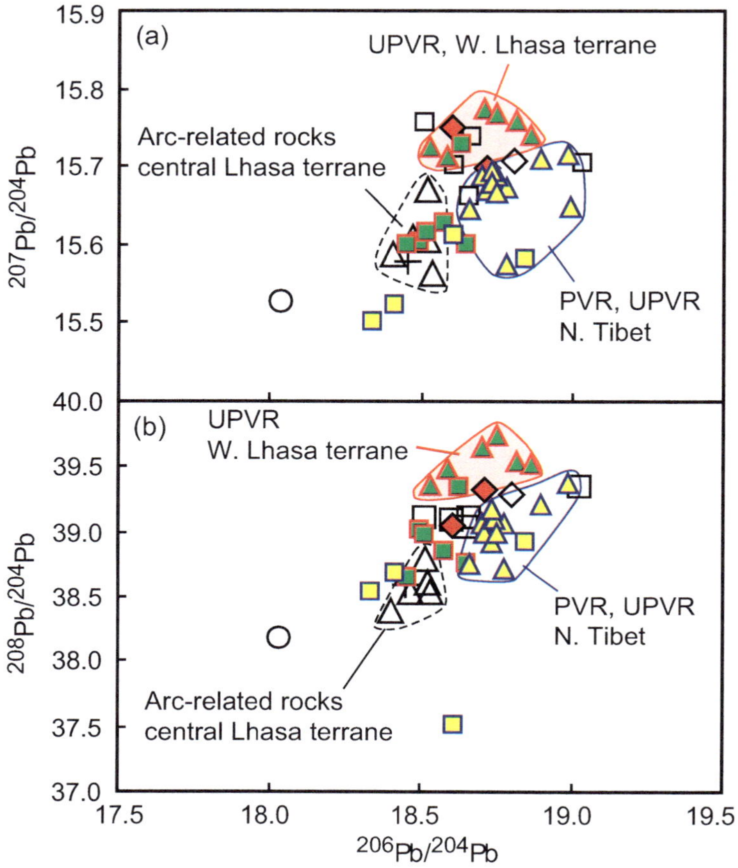

Fig. 9.4 Plots of $^{207}Pb/^{204}Pb$ versus $^{206}Pb/^{204}Pb$ (**a**) and $^{208}Pb/^{204}Pb$ versus $^{206}Pb/^{204}Pb$ (**b**) of rocks from the Lhasa terrane and northern Tibetan plateau. See Tables 9.1 and 9.2 for data source. Symbols are as in Fig. 9.3

9.3 Geochemical Mixing of Major Geochemical Reservoirs in the Magmatism of the Tibetan Plateau

Combination of Pb-Sr-Nd isotopes of magmatic rocks in the Lhasa terrane, northern Tibetan Plateau (Qiangtang, Hoh Xil and West Kunlun), Himalayan basement and granitoids, and basaltic rocks from the Yarlung Zangbo suture, allows better appreciation of the above-mentioned three geochemical reservoirs recorded in rocks throughout the Tibetan Plateau (Mo et al. 2006a, b; Zhao et al. 2006, 2007, 2009).

The three reservoirs are detailed as follows (Fig. 9.5):

(A) The North Tibet Geochemical Province ($^{87}Sr/^{86}Sr = 0.7071–0.7105$, $\varepsilon_{Nd} = -2$ to -8, $T_{DM} = 0.7–1.3$ Ga, $^{206}Pb/^{204}Pb = 18.662–18.996$, $^{207}Pb/^{204}Pb = 15.575–15.716$,

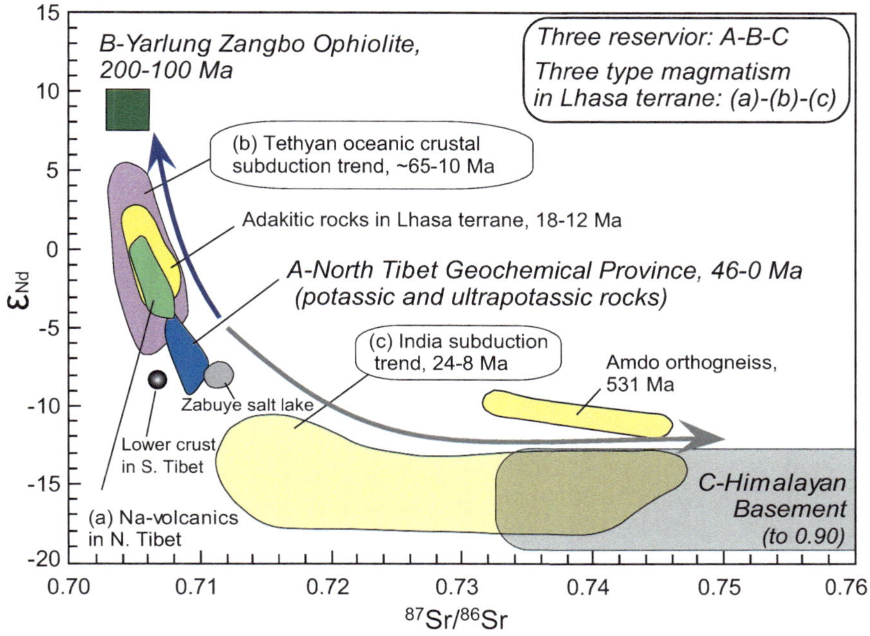

Fig. 9.5 Plot of ε_{Nd} versus $^{87}Sr/^{86}Sr$ showing Sr-Nd features of rocks on the Tibet plateau. Modified from Zhao et al. (2009). Data sources listed in Tables 9.1, 9.2, and references in Zhao et al. (2009)

$^{208}Pb/^{204}Pb$ = 38.712–39.374) is represented by magmatic rocks of both crust and upper mantle derivation since ~46 Ma. Rocks in this province have relatively uniform composition with narrow ranges of isotopic compositions as shown by shoshonitic, ultrapotassic and adakitic rocks in Qiangtang, Hoh Xil, and West Kunlun orogenic belts.

(B) The Neo-Tethyan mantle reservoir is represented by remnant oceanic lithosphere preserved in the Yarlung Zangpo ophiolite ($^{87}Sr/^{86}Sr$ = 0.703–0.706, ε_{Nd} = +7.8 to +10, $^{206}Pb/^{204}Pb$ = 17.707–18.164, $^{207}Pb/^{204}Pb$ = 15.415–15.543, $^{208}Pb/^{204}Pb$ = 37.396–38.222). This reservoir ultimately represents MORB-like depleted oceanic upper mantle asthenosphere.

(C) The Himalayan continental crust component, represented by the crystalline basement and granitoids from the Higher and Lesser Himalayas. Rocks of this geochemical component have the most enriched Sr and Nd isotopic compositions, and the oldest Nd model ages (ε_{Nd} = −12 to −25, $^{87}Sr/^{86}Sr$ = 0.733–0.760, or even as high as 0.90; T_{DM} = 1.9–2.9 Ga), among the three reservoirs. Importantly, these rocks have the most enriched Pb isotopes ($^{206}Pb/^{204}Pb$ = 19.500–19.736, $^{207}Pb/^{204}Pb$ = 15.833–15.843, $^{208}Pb/^{204}Pb$ = 40.212–40.241). These typical continental materials in the northern margin of Indian continent may have subducted under the Lhasa terrane along the Yarlung Zangbo suture zone.

The three reservoirs outlined above control the composition of the magmatism in the Lhasa terrane by means of material mixing. The three magmatic types are well identified in rocks of the southern Lhasa terrane (Fig. 9.5).

(a) The first magmatic type is inherent to the Lhasa terrane and has similar Sr-Nd compositions to those of North Tibet Geochemical Province. These shoshonitic rocks cropped out in Xungba (Miller et al. 1999) and Yangying (Nomade et al. 2004) are similar to those from north Tibet, suggesting that the Lhasa terrane is geochemically related to the North Tibet Geochemical Province.
(b) The second type of magmatism, showing mixing between North Tibet and the mantle component of Yarlung Zangpo ophiolite. This type includes the Linzizong volcanic succession (44–65 Ma, Zhou et al. 2004; Mo et al. 2007, 2008; Zhu et al. 2015), I-type Gangdese granitoid batholiths (45–55 Ma; Mo et al. 2005; Dong et al. 2005, 2006; Wen et al. 2008; Alexander et al. 2019; DePaolo et al. 2019; Ji et al. 2009, 2012, 2016; Liu et al. 2017; Wang et al. 2020, 2021, Wang et al. 2022; Xu et al. 2019; Zhu et al. 2015, 2017a, b, 2019), the ore-bearing adakitic porphyries (12–18 Ma, Chung et al. 2003; Hou et al. 2004; Liu et al. 2017), and the shoshonitic rocks in the Wuyu basin (Zhao et al. 2001). Mantle input or subduction related processes significantly contribute to the petrogenesis of these rocks.
(c) The third type of magmatism, represented by the postcollisional ultrapotassic volcanic rocks in the western Lhasa terrane, shows clear affinity to the Himalayan basement (Zhao et al. 2009). These rocks show mixing between the North Tibet Geochemical Province and the Himalayan continental crust reservoir, providing evidence for Indian continental slab subduction beneath southern Tibet.

9.4 Indian Continent Subducting Underneath Southern Tibet

The ultrapotassic rocks with Sr-Nd-Pb-O isotopic affinity with the Indian continents, provide an important clue for the India subduction beneath Asia (Zhao et al. 2009). The outcrops of the ultrapotassic rocks along the line of Shiquanhe-Xungba-Dangreyong Lake may mark the frontline of the northward subducting India plate in the Miocene, which agrees very well with the helium isotopic study (Hoke et al. 2000) and geophysical models (Jin et al. 1996; Kosarev et al. 1999). A simple calculation using the northward movement rate of Indian plate (40–50 mm/year using the present GPS data; Wang et al. 2001) can help estimate how far to the north has the Indian continent been subducted. India could have moved northward for ~1200 km from the suture during the past 24 Myrs, which implies that the northern edge of the Indian continent may have crossed the Bangong-Nujiang suture (Zhao et al. 2009).

In the northern Tibetan Plateau, however, the magmatism has maintained uniform Sr-Nd-Pb isotopic compositions since 46 Ma. This implies that their source regions may not have been affected by India continental materials. Therefore, the Indian plate subducted beneath the Lhasa terrane, but may not have surpassed the Bangong-Nujiang suture zone. This petrogeochemistry-constrained Indian continental subduction model is consistent with other geophysical data (Jin et al. 1996; Owens and Zandt 1997; Kosarev et al. 1999; Chemenda et al. 2000; Wei et al. 2001; Tilmann et al. 2003; Priestley et al. 2008).

Acknowledgements This work was supported by Ministry of Science and Technology of China (grants: 2019QZKK0702), National Natural Science Foundation of China (grants: 42121002, 42073035), and the 111 project (B18048).

References

Alexander EW, Wielicki MM, Harrison TM, DePaolo DJ, Zhao ZD, Zhu DC (2019) Hf and Nd isotopic constraints on pre- and Syn-collisional crustal thickness of southern Tibet. J Geophys Res Solid Earth 124:11038–11054

Arnund NO, Vidal P, Tapponnier P et al (1992) The high K2O volcanism of northwestern Tibet: Geochmistry and tectonic implications [J]. Earth Planet Sci Lett 111:351–367

Chemenda AI, Burg J-P, Mattauer M (2000) Evolutionary model of the Himalaya–Tibet system: geopoem based on new modelling, geological and geophysical data. Earth Planet Sci Lett 174:397–409

Chung SL, Liu D, Ji J, Chu MF, Lee HY, Wen DJ, Lo CH, Lee TY, Qian Q, Zhang Q (2003) Adakite from continental collision zones: melting of thickened lower crust beneath southern Tibet. Geology 31:1021–1024

Cooper KM, Reid MR, Dunbar NW et al (2002) Origin of mafic magmas beneath northwestern Tibet: constraints from 230Th-238U disequilibriav. Geochem Geophys Geosyst 3(11):1065–1023. https://doi.org/10.1029/2002GC000332

Deng W (1998) Cenozoic intraplate volcanic rocks in the northern Qinghai-Xizang Plateau. Geological Publishing House, Beijing

DePaolo DJ, Harrison TM, Wielicki M, Zhao Z, Zhu D-C, Zhang H, Mo X (2019) Geochemical evidence for thin syn-collision crust and major crustal thickening between 45 and 32 Ma at the southern margin of Tibet. Gondwana Res 73:123–135

Ding L, Zhang J, Zhou Y, Deng W, Xu R, Zhong D (1999) Tectonic implication on the lithosphere evolution of the Tibetan Plateau: petrology and geochemistry of sodic and ultrapotassic volcanism in Northern Tibet. Acta Petrol Sin 15:408–421. (in Chinese, with English abstract)

Ding L, Kapp P, Zhong DL et al (2003) Cenozoic volcanism in Tibet: evidence for a transition from oceanic to continental subduction [J]. J Petrology 44:1835–1865

Dong GC, Mo XX, Zhao Z, Guo T, Wang LL, Chen T (2005) Geochronologic constraints by SHRIMP II zircon U-Pb dating on magma underplating in the Gangdise belt following India-Eurasia collision. Acta Geol Sin 79:787–794

Dong GC, Mo XX, Zhao Z, Zhu DC, Wang LL, Chen T, Li B (2006) Magma mixing in the middle part of the Gangdise magmatic belt: evidence from granitoid complexes. Acta Petrol Sin 24:835–844

Foley SF, Andronikov AV, Jacob DE, Melzer S (2006) Evidence from Antarctic mantle peridotite xenoliths for changes in mineralogy, geochemistry and geothermal gradients beneath a developing rift. Geochim Cosmochim Acta 70:3096–3120

Gao Y, Hou Z, Wei R, Meng X, Hu H (2006) The geochemistry and Sr-Nd-Pb isotopes of basaltic sub-volcanics from the Gangdese: constraints on depleted mantle source for post-collisional volcanisms in the Tibetan plateau. Acta Petrol Sin 22(3):547–557. (in Chinese, with English abstract)

Gariépy C, Allègre C, Xu R (1985) The Pb isotope geochemistry of granitoids from the Himalaya-Tibet collision zone: implications for crustal evolution [J]. Earth Planet Sci Lett 74:220–234

Göpel C, Allègre CJ, Xu R-H (1984) Lead isotopic study of the Xigaze ophiolite (Tibet): the problem of the relationship between magmatites (gabbros, dolerites, lavas) and tectonites (harzburgites). Earth Planet Sci Lett 69:301–310

Guo ZF, Wilson M, Liu J et al (2006) Post-collisional, potassic and ultrapotassic magmatism of the characteristics of the mantle source, geodynamic setting and uplift mechanisms [J]. J Petrol 47(6):1177–1220

Harrison TM, Grove M, McKeegan KD et al (1999) Origin and episodic emplacement of the Manaslu instrusive complex, Central Himalaya [J]. J Petrol 40:3–19

Hoke L, Lamb S, Hilton DR, Poreda RJ (2000) Southern limit of mantle-derived geothermal helium emissions in Tibet: implications for lithospheric structure. Earth Planet Sci Lett 180:297–308

Hou ZQ, Gao YF, Qu XM, Rui ZY, Mo XX (2004) Origin of adakitic intrusives generated during mid-Miocene east–west extension in southern Tibet. Earth Planet Sci Lett 220:139–155

Huang X-L, Xu Y-G, Liu D-Y (2004) Geochronology, petrology and geochemistry of the granulite xenoliths from Nushan, east China: implication for a heterogeneous lower crust beneath the Sino-Korean Craton. Geochim Cosmochim Acta 68(1):127–149

Ji W-Q, Wu F-Y, Chung S-L, Li J-X, Liu C-Z (2009) Zircon U-Pb geochronology and Hf isotopic constraints on petrogenesis of the Gangdese batholith, southern Tibet. Chem Geol 262(3–4):229–245

Ji W-Q, Wu F-Y, Liu C-Z, Chung S-L (2012) Early Eocene crustal thickening in southern Tibet: new age and geochemical constraints from the Gangdese batholith. J Asian Earth Sci 53:82–95

Ji W-Q, Wu F-Y, Chung S-L, Wang X-C, Liu C-Z, Li Q-L, Liu Z-C, Liu X-C, Wang J-G (2016) Eocene Neo-Tethyan slab breakoff constrained by 45 Ma oceanic Island basalt–type magmatism in southern Tibet. Geology 44(4):283–286

Jin Y, McNutt MK, Zhu Y (1996) Mapping the descent of Indian and Eurasian plates beneath the Tibetan Plateau from gravity anomalies. J Geophys Res 101:11275–11290

Kosarev G, Kind R, Sobolev SV, Yuan X, Hanka W, Oreshin S (1999) Seismic evidence for a detached Indian lithosphere mantle beneath Tibet. Science 283:1306–1309

Li X, Gui X, Liu J (1987) Three-cmponent mixting model of Pb and Sr isotopic compositions for the origin of Quxu batholith, Xizang. Geochimica 1:60–66. (in Chinese, with English abstract)

Liao Z (2003) The aluminum granite in southern Tibet: characteristics, origin, and tectonic significance. Doctoral Thesis of Chinese Academy of Geological Sciences

Liu C-Z, Wu F-Y, Chung S-L, Zhao Z-D (2011) Fragments of hot and metasomatized mantle lithosphere in Middle Miocene ultrapotassic lavas, southern Tibet. Geology 39(10):923–926. https://doi.org/10.1130/G32172.1

Liu C-Z, Wu F-Y, Chung S-L, Li Q-L, Sun W-D, Ji W-Q (2014) A 'hidden' ^{18}O-enriched reservoir in the sub-arc mantle. Sci Rep 4(4232):2014. https://doi.org/10.1038/srep04232

Liu D, Zhao Z, DePaolo DJ, Zhu D-C, Meng F-Y, Shi Q, Wang Q (2017) Potassic volcanic rocks and adakitic intrusions in southern Tibet: insights into mantle-crust interaction and mass transfer from Indian plate. Lithos 268-271:48–64

Mahoney JJ, Frei R, Tejada MLG et al (1998) Tracing the Indian Ocean mantle domain through time: isotopic results from old West Indian, East Tethyan, and South Pacific seafloor. J Petrol 39(7):1285–1306

McKenna LW, Walker JD (1990) Geochemistry of crustally derived leucocratic igneous rocks from the Ulugh Muztagh area, northern Tibet and their implications for the formation of the Tibetan plateau [J]. J Geophy Res B (13):21483–21502

Miller C, Schuster R, Klotzli U et al (1999) Post-collisional potassic and ultrapotassic magmatism in SW Tibet: geochemical and Sr-Nd-Pb-O isotopic constraints for mantle source characteristics and petrogenesis [J]. J Petrol 40(9):1399–1424

Mo X, Dong G, Zhao Z, Guo T, Wang L, Chen T (2005) Timing of magma mixing in the Gangdisê magmatic belt during the India-Asia collision: zircon SHRIMP U-Pb dating. Acta Geologica Sinica – English Edition 79(1):66–76. https://doi.org/10.1111/acgs.2005.79.issue-1. https://doi.org/10.1111/j.1755-724.2005.tb00868.x

Mo X, Zhao Z, Deng J, Flower M, Yu X, Luo Z, Li Y, Zhou S, Dong G, Zhu D, Wang L (2006a) Petrology and geochemistry of postcollisional volcanic rocks from the Tibetan plateau: implications for lithosphere heterogeneity and collision-induced asthenospheric mantle flow. In: Dilek Y, Pavlides S (eds) Postcollisional tectonics and magmatism in the Mediterranean region and Asia: Geological Society of America Special Paper, vol 409, pp 507–530. https://doi.org/10.1130/2006.2409(24)

Mo XX, Zhao ZD, DePaolo DJ, Zhou S, Dong GC (2006b) Three types of collisional and post-collisional magmatism in the Lhasa block, Tibet and implications for India continental subduction and mineralization: evidence from Sr-Nd isotopes. Acta Petrol Sin 22:795–803

Mo X, Hou Z, Niu Y, Dong G, Qu X, Zhao Z, Yang Z (2007) Mantle contributions to crustal thickening during continental collision: evidence from Cenozoic igneous rocks in southern Tibet. Lithos 96:225–242

Mo X, Niu Y, Dong G, Zhao Z, Hou Z, Zhou S, Ke S (2008) Contribution of syncollisional felsic magmatism to continental crust growth: a case study of the Paleogene Linzizong Volcanic Succession in southern Tibet. Chem Geol 2008(250):49–68

Niu XL, Zhao ZD, Mo XX, DePaolo DJ, Dong GC, Zhang SQ, Zhu DC, Guo TY (2006) Elemental and Sr-Nd-Pb isotopic geochemistry for basic rocks from Decun-Angren ophiolites in the Xigaze area, Tibet: implications for the characteristics of the Tethyan upper mantle domain. Acta Petrol Sin 22:2875–2888. (in Chinese, with English abstract)

Nomade S, Renne PR, Mo XX et al (2004) Miocene volcanism in the Lhasa block, Tibet: spatial trends and geodynamic implications [J]. Earth Planet Science Letters 221:227–243. (SCI收录)

Owens TJ, Zandt G (1997) Implications of crustal property variations for models of Tibetan plateau evolution. Nature 387:37–43

Priestley K, Jackson J, McKenzie D (2008) Lithospheric structure and deep earthquakes beneath India, the Himalaya and southern Tibet. Geophys J Int 172:345–362

Qu X, Hou Z, Li Y (2002) Implications of S and Pb isotopic compositions of the Gangdise porphyry copper belt for the ore-forming material source and material recycling within the orogenic belt. Geol Bull China 21(11):768–776. (in Chinese, with English abstract)

Schärer U, Xu RH, Allègre CJ (1984) U-(Th)-Pb geochronology of Gangdese (Transhimalaya) plutonism in the Lhasa-Xigaze region, Tibet [J]. Earth Planet Sci Lett 69:311–320

Schärer U, Xu RH, Allègre CJ (1986) U-(Th)-Pb systematic and ages of Himalayan leucogranites, South Tibet [J]. Earth Planet Sci Lett 77:35–48

Tilmann F, Ni J, INDEPTH III Seismic Team (2003) Seismic imaging of the downwelling Indian lithosphere beneath Central Tibet. Science 300:1424–1427

Turner S, Arnaud N, Liu J et al (1996) Post-collision, shoshonitic volcanism on the Tibetan plateau: implications for convective thinning of the lithosphere and the source of ocean Island basalts [J]. J Petrol 37(1):45–71

Vidal P, Cocherie A, Le Fort P (1982) Geochemical investigations of the origin of the Manaslu leucogranite (Himalaya, Nepal) [J]. Geochim Cosmochim Acta 46:2279–2292

Wang Q, Zhang P-Z, Freymueller JT, Bilham R, Larson KM, Lai X, You X, Niu Z, Wu J, Li Y, Liu J, Yang Z, Chen Q (2001) Present-day crustal deformation in China constrained by global positioning system measurements. Science 294:574–577

Wang Z, Zhao Z, Asimow PD, Liu D, Zhu D-C, Mo X, Wang Q, Zhang L, Sheikh L (2020) Shoshonitic enclaves in the high Sr/Y Nyemo pluton, southern Tibet: implications for Oligocene magma mixing and the onset of extension of the southern Lhasa terrane. Lithos 362-363:105490

Wang Z, Zhao Z, Li X, Asimow PD, Liu D, Mo X, Qi N (2021) Late Cretaceous adakitic and A-type granitoids in Chanang, southern Tibet: implications for Neo-Tethyan slab rollback. Gondwana Res 96:89–104

Wang Z, Zhao Z, Asimow PD, Li X, Meng Y, Liu D, Mo X, Zhu D-C, Tang Y, Cong F (2022) Two episodes of Eocene mafic magmatism in the southern Lhasa terrane imply an eastward propagation of slab breakoff. Gondwana Res 110:31–43

Wei W, Unsworth M, Jones A, Booker J, Tan H, Nelson D, Chen L, Li S, Solon K, Bedrosian P, Jin S, Deng M, Ledo J, Kay D, Roberts B (2001) Detection of widespread fluids in the Tibetan crust by magnetotelluric studies. Science 292:716–718

Wen D-R, Chung S-L, Song B, Iizuka Y, Yang H-J, Ji J, Liu D, Gallet S (2008) Late Cretaceous Gangdese intrusions of adakitic geochemical characteristics, SE Tibet: petrogenesis and tectonic implications. Lithos 105:1–11

Xie G (1992) Geochemistry of the Cenozoic volcanic rocks in the margins of Tibetan Plateau: evidence for an ancient enriched mantle. In: Liu R (ed) Geochronology and geochemistry of Cenozoic Volcanic Rock in China. Seismology Press, Beijing, pp 400–427

Xu XS, O'Reilly SY, Griffin WL, Zhou XM, Huang XL (1998) The nature of the Cenozoic lithosphere beneath Nushan, east Central China. In: MFJ F, Chung SL, Lo CH, Lee TY (eds) Mantle dynamics and plate interactions in East Asia, AGU Geodynamics Series, vol 27, pp 167–196

Xu Y, Lin C, Shi L (1999) The geotherm of the lithosphere beneath Qilin, SE China: a re-appraisal and implications for P–T estimation of Fe-rich pyroxenites. Lithos 47:181–193

Xu W, Zhu DC, Wang Q, Weinberg RF, Wang R, Li SM, Zhang LL, Zhao ZD (2019) Constructing the Early Mesozoic Gangdese crust in southern Tibet by hornblende-dominated magmatic differentiation. J Petrol 60:515–555

Yang JS, Wu CL, Shi RD, Li HB, Xu ZQ, Meng FC (2002) Miocene and Pleistocene shoshonitic volcanic rocks in the Jingyuhu area, northern part of the Qinghai-Tibet Plateau. Acta Petrol Sin 18:161–176. (in Chinese, with English abstract)

Yi H, Lin J, Li B (2004) Cenozoic high-potassic calc-alkaline series volcanic rocks in northern Tibet and their implications for crust-mantle interaction. Geological Science Press, Beijing, pp 1–95

Zhang SQ, Mahoney JJ, Mo XX et al (2005) Evidence for a widespread Tethyan upper mantle with Indian-Ocean-type isotopic characteristics [J]. J Petrology 46(4):829–858

Zhao Z, Mo X, Zhang S, Guo T, Zhou S, Dong G, Wang Y (2001) Post-collsional magmatism in Wuyu basin, central Tibet: evidence for recycling of subducted Tethyan oceanic crust. Sci China (Series D) 44:27–34

Zhao Z, Mo X, Nomade S, Renne PR, Zhou S, Dong G, Wang L, Zhu D, Liao Z (2006) Post-collisional ultrapotassic rocks in the Lhasa Block, the Tibetan Plateau: special and temporal distribution and its implications. Acta Petrol Sin 22:787–794. (in Chinese, with English abstract)

Zhao Z, Mo X, Dong G, Zhou S, Zhu D, Liao Z, Sun C (2007) Pb isotopic geochemistry of Tibetan plateau and its implications. Geoscience 21:269–278. (in Chinese, with English abstract)

Zhao ZD, Mo XX, Sun CG, Zhu DC, Niu YL, Dong GC, Zhou S, Dong X, Liu YS (2008a) Mantle xenoliths in southern Tibet: geochemistry and constraints for the nature of the mantle. Acta Petrol Sin 24(2):193–202

Zhao Z, Mo X, Niu Y, Zhu D, Sun C, Dong G (2008b) Nature of the subcontinental mantle beneath southern Tibet revealed by mantle xenoliths entrained by ultrapotassic rocks. Geochim Cosmochim Acta 72(12):A1095

Zhao Z, Mo X, Dilek Y, Niu Y, DePaolo DJ, Robinson P, Zhu D, Sun C, Dong G, Zhou S, Luo Z, Hou Z (2009) Geochemical and Sr-Nd-Pb-O isotopic compositions of the post-collisional ultrapotassic magmatism in SW Tibet: Petrogenesis and implications for India intra-continental subduction beneath southern Tibet. Lithos 2009(113):190–212

Zhao Z, Dale CW, Pearson DG, Niu Y, Zhu D, Mo X (2011) Os and HSE of the hot upper mantle beneath southern Tibet: Indian mantle affinity? American Geophysical Union, Fall Meeting 2011, abstract #T44B-02

Zhou, S., 2002. Study on the geochronology of pivotal regions of Gangdese magmatic and Yarlung Zangpo ophiolite belts, Tibet. Ph. D Dissertation, China University of Geosciences, Beijing (in Chinese, with English abstract)
Zhou S, Mo X, Dong G, hoaxing Zhao Z, Qiu R, Guo T, Wang L (2004) 40Ar-39Ar geochronology of Cenozoic Linzizong volcanic rocks from Linzhou Basin, Tibet, China, and their geological implications. Chin Sci Bull 49:1970–1979
Zhu D. (2003) The volcanic rocks since the late Paleozoic in central Tethyan Himalaya, and it's implication. Doctoral Thesis of Chinese Academy of Geological Sciences
Zhu DC, Wang Q, Zhao ZD, Chung SL, Cawood PA, Niu Y, Liu SA, Wu FY, Mo X (2015) Magmatic record of India-Asia collision. Scientific Report 5:14289
Zhu DC, Wang Q, Cawood PA, Zhao ZD, Mo XX (2017a) Raising the Gangdese Mountains in southern Tibet. J Geophys Res Solid Earth 122:214–223
Zhu DC, Wang Q, Zhao ZD (2017b) Constraining quantitatively the timing and process of continent-continent collision using magmatic record: method and examples. SPECIAL TOPIC: initial collision between India and Asia. Sci China Earth Sci 60(6):1040–1056
Zhu D-C, Wang Q, Chung S-L, Cawood PA, Zhao Z-D (2019) Gangdese magmatism in southern Tibet and India-Asia convergence since 120 Ma. Geological Society, London, Special Publications 483(1):583–604

Chapter 10
Formation, Preservation, and Metallogenic Fertility of Juvenile Crust in the Eastern Tethyan Collisional Orogens

Abstract Genesis and growth of continental crust is fundamental to maintain life and supply resources to our society. It is widely accepted that continental collision zones are favored locations of crustal preservation, yet it remains unclear when and how continental collision generates juvenile crust and maintains net crustal growth. Here we carry out hafnium isotopic mapping of three collisional orogens with distinct maturity in the Tethyan domain, i.e., the Zagros, Himalaya, and Qinling, to reveal evolution of juvenile crust in collisional zones. Our results indicate that the amount of juvenile crust generated and preserved in collisional orogens far exceeds previous estimates. The pre–collisional juvenile crust generated by arc magmatism was mostly preserved in Zagros, slightly reworked in Himalaya, and partly lost in the Qinling orogen. In growing orogens such as Zagros and Himalaya, the collision produces juvenile crust and leads to net crustal growth by input of juvenile mafic magmas derived from partial melting of both subduction–modified lithospheric mantle and asthenosphere. In decayed orogen such as Qinling, there is little juvenile crust produced during collision due to limited mantle melting. The juvenile lower crust is fertilized in ore–forming components, which can be released into hydrous felsic magmas during later remelting, and thus plays a significant role in the genesis of magmatic–hydrothermal ore deposits in collision zones.

Keywords Juvenile crust · Crustal growth · Collisional orogen · Tethyan domain

---

Prof. Zengqian Hou has fully contributed to this chapter.

Supplementary Information The online version contains supplementary material available at https://doi.org/10.1007/978-981-97-2109-2_10.

X. Mo et al., *Magmatism and Crustal Evolution of the Tibetan Plateau*, Modern Approaches in Solid Earth Sciences 25,
https://doi.org/10.1007/978-981-97-2109-2_10

10.1 Introduction

Two fundamental types of orogenic systems occur on Earth, i.e., accretionary and collisional (Sengor et al. 1988; Murphy and Nance 1991; Collins et al. 2011). Accretionary orogens mainly distribute around the Pacific rim (Cawood et al. 2009), where oceanic lithosphere subducts beneath continental lithosphere and the juvenile materials are either magmatically underplated or tectonically accreted to these orogens, leading to the generation of continental crust (Jahn 2004; Hawkesworth et al. 2010; Collins et al. 2011). Collisional orogens are widely developed in the Tethyan domain, where continental fragments of varying size were accreted to the southern margins of Eurasia following successive closure of the Tethyan Oceans, which eventually led to collision between Eurasia and Gondwana–derived continental blocks such as Arabia, India and South China (Fig. 10.1; Sengor et al. 1988; Metcalfe 1996). Although subduction zones are favorable settings for the formation of new continental crust as arcs, the low preservation potential of arcs make subduction zones inefficient sites of crustal growth (Clift and Vannucchi 2004; Stern 2011). By contrast, collision may lead to net crustal growth by input of juvenile mantle components that can be better preserved in collisional orogens (Niu et al. 2013; Moyen et al. 2017; Hou et al. 2020). Yet, the size, genesis, and preservation of juvenile crust in collisional orogens remain poorly understood.

Collisional orogens often underwent a complex evolution history from syn–collision through post–collision to final collapse. However, these processes are not

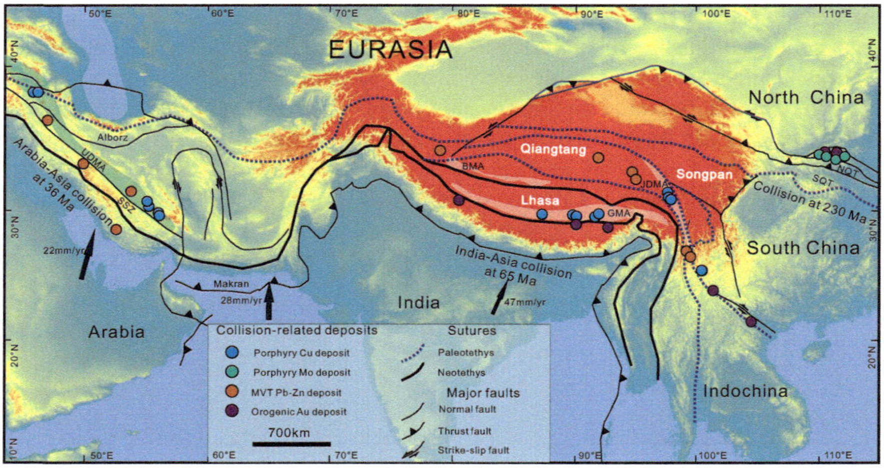

Fig. 10.1 Tectonic framework of the Tethyan domain showing three collisional orogens (i.e., Zagros, Himalaya, and east Qinling) and large ore deposits (generally metal Cu, Pb, Zn > 500,000 t; Mo > 100,000 t, and Au > 20 t). The background image is from SRTM Data (National Aeronautics and Space Administration, NASA, http://srtm.csi.cgiar.org/srtmdata/). Abbreviations: *BMA* Bangonghu magmatic arc, *GMA* Gangdese magmatic arc, *JDMA* Jiangda magmatic arc, *SSZ* Sanandaj–Sirjan magmatic arc zone, *UDMA* Urumieh–Dokhtar magmatic arc, *Terrane in Tibet* Lhasa, Qiangtang, and Songpan, *NQT* North Qinling terrane, *SQT* South Qinling terrane

always completely recorded by a single orogen. Therefore, comparison of the lithospheric architectures of collisional orogens with distinct maturities can be a new approach to understanding the formation and preservation of juvenile crust during collision. In this chapter, we compare evolution of lithospheric architectures of three collisional orogens including Zagros, Himalaya, and Qinling, which likely recorded orogenic evolution from preliminary growth to large–scale growth, and to final collapse, respectively. Their lithospheric architectures are imaged by terrane–scale zircon Hf isotopic mapping, which shed new lights on the formation, reworking and destruction of juvenile crust during orogenic evolution and its control on distinct mineralization. Our study also proved that zircon multi–isotopic mapping is an effective approach to evaluating mass exchanges in orogens through time and to rapidly screen out perspective areas for mineral exploration on continental scales.

10.2 Evolution of Collisional Orogens

There is a wide variety of collisional orogens on Earth. They can be grouped into two dominant types accordingly to different criteria (Hou and Zhang 2015), including kinematically orthogonal (e.g., Himalaya, Pyrenees, and Qinling) and oblique (e.g., Alpine, Zagros, and Variscan) types, or morphologically symmetric (e.g., Alpine, Variscan, Ural, Pyrenean, and Qinling) and asymmetric (e.g., Himalaya and Zagros) types (Sengor 1991; Kerrich et al. 2005). These orogens, despite of various types, usually experienced a three–stage evolution process from an infancy to an adulthood, and finally to a senility, in which the former can be partly disturbed by later. Therefore, collisional orogens can be also divided into infantile, adult and senile orogens. Zagros, Himalayan, and Qinling probably recorded orogenic evolution from preliminary growth (i.e., infantile, small–hot orogen), large–scale growth (i.e., adult, large–hot orogen), and finally to decay (i.e., senile, relatively–cold orogen), respectively, in terms of their magnitude, architecture, and thermal status, as demonstrated below.

10.2.1 Zagros Orogen

Zagros is the youngest active collisional orogen on Earth. It was formed by oblique collision between Arabia and Eurasia in the Paleogene, following the protracted continuous subduction of Neo–Tethyan Ocean since the Mesozoic (Figs. 10.1, 10.2; Agard et al. 2011). The successive subduction generated two parallel, NW–trending, 1500 km long, volcanoplutonic arcs, which migrates toward the central Iran block from early Jurassic–Cretaceous to early Eocene (Verdel et al. 2011; Chiu 2013; Zhang et al. 2018). The collision experienced a three–stage process (Ballato et al. 2011). The syn– or "soft" collision started at ~36 Ma (McQuarrie and van Hinsbergen 2013) and resulted in potassic/calc–alkaline basaltic to felsic magmatism along the Eocene volcanoplutonic arc (i.e., Urumieh–Dokhtar magmatic arc

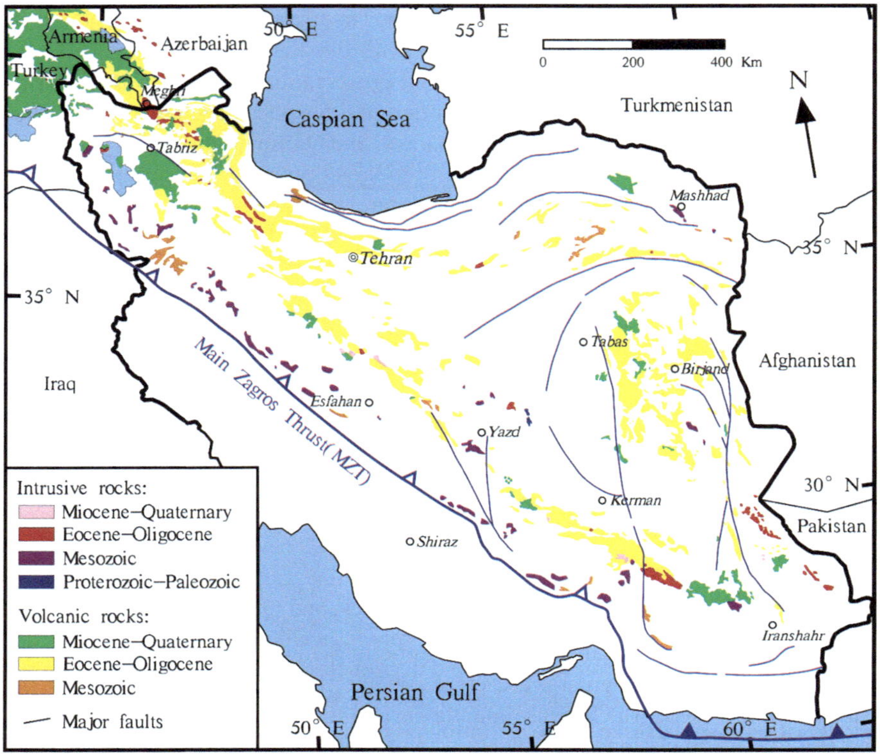

Fig. 10.2 Geological map of the Zagros orogen (modified from Aghazadeh et al. 2015a, b)

(UDMA); Verdel et al. 2011) and tectonic uplift along the collisional front (i.e., SSZ: Sanandaj–Sirjan magmatic arc zone (SSZ)) (Fig. 10.1; Moutherea 2011), whereas the late– or "hard" collision occurred at 20–17.5 Ma, and led to crustal thickening and accelerated deformation along the SSZ– with associated potassic basaltic–andesitic rocks along the UDMA (Verdel et al. 2011; Ballato et al. 2011; Horton et al. 2008). The post–collision, though lacking widespread crustal extension, produced a two–stage magmatism, characterized by the isolated granitoid plutons (17–10 Ma) and small–volume potassic–ultrapotassic lavas (9.4–2 Ma) along the UDMA (Chiu 2013;Aghazadeh et al. 2015a). The Arabia plate is moving northwards by 22 mm/year and still colliding with Eurasia (Fig. 10.1; Reilinger et al. 2006). High terrestrial heat flow (av. 90 mW•m^{-2}) in Zagros suggests it as a small–hot ongoing infantile orogen (Global Heat Flow Compilation Group 2013).

10.2.2 Himalayan Orogen

The Himalayan orogen, formed by India–Asia collision at 65–55 Ma (Mo et al. 2003; DeCelles et al. 2004), is the largest collisional orogen with the thickest crust (60–80 km) and the highest elevation (>4000 m) on the present-day Earth (Figs. 10.1,

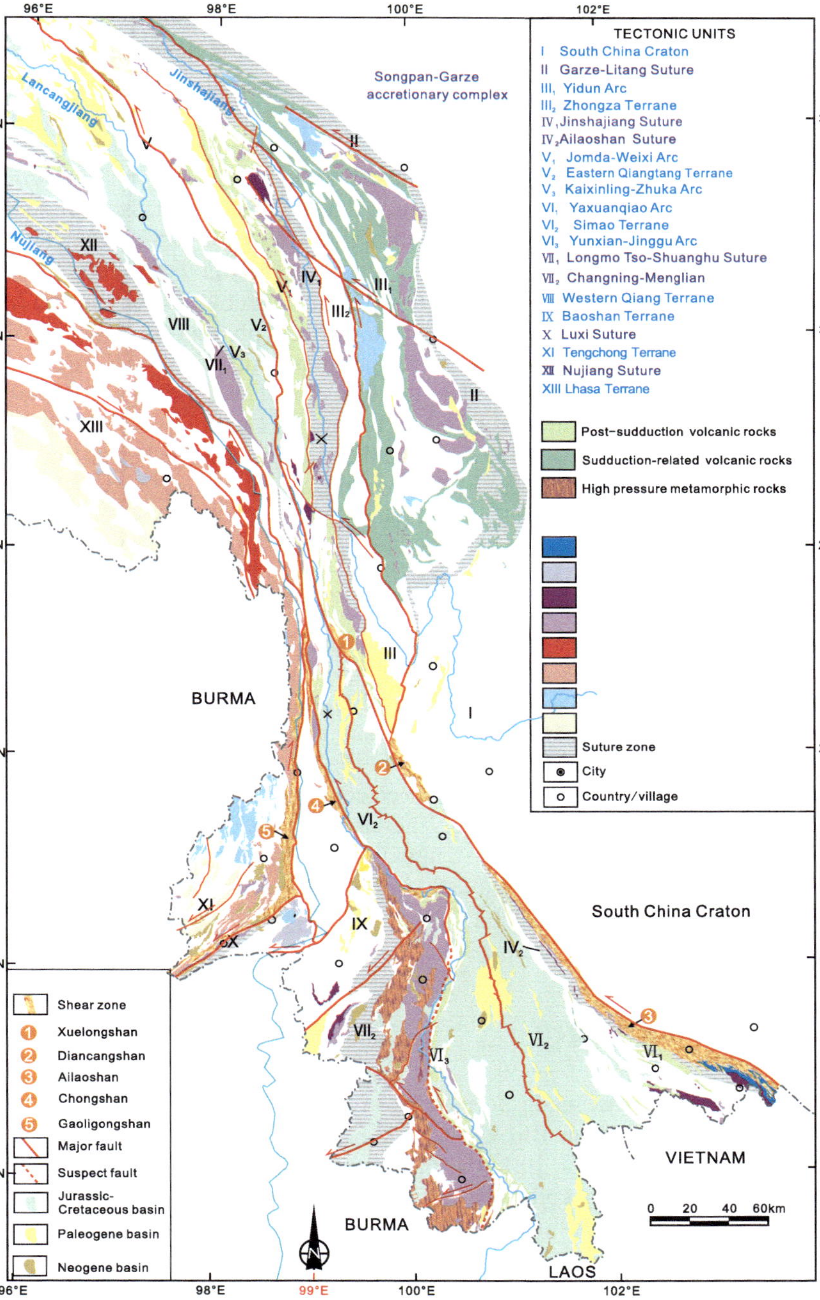

Fig. 10.3 Geological map of the in eastern Himalayan orogen (modified from Wang et al. 2016)

10.3). It is a complex collage of several terranes (e.g., Lhasa, Qiangtang, Songpan) that accreted to the southern margin of Asia by successive subduction of Paleo– and Neo–Tethyan Oceans since the late–Paleozoic, which generated numerous volcano-plutonic arcs, such as the Permian Zaduo–Jiangda–Weixi arc in east Tibet (Wang et al. 2014a, b; Zhang et al. 2017), Triassic Yidun arc in northeast Tibet (Yang et al. 2014a, b), Jurassic–Cretaceous Gangdese arc and Banggong arc on both margins of the Lhasa terrane (Fig. 10.1; Yin and Harrison 2000). This orogen also underwent a three–stage collisional process, i.e., syn–(or "soft") collisional (65–41 Ma), late–(or "hard") collisional (40–26 Ma), and post–collisional (25 Ma to present) (Hou and Cook 2009), leading to widespread multiple magmatic activities for over 50 Myrs (65–10 Ma), crustal thickening up to 80 km, and the plateau uplift since 36 Ma (Chung et al. 2005). The syn–collisional magmatism forms a ~ 5000 m thick potassic volcanic sequence (64–43 Ma) and the isolated gabbro and granitoid batholiths (55–50 Ma), emplaced along the southern margin of Lhasa terrane, a collisional front (Fig. 10.1; Mo et al. 2003). The late–collision is characterized by crustal shorting in the collisional front and the block–rotating, thrust–folding, strike–slip faulting, and dextral shearing in the hinterland (Hou and Cook 2009). Associated Eocene magmatism dominantly generated a 1500 km long semi–continuous belt consisting of potassic–ultrapotassic mafic to felsic volcanic rocks and numerous granitoids in the Qiangtang terrane (Fig. 10.1; Wang et al. 2001; Ou et al. 2017). The post–collision is associated with crustal extension that forms NS–strike normal fault system across the orogen (Blisniuk et al. 2001), and relevant magmatism dominantly produced mid–Miocene granitoids with minor ultrapotassic mafic rocks in the Lhasa terrane (Miller et al. 1999; Hou et al. 2004), and numerous coeval leucogranites in the Indian block (Fig. 10.1; Harrison et al. 1997). All of these suggest that the Himalaya is an adult orogen that experienced complete orogenic process and likely initiated a transform toward collapse. The large orogenic magnitude (Fig. 10.1), great convergent amounts (1500 km), high terrestrial heat flow (av. 138 mW•m^{-2}) and hot lower crust (>800 °C) indicate that Himalaya is a large–hot orogen (Hu et al. 2000; Yin and Harrison 2000).

10.2.3 Qinling Orogen

The Qinling orogen was finally formed by strong collision of South China Block (SCB) with the North China Block (NCB) at ~230 Ma, following the closure of the Paleo–Tethyan Ocean (Figs. 10.1, 10.4; Meng and Zhang 1999; Dong and Santosh 2016). It underwent a syn–collisional crustal shorting and thickening, characterized by a flower–style thrust system and 228–215 Ma granitoid batholiths, and a post–collisional crustal extension, characterized by rapakivi–textured granitoids (210 Ma), lamprophyre dykes and A–type granites (200–180 Ma) (Wang et al. 2013b), as well as early–Jurassic fault–bounded basins (Dong and Santosh 2016). The collision ceased at 180–160 Ma, and then extensive multiple magmatic activities generated widespread granitoid batholiths of varying volume at 155–100 Ma,

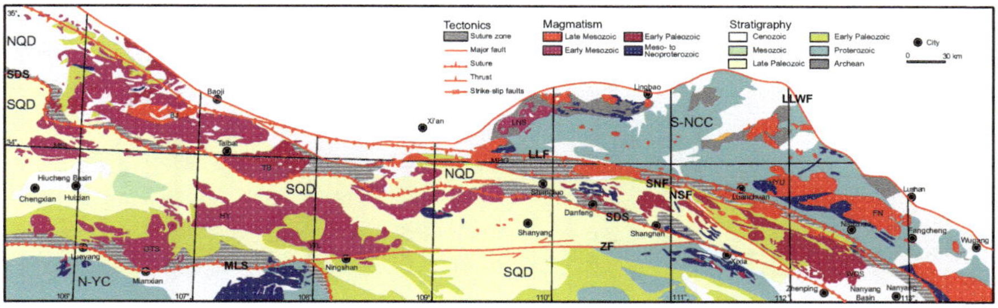

Fig. 10.4 Geological map of the Eastern Qinling orogen (modified after Dong and Santosh 2016; Wang et al. 2017). Abbreviations for tectonics: *LLWF* Lingbao–Lushan–Wuyang fault, *LLF* Luonan–Luanchuan fault, *MLS* Mianlue Suture, *SDS* Shangdan Suture, *NQT* North Qinling terrane, *SQT* South Qinling terrane, and *S–NCC* Southern North China Block

formed by crustal melting related to orogenic collapse (Dong and Santosh 2016). Available geophysical data show that the orogen lacks mountain root, and has normal crust thickness (~40 km; Yuan 1996). Weak collisional magmatism, strong tectonic exhumation, and low heat flow (av. 68 mW•m^{-2}) indicate that Qinling is a slab–stacked, nearly–decayed senile orogen (Hu et al. 2001).

10.3 Major Collision–Related Ore Depositsof Collisional Orogens

Four main styles of mineralization have been recognized in the Zagros, Himalayan and East Qinling orogens (Fig. 10.1). These are porphyry Cu deposits (PCD), porphyry Mo deposits (PMD), Pb–Zn Mississippi Valley–type (MVT) and orogenic Au deposits. The geological characteristics and ages of representative ore deposits in the Tethyan domain are summarized in Table 10.1.

10.3.1 Zagros Orogen

In Iran, the Arasbaran–Kerman porphyry Cu belt within the UDMA extends for ~1700 km from Armenia to southern Iran, and contains a series of large–scale PCDs, such as the Sar Cheshmeh and Sungun (Fig. 10.1; Table 10.1; Aghazadeh et al. 2015a; Asadi 2018; Zürcher et al. 2019). The Miocene porphyritic granodiorites and monzogranites are associated with several world–class Cu deposits. PCDs predominate in the Zagros orogen and yield molybdenite Re–Os ages between 16 and 12 Ma hosted by granitoids with zircon U–Pb ages ranging from 21 to 12 Ma (Perelló et al. 2008; Razique et al. 2014; Aghazadeh et al. 2015a; Asadi 2018; Zürcher et al. 2019). The Sar Cheshmeh deposit contains a resource of ~1200 mil-

Table 10.1 A summary of the geological characteristics of major large ore deposits in the Tethyan orogen

District	Deposit	Long. lat.	Type	Metallic comm.	Tonnage	Grade	Host rock	Alteration	Mineralization	Igneous age (Ma)	Mineralization age (Ma)	Data source
Tibet	Machangqing	100.45E 25.53 N	Porphyry	Cu-Mo-Au	Cu: 81,258 t; Mo: 44,525 t; Au: 26 t	Cu: 0.50%; Mo: 0.08%; Au: 4.01–8.70 g/t	Syenite porphyry, monzonite porphyry, granite porphyry, limestone and sandstone	Potassic alteration, silicification,sericitization,chloritization	Chalcopyrite, bornite, molybdenite, pyrite, pyrrhotite	LA-ICP-MS zircon U-Pb: 33.78 ± 0.21, 35.6 ± 0.3, 35.0 ± 0.2, 37.93 ± 0.82	Molybdenite re-Os: 35.3 ± 0.7, 35.8 ± 1.6, 33.9 ± 1.1, 34.72 ± 0.5, 34.94 ± 0.38	Guo et al. (2009, 2011); He et al. (2011)
	Beiya	100.20E 26.16 N	Porphyry	Cu-Au-Fe	Cu: 0.65 Mt.; Au: 323 t; Fe: 170 Mt	Cu: 0.52%; Au: 2.47 g/t; Fe: 33.3%	Quartz-albite porphyry, quartz-K-feldspar porphyry, biotite-K-feldspar porphyry	Potassic alteration, silicification, sericitization, chloritization, carbonization	Pyrite, quartz, chalcopyrite, bornite, and chalcocite	LA-ICP-MS zircon U-Pb: 34.72 ± 0.94, 37.67 ± 0.97	Molybdenite Re-Os: 36.82 ± 0.48	He (2014); Jiang (2014); Deng et al. (2015a); He et al. (2016)
	Yulong	97.74E 31.40 N	Porphyry	Cu-Mo	Cu: 6.5 Mt	Cu: 0.99%; Mo: 0.03%	Monzonitic granite porphyry, quartz monzonitic porphyry, Triassic sandstone and mudstone	Potassic alteration quartz sericite zone argillization propylitization	Chalcopyrite, molybdenite, pyrite, bornite, tetrahedrite, cubanite, sphalerite, galena, native gold, native silver	LA-ICP-MS zircon U-Pb: 43.9 ± 0.6 to 40.9 ± 0.3; SHRIMP zircon U-Pb: 40.9 ± 0.1	Molybdenite Re-Os: 40.1 ± 1.8, 42.28 ± 0.17 to 40.92 ± 0.17	Hou et al. (2003b); Liang et al. (2009); Chang et al. (2017); Huang et al. (2019)
	Duoxiasongduo	97.92E 31.17 N	Porphyry	Cu-Mo	Cu: 0.5 Mt	Cu: 0.38%; Mo: 0.04%	Alkali-feldspar granite porphyry, monzonitic granite porphyry	Potassic alteration quartz sericite zone propylitization	Chalcopyrite, pyrite, molybdenite, magnetite, galena, sphalerite, bornite	SHRIMP zircon U-Pb: 37.5 ± 0.2	Molybdenite Re-Os: 36.0 ± 0.4	Hou et al. (2003a)
	Malasongduo	98.00E 31.00 N	Porphyry	Cu-Mo	Cu: 1.0 Mt	Cu: 0.44%; Mo: 0.14%	Monzonitic granite porphyry, syenitic granite porphyry	Potassic alteration quartz sericite zone argillization propylitization	Chalcopyrite, pyrite, molybdenite, galena, sphalerite, bornite, tetrahedrite	LA-ICP-MS zircon U-Pb: 36.9 ± 0.4	Molybdenite Re-Os: 38.5 ± 0.5	Hou et al. (2003a); Liang et al. (2009)
	Zhunuo	87.47E 29.65 N	Porphyry Cu-Mo	Cu-Mo	Cu: 2.4 Mt	Cu: 0.83%	Cretaceous monzogranite, Paleogene volcanic rocks	Potassic alteration, quartz-sericite alteration, argillic alteration, propylitic alteration	Chalcopyrite, pyrite, molybdenite, malachite, azurite	LA-ICP-MS zircon U-Pb: 15.6 ± 0.6, 14.7 ± 0.3 and 14.6 ± 0.2	Molybdenite Re-Os: 13.7 ± 0.6, 14.78 ± 0.13 and 14.44 ± 0.13 (B), 14.23 ± 0.13 and 13.51 ± 0.07 (D)	Zheng et al. (2007); Sun et al. (2018)

	Chongjiang	89.96E 29.62 N	Porphyry Cu-Mo	Cu-Mo	Cu: 3.3 Mt	Cu: 0.45%	Mid-Miocene monzogranite	Potassic alteration, quartz-sericite alteration, argillic alteration, propylitic alteration	Chalcopyrite, molybdenite, pyrite	LA-ICP-MSzircon U-Pb: 14.9 ± 0.3 and 12.9 ± 0.3	Molybdenite Re-Os: 14.04 ± 0.16	Hou et al. (2003b); Hu et al. (2017)
	Tinggong	90.05E 29.58 N	Porphyry Cu-Mo	Cu-Mo	Cu: >2.0 Mt	Cu: 0.5%	Mid-Miocene quartz monzogranite, granite	Potassic alteration, quartz-sericite alteration, argillic alteration, propylitic alteration	Chalcopyrite, molybdenite, pyrite	LA-ICP-MS zircon U-Pb: 15.54 ± 0.28, 15.02 ± 0.25 and 14.74 ± 0.22	Molybdenite Re-Os: 15.49 ± 0.36	Rui et al. (2004); Hou et al. (2013); Chen et al. (2014b)
	Qulong	91.60E 29.62 N	Porphyry Cu-Mo	Cu-Mo	Cu: 14.2 Mt	Cu: 0.5%; Mo: 0.03%	Mid-Miocene monzogranitic porphyry and granite	Potassic alteration, quartz-sericite alteration, argillic alteration, propylitic alteration	Chalcopyrite, molybdenite, pyrite	LA-ICP-MS zircon U-Pb: 16.2 ± 0.4, 15.9 ± 0.4 and 15.7 ± 0.2; (CA-IDTIMS) U-Pb: 17.142 ± 0.014/0.014/0.023,16.009 ± 0.016/0.017/0.024 and 15.166 ± 0.010/0.011/0.020; biotite and sericite $^{40}Ar/^{39}Ar$: 15.7 ± 0.2 and 15.7 ± 0.2.	Molybdenite Re-Os: 16.41 ± 0.48; Molybdenite (ID-N-TIMS) Re-Os: 16.126 ± 0.008/0.060/0.077 and 15.860 ± 0.010/0.058/0.075	Yang et al. (2009); Hou et al. (2013); Zhao et al. (2016); Li et al. (2017)
	Jiama	91.75E 29.70 N	Porphyry Cu-Mo	Cu-Mo Pb-Zn	Cu: 10.55 Mt.; Pb-Zn: 58 Mt.; Au: 106 t; Ag: 6675 t	Cu: 0.44%; Pb: 1.01%; Zn: 0.75%; Au: 0.34 g/t; Ag: 14.1 g/t	Lower cretaceous clastic rocks and upper Jurassic limestone, mid-Miocene monzogranite	Skarn alteration, biotitization, quartz-sericite alteration, chloritization, marbleization	Chalcopyrite, bornite, chalcocite, galena, sphalerite, molybdenite, tendrahedrite, pyrite, native gold	LA-ICP-MS zircon U-Pb: 15.96 ± 0.5, 15.72 ± 0.14, 15.59 ± 0.09, and 15.48 ± 0.08	Molybdenite Re-Os: 15.37 ± 0.15, 14.67 ± 0.37, and 14.66 ± 0.27, 15.34 ± 0.10	Zheng et al. (2016); Gong et al. (2013)
	Zhenyuan	101.45E 23.90 N	Orogenic	Au	Au: 50 t	Au: 4.2 g/t	Paleozoic greywacke	Sericitization, silicification, chloritization, carbonatization	Pyrite, chalcopyrite, arsenopyrite, stibnite	No data	Phlogopite Ar-Ar age: 26.4 ± 0.2	Deng et al. (2015b); Li et al. (2019b, 2020)
	Daping	102.98E 22.85 N	Orogenic	Au	Au: 70 t	Au: 10.8 g/t	Diorite, granite porphyry, lamprophyre, siliceous shale, sandstone, limestone	Sericitization, carbonatization, chloritization, silicification, tremolitization, epidotization	Sphalerite, pyrite, galena, chalcopyrite, tetrahedrite, pyrrhotite, chalcocite	No data	Sericite Ar-Ar age: 33.76 ± 0.65; Phlogopites of lamprophyres Ar-Ar age: 36.8 ± 0.2, 29.6 ± 0.2	Chen et al. (2014a, c)

(continued)

Table 10.1 (continued)

District	Deposit	Long. lat.	Type	Metallic comm.	Tonnage	Grade	Host rock	Alteration	Mineralization	Igneous age (Ma)	Mineralization age (Ma)	Data source
	Mayum	82.24E 30.28 N	Orogenic	Au	Au: 28 t	Au: 31.84 g/t	Mica–quartz schist, limestone	Sericitization, silicification, carbonatization, argillization	Pyrite, galena, stibnite, minor chalcopyrite	No data	Sericite Ar-Ar age:44.08 ± 0.39; 59.34 ± 0.62	Wen et al. (2006); Jiang et al. (2009)
	Bangbu	92.50E 29.00 N	Orogenic	Au	Au: 20 t	No data	Phyllite, siltstone	Carbonatization, silicification, sericitization	Pyrite, galena, sphalerite, chalcopyrite	No data	Sericite Ar-Ar age:44 ± 3, 44.8 ± 1.0, and 49.52 ± 0.52	Pei et al. (2016); Sun et al. (2016)
	Nianzha	90.00E 29.33 N	Orogenic	Au	Au: 25 t	Au: 4.2 g/t	Greywacke, metasandstone, silty slate	Silicification, Serpentinization, carbonatization, chloritization	Pyrite	No data	No data	Yang et al. (2017)
	Dongmozazhua	95.68E 33.10 N	MVT	Pb-Zn	Pb-Zn: >1 Mt	No data	Limestone	Silicification, Pyritization, dolomitization	Sphalerite, galena, pyrite	No data	Sphalerite Rb-Sr: 35.0 ± 0.3	Tian et al. (2014)
	Mohailaheng	95.77E 33.42 N	MVT	Pb-Zn	Pb-Zn: >1 Mt	Pb-Zn: 2.18–4.23%	Limestone	Silicification, dolomitization	Galena, sphalerite	No data	Fluorite Sm-Nd: 31.8 ± 0.3	Tian et al. (2014); Liu et al. (2015)
	Chaqupacha	91.90E 34.08 N	MVT	Pb-Zn	Pb-Zn: 4.8 Mt	Pb: 2.2%; Zn: 0.3%	Limestone	Silicification, dolomitization	Galena, pyrite, smithsonite	No data	No data	Song et al. (2015)
	Huoshaoyun	78.98E 34.60 N	MVT	Pb-Zn	14 Mt	Pb: 4.98%; Zn: 20.85%	Mesozoic clastic, carbonate rock	Silicification	Sphalerite, galena, pyrite	No data	No data	Li et al. (2019a)
	Jinding	99.42E 26.45 N	MVT	Pb-Zn	15 Mt	Zn: 6.1%; Pb: 1.3%	Limestone clast-bearing clast-free sandstones brecciated limestones	Dolomitization	Pyrite, marcasite	Paleocene	Late Oligocene to early Miocene	Wang et al. 2018; Song et al. (2020)
	Baiyangping	99.25E 26.87 N	MVT	Pb-Zn-Cu-Ag	No data	No data	Cretaceous transitional zone between porous sandstone and low permeability carbonaceous bargillite	Silicification, carbonization, chloritization, epidotization	Tetrahedrite, pyrite, chalcopyrite, galena, freibergite	No data	Calcite Sm-Nd ages: 29.9 ± 1.1 and 27.4 ± 1.8; sphalerite Rb-Sr age, 28.9 ± 0.6	He et al. (2004); Wang et al. (2015a, b); Zou et al. (2015)

Zagros	Sungun	46.70E 38.70 N	Porphyry	Cu-Mo	Cu: 5.1 Mt.; Mo: 0.2 Mt	Cu: 0.6%; Mo: 0.024%	Monzonite–quartz monzonite	Potassic, phyllic, propylitic argillic, skarn	Magnetite, chalcopyrite, molybdenite	LA-ICP-MS zircon U-Pb: 21.13 ± 0.23 to 19.28 ± 0.33	No data	Aghazadeh et al. (2015a)
	Masjed Daghi	45.94E 38.87 N	Porphyry	Cu-Mo	Cu: 0.9 Mt.; Mo: 20400 t	Cu: 0.27%; Mo: 0.006%	Monzonite	Potassic, phyllic, propylitic argillic	Chalcopyrite, molybdenite, pyrite	No data	Molybdenite Re-Os: 20.46 ± 3.55	Aghazadeh et al. (2015a)
	Meiduk	55.17E 30.33 N	Porphyry	Cu-Mo	Cu: 1.07 Mt.; Mo: 12,000 t	Cu: 0.86%; Mo: 0.007%	Quartz–monzodiorite, granodiorite	Potassic, phyllic, argillic, propylitic	Chalcopyrite, pyrite, molybdenite, bornite	LA-ICP-MS zircon U-Pb: 12.5 ± 0.1,11.58 ± 0.21	Biotite $^{40}Ar/^{39}Ar$: 11.2 ± 0.5; Molybdenite Re-Os: 12.23 ± 0.07	Aghazadeh et al. (2015a); Asadi (2018)
	Sar-cheshmeh	55.91E 29.93 N	Porphyry	Cu-Mo	Cu: 8.9 Mt.; Mo: 0.46 Mt	Cu: 1.2%; Mo: 0.03%; Au: 0.27 g/t	Granodiorite	Potassic, phyllic, argillic, propylitic	Chalcopyrite, molybdenite, pyrite,pyrrhotite	LA-ICP-MS zircon U-Pb: 13.6 ± 0.1, 12.97 ± 0.24 to 12.15 ± 0.19	Biotite $^{40}Ar/^{39}Ar$: 11.8 ± 0.6; Biotite$^{87}Rb/^{86}Sr$: 12.2 ± 1.2; Molybdenite Re-Os: 13.2 ± 0.2 to 13.4 ± 0.19	Aghazadeh et al. (2015a); Asadi (2018); Zürcher et al. (2019)
	Darreh Zar	55.88E 29.85 N	Porphyry	Cu-Mo	Cu: 1.07 Mt.; Mo: 11,000 t	Cu: 0.36%; Mo: 0.0052%	Diorite, granodiorite	Potassic, phyllic, argilli, propylitic	Chalcopyrite, pyrite, molybdenite	LA-ICP-MS zircon U-Pb: 15.71 ± 0.23, 16.77 ± 0.18 and 16.40 ± 0.22	Molybdenite Re-Os: 15.66 ± 0.23	Aghazadeh et al. (2015a); Asadi (2018); Zürcher et al. (2019)
	Chah Firouzeh	55.03E 30.40 N	Porphyry	Cu	Cu: >1 Mt	Cu: 0.35%	Quartz–diorite, tonalite	Potassic, phyllic, argillic, propylitic	Chalcopyrite, pyrite, bornite	LA-ICP-MS zircon U-Pb: 16.19 ± 0.23, 16.91 ± 0.37	Molybdenite Re-Os: 16.19 ± 0.23	Aghazadeh et al. (2015a); Asadi (2018)
	Saindak	61.62E 29.25 N	Porphyry	Cu-Au-Mo	Cu: 1.8 Mt.; Mo: 9000 t; Au: 200 t	Cu: 0.41%; Mo: 0.002%; Au: 0.5 g/t	Granodiorite, tonalite and quartz monzonite	K-silicate, weak quartz-sericite	Chalcopyrite, pyrite, magnetite	LA-ICP-MS zircon U-Pb: 22.2	Hydrothermal biotite K-Ar: 22.4 ± 0.4	Perelló et al. (2008); Zürcher et al. (2019)
	Reko Diq (H14-H15)	62.05E 29.12 N	Porphyry	Cu-Au	Cu: 20.5 Mt.; Au: 1189 t	Cu: 0.5%; Au: 0.29 g/t	Quartz diorite, granodiorite	Potassic, sericite-clay-chlorite, seritic, propylitic	Chalcopyrite, bornite	LA-ICP-MS zircon U-Pb:12.22	Hydrothermal biotite K-Ar: 12.0 ± 0.2	Perelló et al. (2008); Razique et al. (2014)
	Angouran	47.24E 36.59 N	MVT	Pb-Zn-Ag	Pb-Zn: >7.62 Mt	Pb: 1.9%; Zn: 40%	Schist marble	Skarnization, epidotization	Galena, sphalerite	Cenozoic	^{40}Ar-^{39}Ar: 20	Zhuang et al. (2019)
	Irankuh	51.51E 32.51 N	MVT	Pb-Zn	Pb-Zn: >2.7 Mt	No data	Limestone	Dolomitization, silicification	Sphalerite, galena	No data	No data	Liu et al. (2019)
	Mehdiabad	54.56E 31.37 N	MVT	Pb–Zn-Ag	Pb + Zn: 20.7 Mt	Zn: 4.2% Pb: 1.6%, Ag: 36 g/t	Limestone Shale	Dolomitization, silicification	Sphalerite, galena,	No data	No data	Maghfouri et al. (2019, 2020)
	Emarat	49.63E 33.84 N	MVT	Pb-Zn	Pb-Zn: 0.83 Mt	No data	Shale	Silicification, chloritization	Sphalerite, galena	No data	No data	Maghfouri et al. (2020)
	Robat Kuhkolangeh	51.88E 33.12 N	MVT	Pb-Zn	Pb-Zn: 0.9 Mt	Zn-Pb: 8.6%	Shale limestone	No data	Sphalerite, galena	No data	No data	Rajabi et al. (2012)

(continued)

Table 10.1 (continued)

District	Deposit	Long. lat.	Type	Metallic comm.	Tonnage	Grade	Host rock	Alteration	Mineralization	Igneous age (Ma)	Mineralization age (Ma)	Data source
Qinling	Jinduicheng	109.97E 34.34 N	Porphyry-skarn	Mo(-W-Au)	Mo: 1.08 Mt	Mo: 0.1%	Porphyritic monzogranite, porphyritic andesite, and minor andesite	Silica, K-feldspar, fluorite, chlorite, calcite	Molybdenite, pyrite, galena, bismuthinite, chalcopyrite, sphalerite	LA-ICP-MS zircon U-Pb: 146 ± 1	Molybdenite Re-Os: 140 ± 4; LA-ICP-MS zircon U-Pb: 141 ± 1	Mao et al. (2008, 2011b); Du et al. (2004); Zhu et al. (2008)
	Huanglongpu	111.03E 34.37 N	Hydrothermal vein	Mo-Pb(-W-Au)	Mo: 89,400 t; Pb: 0.19 Mt	Mo: 0.08–0.1%	Mesoproterozoic volcanic rocks of the Xiong'er group	Silica, sericite, chlorite, fluorite, epidote, biotite	Molybdenite, pyrite, galena	LA-ICP-MS zircon U-Pb: 228 ± 1, 238 ± 5, 217 ± 2, 215 ± 4	Molybdenite Re-Os: 221.5 ± 0.3, 222 ± 4	Du et al. (2004); Huang et al. (1994)
	Yechangping	110.73E 34.06 N	Porphyry-skarn	Mo(-W-Au)	Mo: 0.19 Mt.; WO_3: 0.15 Mt	Mo: 0.13%; W: 0.1%	Porphyritic K-feldspar granodiorite and skarn at the contact with dolomite of the Guandaokou group	Silica, calcite, chlorite	Magnetite, molybdenite, pyrite, scheelite	LA-ICP-MS zircon U-Pb: 158 ± 2, 145 ± 2	Molybdenite Re-Os: 144.89 ± 0.96, 145 ± 4	Mao et al. (2011a, b); Yan et al. (2012)
	Leimengou	111.92E 34.21 N	Porphyry-skarn	Mo(-W-Au)	Mo: 0.34 Mt	Mo: 0.07%	Monzogranite in contact with gneiss	Silica, calcite, chlorite, K-feldspar, biotite, sericite, fluorite	Molybdenite, pyrite, chalcopyrite, galena, scheelite	SHRIMP zircon U-Pb: 136 ± 2	Molybdenite Re-Os: 132 ± 2	Mao et al. (2008, 2011b); Li et al. (2006)
	Yuchiling	111.87E 32.96 N	Porphyry-skarn	Mo(-W-Au)	Mo: 0.55 Mt	Mo: 0.06%	Porphyritic monzogranite	Silica, calcite, chlorite, K-feldspar, biotite, sericite, fluorite, epidote	Molybdenite, pyrite, chalcopyrite, galena	LA-ICP-MS zircon U-Pb: 143 ± 2	Molybdenite Re-Os: 144 ± 5, 131 ± 1, 134 ± 1	Li et al. (2009); Zhou et al. (2009)
	Donggou	112.38E 33.93 N	Porphyry-skarn	Mo(-W-Au)	Mo: 0.63 Mt	Mo: 0.11%	Porphyritic monzogranite and andesite and rhyolite of the Xiong'er group	Silica, K-feldspar, fluorite, sericite, chlorite, epidote, calcite	Molybdenite, pyrite, galena, sphalerite, chalcopyrite, scheelite	SHRIMP zircon U-Pb: 112 ± 1	Molybdenite Re-Os: 116 ± 2	Ye (2006); Mao et al. (2011a)
	Nannihu-Sandao-Zhuang	111.49E 33.91 N	Porphyry-skarn	Mo-W(-Au)	Mo: 1.34 Mt.; WO_3: 0.64 Mt	Mo: 0.08–0.14%	Skarn at the contact with porphyritic monzogranite and carbonate in the Luanchuan group	Silica, K-feldspar, fluorite, sericite, chlorite, epidote	Molybdenite, pyrite, galena, sphalerite, chalcopyrite, scheelite	LA-ICP-MS zircon U-Pb: 147 ± 1	Molybdenite Re-Os: 142 ± 2.1, 146 ± 1	Mao et al. (2008, 2011b); Li et al. (2003)
	Wenyu	111.44E 34.43 N	Orogenic	Au	Au: 32.7 t	Au: 7.0 g/t	Metamorphic rocks of the Taihua group	Silica, sericite, chlorite, graphite, epidote, biotite, K-feldspar	Pyrite, galena, chalcopyrite, calaverite	No data	No data	Mao et al. (2002)

	Dongchuang	110.45E 34.43 N	Orogenic	Au	Au: 27.6 t	Au: 8.2 g/t	Neoarchean rocks of the Lujiayu formation	Silica, sericite, chlorite, graphite, epidote, biotite	Pyrite, galena, pyrrhotite, chalcopyrite, chalcocite, bornite, scheelite, wolframite, calaverite, petzite	No data	No data	Mao et al. (2002)
	Yangzhaiyu	110.54E 34.41 N	Orogenic	Au	Au: 34.4 t	Au: 11.4 g/t	Migmatite and hornblende-plagioclase gneiss of the Taihua group	Silica, K-feldspar, calcite, sericite, biotite, kaolin	Pyrite, galena, chalcopyrite, sphalerite, magnetite, scheelite, wolframite, silver, pyrrhotite, ilmenite, telluro-bismuthite	No data	No data	Mao et al. (2008)
	Dahu	110.61E 34.43 N	Orogenic	Au-Mo	Au: 38 t; Mo: 20,000 t	Au: 6.8 g/t; Mo: 0.23 g/t	Neoarchean biotite–Plagioclase gneiss	Silica, K-feldspar, calcite, sericite, biotite	Molybdenite, pyrite, galena, bismuthinite, chalcopyrite, gold, pyrrhotite, bornite	No data	Molybdenite Re-Os: 223 ± 2 to 233 ± 3, 218 ± 41	Mao et al. (2002, 2008, 2011a); Li et al. (2008)
	Linghu	110.64E 34.38 N	Orogenic	Au	Au:20 t	Au: 5.6 g/t	Neoarchean rocks of the Lujiayu formation	Silica, K-feldspar, calcite, sericite	Pyrite, chalcopyrite, galena, sphalerite, pyrrhotite, bornite, covellite, bismuthinite, arsenopyrite, magnetite	No data	No data	Mao et al. (2002)
	Shang-gong	111.56E 34.18 N	Orogenic	Au-Ag	Au: 29.1 t	Au: 6.9 g/t	Porphyritic andesite and andesite in the Proterozoic Xiong'er group	Silica, chlorite, sericite, fluorite, barium	Pyrite, galena, sphalerite, chalcopyrite, tetra, hedrite, scheelite, calaverite	No data	No data	Chen et al. (2008)
	Qiyugou	111.95E 34.23 N	Orogenic	Au	Au: 40 t	Au: 7.4 g/t	Volcanic breccia pipe	Silica, K-feldspar, calcite, chlorite, sericite	Pyrite, chalcopyrite, galena, bismuthinite, native gold, chalcocite	LA-ICP-MS zircon U-Pb: 134 ± 2	Molybdenite Re-Os: 136 ± 6	Mao et al. (2002)
	Huaishuping	111.93E 34.03 N	Orogenic	Au	Au: 32 t	Au: 5.5 g/t	Dacite and andesite of the Xiong'er group	Silica, chlorite, K-feldspar, sericite, epidote, kaolin, graphite	Pyrite, sphalerite, galena	No data	Molybdenite Re-Os: 202 ± 8	Wang et al. (2021)

lion tons with an average of 0.69% Cu, 0.03% Mo, 0.06 g/t Au, and is presently the largest producer of Cu and Mo in Iran (Shafiei and Shahabpour 2012). This deposit is situated within a folded and faulted Tertiary igneous–sedimentary complex which consists of Eocene volcano–sedimentary successions, Oligo-Miocene intrusions and dikes, and Mio-Pliocene subvolcanic and volcanic rocks. The Sar Cheshmeh ore body is oval shaped, about 2000 m long by 1000 m wide with a northeast–southwest trend and is centered on the Sar Cheshmeh porphyry within Eocene fine-grained porphyritic andesite (Fig. 10.5). Granodiorite–quartz monzonite porphyry is the main component of Sar Cheshmeh porphyry and yields zircon U–Pb ages of 13.6 ± 0.1 to 13.0 ± 0.2 Ma (McInnes et al. 2003; Aghazadeh et al. 2015a). The late fine-grained porphyry with zircon U–Pb age of 12.4 ± 0.2 Ma crops out in the eastern part of the mine (Aghazadeh et al. 2015a). All porphyries and andesitic rocks were cut by a series of intra- to post-mineralization dikes (Fig. 10.5), including early hornblende porphyry, late hornblende porphyry, feldspar porphyry and biotite porphyry, which have variable composition and age (from 12.2 ± 0.2 to 3.4 ± 1 Ma, K–Ar on biotite and U–Pb on zircon; Haynes and Ghorashi-Zadeh 1978; McInnes et al. 2003; Aghazadeh et al. 2015a). Molybdenite Re–Os dating yielded an age of 13.4 ± 0.19 to 13.2 ± 0.2 Ma of the deposit (Aghazadeh et al. 2015a). The granodiorite in Sar Cheshmeh deposit with positive εHf(t) values (+6.3 to +10.6) indicate that the primary magma was likely originated from a juvenile lower crust with negligible sedimentary input (Asadi 2018). The highest values of Mo and precious metals are concentrated into Cu-rich potassic core of the deposit and also associated with strong phyllic ores which are developed in upper parts of the porphyry system (Shafiei and Shahabpour 2012).

The SSZ in Iran hosts the most economically significant MVT Pb–Zn deposits, such as Angouran and Irankuh. The Pb–Zn mineralization in this area is restricted to Early Cretaceous limestones, dolomites, or along contacts between Cretaceous shales and limestones (Rajabi et al. 2012). Apart from the SSZ, MVT Pb–Zn deposits can also be found in other units within the Zagros collisional orogen, such as Anjireh–Tiran in the Zagros fold and thrust belt (Rajabi et al. 2012).

10.3.2 Himalayan Orogen

PCDs in the Himalayan orogen are mainly distributed along within the southern margin of the Lhasa Block in southern Tibet and along the Jinshajiang–Ailaoshan tectonic belt (Fig. 10.1). The Qulong, Jiama, Chongjiang, and Tinggong PCDs have been discovered in the Gangdese porphyry Cu belt, which are related to the Miocene post–collisional magmatic suites. The mid–Miocene porphyry Cu–Mo mineralization yields molybdenite Re–Os ages between 17 and 14 Ma hosted by monzogranitic and quartz monzonitic stocks with zircon U–Pb ages ranging from 26 to 13 Ma (Hou and Zhang 2015). The Jinshajiang–Ailaoshan tectonic belt contains the Yulong, Malasongduo, Duoxiasongdui and Narigongma PCDs, which are associated with the Miocene post–collisional suites constituted by numerous

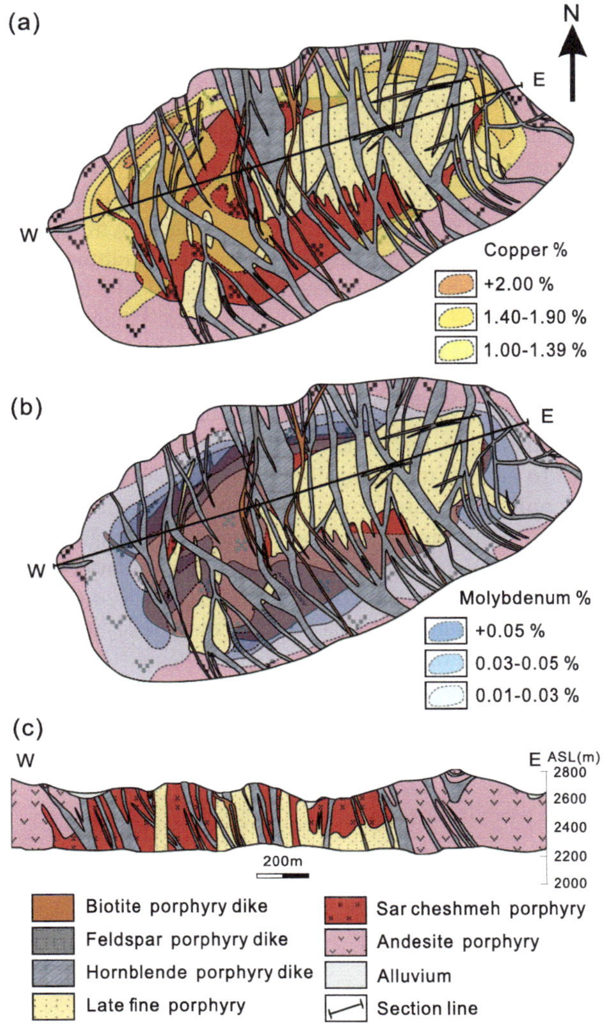

Fig. 10.5 (**a–b**) Maps showing distribution of hypogene Cu and Mo grades within 0.40% Cu cutoff at 2400 m elevation; and (**c**) East–west cross section of the Sar Cheshmeh deposit (after Shafiei and Shahabpour 2012)

small–volume monzogranite stocks (Hou and Cook 2009). Geochronological data suggest that the magmatism occurred between 43 and 33 Ma, broadly coincident with 40–32 Ma mineralization (Yang et al. 2014a, b; Hou and Zhang 2015). The giant Yulong porphyry Cu-Mo deposit, located in eastern Tibet, is the third largest porphyry–type Cu-Mo deposit in China, containing a resource of over 1.0 billion tons with an average grade of 0.62% Cu, 0.042% Mo and 0.05 g/t Au (Yang and Cooke 2019). The area in and around the Yulong deposit includes multiphase intrusions intruding Triassic carbonate and clastic sedimentary sequences and has relatively shallow emplacement depths (Fig. 10.6a; Hou et al. 2003a). The intrusive body is controlled by several ring-shaped, steeply dipping faults and the southern end of the N- to NW-trending Hengxingcuo anticline (Fig. 10.6a). The orebody in Yulong deposit is steeply dipping, 1000 m long, 600 m wide and at least 500 m deep, and the highest Cu-Mo grades are hosted in ring–shaped orebody surrounding the pipelike orebody (Fig. 10.6b, c). Hydrothermal alteration at Yulong includes potassic, phyllic, argillic, propylitic, and calc-silicate assemblages. The potassic core has been overprinted by intense phyllic and argillic alteration assemblages. The porphyritic monzogranite hosts the majority of Cu and Mo mineralization of the deposit yielding zircon U–Pb dating of ca. 43–41 Ma (Chang et al. 2017; Huang et al. 2019; Chen et al. 2023).Zircon Hf isotopes from the ore-bearing porphyries has positive εHf(t) values (+0.5 to +8), indicating that the primary magma was likely originated from a juvenile lower crust (Huang et al. 2019). Microtextural petrography, fluid inclusion microthermometry and LA–ICP–MS microanalysis demonstrate that Mo and Cu mineralization were introduced by single-phase intermediate-density magmatic–hydrothermal fluids exsolved from evolved magma reservoirs in the underlying magma chamber (Chang et al. 2018). The bulk of the Cu-Mo mineralization ages constrained by high-precision molybdenite Re–Os dating range from ca. 42.28 to 40.92 Ma (Chang et al. 2017). Ti diffusion modeling for different quartz generations indicate that multiple magmatic-hydrothermal pulses at Yulong formed in a short period of 880,000 to 3500 years (Chen et al. 2021).

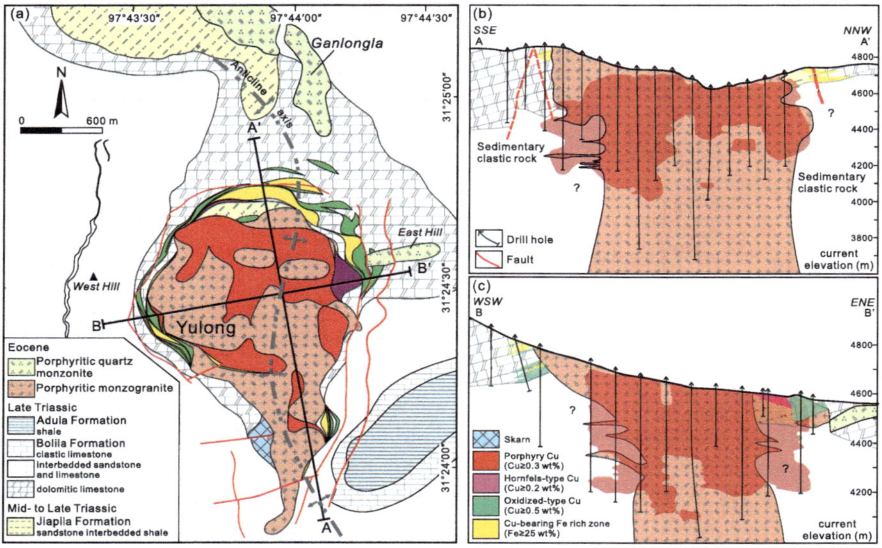

Fig. 10.6 Simplified geological maps of: (**a**) Yulong deposit in the Himalayan orogen; and (**b–c**) section A–B showing distribution and zonation of hydrothermal alteration and mineralization (after Chang et al. 2017)

MVT Pb–Zn deposits in the Himalayan orogen are mainly distributed in northeast Tibet, such as Jinding, Chaqupacha, and Dongmozhazhua (Fig. 10.1; Table 10.1). These deposits are hosted in dolomitized limestones or sandstones within slices of Mesozoic rocks, which are controlled by Cenozoic fault systems associated with the India-Asian collision (Hou and Cook 2009). There are also two main stages of Pb–Zn mineralization during the Eocene and late Oligocene–early Miocene constrained by sphalerite Rb–Sr isochron age of 35.0 ± 0.3 Ma for the Dongmozhazhua deposit (Tian et al. 2014), calcite Sm–Nd and sphalerite Rb–Sr isochron ages of 29.9 ± 1.1, 28.9 ± 0.6, and 27.4 ± 1.8 Ma for the Baiyangping deposit (He et al. 2004; Wang et al. 2015a; Zou et al. 2015), and Paleomagnetic age of 23 ± 3 Ma for the Jinding deposit (Yalikun et al. 2018). The Jinding deposit is regarded as the first largest MVT deposit in China, structurally controlled by a thrust–diapiric dome system (Leach et al. 2017). The MVT Pb–Zn mineralization is hosted by evaporite diapir-related rocks, such as diapir breccias, diapir-related breccia-bearing sandstones, and dilatancy zones in thrust belts as well as footwall sandstone (Song et al. 2020; Fig. 10.7). The ore types have disseminated breccia, and vein-like ores. The ore mineral assemblages consist of sphalerite and galena. The gangue minerals are pyrite, marcasite, barite, calcite, celestine, gypsum, anhydrite and bitumen. Metal minerals are derived from metal-rich evaporite formation and basement rocks (Deng et al. 2017, 2022; Wang et al. 2018). Metal minerals precipitated when metal-rich brines mixed H_2S–rich hydrocarbon-bearing fluid, which was produced by bacteria-assisted reaction of hydrocarbons with gypsum-bearing evaporites (Mu et al. 2021).

Gold deposits in the Himalayan orogen are mainly distributed along the large strike–slip Ailaoshan Red–River shear zone and the Indus–Yarlung Tsangpo suture zone. The Mayum, Bangbu and Nianzha gold deposits have been discovered in the Indus–Yarlung.

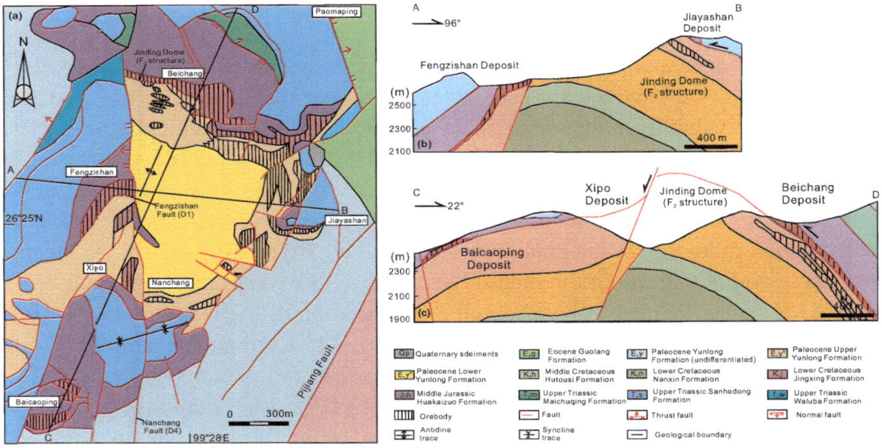

Fig. 10.7 Simplified geological maps of: (**a**) Jinding deposit in the Himalayan orogen; and (**b**) section A–B and C–D (after Wang et al. 2018)

Tsangpo suture zone, which are characterized by mesozonal Au lode (Fig. 10.1; Sun et al. 2016). Recent high precision $^{40}Ar/^{39}Ar$ dating yielded a limited mineralization age in the range of 44 to 59 Ma constrained by sericite isochron ages of 44 ± 3, 44.8 ± 1.0, and 49.52 ± 0.52 Ma for the Bangbu deposit (Pei et al. 2016; Sun et al. 2016), and sericite isochron ages of 44.08 ± 0.39 and 59.34 ± 0.62 Ma for the Mayum deposit (Wen et al. 2006; Jiang et al. 2009), which are broadly coincident with the timing of the India-Asia collision (Hou and Cook 2009).

10.3.3 Qinling Orogen

PMDs associated with magmatic rocks formed during the Late Jurassic to Early Cretaceous in the Qinling orogen (Table 10.1; Fig. 10.1). The PMDs predominate in the Qinling orogen and yield molybdenite Re–Os ages between 156 and 116 Ma hosted by porphyritic monzogranite with zircon U–Pb ages ranging from 156 to 112 Ma (Li et al. 2003, 2006, 2009; Mao et al. 2008; Wang et al. 2008; Mai et al. 2011). The Late Jurassic porphyry Cu–Mo mineralization is minor in the Qinling orogen, and yields a molybdenite Re–Os age of 148 ± 2 Ma hosted by porphyritic quartz diorite with a zircon U–Pb age of 145 ± 1 Ma (Xie et al. 2012; Ren et al. 2014). The Qinling orogen is the second most endowed auriferous region in China, with world–class gold deposits. Although the genesis of the Au mineralization in this region has been debated, the deposits have been either classified as orogenic or intrusion–related types (Mao et al. 2002; Chen et al. 2008; Wang et al. 2013a, 2017, 2021).

10.4 Methodology

In order to produce a dataset that is compatible with the previously published εHf–isotope studies, we have used a consistent method to recalculate all the data and took a median value from the range of Hf isotopic values of individual sample. The chondritic data of $^{176}Lu/^{177}Hf = 0.0336$ and $^{176}Hf/^{177}Hf = 0.282785$ were used to calculate the εHf (t) and fLu/Hf values (Griffin et al. 2000; Bouvier et al. 2008). Contour maps can be produced using the kriging and inverse distance weighted interpolation methods (Wang et al. 2016; Deng et al. 2018). We employ the inverse distance weighted interpolation method using ArcGIS to contour the Hf dataset. In order to produce the most robust spatial representation of the isotopic dataset, this method used 12 nearest neighbours at a 'power' of 2 following the examples. Using the median for a range of Hf isotopic values from an individual sample helped to exclude obvious outlier. Probability density curves (Fig. 10.8) demonstrate that the interpolation method is reliable and representative of the full information contained within the dataset.

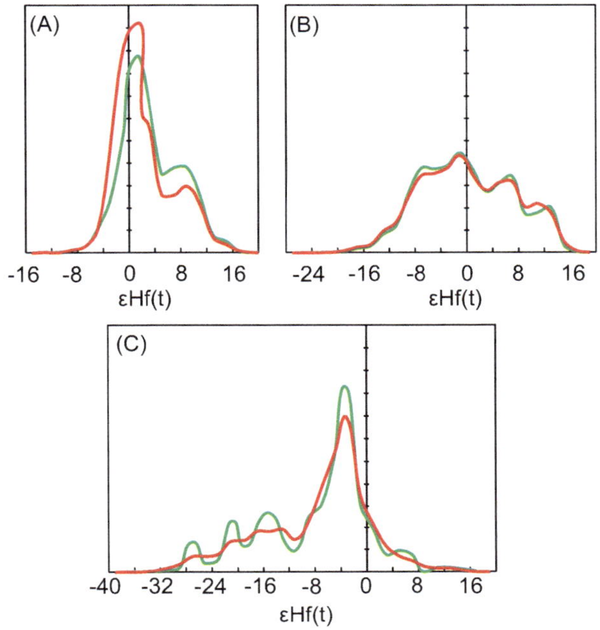

Fig. 10.8 Probability density curves for Hf values in the: (**a**) Zargos, (**b**) Himalaya, and (**c**) Qinling. The green curve represents the median Hf for discrete temporal groups, whereas the red curve represents all of the individual grain analyses. Data are from this study and literature (see Supplementary Table S1)

Here we consider that "juvenile" Hf isotopic features ($_{\varepsilon}$Hf > 0) as suggesting derivation from a depleted mantle source. In contrast, "old" and "reworked" Hf isotopic features ($_{\varepsilon}$Hf < 0) suggest reworking of pre–existing crust by partial melting or processes of erosion and sedimentation (Belousova et al. 2010; Hawkesworth et al. 2010). In order to estimate the contribution of magmas to crustal growth (R1), we have taken into account a ratio (R1) of the whole current area of juvenile domains to the study area. We can calculate a ratio (R2) of pre–collisional arc magmas to crustal growth. Finally, the ratio of post–collisional magmas to crustal growth (R3) is equal to R1–R2.

10.5 Result

The lithospheric architecture and evolution processes of orogens can be revealed by Hf isotopic mapping using igneous rock–hosted zircons (Hou et al. 2015a; Wang et al. 2015b; Wang et al. 2016). We integrate geologic and geochemical data for Mesozoic–Cenozoic igneous rocks from the three orogens, including 20,000 zircon Lu–Hf and U–Pb isotope analyses from 1212 igneous samples and 574 bulk–rock

Sr–Nd isotope analyses (Supplementary Tables S1 and S2; Figs. 10.9, 10.10, 10.11, 10.12).

The three collisional orogens show distinct Hf isotope evolution arrays (Fig. 10.13a). The.

Qinling orogen is characterized by progressively more negative $_{\varepsilon}$Hf after collision at 230 Ma, indicating increasing reworking of existing crust and decreasing addition of juvenile components, similar to secular trend of Eo– to Paleo–Archean TTG crust in Greenland and Pilbara Craton (Næraa et al. 2012; Gardiner et al. 2017). By contrast, the Zagros and Himalayan orogens show converging trends of $_{\varepsilon}$Hf after collision with downward and upwards shifts in maximum and minimum $_{\varepsilon}$Hf, reflecting decreased reworking of ancient crust and homogenizing, similar to trend of Meso–Archean TTG crust in Greenland (Næraa et al. 2012), Paleoproterozoic magmatism in North Australian Craton (Laccheri et al. 2018), and Paleozoic magmatism in peri–Laurentia (Henderson et al. 2018). These distinct trends reflect fundamentally different geodynamic processes between the Qinling and the other two orogens (Fig. 10.13a).

The zircon εHf contour maps are used to image lithospheric architectures of studied orogens, which demonstrate that the three orogens show differing lithospheric architectures through time (Figs. 10.10, 10.11 and 10.14). The Zagros orogen is dominated by positive zircon εHf values with a peak at +2 (Figs. 10.9 and 10.14a). High–εHf regions (εHf > 0) generally overlap with the Jurassic and Eocene UDMA arcs, whereas low–$_{\varepsilon}$Hf regions ($_{\varepsilon}$Hf < 0) mostly cover the central Iran block with a few isolated low–$_{\varepsilon}$Hf patches distributed along the Jurassic arc (Fig. 10.14a). The larger $_{\varepsilon}$Hf regions (<36 Ma, Fig. 10.11a) than those on the pre–collisional time slice (>36 Ma, Fig. 10.10a) reflect more widespread generation of juvenile materials since collision in Zagros (Chiu et al. 2017).

The Himalayan orogen is characterized by a much wider range of zircon $_{\varepsilon}$Hf values (c. −17 to +19, Fig. 10.13b). There are at least four high–$_{\varepsilon}$Hf (>0) regions, surrounded by large areas of low–$_{\varepsilon}$Hf (<0) domains (Fig. 10.14b). The high–$_{\varepsilon}$Hf regions occur dominantly along terrane boundaries, and correspond to the three volcanoplutonic arcs (Fig. 10.1). The orogen is dominated by low–$_{\varepsilon}$Hf domains, which correspond to the reworked old crust of the Lhasa, Qiangtang and Songpan terranes (Fig. 10.14b). The collisional time–slice (<65 Ma, Fig. 10.11b) generally shows similar features to pre–collisional one (Fig. 10.10b), suggesting similar lithospheric architecture.

The Qinling orogen is dominated by widespread, negative, low–$_{\varepsilon}$Hf values with peaks at −4 and − 16 (Figs. 10.9 and 10.14c). Five isolated small positive $_{\varepsilon}$Hf domains and a large–area of negative $_{\varepsilon}$Hf region occur in the southern and northern parts of the orogen, respectively, reflecting different crustal components and reworking. The pre–collisional time–slice is characterized by a few patches of high–$_{\varepsilon}$Hf values embedded within a large area of negative $_{\varepsilon}$Hf values (Fig. 10.10c), whereas the collisional time–slice is dominated by more negative $_{\varepsilon}$Hf values (Fig. 10.11c), indicating substantial decrease of juvenile components since collision.

According to the surface distribution of high–$_{\varepsilon}$Hf (>0) regions, we estimate the proportion of the juvenile materials to be the most in Zagros, less in Himalaya, and the least in Qinling (Figs. 10.10, 10.14). Considering potential ponding of

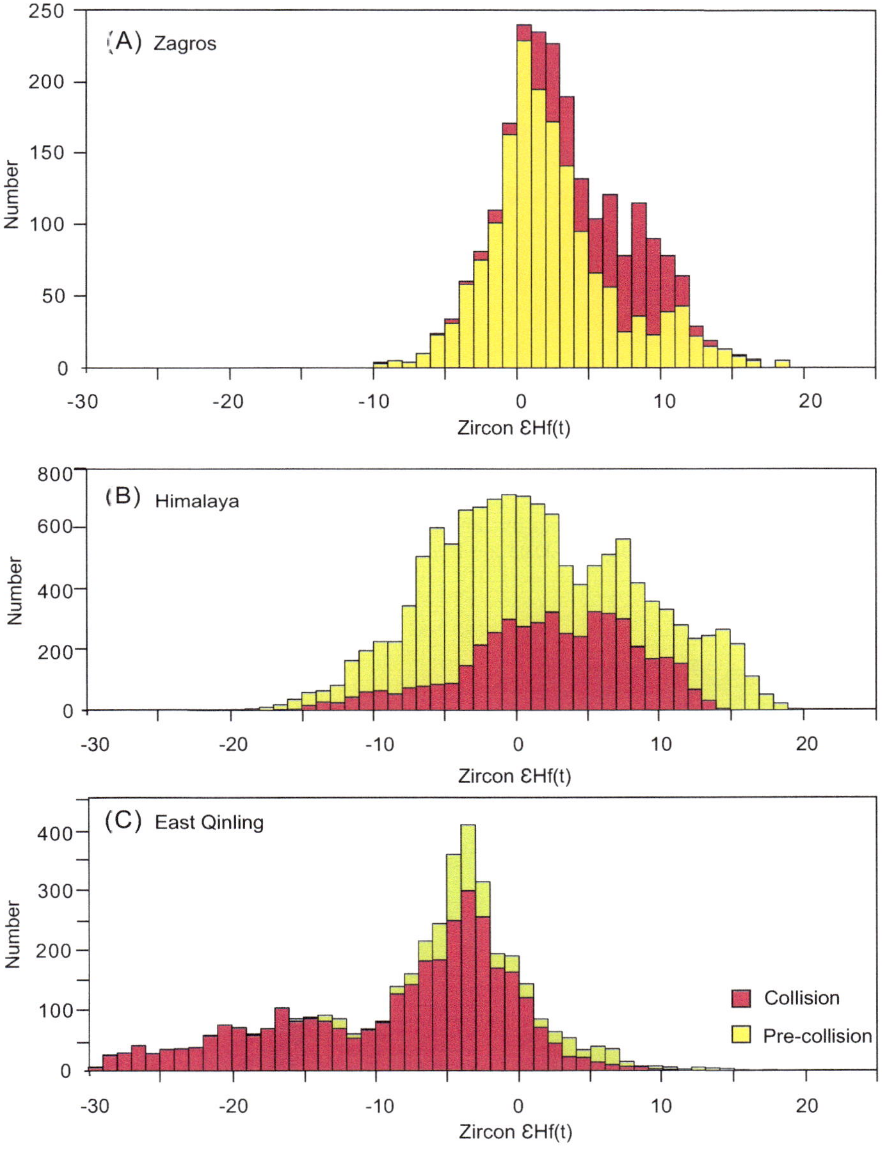

Fig. 10.9 Histograms of εHf values of magmatic zircons from the Mesozoic–Cenozoic granitoid rocks and felsic volcanic rocks in Zagros (**a**), Himalaya (**b**), and Qinling (**c**). Data are from this study and literature (see Supplementary Table S1)

mantle–derived mafic magmas at deep crust, these estimates based on exposed rocks represent minimum proportions of juvenile crust in these orogens. Such estimates are also similar to the minimum contribution of mantle components to continental crust in these orogens (Fig. 10.12), derived by a binary mixing calculation

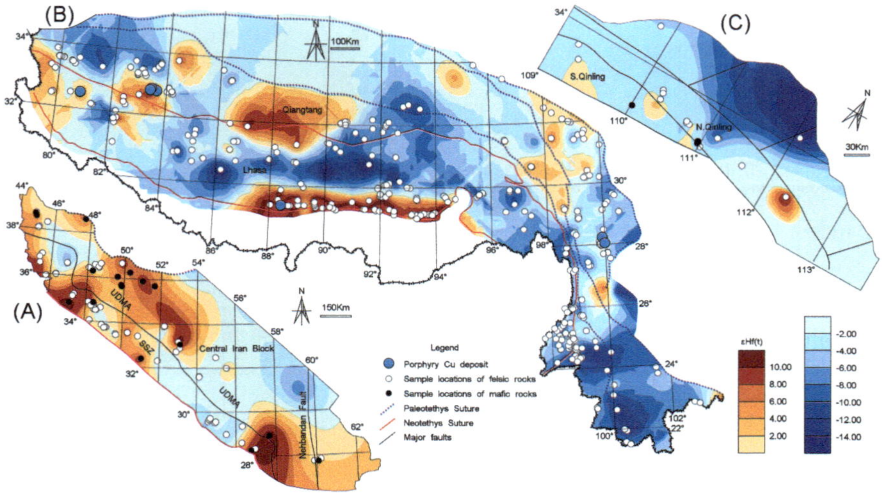

Fig. 10.10 Pre–collisional time slices for lithospheric architecture of the collisional orogens in Iran (**a**), Tibet (**b**) and Qinling (**c**). Data are from this study and literature (see Supplementary Table S1)

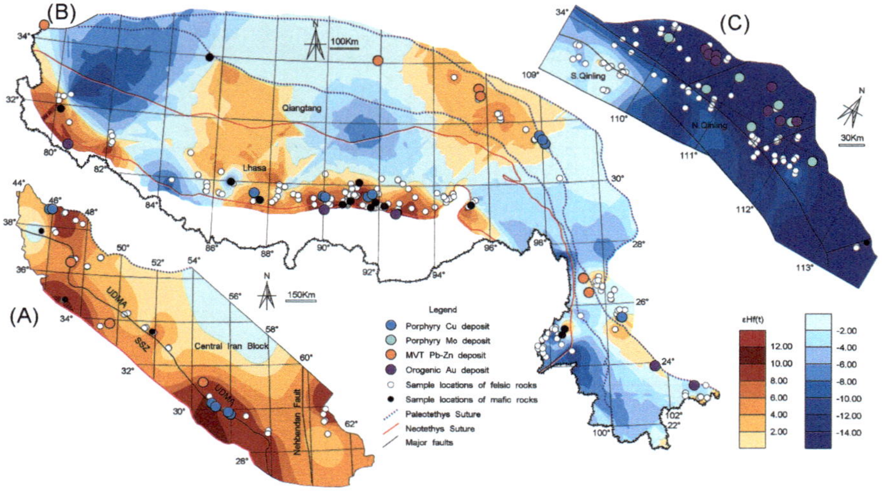

Fig. 10.11 Collisional time slices for lithospheric architecture of the collisional orogens in Iran (**a**), Tibet (**b**) and Central China (**c**). Data are from this study and literature (see Supplementary Table S1)

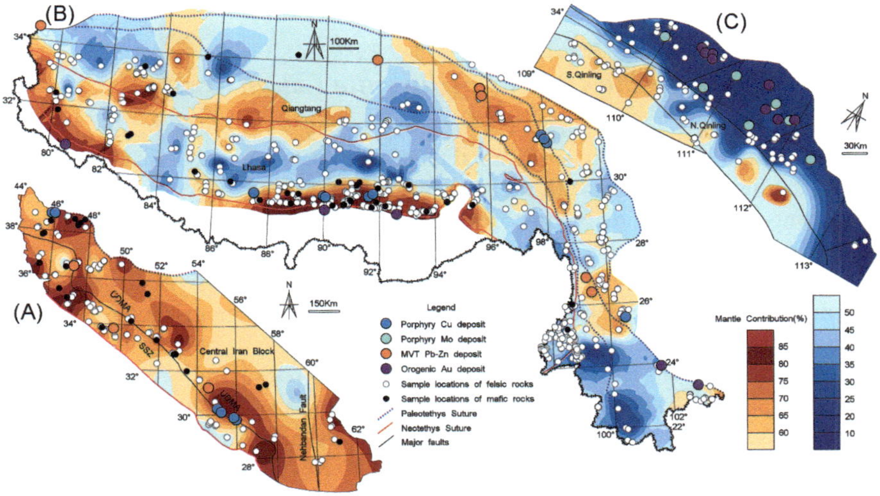

Fig. 10.12 Contour map of the quantitative proportions of the juvenile mantle components adding to the sources for the Mesozoic–Cenozoic granitoid and felsic volcanic rocks in the Zagros (**a**), Himalaya (**b**) and Qinling (**c**). The percentage of mantle component in the felsic melt was calculated following Mišković and Schaltegger (2009), and represents the maximum mantle contribution. Zircons from a Jurassic basaltic andesite sample (sample YH06–3) with $_{\varepsilon}$Hf = 18.8 and Hf = 8811 ppm are assumed to represent a depleted mantle–derived magma (Zhu et al. 2011). The zircons from a 194 Ma strongly peraluminous granite with εHf of −20.5 and Hf content of 12,761 ppm are assumed to represent compositions of continental crust of the Lhasa terrane (Liu et al. 2006). The results calculated for mixing between mantle and crustal components are listed in Supplementary Table S1. All sample points are given in Supplementary Table S1

assuming an old crust with $_{\varepsilon}$Hf = −20.5 and a juvenile component with $_{\varepsilon}$Hf = +18.8 (Figs. 10.10, 10.14; Hou et al. 2015a). We thus suggest that the amount of juvenile materials involved in collisional orogens far exceeds previous estimates (Kerrich et al. 2005), but gradually decreases from growing (Zagros and Himalaya) to decayed orogen (Qinling).

It is estimated that the contribution of juvenile domains (including subduction and collision) to the current area of orogens is 73.9% in Zagro, 28.2% in Himalaya, and 6.5% in Qinling (Fig. 10.14), the contribution of pre-collisional arc magmas to crustal growth is 60.7% in Zagro, 21.3% in Himalaya, and 5.8% in Qinling (Fig. 10.10), thus the contribution of syn-collisional magmas to crustal growth is 13.2% in Zagro, 6.9% in Himalaya, and 0.7% in Qinling.

10.6 Discussion

10.6.1 Pre–collision Genesis of Juvenile Lower Crust

The Hf contour maps for pre–collisional magmatic rocks show that juvenile crustal domains coincide spatially with the volcanoplutonic arcs in Zagros and Himalaya, supporting new crust generation from arc magmatism (Fig. 10.10; Hawkesworth et al. 2010).However, these Hf samples are from exposed felsic magmatic rocks, which only reflect crustal growth at their emplacement levels in shallow crust. Several lines of evidence suggest that pre–collisional arc magmatism also produced juvenile lower crust by mafic underplating.

Firstly, there is an exposed lower crustal section consisting of meta–gabbros (garnet amphibolites) with minor cumulate hornblendites with U–Pb ages of 200 Ma to 85 Ma (Xu et al. 2019; Zhang et al. 2020)in the juvenile domain around the eastern Himalayan syntaxis (Fig. 10.14b). The ca. 200 Ma Cuijiu igneous complex in the eastern Gangdese arc includes a wide variety of rocks from ultramafic–mafic and felsic cumulates to mafic to felsic non-cumulate plutonic rocks. The cumulate hornblende gabbros have a wide zircon εHf(t) values from +0.37 to +9.5 and the positive bulk-rock εNd(t) from +3.41 to +3.96 (Xu et al. 2019). These isotopic data combine with mineral and bulk-rock composition, showing that the subduction-related magmatic arc might be derived from mantle-derived hydrous basaltic magmas that underwent high- to intermediate-pressure fractional crystallization (<10 kbar) and crustal assimilation (Xu et al. 2019). In addition, the voluminous migmatitic garnet amphibolites (meta-gabbro) in the eastern Gangdese arc (ca. 95–82 Ma) forming as the major component of the Gangdese lower crust, have the positive and high zircon εHf(t) values ranging from +14.1 to +10.6. These isotopic compositions also indicate that the Gangdese arc magmas contain a significant mantle component and represent a juvenile continental crust with variable input of ancient crustal material (Zhang et al. 2020). These ultramafic–mafic cumulates, derived from metasomatized asthenospheric mantle evidenced by their arc–like signatures, are direct evidence for the presence of mafic rocks underplated in the arc roots of Himalaya. In addition, there are potassium volcanic rocks and ore bearing porphyry (35–33 Ma) in the eastern part of the Tibet plateau, which contains lower crustal amphibolite and garnet amphibolite xenoliths.

These xenoliths are thought to represent cumulates or residues of Neoproterozoic arc magmas ponded at the base of arc that subsequently underwent high-pressure metamorphism ca. 738 Ma (Hou et al. 2017). They have relatively high Y (>15 ppm) and low Sr/Y (< 30), and yield a wide zircon εHf(t) values of +0.7 to +14.8, which are consistent with the Neoproterozoic arc magmatism in the eastern part of the present-day Tibetan plateau, indicating that the earlier oceanic subduction triggered asthenosphere upwelling and induced large-scale underplating of mafic arc magmatism, leading to the formation of a mafic juvenile crust (Hou et al. 2017).

Secondly, the juvenile syn–collisional granite batholiths (55–50 Ma) with high $_\varepsilon$Hf and Nd in the Gangdese magmatic arc of the Lhasa terrane have been interpreted to be derived by remelting of pre–collisional Jurassic mafic arc roots (Chu et al. 2006). This is supported by the observation that the juvenile domain in the Lhasa terrane on the collisional time–slice spatially overlaps with that on the pre–collisional one (Figs.10.10, and 10.11).

Thirdly, many post–collisional Miocene felsic rocks in Zagros and Himalaya, characterized by arc–like signature, adakite affinity and juvenile Sr–Nd–Hf isotopic compositions, have been demonstrated to be derived from re–melting of low–crustal mafic underplated rocks formed during the pre–collisional subduction setting (Hou et al. 2004, 2015b; Shafiei et al. 2009).In Zagros, pre-collisional Eocene–OligoceneKerman arc segments are composed of diorites, quartzdiorites, granodiorites, and volcanic equivalents (Shafiei et al. 2009). The ore-hosting porphyries have high Sr (>400 ppm), low Y (<12 ppm) contents, and non-radiogenic $^{87}Sr/^{86}Sr$ ratios (0.7042–0.7047) relative to barren granitoids ($^{87}Sr/^{86}Sr = 0.7053–0.7068$), reflecting a thickening arc crust by recycling melting of the arc crustal keel (Shafiei et al. 2009). In Himalaya, the mid-Miocene Gangdese adakiticintrusives are characterized by relatively high K and Mg, highinitial $(^{87}Sr/^{86}Sr)_i$ between 0.7049 and 0.7079, and a wide range εNd(t) values of 6.18 to +5.52, indicating the adakitic magmas were formed by a complex mechanism involving partial melting of mafic materials in a thickened lower crust with input of enriched mantle and/or upper crust components (Hou et al. 2004). In addition, the Miocene (ca. 15 Ma) ore-forming porphyries have large isotopic variations with $(^{87}Sr/^{86}Sr)_i$ from 0.7048 to 0.7062, εNd(t) from −4.8 to +2.2, and zircon εHf(t) from +2 to +12, suggesting that remelting of the lower crustal sulfide-bearing Cu-rich Jurassic cumulates, triggered by Cenozoic crustal thickening and/or subsequent slab break-off, leading to the formation of giant Miocene porphyry Cu deposits (Hou et al. 2015b).

Fourthly, geophysical imaging identified a 14 to 20 km–thick high–velocity layer with Vp of 7.2 to 7.5 km/s and Vs of 6 km/s at the 60–km depth in Tibet (Kind et al. 1996), which has been interpreted as a high–density (>3.0 g/cm^3) garnet–bearing mafic rock (Nábelek et al. 2009), formed by underplating of mantle–derived magmas (Owens and Zandt 1997).

For the Tibetan plateau, the amount of juvenile crust (subduction + collision) accounts for about 28% of crustal growth, in which pre-collisional arc magmas contribute about 21% to crustal growth (Figs. 10.10 and 10.14), indicating 75% contribution from mantle arc magma underplating or emplacement. According to the deep crustal high-speed layer revealed by geophysical exploration (14 km) and 75% ratio, the low crust has thickened by10 km from contribution of pre-collisional arc magmas. The arc magma (La/Yb)n ratios reveal that the Gangdese crust in Tibet was likely of normal thickness prior to approximately 70 Ma (~37 km) but has been thickened locally to 47 km at approximately 70 Ma (Zhu et al. 2017), which is consistent with the estimate of 10 km thickening.

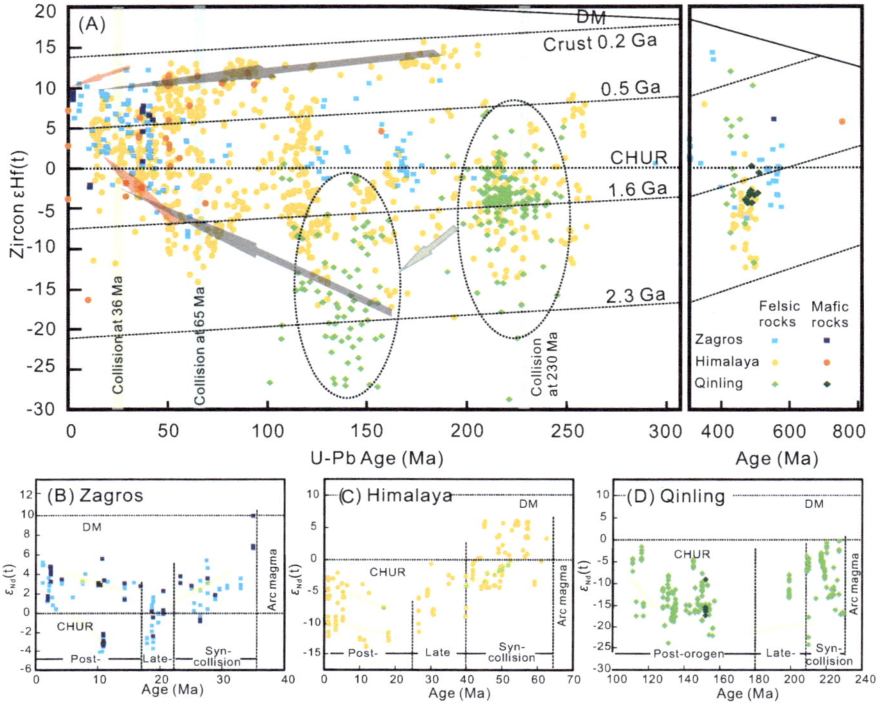

Fig. 10.13 Zircon hafnium (**a**) and whole–rock Nd isotopic compositions (**b–d**) of the Mesozoic–Cenozoic felsic and mafic igneous rocks from the collisional orogens in Iran (Zagros), Tibet (Himalaya) and Central China (Qinling). (**a**) DM represents a generalized 'Depleted Mantle' growth curve, whereas CHUR denotes the chondritic reference. The three crustal evolution lines labelled 0.5 Ga, 1.6 Ga and 2.3 Ga are presumably average continental crust with $^{176}Lu/^{177}Hf = 0.015$. The evolution line of mafic lower crust generated at 0.2 Ga with $^{176}Lu/^{177}Hf = 0.022$ was shown for comparison (Hou et al. 2015b). (**b–d**) The depleted mantle growth curves of Nd were shown, using linear depletion from $_{\varepsilon}Nd = 0$ at 4.56 Ga to $_{\varepsilon}Nd = 10$ at present. Data are from this study and literature (see Supplementary Tables S1 and S2 and references)

10.6.2 Formation and Preservation of Juvenile Crust during Collision

How and when collisional processes create juvenile crust and cause crustal growth remains unclear. However, there is increasing evidence that mantle–derived mafic magmas produced during collision contribute to the formation of collisional juvenile crust (Mo et al. 2007; Niu et al. 2013; Moyen et al. 2017; Hou et al. 2017, 2020).

Mantle–derived mafic magmas were produced continuously since the collision at Himalaya and Zagros orogens and their isotopic compositions vary systematically through time (Fig. 10.13). In Himalaya, the syn–collisional gabbros (55–50 Ma) and basalts (58–45 Ma) were emplaced along the collisional front (Mo et al. 2007;

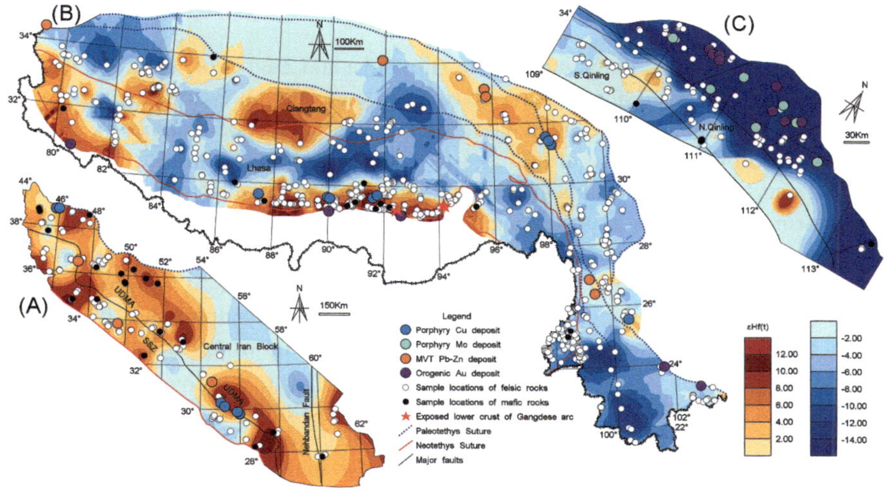

Fig. 10.14 Hafnium isotopic mapping showing lithospheric architecture of the collisional orogens of Zagros (**a**), Himalaya (**b**) and Qinling (**c**). The $_{\varepsilon}$Hf(t) data of igneous rocks were contoured using the geometric interval (GI) method in ArcGIS, where all groups are sized using a geometric series that highlights the differences between areas. The color bands indicate the ranges of zircon $_{\varepsilon}$Hf(t) values at the time of crystallization of the felsic igneous magmas. Data are from this study and literature (see Supplementary Tables S1 and S2 and references)

Ji et al. 2016) and late–collisional basaltic rocks and lamprophyres (41–26 Ma) were emplaced in the hinterland (Fig. 10.14b; Wang et al. 2010). The Eocene (ca. 45 Ma) gabbros in Himalaya exhibit similar geochemical characteristics with HIMU (high μ)–type oceanic island basalt, and have a narrow range of initial $(^{87}Sr/^{86}Sr)_i$ ratios (0.70312–0.70317) and positive εNd(t) values of +4.9 to +5.0, indicating that thegabbrosare formed by partial melting of the asthenosphere followed by rapid magma ascent with negligible crustal contamination (Ji et al. 2016). By contrast, both the syn– and late–collisional mafic magmas (36–17.5 Ma) overprinted the Eocene volcanoplutonic arc in Zagros(Fig. 10.14a; Verdel et al. 2011; Chiu et al. 2017). The Eocene-latest Miocene magmatic rocks, which were triggered in late–collisionaltectonic setting in Zagros, yield positive zircon εHf(t) values of +6 to +15, reflecting significant juvenile component (Chiu et al. 2017). These syn– to late– collisional mafic magmas in both orogens are dominantly potassic, calc–alkaline, and arc–like (Clift and Vannucchi 2004; Mo et al. 2007; Wang et al. 2010; Verdel et al. 2011; Ji et al. 2016), with minor Na–rich basalts showing OIB–like character (Ji et al. 2016), and yielded variable $_{\varepsilon}$Hf values (−4 to +12), and show decreasing $_{\varepsilon}$Nd trends since the collision (Fig. 10.13b,c). The syn–collisional mafic melts with high $_{\varepsilon}$Hf and $_{\varepsilon}$Nd were derived by partial melting of the upwelling asthenosphere induced by slab roll-back and subsequent slab break-off (Motavalli-Anbaran et al. 2011; Verdel et al. 2011; Ji et al. 2016). The late–collision led to further upwelling of the asthenosphere (Tilmann et al. 2003) and melting of SCLM

(Wang et al. 2010), which produced mafic magmas with low εHf and εNd (Fig. 10.13).

The post–collisional mafic magmas in Himalaya and Zagros occur as small–volume potassic–ultrapotassic and Na–rich basaltic–andesitic lavas emplaced in transtensional settings (Aghazadeh et al. 2015b; Chiu et al. 2017; Xu et al. 2017), and are characterized by large range of εHf (−4 to +10), and εNd (−14 to +5). In Zagros, their Hf and Nd isotopic arrays gradually contract towards more juvenile compositions during post–collision, reflecting shift of magmatic sources from subduction–modified SCLM to the upwelling asthenosphere (Fig. 10.13a,b; Verdel et al. 2011). However, post–collisional mafic magmas in Himalaya only exhibit a slight increase in $_{\varepsilon}$Hf and $_{\varepsilon}$Nd through time, indicating that their sources were dominated by SCLM (Xu et al. 2017) albeit with slightly increasing contribution from the asthenosphere (Fig. 10.13c).

Theabove observations indicate that continuous continental subduction from syn– to post–collisional stages drives upwelling of the asthenosphere and thermal erosion of SCLM to produce mafic magmas. These magmas either underplated at the lower continental crust and undergone MASH process to form hybrid andesitic melts (Hildreth and Moorbath 1988), differentiation of which eventually formed calc–alkaline felsic magmas emplaced into the upper crust (Richards 2009), or directly injected into molten felsic magmas to form microgranular mafic enclaves in granitoid batholiths (Mo et al. 2007), and both processes represent net growth of continental crust.

The syn– to post–collisional felsic magmas from Zagros and Himalaya have similar Hf and Nd isotopic compositions and evolution arrays to mafic magmas, which support that these felsic intrusions in the upper crust are genetically linked to mafic magmas through the MASH process, hence contribute to net crustal growth (Fig. 10.13). Moreover, compared with pre–collisional time–slices (Fig. 10.10), the collisional time–slices for Zagros and Himalaya exhibit larger area of high–$_{\varepsilon}$Hf domains spatially correlated with locations of mafic magmas (Fig. 10.11), consistent with contributions of the collision–related mantle– derived magmas to juvenile crust.

By contrast, there is a lack of asthenospheric mantle–derived mafic magma in Qinling, where the collisional mafic magmas are dominated by ultrapotassic lamprophyres derived by low–degree partial melting of SCLM (Fig. 10.13d; Wang et al. 2013b).The Qinlingultrapotassic lamprophyres are characterized by high Th/Hf (>1), and yield a negative zircon εNd(t) values of −0.5 to −3.3 and initial Sr isotope ratios from 0.7036 to 0.7058, indicating the generation of the lamprophyre by partial melting of a metasomatized garnet peridotite mantle source (Wang et al. 2006). The limited input of juvenile mantle melts in Qinling is likely related to minimum upwelling of the asthenosphere inhibited by thickened lithosphere during collision (Yuan 1996; Wang et al. 2013b).

The Hf isotopic time–slice maps for the three orogens also record crustal preservation and destruction through time. In Zagros, the pre–collisional juvenile crust was preserved during collision, whereas the ancient crusts in NW and central Iran have been completely replaced by juvenile mantle components during collision

(Figs. 10.10 and 10.11). In Himalaya, the collisional time–slice map is mostly similar to the pre–collisional one, suggesting limited destruction of the Asian lithospheric architecture during collision, and that pre–collisional juvenile crust is mostly preserved in the orogen (Figs. 10.10 and 10.11).

The lithospheric architecture in Qinling reflects the final status of this orogen after collisional collapse and later tectonism (Fig. 10.14c). Despite several pre–collisional high–$_{\varepsilon}$Hf domains (Fig. 10.10c), the collisional time–slice for Qinling only exhibit one high–εHf domain (Fig. 10.11c), suggesting that either the ancient crust there was strongly reworked by collision or the pre–collisional juvenile crust was partially lost due to later delamination. The delamination is supported by its thin crust (40 km) and lack of continental root (Yuan 1996) as well as a Jurassic–Cretaceous magma flare–up in Qinling (Dong and Santosh 2016). The collision at 230 Ma in Qinling only caused the thickening of continental crust rather than destruction of the former architecture, because the Triassic syn–collisional granitoids (228–215 Ma) show adakite affinity indicating thickened crust (Wang et al. 2015b) with inherited Hf isotopic characteristics of Paleozoic intrusions (Fig. 10.13a). However, significant modification of the Qinling lithospheric architecture occurred since post–collision, when widespread late–Jurassic granitoids were emplaced with negative $_{\varepsilon}$Hf values and absence of adakite affinity (Fig. 10.13a; Wang et al. 2015b). These observations suggest that the post–collisional granitiods tapped a more ancient crustal source within a thinned crust, which can be explained by pre–collisional juvenile mafic lower crust that was lost by delamination, and the overlying ancient crust was subsequently underthrust by continental crust from the South China Block.

For the Tibetan plateau, the amount of juvenile crust (subduction + collision) accounts for about 28% of crustal growth, in which syn-collisional magmas contribute about 7% to crustal growth (Figs. 10.10 and 10.14), indicating 25% contribution from syn–collisional felsic magmas. The syn–collisional felsic magma (La/Yb)n ratios reveal that the Gangdese crust in Tibet was thickened by 5 to 10 km at 65–45 Ma (Zhu et al. 2017). According to the deep crustal high-speed layer revealed by geophysical exploration (14 km) and 25% ratio, the lower crust has thickened by 4 km from syn–collisional felsic magmas.Together with the 5 km thickness of the Linzizong volcanic succession, it is estimated that the mantle derived mafic magma since the collision has thickened the crust of the southern Qinghai Tibetan Plateau by at least 9 km, including the contribution of the syn–collisional felsic magmas.

10.6.3 Evolution of Collisional Orogens

Continental collisional orogens usually experienced evolution from their growth to decay (Jamieson and Beaumont 2013). The collisional orogenesis is driven largely by subduction and accretion of materialat convergent margins, accompanied by shortening, thickening, and heating of deformed crust. By using numerical geodynamic models, the evolution of collisional orogens include small-cold orogens,

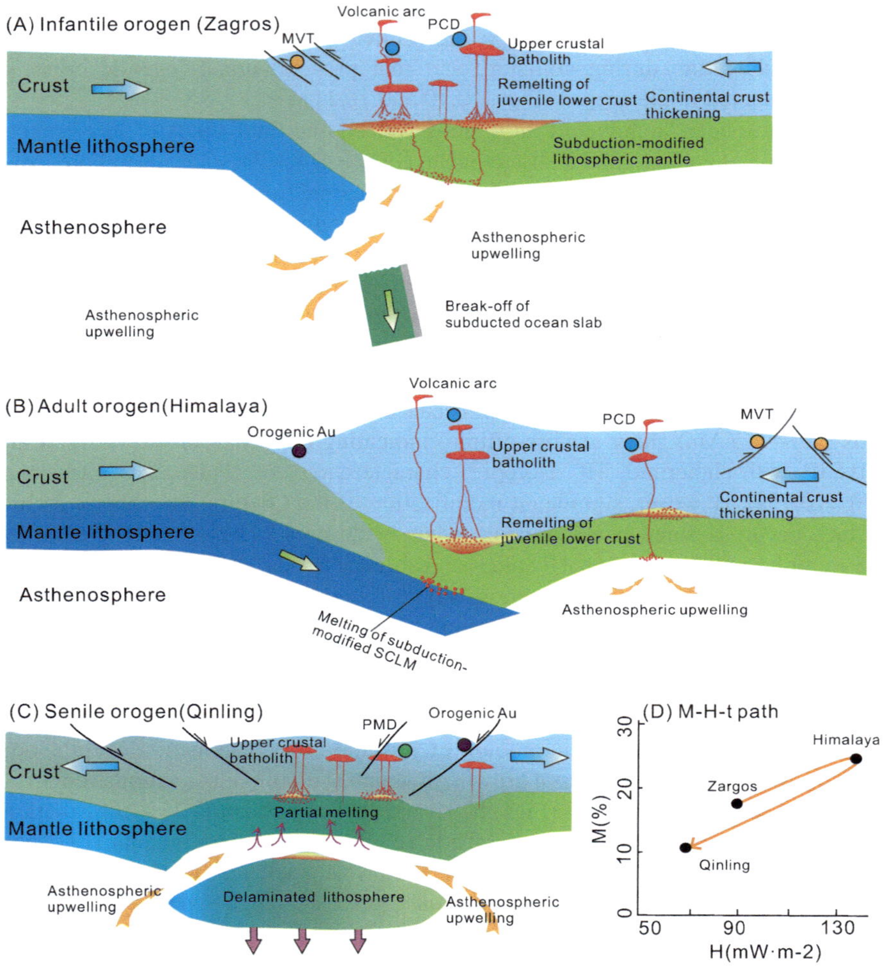

Fig. 10.15 Geodynamic evolution of collisional orogen and associated metallogeny in the context of terrestrial heat flow and juvenile crust generation during collision. (**a**) In infantile orogen (i.e. Zagros), breaking off of subducted oceanic–slab following continental collision triggered asthenospheric upwelling and melting of subduction–modified lithospheric mantle and juvenile mafic lower–crust, resulting in net growth of continental crust in upper plate. (**b**) in adult orogen (i.e. Himalaya), continuous underthrusting of a continental lithosphere caused fluid–fluxed re–melting of pre–collisional juvenile lower crust in collisional front and partial melting of subduction–modified lithospheric mantle in hinterland, triggered by fluid flux from underthrusting continental lithosphere and upwelling of the asthenosphere, respectively. (**c**) In senile orogen (i.e. Qinling), prolonged collision resulted in over–stacking of crustal slabs and shortening of the lithosphere, which eventually delaminated due to gravitational instability, leading to partial loss of the juvenile lower crust and anataxis of old continental crust. (**d**) M–H–t path for collisional orogeny. "M" here refers to the ratio of the amount of juvenile crust generated during collision to the total amount of juvenile crust in the orogen, which is 18% in Zagros, 25% in Himalaya and 11% in Qinling; "H" refers to terrestrial heat flow value; "t" refers to the time when orogen evolved

transition from oceanic subduction to continental collision, and the subsequent evolution of large-hot orogens (Jamieson and Beaumont 2013). Based on the above work on the three orogens, we propose that the evolution of collisional orogens can be semi–quantitatively described by two factors including the amount of collision–produced juvenile crust and terrestrial heat flow. The former factor controls the orogenic magnitude and maturity, and the latter affects temperature and thus crustal rheology (Fig. 10.15d).

During the initial growth stage of collisional orogen (e.g., Zagros), the oceanic slab breakoff following continental collision induced upwelling of the asthenosphere (Motavalli-Anbaran et al. 2011; Mouthereau et al. 2012), which resulted in vertical crustal growth by underplating of mantle–derived mafic magmas (Fig. 10.15a). In addition, the high heat flow (av. 90 $mW•m^{-2}$; GHFCG 2013) due to input of hot mantle–derived melts is responsible for minor thrust sheets and pervasive deformation in such infantile orogen (Fig. 10.15a).

During the mature stage of collision such as in Himalaya, the asthenosphere continued to upwell, and thus mantle–derived magmas continued to inject into the orogen to maintain vertical crustal growth (Hou et al. 2015a, b, 2020; Zhu et al. 2017). Meanwhile, continuous increase in the heat flow (av. 138 $mW•m^{-2}$; Hu et al. 2000) resulted in crustal melting, ductile deformation and lateral crustal flow (Jamieson and Beaumont 2013), which contributed to increasing scale of the orogen (Fig. 10.15b).Deep heat was provided by a serial of "mantle flow channels" and sustained the uplift of the Tibetan Plateau (Hou et al. 2021).

Finally during the orogenic decay stage such as at Qinling, delamination lead to lithospheric extension, crustal thinning, and enhanced erosion with limited accretion of juvenile crust (Wang et al. 2013b; Wu and Zheng 2013; Dong and Santosh 2016; Wang et al. 2017). Therefore, heat flow (av. 68 $mW•m^{-2}$; Hu et al. 2001) decreased accompanied by limited mantle melting (Fig. 10.15c).

10.6.4 Fertility of Juvenile Crust and Its Control on Ore Deposit Formation

It was previously accepted that collisional orogens have a limited metallogenic potential, due to the assumption that collision rarely produces juvenile crust (Kerrich et al. 2005). However, the recognition of large amounts of juvenile mantle materials (Fig. 10.14), and the discovery of numerous world–class ore deposits in collisional orogens (Fig. 10.1) challenge the status quo (Hou and Cook 2009). Moreover, our Hf isotope mapping clearly shows that the lithospheric architecture and specific crustal types correspond to the spatial localization of some giant ore deposits, reflecting lithosphere–scale source control on these deposits (Fig. 10.1).

One example is porphyry Cu deposits (PCDs) in the Zagros and Himalayan orogens (Fig. 10.1). These PCDs formed in late–collisional transtentional and post–collisional extensional settings (Hou et al. 2003a, 2015b, 2017), and are spatially localized within juvenile crustal blocks (Fig. 10.14). The ore–forming magmas are high–K calc–alkaline, and show arc–like signatures and adakite affinity,previously interpreted to be entirely derived from re–melting of the juvenile mafic lower crust (Hou et al. 2015b, 2017). However, the high water content of Cu–ore–forming magmas (>10 wt% H_2O; Lu et al. 2015), close temporal–spatial association of these PCDS with juvenile crust blocks and coeval mafic magmas derived from subduction–modified SCLM (Fig. 10.14) together suggest that the juvenile hydrous mantle melts have likely provided H_2O, S, and Cl (Lu et al. 2015; Yang et al. 2015; Sun et al. 2018) and triggered fluid–fluxed remelting of juvenile lower crust to form the PCD in the Himalaya orogen. The Qulong is the largest deposit in the Oligocene-Miocene Gangdese porphyry Cu belt (Yang et al. 2015). The high-Mg diorite porphyry exhibits geochemical characteristics of both adakite-like and ultrapotassic magmas, and confirms a magma mixing of the adakite-like melt with underplated ultrapotassic magmas. Due to the large decrease in H_2O solubility in the ultrapotassic mafic melt upon ascent and decompression, this magma mixing model supports the hypothesis that fluid exsolution would take place upon the ascent of mafic ultrapotassic magmas from their source (~100 km) into the lower part (~70 km) of the crust (Yang et al. 2015).

It has been demonstrated that the juvenile lower–crustal rocks in Tibet are fertile source for Cu–rich magmas because these lower crustal garnet amphibolites and cumulates contain abundant primary magmatic Cu–sulfides (Hou et al. 2017) and have bulk–rock Cu contents up to 1326 ppm (Xu et al. 2019), much higher than that of ocean crust (60 ppm Cu on average). The source fertility is probably controlled by crustal thickness (Chiaradia 2013), because the thick crust prompts magnetite crystallization, which in turn triggers the precipitation of Cu–rich sulphides in hydrous fertile magmas. In addition, excess aluminum may exclude copper from plagioclase, leading to the remaining meltsenriched in copper (Williamson et al. 2016). High-precision copper isotopes for the collision-related PCDs in southern Tibet reveal that initial Cu enrichment in magma sources could be a key step in the formation of giant PCDs in continental collision zones (Zheng et al. 2019). Lee and Tang (2020) also proposed that the garnet fractionation drives oxidation of the residual magma, releasing Cu from sulfides into solution and increasing the ability of endogenic magmatic fluids to transport Cu to the tops of the plutons. This means that the hot collisional orogens with more juvenile mantle components usually contain chalcophile–rich ore–forming components, thus having great potential for PCDs. However, the weak positive correlation between zircon $_{\varepsilon}$Hf of Cu–porphyries and Cu tonnages implies that other factors apart from juvenile crust also played roles in determining Cu endowments such as deep crustal magma differentiation, sulphide saturation, fluid exsolution and precipitation efficiency (Fig. 10.16; Wilkinson 2013).

The juvenile crust played a first order control on the locations of PCDs. This is supported by the lack of PCDs in Qinling orogen where there is limited juvenile crust but abundant ancient crust. Six world–class early–Cretaceous porphyry Mo deposits (Li et al. 2013) all cluster within the oldest crustal block in Qinling (Fig. 10.9c). Geochemical data indicate that these Late Mesozoic Mo-bearing

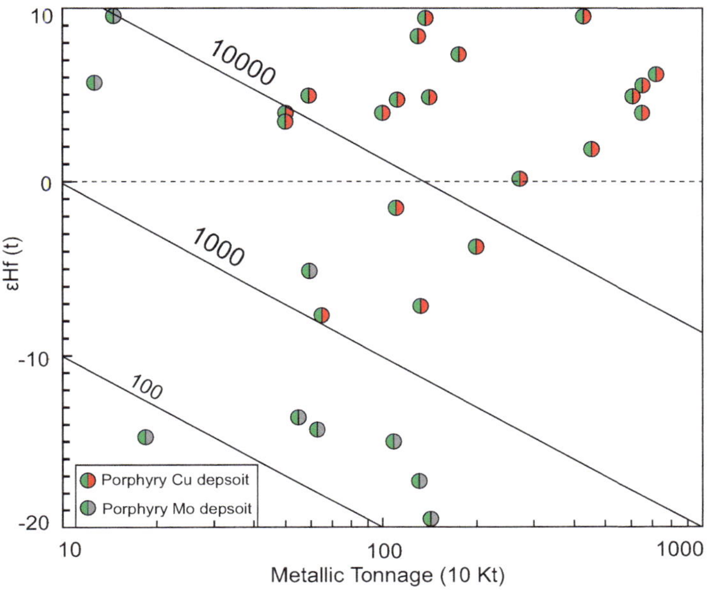

Fig. 10.16 Correlation between zircon εHf values of the host rocks and metallic (Cu, Mo) tonnage of porphyry deposits in the Zagros, Himalayan and East Qinling orogens. All data are given in Supplementary Table S1 and references

intrusions in Qinling show affinity to high K calc-alkaline to shoshonite, metaluminous to peraluminous series (Li and Pirajno 2017). Combined bulk-rock Sr–Nd with zircon Lu–Hf isotopic features, the Mo-bearing intrusionsare likely generated by partial melting of thickened lower continental crust with some contribution of mantle material (Li and Pirajno 2017). A negative correlation between Mo tonnages and $_{\varepsilon}$Hf values of Mo–porphyries indicate that melting of ancient crustal source triggered by delamination generated the low–$_{\varepsilon}$Hf causative magmas and Mo deposits in Qinling (Figs. 10.11 and 10.16).

Our Hf isotopic mapping also provides new constraints on the genesis of other significant deposits in collisional orogens (Fig. 10.14). The majority of giant Mississippi valley type (MVT) lead–zinc deposits are located on the isotopic boundary between the ancient and juvenile crustal blocks in Zagros and Himalaya (Fig. 10.14a, b). Hoggard et al. (2020) also have highlighted that the transition between thick and thin lithosphere controlled global distribution of sediment-hosted metals, including all the giant deposits, which is consistent with Hf isotopic mapping of this study. The formation of these MVT deposits is most likely related to processes that ancient crustal blocks, when thermally eroded or reworked by juvenile materials, released metals (Pb, Zn, Ag) which were swept by crustal fluids migrating towards the front zone of thrust–fold systems (Wang et al. 2014a, b; Leach and Song 2019). The weakly–metamorphosed ancient crustal blocks contain gold, entrained by greenstone basement (e.g.,

Qinling) or ophiolitic complex relicts (e.g., Himalaya) (Goldfarb et al. 2007). The gold was easily mobilized by collisional events and then leached by fluids migrating through translithospheric faults (Goldfarb et al. 2007), thus the resultant gold mineralization mostly concentrated in the interiors or edges of the ancient crustal blocks cut by fault network. These gold deposits are different from traditional orogenic Au deposits which are related to metamorphism-derived ore fluids and materials formed in oceanic subduction (Groves et al. 1998; Tomkins 2013).

We hence suggest that the lithospheric architecture and its crustal type (juvenile and ancient) control the genesis and localization of distinct ore deposits, and should be considered as a key criterion for further exploration of mineral deposits.

Acknowledgements Z.Q.H thanks Dr. Changming Wang for collecting geochemical data, drawing maps and interpreting data. Many thinks to Drs.Tao Wang, Hongrui Zhang, and Yongjun Lu for analysis and discussions, and Zhiming Yang, Tiannan Yang, Zhusen Yang, and Xiaoxia Wang for help with field–work. This study was financially supported by National Key Research and Development Program of China (2022YFF0800903,2016YFC0600300) and NSFC (92162101, 41872080).

References

Agard P, Omrani J, Jolivet L, Whitechurch H (2011) Zagros orogeny: a subduction–dominated process. Geol Mag 148:692–725

Aghazadeh M, Hou Z, Badrzadeh Z, Zhou L (2015a) Temporal–spatial distribution and tectonic setting of porphyry copper deposits in Iran: constraints from zircon U– Pb and molybdenite re–Os geochronology. Ore Geol Rev 70:385–406

Aghazadeh M, Prelević D, Badrzadeh Z, Braschi E, Bogaard PV, Conticelli S (2015b) Geochemistry, Sr–Nd–Pb isotopes and geochronology of amphibole– and mica–bearing lamprophyres in northwestern Iran: implications for mantle wedge heterogeneity in a paleo–subduction zone. Lithos 216–217:352–369

Asadi S (2018) Triggers for the generation of post–collisional porphyry cu systems in the Kerman magmatic copper belt, Iran: new constraints from elemental and isotopic (Sr–Nd–Hf–O) data. Gondwana Res 64:97–121

Ballato P, Uba CE, Landgraf A, Strecker MR, Sudo M, Stockli DF, Friedrich A, Tabatabaei SH (2011) Arabia–Eurasia continental collision: Insights from late Tertiary foreland-basin evolution in the Alborz Mountains, Northern Iran. Geol Soc Am Bull 123:106–131

Belousova EA, Kostitsyn YA, Griffin WL, Begg GC, O'Reilly SY, Pearson NJ (2010) The growth of the continental crust: constraints from zircon Hf–isotope data. Lithos 119:457–466

Blisniuk PM, Hacker B, Glodny J, Ratschbacher L, Bill S, Wu ZH, McWilliams MO, Calvert A (2001) Normal faulting in Central Tibet since at least 13.5 myr ago. Nature 412:628–632

Bouvier A, Vervoort JD, Patchett PJ (2008) The Lu–Hf and Sm–Nd isotopic composition of CHUR: constraints from unequilibrated chondrites and implications for the bulk composition of terrestrial planets. Earth Planet Sci Lett 273:48–57

Cawood, P.A., Kroner, A., Collins, W.J., Kusky, T.M., Mooney, W.D, Windley, B.F., 2009. Accretionary orogens through earth history. Geol Soc Spec Publ 318, 1–36

Chang J, Li JW, Selby D, Liu JC, Deng XD (2017) Geological and chronological constraints on the long–lived Eocene Yulong porphyry Cu–Mo deposit, eastern Tibet, China: implications

for lifespan of magmatic–hydrothermal processes forming giant and supergiant porphyry Cu deposits. Econ Geol 112:1719–1746
Chang J, Li JW, Audétat A (2018) Formation and evolution of multistage magmatichydrothermal fluids at the Yulong porphyry Cu-Mo deposit, eastern Tibet: insights from LA-ICP-MS analysis of fluid inclusions. Geochim Cosmochim Acta 232:181–205
Chen YJ, Pirajno F, Qi JQ (2008) The Shanggong gold deposit, Eastern Qinling Orogen, China: isotope geochemistry and implications for ore genesis. J Asian Earth Sci 33:252–266
Chen JL, Xu JF, Ren JB, Huang XX, Wang BD (2014a) Geochronology and geochemical characteristics of late Triassic porphyritic rocks from the Zhongdian arc, eastern Tibet, and their tectonic and metallogenic implications. Gondwana Res 26:492–504
Chen R, Liu YL, Guo LS, Wang ZH, Liu HF, Xu KF, Zhang JS (2014b) Geochronology and geochemistry of the Tinggong porphyry copper ore deposit, Tibet. Acta Geol Sin 88:780–800
Chen XC, Hu RZ, Bi XW, Li HM, Lan JB, Zhao CH, Zhu JJ (2014c) Cassiterite LA–MCICP–MS U/Pb and muscovite $^{40}Ar/^{39}Ar$ dating of tin deposits in the Tengchong–Lianghe tin district, NW Yunnan, China. MineralDeposita 49:843–860
Chen Q, Wang CM, Bagas L, Zhang ZC, Du B (2021) Time scales of multistage magmatic-hydrothermal fluids at the giant Yulong porphyry cu-Mo deposit in eastern Tibet: insights from titanium diffusion in quartz. Ore Geol Rev 139:104459
Chen Q, Wang CM, Bagas L, Du B, Shi KX, Zhu JX (2023) Hydrothermal fluid signatures of the Yulong porphyry Cu-Mo deposit: clues from the composition and U-Pb dating of W-bearing rutile. Am Mineral 108:1092–1108. https://doi.org/10.2138/am-2022-8453
Chiaradia M (2013) Copper enrichment in arc magmas controlled by overriding plate thickness. Nature Geosci. 7:43–46
Chiu HY (2013) Zircon U–Pb age constraints from Iran on the magmatic evolution related to Neotethyan subduction and Zagros orogeny. Lithos 162–163:70–87
Chiu HY, Chung SL, Zarrinkoub MH, Melkonyan R, Pang KN, Lee HY, Wang KL, Mohammadi SS, Khatib MM (2017) Zircon Hf isotopic constraints on magmatic and tectonic evolution in Iran: implications for crustal growth in the Tethyan orogenic belt. J Asian Earth Sci 145:652–669
Chu MF, Chung SL, Song B, Liu D, O'Reilly SY, Pearson NJ, Ji J, Wen DJ (2006) Zircon U–Pb and Hf isotope constraints on the Mesozoic tectonics and crustal evolution of southern Tibet. Geology 34:745–748
Chung SL, Chu MF, Zhang YQ, Xie YW, Lo CH, Lee TY, Lan CY, Li XH, Zhang Q, Wang YZ (2005) Tibetan tectonic evolution inferred from spatial and temporal variations in post–collisional magmatism. Earth–Sci. Rev 68:173–196
Clift P, Vannucchi P (2004) Controls on tectonic accretion versus erosion in subduction zones: implications for the origin and recycling of the continental crust. Rev Geophys 42:1–31
Collins WJ, Belousova EA, Kemp AIS, Murphy JB (2011) Two contrasting Phanerozoic orogenic systems revealed by hafnium isotope data. Nat Geosci 4:333–337
DeCelles PG, Gehrels GE, Najman Y, Martin AJ, Carter A, Garzanti E (2004) Detrital geochronology and geochemistry of Cretaceous early Miocene strata of Nepal: Implications for timing and diachroneity of initial Himalayan orogenies. Earth Planet. Sci. Lett 227:313–330
Deng J, Wang QF, Li GJ, Hou ZQ, Jiang CZ, Danyushevsky L (2015a) Geology and genesis of the giant Beiya porphyry–skarn gold deposit, northwestern Yangtze Block, China. Ore Geol. Rev. 70:457–485
Deng J, Wang QF, Li GJ, Zhao Y (2015b) Structural control and genesis of the Oligocene Zhenyuan orogenic gold deposit, SW China. Ore Geol. Rev. 65:42–54
Deng J, Wang CM, Bagas L, Selvaraja V, Jeon H, Wu B, Yang LF (2017) Insights into ore genesis of the Jinding Zn–Pb deposit, Yunnan Province, China: Evidence from Zn and in-situ S isotopes. Ore Geol. Rev. 90:943–957
Deng J, Wang CM, Bagas L, Santosh M, Yao EY (2018) Crustal architecture and metallogenesis in the south-eastern North China Craton. Earth Sci. Rev. 182:251–272

Deng J, Wang QF, Sun X, Yang L, Gao L, Groves DI, Shu QH, Yang LQ, Qiu QF, Wang CM (2022) Tibetan ore deposits: a conjunction of accretionary orogeny and continental collision. Earth Sci. Rev. 235:104245

Dong YP, Santosh M (2016) Tectonic architecture and multiple orogeny of the Qinling orogenic belt, Central China. Gondwana Res. 29:1–40

Du AD, Wu SQ, Sun DZ, Wang SX, Qu WJ, Markey R, Stain H, Morgan J, Malinovskiy D (2004) Preparation and certification of Re–Os dating reference materials: molybdenites HLP and JDC. Geostand Geoanal Res. 28:41–52

Gardiner NJ, Hickman AH, Kirkland CL, Lu Y, Johnson T, Zhao JX (2017) Processes of crust formation in the early Earth imaged through Hf isotopes from the East Pilbara Terrane. Precambrian Res. 297:56–76

Global Heat Flow Compilation Group (2013) Component parts of the World Heat Flow Data Collection. PANGAEA. https://doi.org/10.1594/PANGAEA.810104

Goldfarb R, Hart C, Davis G, Groves D (2007) East Asian gold: deciphering the anomaly of phanerozoic gold in Precambrian cratons. Econ. Geol. 102:341–345

Gong P, Hu XM, Li J, Chou YF (2013) The construction of the geological and geochemical prospecting model: a case study of the Jiama copper–polymetallic deposit in Tibet. Geol. Bull. China 32:1601–1612. (in Chinese with English abstract)

Griffin WL, Pearson NJ, Belousova E, Jackson SE, Van Achterbergh E, O'Reilly SY, Shee SR (2000) The Hf isotope composition of cratonic mantle: LA–MC–ICPMS analysis of zircon megacrysts in kimberlites. Geochim.Cosmochim. Acta 64:133–147

Groves DI, Goldfarb RJ, Gebre-Mariam M, Hagemann SG, Robert F (1998) Orogenic gold deposits: a proposed classification in the context of their crustal distribution and relationship to other gold deposit types. Ore Geol. Rev. 13:7–27

Guo XD, Wang ZH, Chen X, Wang X, Wang XH (2009) Machangqing Cu–Mo–Au deposit, Yunnan Province: geological characteristics and genesis. Acta Geol.Sin. 83:1901–1914. (in Chinese with English abstract)

Guo XD, Wang ZH, Wang L, Yang YX, Chen XW (2011) LA–ICP–MS zircon U–Pb ages of porphyritic granite in Machangqing complex of Yunnan Province and their geological significance. Geol. China 38:610–622. (in Chinese with English abstract)

Harrison TM, Lovera OM, Grove M (1997) New insight into the origin of two contrasting Himalayan granite belts. Geology 25:899–902

Hawkesworth CJ, Dhuime B, Pietranik AB, Cawood PA, Kemp AIS, Storey CD (2010) The generation and evolution of the continental crust. J. Geol. Soc. 167:229–248

Haynes SJ, Ghorashi-Zadeh M (1978) Hydrothermal alteration, Cu mineralization, and supergene pattern, Sar Cheshmeh, Iran. Econ. Geol. 73:1391–1392

He, W.Y., 2014. The Beiya Giant Gold–polymetallic deposit: magmatism and metallogenic model (China University of Geosciences. PhD Thesis, Beijing, pp 150 (in Chinese with English abstract)

He MQ, Liu JJ, Li CY, Li ZM, Liu YP, Yang AP, Sang HQ (2004) $^{40}Ar-^{39}Ar$ dating of quartz from ore in the Baiyangping Cu–Co polymetallic ore–concentrated area, Lanping Basin, Yunnan Chin. J. Geochem. 23:342–348

He WY, Mo XX, Yu XH, Li Y, Huang XK, He ZH (2011) Geochronological study of magmatic intrusions and mineralization of Machangqing porphyry Cu–Mo– Au deposit, Western Yunnan Province. Earth Sci. Front. 18:207–215. (in Chinese with English abstract)

He ZH, Guan DR, He WY, Zhou YM, Yang SW, Li WH, Lu YZ, Su GS, Yang R (2016) Exploration model of Beiya supergiant gold–polymetal deposit, northwestern Yunnan. Mineral Deposits 35:261–282. (in Chinese with English abstract)

Henderson BJ, Collins WJ, Murphy JB, Hand M (2018) A hafnium isotopic record of magmatic arcs and continental growth in the Iapetus ocean: the contrasting evolution of Ganderia and the peri–Laurentian margin. Gondwana Res. 58:141–160

Hildreth W, Moorbath S (1988) Crustal contributions to arc magmatism in the Andes of central Chile. Contrib. Mineral. Petrol. 98:455–489

Hoggard MJ, Czarnota K, Richards FD, Huston DL, Jaques AL, Ghelichkhan S (2020) Global distribution of sediment-hosted metals controlled by craton edge stability. Nature Geosci 13:504–510

Horton BK, Hassanzadeh J, Stockli DF, Axen GJ, Gillis RJ, Guest B, Amini A, Fakhari MD, Zamanzadeh SM, Grove M (2008) Detrital zircon provenance of Neoproterozoic to Cenozoic deposits in Iran: implications for chronostratigraphy and collisional tectonics. Tectonophysics 451:97–122

Hou ZQ, Cook N (2009) Metallogenesis of the Tibetan collisional orogen: a review. Ore Geol. Rev. 36:1–24

Hou ZQ, Zhang HR (2015) Geodynamics and metallogeny of the eastern Tethyan metallogenic domain. Ore Geol. Rev. 70:346–384

Hou ZQ, Ma HW, Zaw K, Zhang YQ, Wang MJ, Wang Z, Pan GT, Tang RL (2003a) The Himalayan Yulong porphyry copper belt: product of large-scale strike-slip faulting in Eastern Tibet. Econ. Geol. 98:125–145

Hou ZQ, Qu XM, Wang SX, Gao YF, Du AD, Huang W (2003b) Re–Os ages of molybdenite in the Gangdese porphyry copper belt in south Tibet: duration of mineralization and application of the dynamic setting. Sci. China 33:509–618. (in Chinese with English abstract)

Hou ZQ, Gao YF, Qu XM, Rui ZY, Mo XX (2004) Origin of adakitic intrusives generated during mid–Miocene east–west extension in southern Tibet. Earth Planet. Sci. Lett. 220:139–155

Hou ZQ, Zheng YC, Yang ZM, Rui ZY, Zhao ZD, Qu XM, Jiang SH, Sun QZ (2013) Contribution of mantle components within juvenile lower–crust to collisional zone porphyry Cu systems in Tibet. Mineral. Deposita 48:173–192

Hou ZQ, Duan LF, Lu YJ, Zheng YC, Zhu DC, Yang ZM, Yang ZS, Wang BD, Pei YR, Zhao ZD, McCuaig TC (2015a) Lithospheric architecture of the Lhasa terrane and its control on ore deposits in the Himalayan–Tibetan Orogen. Econ. Geol. 110:1541–1575

Hou ZQ, Yang ZM, Lu YJ, Kemp T, Yang ZS, Tang JX, Qu XM, Duan LF, Zheng YC (2015b) A genetic linkage between subduction– and continental collision–related porphyry Cu deposits in Tibet. Geology 43:247–250

Hou ZQ, Zhou Y, Wang R, Zheng YC, He WY, Zhao M, Evans NJ, Weinberg RF (2017) Recycling of metal–fertilized lower continental crust: Origin of non–arc Au–rich porphyry deposits at cratonic edges. Geology 45:563–566

Hou ZQ, Zheng YC, Lu ZW, Xu B, Wang CM, Zhang HR (2020) Growth, thickening and evolution of the thickened crust of the Tibet Plateau. Acta Geol. Sin. 94:2797–2815. (in Chinese with English abstract)

Hou ZQ, Xu B, Zheng YC, Zheng HW, Zhang HR (2021) Mantle flow: the deep mechanism of large-scale growth in Tibetan Plateau (in Chinese). China Sci. Bull. 66:2671–2690

Hu SB, He LJ, Wang JY (2000) Heat flow and thermal regimes in the continental area of China: a new data set. Earth Planet. Sci. Lett. 179:407–419

Hu SB, He LJ, Wang JY (2001) Compilation of heat flow data in the China continental area (3rd Edition). Chinese J. Geophysics 44:611–626

Hu YB, Liu JQ, Ling MX, Liu Y, Ding X, Liu DY, Sun WD (2017) Constraints on the origin of adakites and porphyry Cu–Mo mineralization in Chongjiang, Southern Gangdese, the Tibetan plateau. Lithos 292–293:424–436

Huang DH, Wu CY, Du AD, He HL (1994) Re–Os isotope ages of molybdenum deposits in east Qinling and their significance. Mineral Deposits 13:221–230. (in Chinese with English abstract)

Huang ML, Bi XW, Richards JP, Hu RZ, Xu LL, Gao JF, Zhu JJ, Zhang XC (2019) High water contents of magmas and extensive fluid exsolution during the formation of the Yulong porphyry Cu–Mo deposit, eastern Tibet. J. Asian Earth Sci. 176:168–183

Jahn BM (2004) The Central Asian Orogenic Belt and growth of the continental crust in the Phanerozoic. Geol. Soc. Spec. Publ. 226:73–100

Jamieson RA, Beaumont C (2013) On the origin of orogens. Geol. Soc. America Bulletin 125:1671–1702

Ji WQ, Wu FY, Chung SL, Wang XC, Liu CZ, Li QL, Liu ZC, Liu XC, Wang JG (2016) Eocene Neo–Tethyan slab breakoff constrained by 45 Ma oceanic island basalt–type magmatism in southern Tibet. Geology 44:283–286

Jiang, C.Z., 2014. Potassic Magmatism and Ore–forming Processes of the Beiya Gold Deposit in Western Yunnan ((China University of Geosciences (Wuhan) PhD Thesis, Wuhan, pp 150 (in Chinese with English abstract)).

Jiang SH, Nie FJ, Hu R, Lai XR, Liu YF (2009) Mayum: an orogenic gold deposit in Tibet, China. Ore Geol. Rev. 36:160–173

Kerrich R, Goldfarb RJ, Richards J (2005) Metallogenic provinces in an evolving geodynamic framework. Econ. Geol. 100:1097–1136

Kind R, Ni J, Zhao W, Wu J, Yuan X, Sandvol E, Reese C, Nabelek J, Hearn T (1996) Evidence from earthquake data for a partially molten crustal layer in southern Tibet. Nature 274:1692–1694

Laccheri LM, Kemp AIS, EIMF (2018) Detrital zircon age, oxygen and hafnium isotope systematics record rigid continents after 2.5 Ga. Gondwana Res. 57:90–118

Leach DL, Song YC (2019) Sediment–hosted zinc–lead and copper deposits in China. Econ. Geol. Special Publication 22:325–409

Leach DL, Song YC, Hou ZQ (2017) The world-class Jinding Zn-Pb deposit: Ore formation in an ecaporite dome, Lanping Bsin, Yunnan, China. Mineral Deposita 52:281–296

Lee CT, Tang M (2020) How to make porphyry copper deposits. Earth Planet Sci. Lett. 529:115868

Li N, Pirajno F (2017) Early Mesozoic Mo mineralization in the Qinling Orogen: an overview. Ore Geol. Rev. 81:431–450

Li YF, Mao JW, Bai FJ, Li JP, He ZJ (2003) Re–Os isotopic dating of molybdenites in the Nannihu molybdenum (tungsten) orefield in the Eastern Qinling and its geological significance. Geol Rev 49:652–659. (in Chinese with English abstract)

Li YF, Mao JW, Liu DY, Wang YB, Wang ZL, Wang YT, Li XF, Zhang ZH, Guo BJ (2006) SHRIMP zircon U–Pb and molybdenite Re–Os dating for the Leimengou porphyry molybdenum deposit, Western Henan and its geological implication. Geol. Rev. 52:122–131. (in Chinese with English abstract)

Li N, Sun YL, Li J, Xue LW, Li WB (2008) Molybdenite Re–Os isotope age of the Dahu Au–Mo deposit, Xiaoqinling and the Indosinian mineralization. Acta Petrol.Sin. 24:810–816. (in Chinese with English abstract)

Li N, Chen YJ, Sun YL, Hu HZ, Li J, Zhang H, Ni ZY (2009) Molybdenite Re– Os isochron age of the Yuchiling porphyry Mo deposit, Henan Province and its geological implications. Acta Petrol.Sin. 25:413–421. (in Chinese with English abstract)

Li N, Chen YJ, Pirajno F, Ni ZY (2013) Timing of the Yuchiling giant porphyry Mo system, and implications for ore genesis. Mineral. Deposita 48:505–524

Li, Y., Selby, D., Condon, D., and Tapster, S., 2017. Cyclic magmatic–hydrothermal evolution in porphyry systems: high–precision U–Pb and Re–Os geochronology constraints on the Tibetan Qulong porphyry Cu–Mo deposit. Econ. Geol. 112, 1419–1440.

Li H, Xu XW, Borg G, Gilg HA, Dong LH, Fan TB, Zhou G, Liu RL, Hong T, Ke Q, Wu C, Zhang GL, Li H (2019a) Geology and Geochemistry of the giant Huoshaoyun zinc–lead deposit, Karakorum Range, northwestern Tibet. Ore Geol. Rev. 106:251–272

Li HJ, Wang Q, Deng J, Yang L, Dong CY, Yu HZ (2019b) Alteration and mineralization styles of the orogenic disseminated Zhenyuan gold deposit, southeastern Tibet: contrast with carlin gold deposit. Geosci. Front. 10:1849–1862

Li HJ, Wang Q, Deng J, Yu HZ, Sun PF, Dong CY, Yang L (2020) Source and evolution of ore fluids in the Zhenyuan orogenic gold deposit, SE Tibet: Constraints from the S-C-O isotopes. Ore Geol. Rev. 121:103534

Liang HY, Mo JH, Xie YW, Zhang YQ, Zeng T, Hu GQ, Charllote MA (2009) Study on geochemical composition and isotope ages of the Malasongduo porphyry associated with Cu–Mo mineralization. Acta Petrol.Sin. 25:385–392. (in Chinesewith English abstract)

Liu QS, Jiang W, Jian P, Yem P-S, Wu ZH, Chen DG (2006) Zircon SHRIMP U–Pb age and petrochemical and geochemical features of Mesozoic muscovite monzonitic granite at Ningzhong, Tibet. Acta Petrol.Sin. 22:643–652. (in Chinese with English abstract)

Liu YC, Yang ZS, Tian SH, Song YC, Zhang HR (2015) Fluid origin of fluorite– rich carbonate–hosted Pb–Zn mineralization of the Himalayan–Zagros collisional orogenic system: a case study of the Mohailaheng deposit, Tibetan Plateau, China. Ore Geol. Rev. 70:546–561
Liu YC, Song YC, Fard M, Zhou LM, Hou ZQ (2019) Pyrite Re–Os age constraints on the Irankuh Zn–Pb deposit, Iran, and regional implications. Ore Geol. Rev. 104:148–159
Lu YJ, Loucks RR, Fiorentini ML, Yang ZM, Hou ZQ (2015) Fluid flux melting generated post-collisional high Sr/Y copper ore–forming water–rich magmas in Tibet. Geology 43:583–586
Maghfouri S, Hosseinzadeh MR, Choulet F, Alfonso P, Zadeh AMA, Rajabi A (2019) Vent-proximal sub-seafloor replacement clastic-carbonate hosted SEDEXtype mineralization in the Mehdiabad world-class Zn-Pb-Ba-(Cu-Ag) deposit, southern Yazd Basin, Iran. Ore Geology Rev 113:103047
Maghfouri S, Hosseinzadeh MR, Choulet F (2020) Supergene nonsulfide Zn–Pb mineralization in the Mehdiabad world–class sub–seafloor replacement SEDEX–type deposit, Iran. Int. J. Earth Sci. 109:2531–2555
Mai B, Ye HS, Li C, Xiao ZJ, Yang GQ (2011) Molybdenite Re–Os isochron age of Yechangping Mo deposit in western Henan Province and its geological implications. Mineral Deposits 30:1069–1074. (in Chinese with English abstract)
Mao JW, Goldfarb RJ, Zhang ZW, Xu WY, Qiu YM, Deng J (2002) Gold deposits in the Xiaoqinling–Xiong'ershan region, Qinling Mountains, central China. Mineral. Deposita 37:306–325
Mao, J.W., Xie, G.Q., Bierlein, F., Ye, H.S., Qu, W.J., Du, A.D., Pirajno, F., Li, H.M., Guo, B.J., Li, Y.F., Yang, Z.Q., 2008. Tectonic implications from Re–Os dating of Mesozoic molybdenum deposits in the East Qinling–Dabie orogenic belt. Geochim.Cosmochim. Acta 72, 4607–4626.
Mao B, Ye HS, Li C, Xiao ZJ, Yang GQ (2011a) Molybdenite Re–Os isochron age of Yechangping Mo deposit in western Henan Province and its geological implications. Mineral Deposit 30:1069–1074. (in Chinese with English abstract)
Mao JW, Pirajno F, Xiang JF, Gao JJ, Ye HH, Li YF, Guo BJ (2011b) Mesozoic molybdenum deposits in the east Qinling–Dabie orogenic belt: characteristics and tectonic settings. Ore Geol. Rev. 43:264–293
McInnes, B.I.A., Evans, N.J., Belousova, E., Griffin, W.T., Andrew, R.L., 2003. Timing of mineralization and exhumation processes at the Sar Cheshmeh and Meiduk porphyry Cu deposits, Kerman belt, Iran. In: Eliopoulos, et al. (Eds.), Mineral exploration and sustainable development (7th Biennial SGA Meeting, Athens, August 24–28). Millpress, Rotterdam, pp 1197–1200.
McQuarrie N, van Hinsbergen DJJ (2013) Retrodeforming the Arabia–Eurasia collision zone: age of collision versus magnitude of continental subduction. Geology 41:315–318
Meng QR, Zhang GW (1999) Timing of collision of the North and South China blocks: controversy and reconciliation. Geology 27:123–126
Metcalfe I (1996) Gondwanaland dispersion, Asian accretion and evolution of eastern Tethys. Aus. J. Earth Sci. 43:605–623
Miller C, Schuster R, Klötzli U, Frank W, Purtscheller F (1999) Post–collisional potassic and ultrapotassic magmatism in SW Tibet: geochemical and Sr–Nd–Pb–O isotopic constraints for mantle source characteristics and petrogenesis. J. Petrol. 40:1399–1424
Mišković A, Schaltegger U (2009) Crustal growth along a non–collisional cratonic margin: a Lu–Hf isotopic survey of the Eastern Cordilleran granitoids of Peru. Earth Planet. Sci. Lett. 279:303–315
Mo, X.X., Zhao, Z.D., Deng, J.F., Dong, G.C., Zhou, S., 2003, Response of volcanism to the India–Asia collision. Earth Sci. Front.10, 135–148.
Mo XX, Hou ZQ, Niu YL, Dong GC, Qu XM, Zhao ZD, Yang ZM (2007) Mantle contributions to crustal thickening during continental collision: evidence from Cenozoic igneous rocks in southern Tibet. Lithos 96:225–242
Motavalli-Anbaran SH, Zeyen H, Brunet MF, Ardestani VE (2011) Crustal and lithospheric structure of the Alborz Mountains, Iran, and surrounding areas from integrated geophysical modeling. Tectonics 30:1–16

Moutherea F (2011) Timing of uplift in the Zagros belt/Iranian plateau andaccommodation of late Cenozoic Arabia–Eurasia convergence. Geol. Mag. 148:726–738
Mouthereau F, Lacombe O, Verges J (2012) Building the Zagros collisional orogen: timing, strain distribution and the dynamics of Arabia/Eurasia plate convergence. Tectonophysics 532:27–60
Moyen JF, Laurent O, Chelle-Michou C, Couzinie S, Vanderhaeghe O, Zeh A, Villaros A, Gardien V (2017) Collision vs. subduction–related magmatism: two contrasting ways of granite formation and implications for crustal growth. Lithos 277:154–177
Mu L, Hu R, Bi X, Tang Y, Lan T, Lan Q, Zhu JJ, Peng JT, Oyebamiji A (2021) New insights into the origin of the world-class Jinding sediment-hosted Zn-Pb deposit, Southwestern China: evidence from LA-ICP-MS Analysis of Individual FluidInclusions. Econ. Geol. 116:883–907
Murphy JB, Nance RD (1991) Supercontinent model for the contrasting character of Late Proterozoic orogenic belts. Geology 19:469–472
Nábelek, J., Hetenyi, G., Vergne, J., Sapkota, S., Kafle, B., Jiang, M., Su, H.P., Chen, J., Huang, B.S., the Hi–CLIMB Team, 2009.Underplating in the Himalaya–Tibet collision zone revealed by the Hi–CLIMB experiment. Science 325, 1371–1374.
Næraa T, Scherstén A, Rosing MT, Kemp AIS, Hoffmann JE, Kokfelt TF, Whitehouse MJ (2012) Hafnium isotope evidence for a transition in the dynamics of continental growth 3.2 Gyr ago. Nature 485:627–630
Niu YL, Zhao ZD, Zhu DC, Mo XX (2013) Continental collision zones are primary sites for net continental crust growth–A testable hypothesis. Earth–Sci. Rev. 127:96–110
Ou Q, Wang Q, Wyman DA, Zhang HX, Yang JH, Zeng JP, Hao LL, Chen YW, Liang H, Qi Y (2017) Eocene adakitic porphyries in the central–northern Qiangtang Block, central Tibet: partial melting of thickened lower crust and implications for initial surface uplifting of the plateau. J. Geophys. Res. Solid Earth 122:1025–1053
Owens T, Zandt G (1997) Implications of crustal property variations for models of Tibetan plateau evolution. Nature 387:37–43
Pei YR, Sun QZ, Zheng YC, Yang ZS, Li W, Huang KX (2016) Genesis of the Bangbu orogenic gold deposit in Tibet: evidence from fluid inclusion, stable isotope and A–Ar geochronology. Acta Geol. Sin. 90:722–737
Perelló J, Razique A, Schloderer J, Ur-Rehman A (2008) The Chagai porphyry copper belt, Baluchistan Province, Pakistan. Econ. Geol. 103:1583–1612
Rajabi A, Rastad E, Canet C (2012) Metallogeny of Cretaceous carbonate–hosted Zn–Pb deposits of Iran: geotectonic setting and data integration for future mineral exploration. Int. Geol. Rev. 54:1649–1672
Razique A, Tosdal RM, Creaser RA (2014) Temporal evolution of the western porphyry Cu–Au systems at Reko Diq, Balochistan, western Pakistan. Econ. Geol. 109:2003–2021
Reilinger, R. et al., 2006. GPS constraints on continental deformation in the Africa– Arabia–Eurasia continental collision zone and implications for the dynamics of plate interactions. J. Geophys. Res. Solid Earth, 111, B05411.
Ren T, Wang RT, Xie GQ, Li JB, Dai JZ, Guo YH, Dang KF, Wu XQ (2014) Geochemistry and rock–forming and ore–forming epochs of Chigou Cu porphyry deposit in Shaanxi Province, and their implications. Mineral Deposits 33:807–820. (in Chinese with English abstract)
Richards JP (2009) Postsubduction porphyry Cu–Au and epithermal Au deposits: products of remelting of subduction–modifi ed lithosphere. Geology 37:247–250
Rui ZY, Li GM, Zhang LS, Wang LS (2004) The response of porphyry copper deposits to important geological events in Xizang. Earth Sci. Front. 11:145–152. (in Chinese with English abstract)
Sengor AMC (1991) Plate tectonics and orogenic research after 25 years: synopsis of a Tethyan perspective. Tectonophysics 187(315–330):337–344
Sengor AMC, Altlner D, Cin A, Ustaiimer T, Hsti KJ (1988) Origin and assembly of the Tethyside collage at the expense of Gondwana Land. Geol. Soc. Spec. Publ. 37:119–181
Shafiei B, Shahabpour J (2012) Geochemical aspects of molybdenum and precious metals distribution in the Sar Cheshmeh porphyry copper deposit, Iran. Mineral Deposita 47:535–543

Shafiei B, Haschke M, Shahabpour J (2009) Recycling of orogenic arc crust triggers porphyry Cu mineralization in Kerman Cenozoic arc rocks, southeastern Iran. Mineral. Deposita 44:265–283
Song YC, Yang TN, Zhang HR, Liu YC, Hou ZQ, Li Z (2015) The Chaqupacha Mississippi Valley–type Pb–Zn deposit, central Tibet: ore formation in a fold and thrust belt of the India–Asia continental collision zone. Ore Geol. Rev. 70:533–545
Song YC, Hou ZQ, Xue CD, Huang SQ (2020) New mapping of the World–Class Jinding Zn–Pb Deposit, Lanping Basin, Southwest China: genesis of Ore Host Rocks and Records of Hydrocarbon–Rock Interaction. Econ.Geol. 115:981–1002
Stern CR (2011) Subduction erosion: rates, mechanisms, and its role in arc magmatism and the evolution of the continental crust and mantle. Gondwana Res. 20:284–308
Sun XM, Wei HX, Zhai W, Shi GY, Liang YH, Mo RW, Han MX, Yi JZ, Zhang XG (2016) Fluid inclusion geochemistry and Ar–Ar geochronology of the Cenozoic Bangbu orogenic gold deposit, southern Tibet. China. Ore Geol. Rev. 74:196–210
Sun X, Lu YJ, McCuaig TC, Zheng YY, Chang HF, Guo F, Xu LJ (2018) Miocene ultrapotassic, high-Mg dioritic, and adakite-like rocks from Zhunuo in Southern Tibet: implications for mantle metasomatism and porphyry copper mineralization in collisional orogens. J Petrol 59:341–386
Tian SH, Gong YL, Yang ZS, Hou ZQ, Liu YC, Song YC, Xue WW, Lu HF, Wang FC, Zhang YB, Zhu T, Yu CJ (2014) Rb–Sr and Sm–Nd isochron ages of the Dongmozhazhua and Mohailaheng Pb–Zn ore deposits in the Yushu area, southern Qinghai and their geological implications. Acta Petrol. Sin. 88:558–569. (in Chinese with English abstract)
Tilmann F, Ni J, INDEPTH III Seismic Team (2003) Seismic imaging of the downwelling Indian lithosphere beneath central Tibet. Science 300:1424–1427
Tomkins AG (2013) On the source of orogenic gold. Geology 41:1255–1256
Verdel C, Wernicke BP, Hassanzadeh J, Guest B (2011) A Paleogene extensional arc flare–up in Iran. Tectonics 30:1–20
Wang JH, Yin A, Harrison TM, Grove M, Zhang YQ, Xie GH (2001) A tectonic model for Cenozoic igneous activities in the eastern Indo–Asian collision zone. Earth Planet. Sci. Lett. 188:123–133
Wang XX, Wang T, Jahn BM, Hu NG, Chen W (2006) Tectonic significance of Late Triassic post-collisional lamprophyre dykes from the Qinling Mountains (China). Geol. Mag. 144:837–848
Wang CM, Cheng QM, Zhang ST, Deng J, Xie SY (2008) Magmatic hydrothermal superlarge systems—a case study of the Nannihu ore field. J. China Univ. Geosci. 19(4):391–403
Wang Q, Wyman DA, Li ZX, Sun WD, Chung SL, Vasconcelos PM, Zhang QY, Dong H, Yu YS, Pearson N, Qiu HN, Zhu TX, Feng XT (2010) Eocene north–south trending dikes in central Tibet: new constraints on the timing of east–west extension with implications for early plateau uplift? Earth Planet Sci. Lett. 298:205–216
Wang CM, Carranza EJM, Zhang ST, Zhang J, Liu XJ, Zhang D, Sun X, Duan CJ (2013a) Characterization of primary geochemical haloes for gold exploration at the Huanxiangwa gold deposit, China. J. Geochem. Explor. 124:40–58
Wang XX, Wang T, Zhang CL (2013b) Neoproterozoic, paleozoic, and mesozoic granitoid magmatism in the Qinling orogen, China: constraints on orogenic process. J. Asian Earth Sci. 72:129–151
Wang CM, Deng J, Carranza EJM, Santosh M (2014a) Tin metallogenesis associated with granitoids in the southwestern Sanjiang Tethyan Domain: nature, deposit types, and tectonic setting. Gondwana Res. 26:576–593
Wang CM, Deng J, Carranza EJM, Lai XR (2014b) Nature, diversity and temporal–spatial distributions of sediment–hosted Pb–Zn deposits in China. Ore Geol. Rev. 56:327–351
Wang XH, Hou ZQ, Song YC, Zhang HR (2015a) Geological, fluid inclusion and isotopic studies of the Baiyangping Pb–Zn–Cu–Ag polymetallic deposit, Lanping basin, Yunnan province, China. J. Asian Earth Sci. 111:853–871
Wang XX, Wang T, Ke CH, Yang Y, Li JB, Li YH, Qi QJ, Lv XQ (2015b) Nd–Hf isotopic mapping of Late Mesozoic granitoids in the East Qinling Orogen, central China: constraint

on the basements of terranes and distribution of Mo mineralization. J. Asian Earth Sci. 103:169–183

Wang CM, Bagas L, Lu YJ, Santosh M, Du B, Campbell MT (2016) Terrane boundary and spatio–temporal distribution of ore deposits in the Sanjiang Tethyan Orogen: insights from zircon Hf–isotopic mapping. Earth-Sci. Rev 156:39–65

Wang CM, Deng J, Bagas L, Wang Q (2017) Zircon Hf–isotopic mapping for understanding crustal architecture and metallogenesis in the Eastern Qinling Orogen. Gondwana Res. 50:293–310

Wang CM, Yang LF, Bagas L, Noreen E, Chen JY, Du B (2018) Mineralization processes at the giant Jinding Zn–Pb deposit, Lanping Basin, Sanjiang Tethys Orogen: evidence from in–situ trace element analysis of pyrite and marcasite. Geol. J. 53:1279–1294

Wang CM, Deng J, Bagas L, He XY, Zhang J (2021) Origin and classification of the Late Triassic Huaishuping gold deposit in the eastern part of the Qinling–Dabie Orogen, China: implications for gold metallogeny. Mineral. Deposita. 56:725–742

Wen CQ, Duo J, Fan XP, Hu XC, Li BH, Sun Y, Liu WZ, Huo Y, Wen Q, Ren WJ (2006) Characteristics of ore–fluids of the Mayum gold deposit, western Tibet, China. Geol. Bull. China 25:261–266. (in Chinese with English abstract)

Wilkinson JJ (2013) Triggers for the formation of porphyry ore deposits in magmatic arcs. Nature Geosci 6:917–925

Williamson BJ, Herrington RJ, Morris A (2016) Porphyry copper enrichment linked to excess aluminium in plagioclase. Nat. Geosci. 9:237–241

Wu YA, Zheng YF (2013) Tectonic evolution of a composite collision orogen: an overview on the Qinling–Tongbai–Hong'an–Dabie–Sulu orogenic belt in central China. Gondwana Res. 23:1402–1428

Xie GQ, Ren T, Li JB, Wang RT, Xian CL, Gou YH, Dai JZ, Shen ZC (2012) Zircon U–Pb age and petrogenesis of ore–bearing granitoid for the Chigou Cu–Mo deposit from the Zhashan basin, Shaanxi Province. Acta Petrol. Sin. 28:15–26. (in Chinese with English abstract)

Xu B, Griffin WL, Xiong Q, Hou ZQ, O'Reilly SY, Guo Z, Pearson NJ, Gréau Y, Yang ZM, Zheng YC (2017) Ultrapotassic rocks and xenoliths from South Tibet: contrasting styles of interaction between lithospheric mantle and asthenosphere during continental collision. Geology 45:51–54

Xu W, Zhu DC, Wang Q, Weinberg RF, Wang R, Li SM, Zhang LL, Zhao ZD (2019) Constructing the early Mesozoic gangdese crust in Southern Tibet by hornblende dominated magmatic differentiation. J. Petrol. 60:515–552

Yalikun Y, Xue C, Symons D (2018) Paleomagnetic age and tectonic constraints on the genesis of the giant Jinding Zn–Pb deposit, Yunnan China. Mineral. Deposita 53:245–259

Yan GL, Wang ZM, Li YQ, Fu YH, Ding GM (2012) Re–Os isotope ages of molybdenite and their geological significance of Yechangping molybdenum deposit, Henan. Miner Explor. 3:184–194. (in Chinese with English abstract)

Yang ZM, Cooke DR (2019) Porphyry copper deposits in China. In: Chang ZS, Goldfarb RJ (eds) Mineral deposits of China. Society of Economic Geologists., Special Publication, the US, pp 133–187

Yang ZM, Hou ZQ, White NC, Chang ZS, Li ZQ, Song YC (2009) Geology of the post–collisional porphyry copper–molybdenum deposit at Qulong, Tibet. Ore Geol. Rev. 36:133–159

Yang TN, Ding Y, Zhang HR, Fan JW, Liang MJ, Wang XH (2014a) Two– phase subduction and subsequent collision defines the Paleotethyan tectonics of the southeastern Tibetan Plateau: evidence from zircon U–Pb dating, geochemistry, and structural geology of the Sanjiang orogenic belt, southwest China. Geol. Soc. Am. Bull. 126:1654–1682

Yang Z, Hou Z, Xu J, Bian X, Wang G, Yang Z, Tian S, Liu Y, Wang Z (2014b) Geology and origin of the post–collisional Narigongma porphyry Cu–Mo deposit, southern Qinghai, Tibet. Gondwana Res. 26:536–556

Yang ZM, Lu YJ, Hou ZQ (2015) High–Mg diorite from Qulong in southern Tibet: implications for the genesis of adakite–like intrusions and associated porphyry Cu deposits in collisional orogens. J. Petrol. 56:227–254

Yang ZS, Hou ZQ, Zheng YC, Liu YC (2017) Genesis of the Gold Deposit in the Indus–Yarlung Tsangpo Suture Zone, Southern Tibet: evidence from Geological and Geochemical Data. Acta Geol. Sin. 91:947–970

Ye, H.S., 2006.The Mesozoic Tectonic Evolution and Pb–Zn–Ag Metallogeny in the South Margin of North China Craton (Chinese Academy of Geological Sciences Doctoral Dessertation, Beijing, pp 217. (in Chinese)).

Yin A, Harrison TM (2000) Geologic evolution of the Himalayan–Tibetan orogen. Annu. Rev. Earth Planet. Sci. 28:211–280

Yuan XC (1996) Velocity structure of the Qinling lithosphere and mushroom cloud model. Sci. China (Ser. D) 39:236–244

Zhang HR, Yang TN, Hou ZQ, Dai MN, Hou KJ (2017) Permian back–arc basin basalts in the Yushu area: new constrain on the Paleo–Tethyan evolution of the north–central Tibet. Lithos 286:216–226

Zhang HR, Chen JL, Yang TN, Hou ZQ, Aghazadeh M (2018) Jurassic granitoids in the northwestern Sanandaj–Sirjan Zone: evolving magmatism in response to the development of a Neo–Tethyan slab window. Gondwana Res. 62:269–286

Zhang ZM, Ding H, Palin RM, Dong X, Tian ZL, Chen YF (2020) The lower crust of the Gangdese magmatic arc, southern Tibet, implication for the growth of continental crust. Gondwana Res. 77:136–146

Zhao J, Qin K, Xiao B, McInnes B, Li G, Evans N, Cao M, Li JX (2016) Thermal history of the giant Qulong Cu–Mo deposit, Gangdese metallogenic belt, Tibet: constraints on magmatic–hydrothermal evolution and exhumation. Gondwana Res. 36:390–409

Zheng YY, Zhang GY, Xu RK, Gao SB, Pang YC, Cao L, Du AD, Shi YR (2007) Geochronologic constraints on magmatic intrusions and mineralization of the Zhunuo porphyry copper deposit in Gangdese, Tibet. Sci. Bull. China 52:3139–3147. (in Chinese with English abstract)

Zheng WB, Tang JX, Zhong KH, Ying LJ, Leng QF, Ding S, Lin B (2016) Geology of the Jiama porphyry copper–polymetallic system, Lhasa Region, China. Ore Geol. Rev. 74:151–169

Zheng YC, Liu SA, Wu CD, Griffin WL, Li ZQ, Xu B, Yang ZM, Hou ZQ, O'Reilly SY (2019) Cu isotopes reveal initial Cu enrichment in sources of giant porphyry deposits in a collisional setting. Geology 47:135–138

Zhou K, Ye HS, Mao JW, Qu WJ, Zhou SF, Meng F, Gao YL (2009) Geological characteristics and molybdenite Re–Os dating of Yuchiling porphyry Mo deposit in western Henan Province. Mineral. Deposit 28:170–184. (in Chinese with English abstract)

Zhu LM, Zhang GW, Guo B, Li B (2008) U–Pb (LA–ICP–MS) zircon dating for the large Jinduicheng porphyry Mo deposit in the East China, and its metallogenic setting. Acta Geol. Sin. 82:204–220. (in Chinese with English abstract)

Zhu DC, Zhao ZD, Niu YL, Mo XX, Chung SL, Hou ZQ, Wang LQ, Wu FY (2011) The Lhasa terrane: record of a microcontinent and its histories of drift and growth. Earth Planet. Sci. Lett. 301:241–255

Zhu DC, Wang Q, Cawood PA, Zhao ZD, Mo XX (2017) The Lhasa terrane: record of a microcontinent and its histories of drift and growth. J. Geophys. Res. Solid Earth 122:214–223

Zhuang LL, Song YC, Liu YC, Fard M, Hou ZQ (2019) Major and trace elements and sulfur isotopes in two stages of sphalerite from the world–class Angouran Zn–Pb deposit, Iran: implications for mineralization conditions and type. Ore Geol. Rev. 109:184–200

Zou ZC, Hu RZ, Bi XW, Wu LY, Feng CX, Tang YY (2015) Absolute and relative dating of Cu and Pb–Zn mineralization in the Baiyangping area, Yunnan Province, SW China: Sm–Nd geochronology of calcite. Geochem. J. 49:103–112

Zürcher L, Bookstrom AA, Hammarstrom JM, Mars JC, Ludington SD, Zientek ML, Dunlap P, Wallis JC (2019) Tectono–magmatic evolution of porphyry belts in the central Tethys region of Turkey, the Caucasus, Iran, western Pakistan, and southern Afghanistan. Ore Geol. Rev:102849

GPSR Compliance

The European Union's (EU) General Product Safety Regulation (GPSR) is a set of rules that requires consumer products to be safe and our obligations to ensure this.

If you have any concerns about our products, you can contact us on ProductSafety@springernature.com

In case Publisher is established outside the EU, the EU authorized representative is:

Springer Nature Customer Service Center GmbH
Europaplatz 3
69115 Heidelberg, Germany

Batch number: 10703867

Printed by Printforce, the Netherlands